Asian Pacific Phycology in the 21st Century:
Prospects and Challenges

Developments in Hydrobiology 173

Series editor

K. Martens

Asian Pacific Phycology in the 21st Century: Prospects and Challenges

Proceedings of The Second Asian Pacific Phycological Forum, held in Hong Kong, China, 21–25 June 1999

Edited by

Put O. Ang, Jr.

Marine Science Laboratory, Department of Biology, The Chinese University of Hong Kong, Shatin, Hong Kong, China

Reprinted from Hydrobiologia, volume 512 (2004)

Springer Science+Business Media, B.V.

Library of Congress Cataloging-in-Publication Data

A C.I.P. Catalogue record for this book is available from the Library of Congress.

ISBN 978-94-010-3748-8   ISBN 978-94-007-0944-7 (eBook)
DOI 10.1007/978-94-007-0944-7

TABLE OF CONTENTS

Hydrobiologia **512:** ix–xi, 2004.
P.O. Ang, Jr. (ed.), Asian Pacific Phycology in the 21st Century: Prospects and Challenges.

Introduction

People from the Asian Pacific region are among the first in the world to utilize algae for various purposes. Some reference to algae could be found in Chinese classic writings some 2500 years ago. It is perhaps not surprising that traditional focus on algae in the region has been on their potential as a resource. Asian Pacific is now the world's largest algal production region. From freshwater to marine environments, from microalgae to macroalgae (seaweeds), through natural harvest or through farming or polyculture, in indoor tanks or outdoor ponds, algal biomass is being produced by millions of tons annually. In addition to all the other associated industries, from food manufacturing, chemical extraction to pharmaceutical, nutraceutical and industrial product development, the entire algae-related industry is certainly one of the most vital in the region. There is continued and sustained interest in the expanded use of algae and the application of algae as a tool in biotechnology.

Notwithstanding the focus on the economic potential of algae, there is also a greater focus on the role of algae in the environment, not simply as the primary producers, but also as structuring forces in the community. There is the question of algae as sources of various toxins during algal blooms, as well as the potential of algae as scavenger of excess nutrients under eutrophication. More and more researchers have also turned to algae as a tool in experimental biology and as a model to understand biological phenomena. All this diversity in interests and focuses could only be linked together simply because they are all related to algae.

It was with this perspective in mind that the Second Asian Pacific Phycological Forum was held at the Chinese University in Hong Kong in 1999, with the theme "Asian Pacific Phycology in the 21st Century: Prospects and Challenges". The forum attracted close to 200 people from the Asian Pacific region and beyond to five days of stimulating discussion. Three keynote speakers presented their valuable insight in three important aspects of algal biology. Professor R. E. De Wreede from the University of British Columbia, Canada, epoused on "The culture of marine ecology". Professor W. Prud'homme van

Reine from Leiden University of the Netherlands explored the extent of utilization of "Useful seaweeds in Southeast Asia". Professor S. Miyachi from the Marine Biotechnological Institute of Japan projected his vision on "Microalgal studies in the 21st century". Ten mini-symposia were organized, together with Student Competition Sessions, Oral Contributed and Poster Sessions. These mini-symposia included "Algal population dynamics and community structure", "Herbivory and algal defenses", "Algal biochemical resources: exploitation and future prospects", "Algal nutraceuticals", "Molecular and morphological approaches to systematics and evolution', "Seaweed mariculture, problems and prospects", "Microalgae and pollution", "Marine macroalgae and the environment", "Green tide", "Algal cell as an experimental system", and "Algal biotechnology". One main and unique characteristic of these mini-symposia was that they were all organized by young scientists as they are the future of Asian Pacific phycology in the 21st century.

Twenty-four students participated in the student competition for best oral presentation which was divided into two main sessions. In the session under Ecology/Cytology/Ecology and Environment, the best student paper was presented by Ms. Corsica Kong Sau Lai of the Chinese University of Hong Kong, Hong Kong, with her paper entitled "Seasonal occurrence and reproduction of *Hypnea charoides* (Rhodophyta) in Ping Chau, N.T., Hong Kong, China". The second best paper was presented by Mr David Ginsburg of the University of Guam, USA, with the paper entitled "The role of diet-derived algal metabolites in the sea hare *Aplysia parvula* Mörch (Opisthobranchia, Anspidea". In the second session on Morphology, Systematics and Genetics/Applied Phycology, the best student paper was presented by Mr Wong Ka Hing of the Chinese University of Hong Kong, Hong Kong, with his paper entitled "Nutritional evaluation of protein concentrates isolated from red seaweeds *Hypnea charoides* and *Hypnea japonica* in growing rats". The second best paper was presented by Chen Ying of the Institute of Genetics, Beijing, China with her paper entitled "Establishment of high frequency gene transformation system and expression of Defensin (Np-1) gene in unicellular green alga (*Chlorella ellipsoidea*) by electroporation". Professor M. Ohno of Kochi University, Japan and Prof. Kwang Young Kim of Chonnam National University, Korea, chaired and co-chaired the first session respectively and members of the judging panel included Dr A. Critchley, Prof. R. De Wreede, Dr G. Kendrick, Prof. S. Murray and Dr C. Towbridge. Professor T. Larkum of the University of Sydney, Australia and Prof. E. Cao of the University of the Philippines, Philippines, chaired and co-chaired the second session respectively and members of the judging panel included Prof. Y. Aruga, Ms. C. A. Borden, Prof. Y. Hara, Prof. R. King, Dr Lawrence Liao, and Prof. W. Prudhomme van Reine. Prizes for the best student paper awards came from the local organizing committee of the meeting and prizes for the second best student paper awards were contributed by judges of the student competitions.

Fifty-two posters were presented at the meeting and the best poster awards went to K. Ishida, B. R. Green and T. Calvalier-Smith of the University of British Columbia, Canada, for their poster entitled "Phylogeny of Chlorarachniophytes: Nuclear and nucleomorph small subunit ribosomal RNA gene trees"; X. H. Yan, Y. Fujita and Y. Aruga of Nagasaki University/Tokyo University of Fisheries, Japan for their poster entitled "Observations on monospore-releasing mutants obtained by treatment with NNG in *Porphyra yezoensis* Ueda (Bangiales, Rhodophyta)"; and K. H. Jeon, Y. K. Lee and G. H. Kim of Kongju National University, Korea, for their poster entitled "Expression patterns of sex-related genes, cyclophilin and heat-shock protein 90 in some red algae". The panel of judges for poster competition was chaired by Prof. Lee Yuan Kun of the National University of Singapore, Singapore and members of the panel included Dr L. Airoldi, Prof. I. K. Lee, Dr J. Lewis, Prof. N. Montaño, Prof. S. Miyachi, Ms. R. Ruangchuay and Prof. J. West.

It was not all talks and discussions. There was also fun time with "Student Night" and conference banquet with Prof. C. K. Tseng of the Chinese Academy of Sciences as the special banquest speaker who talked about "The past, present and future of phycology in China". Some of the conference participants joined the tour to Victoria Peak, a famous tourist destination in Hong Kong, on the last day after the closing ceremony and finished off the conference with a dinner in downtown Hong Kong.

The forum was held under the auspices of the Asian-Pacific Phycological Association, and was sponsored by the Department of Biology, The Chinese University of Hong Kong with financial support from the New Asia College, The Chinese University of Hong Kong, as part of its 50th Anniversary Celebration and the Japan Seaweed Research Association (JASRA, President: Y. Aruga). The meeting benefited from the advice provided by Prof. Y. Aruga (Tokyo University of Fisheries and president of Asian Pacific Phycological Association), Prof. I. J. Hodgkiss (University of Hong Kong), Prof. Vincent Ooi (The Chinese University of Hong Kong), Prof. S. S. M. Sun (The Chinese University of Hong Kong), Prof. C. K. Tseng (The Chinese Academy of Sciences), Prof. P. K. Wong (The Chinese University of Hong Kong), Prof. Y. S. Wong (The City University of Hong Kong), Prof. Norman Y. S. Woo (The Chinese University of Hong Kong), and Prof. Madeline Wu (The Hong Kong University of Science and Technology).

Members of the local organizing committee included Profs P. C. K. Cheung, A. H. Y. Chung, H. M. Lam and Dr L. P. S. Liu.

Many more people made up the secretariat and helped with the running of the meeting. A heartfelt thank to you all who contributed a lot to the success of the meeting!

It was only during the conference that a decision was made to produce a conference proceedings. As there were more papers presented than could be accommodated in a single volume of proceedings, only a limited number of papers could be included. After careful review, 36 papers are collected in this special volume of *Hydrobiologia*. These include seven invited papers and 29 contributed papers. Although there was a considerable time gap between the meeting and the final publication of this special volume, most papers collected in this volume have been updated where necessary.

Many people helped in the production of this special volume. All the contributors did their best and all the reviewers have been most helpful. Both the organization of the meeting and the production of this volume were greatly assisted by the dedication and untiring technical support provided by my assistants and students, especially L. S. Choi, M. M. Choi, C. C. Chiu, S. L. Kong, M. W. Lee, T. P. Lin, L. R. Zhou, L. P. Ding and Prof. Lu Baoren. The latter two of the Institute of Oceanology (the Chinese Academy of Sciences) also helped extensively in many ways. Dr Karla McDermid of the University of Hawaii at Hilo is thanked for helping with the Latin descriptions in several papers collected in this volume. Dr K. Martens, the editor-in-chief of *Hydrobiologia*, and the staff of Kluwer Academic Publishers, especially Ms. Cynthia de Jonge and Mr Anthony Hammick and Ms. Margit Lazar, have been most helpful in providing editorial and technical advice and support.

It is with most sincere hope that this volume may be useful in providing some general picture of the nature of phycological research that is going on in the Asian Pacific region, that this may stimulate further interests and provide an impetus that may lead to future challenges and directions in Asian Pacific Phycology.

PUT O. ANG, Jr.
Chair, Local Organizing Committee
Second Asian Pacific Phycological Forum
and Editor, Conference Proceedings

Hydrobiologia **512:** xiii–xvi, 2004.
P.O. Ang, Jr. (ed.), Asian Pacific Phycology in the 21st Century: Prospects and Challenges.

List of reviewers

The guest editor wishes to acknowledge with most profound appreciation the valuable time these following reviewers contributed in order to help improve the quality of papers collected in this special volume. Those with an * have reviewed at least two manuscripts.

Abbott, Isabella A.
University of Hawaii
U.S.A.

Amano, Hideomi
Mie University
Japan

Airoldi, Laura
Università di Bologna
Italy

Aruga, Yusho
Tokyo University of Fisheries
Japan

Bacic, Tony
University of Melbourne
Australia

Borowitzka, Michael A.
Murdoch University
Australia

Buswell, John A.
Chinese University of H.K.
Hong Kong SAR,
China

Cao, Ernelea P.
University of the Philippines
Philippines

Cheney, Donald P.
Northeastern University
U.S.A.

Cheng, Christopher H. K.
Chinese University of H.K.
Hong Kong SAR,
China

*Cheung, Peter C. K.
Chinese University of H.K.
Hong Kong SAR,
China

*Chirapart, Anong
Kasetsart University
Thailand

Chiu, S. W.
Chinese University of H.K.
Hong Kong SAR,
China

*Chou Hong Nong
National Taiwan University
**Taiwan,
R.O. China**

*Chu Wan Loy
International Medical University
Malaysia

Chung, Anthony H. Y.
Chinese University of H.K.
Hong Kong SAR,
China

Chung, Ik Kyo
Pusan National University
Korea

Clayton, Margaret
Monash University
Australia

Critchley, Alan
University of Namibia
Namibia

*De Wreede, Robert
University of British Columbia
Canada

Fei Xiugeng
Chinese Academy of Sciences
China

Fleurence, J.
IFREMER
France

Forster, Rodney M.
University of Rostock
Germany

Fredericq, Susan
University of SW Louisiana
U.S.A.

Freshwater, D. Wilson
University of North Carolina
U.S.A.

Fukuyo, Yasuwo
The University of Tokyo
Japan

*Ganzon-Fortes, Edna
University of the Philippines
Philippines

Garbary, David J.
St. Francis Xavier University
Canada

*Hermes, Rudolf
PCAAR
Philippines

Hori, Kanji
Hiroshima University
Japan

*Hurtado, Anicia
SEAFDEC
Philippines

Johansen, Jeffrey R.
John Carroll University
U.S.A.

*Kim, Gwang Hoon
Kongju National University
Korea

Kim, Jeong Ha
Sung Kyun Kwan University
Korea

King, Robert J.
University of New South Wales
Australia

Kendrick, Gary
University of Western Australia
Australia

Kusumi, Takenori
Takushima University
Japan

Lam, Hon Ming
Chinese University of H.K.
Hong Kong SAR,
China

*Largo, Danilo
University of San Carlos
Philippines

Larkum, Tony
University of Sydney
Australia

Lau, Arthur P. S.
H.K. University of Sci. & Tech.
Hong Kong SAR,
China

Lee, In Kyu
Seoul National University
Korea

*Lee, Yuan Kun
National University of Singapore
Singapore

Lewis, Jane
National Taiwan Ocean Uniersity
Taiwan,
R.O. China

Li Jun
Ocean University of China
China

Liao, Lawrence M.
San Carlos University
Philippines

Lindstrom, Sandra C.
University of British Columbia
Canada

*Marasigan, B.
University Phil. in the Visayas
Philippines

Marshall, Judith-Anne
University of Tasmania
Australia

McMahon, Brian
University of Calgary
Canada

Miyachi, Shigetoh
Marine Biotech. Institute
Japan

*Montaño, Nemesio
University of the Philippines
Philippines

Motomura, Taizo
Hokkaido University
Japan

Murray, Steve
California State University
U.S.A.

Nagarkar, Sanjay
University of Hong Kong
Hong Kong SAR,
China

Nelson, Wendy
Museum of New Zealand
New Zealand

Nguyen Huu Dinh
Center of Science
Vietnam

Notoya, Masahiro
Tokyo University of Fisheries
Japan

Ohno, Masao
Kochi University
Japan

Paul, Valerie
University of Guam
U.S.A.

Phang Siew Moi
University of Malaya
Malaysia

Pointing, Stephen
University of Hong Kong
Hong Kong SAR,
China

Posten, Clemens
University of Karlsruhe
Germany

Richmond, A.
Ben Gurion University
Israel

Rueness, Jan
University of Oslo
Norway

Shealth, Robert
University of Guelph
Canada

Song Liron
Chinese Academy of Sciences
China

Tam, Nora F. Y.
City University of Hong Kong
Hong Kong SAR,
China

*Tamse, Armando
University Phil. in the Visayas
Philippines

*Trono, Gavino C., Jr.
University of the Philippines
Philippines

Trowbridge, Cynthia D.
Oregon State University
U.S.A.

Tseng, C. K.
Chinese Academy of Sciences
China

Underwood, A. J.
University of Sydney
Australia

Usov, A. I.
Russian Academy of Sciences
Russia

Vonshak, Avigad
Ben-Gurion University of the Negev
Israel

West, John
University of Melbourne
Australia

*Williams, Gray
University of Hong Kong
Hong Kong SAR,
China

Wong, Y. S.
City University of H.K.
Hong Kong SAR,
China

Wu, Chaoyuan
Chinese Academy of Sciences
China

Wu, Madeline
H.K. University of Sci. & Tech.
Hong Kong SAR,
China

Xia, Jinlan
Faculté Polytechinique de Mons
Belgium

Xiang, Jian Hai
Chinese Academy of Sciences
China

Xu, Huai-Shu
Ocean University of China
China

Yoshida, Tadao
Hokkaido University
Japan

*Zhang Ying
Hong Kong Ocean Park
Hong Kong SAR,
China

Group picture taken after the opening ceremony of the Second Asian Pacific Phycological Forum at the Chinese University of Hong Kong, June 1999.

Hydrobiologia **512:** 1–10, 2004.
P.O. Ang, Jr. (ed.), Asian Pacific Phycology in the 21ˢᵗ Century: Prospects and Challenges.
© 2004 *Kluwer Academic Publishers.*

1

The culture of marine ecology

Robert E. DeWreede
Department of Botany, The University of British Columbia, Vancouver, B.C., Canada V6T 1Z4
E-mail: dewreede@science.ubc.ca

Key words: ecology, data analysis, statistics, hypothesis, causality

Abstract

Marine algal ecology today faces many of the same problems as ecology in general, e.g. lack of generality of experimental results, the difficulty of making long-term predictions, and an apparent lack of agreement as to what constitutes the proper or 'acceptable' way of doing this particular component of science. These problems, if real, affect marine algal ecology everywhere but, in different geographical areas, specific problems also occur; science in parts of Asia has some problems different from those in other parts of the world. Since its inception, research in marine algal ecology has been motivated by many factors, ranging from traditional needs, to curiosity, to survival, to new technology, and economic needs. Each of these has shaped the questions that have been asked by, and the level of support society has been willing to supply to, ecology. For example the requisites of tradition pushed marine ecology to ask questions about food and ceremonial biota, and our fears today about loss of biota are pushing for answers to questions about the means of preserving biodiversity. The limitations of many marine ecological studies have been pointed out by different individuals. Their comments have been valuable in forcing us to examine what we are doing as marine ecologists, and how we are doing it. Ecology, and marine algal ecology with it, has been accused of carrying out small-scale studies that have no greater generality than the sites at which the studies were done, and of using statistical procedures that are wrong or inappropriate; also, there is disagreement within the ecological community of how to correct for these 'faults'. Some of the problems arise due to the nature of our particular science, e.g. working with organisms with differing genetic makeup and sensitivity of experimental results to small changes in initial conditions. Other problems are more likely due to the individuals doing the science, e.g. an inability to be an 'expert' on all areas of knowledge required for a modern ecologist (taxonomy, experimental design, data analysis, etc.), and perhaps an unwillingness to recognize that in some instances different methods of data analysis are applicable and valid. As ecologists, we must come to grip with these problems, both for the sake of our science, and for our own sake as practicing ecologists.

Introduction

Today, more ecological research is being published than ever before, and amongst these publications are numerous articles on algal ecology. Yet, despite this outpouring of often very good research, some have argued that ecological research lacks coherence, lacks predictive power, and thus lacks the power to suggest appropriate actions for politicians and managers dealing with the marine environment.

Some who make these claims lay the blame on the ecologists, arguing that most experimental ecological studies are of limited scale, and lack generality. Others have argued that this lack of generality is due to the nature of the science itself, pointing to the complex interactions of many biological and environmental factors that determine the outcome of any ecological process, e.g. competition; as field scientists, we are unable to measure and take into account all of these factors.

Others point out that even if we were able to take all these factors into account, small differences in initial conditions can result in vastly different outcomes, e.g. vastly different numbers of organisms in a population. And since we cannot measure these small differences, we are unable to predict the consequences, even if we know the nature of the interacting processes.

If these difficulties were not enough, ecologists often disagree amongst themselves as to, (1) the appropriate experimental design, or what is aceptable as an experimental design, given the inevitable shortage of funds, researchers, and time, and, (2) the appropriate analytical procedures for examining the data. It is my intent in this paper to: 1. Examine the culture of ecology by discussing what I see as the things we are doing well, the things we are doing badly, and some possible solutions to those things we are not doing well. Clearly these are my personal opinions on these topics. 2. Finally, I hope that we may at least be able to agree on some mechanism for improving what we do, and clearly love to do, research on the ecology of algae.

Definitions of science and ecology

Science in its ideal sense is a method of inquiry that proposes and tests hypotheses, is free to roam the intellectual landscape where it will, and is able to disemminate the information it acquires to all.

Popper (1968), for example, has written and argued that what separates science from non-science is the formulation of a falsifiable hypothesis. Popper (1968) concluded that the criterion that demarcates science from non-science (metaphysics) is that of falsifiability.

Others argue that ecology is too complex to reasonably propose and test single falsifiable hypotheses. Are the upper limits of the distribution of algae due solely to only herbivory, only desiccation, or only dispersal? Or do each of these variables contribute in some degree to the limit of an observed distribution (Underwood, 1985, 1991)? If the latter, no single hypothesis will 'explain' this phenomenon, and the proportional importance of several factors must be examined. Ecology has been defined as "the scientific study of the interactions that determine the distribution and abundance of organisms" (Krebs, 1985) and, in a more cynical sense, as "The science which says what everyone knows in a language that no one understands" (Elton, 1927), and as "The science given over entirely to terminology" (McIntosh, 1976) e.g., a spade = geotome; a rocky sea shore formation = actad; a beach plant = agad.

In some of my darkest moments, when experiments have not gone well, and Hurlbert's Demonic Intrusion (Hurlbert 1984) seems to be rampant, I look at Gertrude Stein's aphorism: (cited in: Fulghum, 1999)

"There ain't no answer.
There ain't going to be any answer.
There never has been an answer.
That's the answer."

and wonder if this is the reality of ecology.

However, a more hopeful view was expressed by Albert Schweitzer: (cited in: Fulghum, 1999)

"To the question whether I am a pessimist or an optimist, I answer that my knowledge is pessimistic, but my willing and hope are optimistic"

Motivations for studying algal ecology

Tradition and natural history

The earliest roots of marine algal ecology probably lie in attempts to find and maintain seaweeds for traditional purposes (e.g. where to find the best seaweed for a particular ceremonial purpose, or where to find seaweeds with the highest agar content). Here, the ability to recognize the characteristics of a suitable habitat for a useful species would become necessary knowledge, and convey status to the individual possessing such knowledge.

The Asia-Pacific Region has a long history of using marine algae, for medicine, for food, and for ceremonial events:

"When the soul of the dead person departs, it travels by sea and stops at the first rock, and is moved because he can still hear the voices of his loved ones crying. At the second rock, he can still see the smoke rising from the yam fields where they are burning weeds. And on the third rock he grabs a little piece of algae and heads towards the island of the dead." (G. Scoditti, quoted in: Stille, 1999).

Still today, the Asia-Pacific region leads the world in the quantity and numbers of seaweed species grown and harvested for industrial use and for food.

Curiosity

Started by tradition, marine algal ecology continued its development by adding curiosity to its motivation. In the context of this presentation, I consider curiosity driven questions those dealing with population, community, and ecosystem processes, as these processes occur in more or less natural (relatively undisturbed by human influence) habitats. In this context, hypotheses are posed about such topics as population dynamics, herbivory, competition, zonation.

Today, the dynamic nature of marine systems is the result of both naturally occurring changes and human induced disturbances, and the resulting dynamics can occur in both obvious and more subtle ways. For example, in Kaneohe Bay in Hawaii, there was an obvious impact resulting from sewage input; the increase in nutrients caused the proliferation of *Dictyosphaeria* (a green algae), which overgrew and killed the corals. Subsequent reduction in nutrients by treatment of the sewage has reversed this process, with a reduction in *Dictyosphaeria*, and a regrowth of corals.

A more subtle dynamic within algal populations has occurred due to oil spills. For example, ongoing studies of the impact of the Exxon Valdez oil spill in Alaska have shown a cycle in the population dynamics of *Fucus* different than found in undisturbed populations of this brown seaweed (Holloway, 1996). Apparently similar cycles were found associated with the Torrey Canyon oil spill, but here it was due to interactions between an active herbivore (a limpet) and *Fucus*. This limpet, and similar herbivore activity, are apparently absent from the Alaska sites.

Survival

The increase in human caused disturbances in marine habitats has added another motivation for doing ecological studies, fuelled this time by fear. This fear arises from both a perceived threat to our own survival, and to the survival of other species. As phycologists, we fear that algal populations, communities and ecosystems, will be destroyed by our growing demand for food and living space.

Hinrichsen (1998) provided the following data on population change and its impact on the marine habitat: "Over 50% of the entire population of the planet (3.2 billion people) lives and works within 200 km of the coast, on about 10% of the earth's land area. By 2025, 75% or about 6.3 billion people will probably live along the coast. South-east Asia has the highest percentage (85%) of coastal dwellers in the world, or about 400 million people."

Hinrichsen (1998) emphasized 3 main points regarding our coastal areas: A. "The world's coastal areas are being overwhelmed with people and pollution."; B. "As a result of the concentration of economic activities along coastlines... critical coastal resources – such as wetlands, mangroves, seagrasses, and coral reefs – are being plundered in the name of development and lost through inertia and neglect."; C. "... the inability of governments, with few exceptions, to craft and implement rational coastal management plans is having far-reaching consequences ...". These consequences occur due to onshore development, increased fishing and seaweed harvesting pressure, and increased run-off from agricultural fields (run-off brought on by the combination of an accelerating rate of removal of terrestrial vegetation and an increased use of fertilizers and pesticides, including in sites far away from the coast).

Yet, in the entire 275 pages of Hinrichsen's book, algae are only mentioned twice, and then only in the context of algal blooms. Are algae not threatened, or are we as phycologists not letting our voices be heard?

Studies generated by this fear of our own and other species' survival include topics such as:

Diversity – What is the diversity of different habitats, how do we best measure diversity, and how do we preserve it?

Function – What is the functional role of different algal species? Are there 'redundant' species that we can afford to loose (e.g. functional redundancy) (Menge et al., 1994; Power & Mills, 1995).

Overharvesting – To what extent can we harvest populations of algae and have them remain viable?

Marine Protected Areas – By what criteria should these be selected? On the basis of high diversity? The presence of one or more rare species? Their functional role as a source of gametes or spores? The viablity of the protected species/populations? Or some optimal combination of criteria such as these?

Introduced Algal Species – How do we recognize an introduced species? What properties makes for a successful invading species? Is it possible to predict the impact of any potential introduced species on a given marine algal community? Should we even be concerned about introduced algal species, e.g. *Sargassum muticum* (Yendo) Fensholt, *Eucheuma*, and *Undaria*, to various oceans?

Can marine algae serve as indicators of global changes? e.g. the sensitivity to temperature of sexual reproduction of many kelps is well known. Similarly, some tropical algae likely live near their temperature maxima. Are some of these algae now found in areas further from the equator?

The combination of the natural dynamism of ecosystems and the added impact of human change has greatly complicated our ability to understand marine

4

systems. For example, the collapse of the West Coast Salmon Fishery has been attributed to overfishing, destruction of habitat, and a long-term increase in ocean temperatures in the NE Pacific. The long-term warming in turn has been attributed to a natural cycle and also as due to global warming resulting from increased CO_2 emissions.

Making use of new technology

Many questions in Science are generated by the development of new technology. The 'answerable questions' in the life sciences changed dramatically, for example, with the advent of the electron microscope, and all its modern permutations, and have changed once again with the development of molecular biology.

For ecologists, molecular tools open the possibility of a much better understanding of, for example, what constitutes an algal population (what is the geographical extent of interbreeding individuals?), e.g. Kusumo (1998). The techniques of molecular ecology also provide powerful tools for understanding the life history of species, e.g. Sussmann et al. (1999) on the identity of green algal unicellular endophytes found in a variety of red algal blades. And of course it has opened our eyes to yet one more level of diversity, that found at the molecular level (e.g. Hommersand et al., 1999).

Economics

Increasingly, funds for research in all areas of science are becoming scarce and, as a result, governments, business, and other private funding agencies, are setting priorities for the allocation of these funds. The research priorities set by these organizations are likely to have a powerful influence on the direction of science and also may slow the rapid dissemination of scientific knowledge. Knowledge gained through research funded by special interest groups, e.g. industry, may be delayed in publication to enable the funding agency to gain some economic profit from the new data. It is ironic that such a possible delay in the sharing of information could come about at the very time that innovations such as the internet are speeding up our access to newly discovered facts.

Also, research funded by special interest groups may suffer from the association. For example, after the massive oil spill in Alaska from the Exxon Valdez in 1989, three different groups (Paine et al., 1996) sponsored research on impact and recovery: A. The

oil company, Exxon; B. The Trustees (an Alaska government interest group) and C. A U.S. Government Science Agency.

The most sound scientific study of recovery from the oil spill was the study carried out by Exxon; it used a random placement of quadrats, and generally larger quadrats. Nevertheless, its conclusions were widely disallowed, in part for good scientific reasons (Exxon's interpretation of recovery was that a similar percent cover had been achieved, even though the species composition and age structure differed markedly from non-oiled sites), but also because of perceived self-interest. Similarly, the results of the Trustees' study was tainted by the perception of self-interest, as it was seen to be in their interest to prove non-recovery, as this might lead to a larger reparation payment by Exxon. It is of interest that each of these two interest groups reached conclusions that favored themselves.

Yet, despite dismay at the Exxon Valdez related events, are we even focussed on the most important component of the problem? Ships off Canada's (East) coast repeatedly dump close to 100 liters of oily bilge to avoid paying pumping fees when in port. "The equivalent of the Exxon Valdez takes place off Newfoundland every 2–3 years" (Globe & Mail, 1999).

In Canada, we see targeted research funding occurring more and more often; the targets (such as the generation of jobs, or knowledge that will be useful in the health sciences) are set both by government priorities, and by industry (such as better ways to grow a particular algal crop). The setting of such guidelines is frequently justified by the argument that there is a limited amount of money available, hence choices must be made.

In the context of this conference, Mervis & Normile (1998) report that most of the S.E. Asian countries have defined common research areas for targetted funding. One of these four areas is biodiversity; this is a target because of the potential for new chemicals, especially bioactive chemicals, among the diversity of life still to be found in many of the S.E. Asian waters. One proposed solution for protecting biodiversity is the establishment of Marine Protected Areas (MPA's). How is this best done? One approach is the establishment of MPA's wherever and whenever they can be established. Another approach is to develop theory, or draw on existing ecological theory, to determine the placement of MPA's. This topic has recently been addressed by Phillips (1998), especially in the context of such actions in Australia, and their relevance to marine macro-algae.

Phillips (1998) identified the following problems related to current efforts in marine conservation:

1. Belief in the inexhaustibility of the oceans;
2. Lack of an inventory of macro-algal species;
3. MPA's exist, but no studies have examined their effect on macro-algae;
4. Transfer of ecological theory from terrestrial systems to marine systems.

Limitations of marine ecological studies

Just as different motivations have fuelled algal ecological studies, critical voices have been raised about how these studies were conducted, and what the generality and predictability of the results are.

Worldwide

Generality of results

Foster (1990) has argued that marine ecologists have frequently over-generalized results, attempting to proclaim ecological laws from experimental results obtained within a relatively limited geographical area. For example, in the northeastern Pacific, the otter is perceived to be both a 'Keystone Species' (e.g. Laur et al., 1988; Van Blaricom & Estes, 1988) in some sites and simply a predator in others (Foster & Schiel, 1988). The otter is a predator of sea-urchins which, in turn and when present, consume subtidal algae in great quantities. In the presence of otter predation, and thus reduced sea urchin herbivory, kelp forests increase greatly in abundance. This scenario has been repeatedly cited to claim its importance in areas where its role has not been studied.

Similarly, the important herbivore in many northeastern Pacific shores, the chiton *Katharina tunicata*, is said to have a similar range of roles, e.g. it is an important structuring agent of low intertidal communities in coastal Washington, U.S.A., where the kelp *Hedophyllum*, is a major constituent of the community (Duggins & Dethier, 1985), whereas elsewhere *Katharina* has only limited importance, e.g. in southeastern Alaska, where *Alaria*, another kelp species, dominates (Dethier & Duggins, 1988).

Another example of over-generalization has occurred in the case of phlorotannins, found in many brown algae, and their attributed widespread role as a herbivore deterrant (Littler et al., 1986; Duffy & Hay, 1990). However, additional research has shown that phlorotannins often do not deter smaller and less mobile herbivores (Hay & Fenical, 1992; Pavia et al., 1997). Furthermore, the suggestion that these chemicals may in general be induced by herbivory (Van Alstyne, 1988; Yates & Peckol, 1993) is negated by numerous studies suggesting this is not the case (Pfister, 1992; Steinberg, 1994, 1995). Phlorotannins are not so induced in, for example, *Sargassum*, and *Hedophyllum*. Instead, their presence has been shown to be more closely correlated to C:N ratios, so that high concentrations of phenolics are generally found in low nutrient waters. In addition, polyphenolics may also play a role in the protection of some algae from UV-B light (Pavia et al., 1997).

In this latter example, a simple and attractive hypothesis has given way to a much more complex picture, and this complexity is undoubtedly much closer to reality. A desirably simple explanation (e.g. that phloroglucinols are only herbivore deterrants, and that they are induced by herbivory) is unlikely ever to be true in ecology, nor is it likely in marine algal ecology. In each of the above examples, the problem was not that the facts were wrong (otters can be a major determinant of community structure, and for at least one species of *Fucus*, herbivory can induce higher phlorotannin levels) but rather that single experiments were over-generalized, or single hypotheses were tested, as discussed earlier. Foster's (1990) point is that the same organism in a different community context may have a quite different role, and that the critical difference is not necessarily easy to discern.

Vague and untestable hypotheses

Peters (1991) and others have commented on the tendency of ecologists to pose vague and possibly untestable hypotheses. Peters (1991) suggests that ecologists too often pose questions: (1) About 'ambiguous entities' such as community and stability. For example, initial questions about the stability of communities have led to a multitude of interpretations of stability itself, which in turn have led to such terms as 'local' and 'global' stability, and 'elasticity'. The ability to determine whether a community has any of these properties has lagged far behind our ability to propose the terms; (2) Answerable only by personal opinions, such as 'Why' questions; (3) Only answerable by an infinitely large research program.

Disagreement re. statistical analysis

Ecologists are called upon to analyze experimental results using appropriate statistics, yet frequently (increasingly?) the opinions of statisticians diverge as to what constitutes appropriate statistical methodology. One recurring problem relates to the fact that ecological data are often unable to meet the assumptions of common statistical tests.

Some marine ecologists/statisticians have provided guidelines for ecological data analysis [e.g. Underwood (1981) on the Use and Mis-use of ANOVA; and Day & Quinn (1989) on the appropriate choice of a post-hoc test]. However, what is the biologically inclined, but often statistically challenged, ecologist to make of the following? (1) Statements on Assumptions of Non-Parametric Analyses: Underwood (1997): "Note, however, that the Kruskal–Wallis procedure is not free from restrictive assumptions. ... Homogeneity of variances and independence of data are, however, also assumptions for the Kruskal–Wallis test, as is the requirement that the distributions are continuous and of the same general shape." (2) Zar (1996: 198–199): The Kruskal–Wallis test... "may be employed in instances where the latter is not applicable, in which cases it may in fact be a more powerful test. The nonparametric analysis is especially desirable... when the k samples do not come from normal populations, and it may also be applied when the k population variances are somewhat heterogeneous." (3) Fowler et al. (1998), referring to the Kruskal–Wallis test: "Non-parametric tests... do not require data to be normally distributed or to have homogeneous variance; i.e. they are distribution free."

Statement 1 above is apparently most correct, the others add confusion to the analytical task ecologists face.

Opinions differ regarding the necessity of incorporating the Bonferoni Correction (on setting the appropriate level of alpha) with multiple tests (Peres-Neto, 1999); defining the limits of the occurrence of pseudoreplication (Hurlbert, 1984), the 'correct' analysis of quadrats measured repeatedly but at different times (is a repeated measures analysis appropriate or not?), the increasing complexity of statistical analysis argued to be necessary for multiple choice feeding experiments (Peterson & Renaud, 1989; Roa, 1992; Manley, 1993), or the impact of Chaos Theory on our ability to draw inferences from any ecological data (Maurer, 1999).

The appropriateness of Bayesian Statistics has also been argued. For example, Ellison (1996) writes "Con-

ceptually, Bayesian inference is the most straightforward way of analyzing and interpreting our hypotheses in light of our data.", whereas Dennis (1996) concludes that "Bayesian and frequentist statistics cannot logically coexist. Until I see some new compelling Bayesian understandings of nature, I will not be a believer."

Let me clarify that it is not the disagreements or the several opinions about a statistical procedure that is the problem. Rather, it is the unwillingness of reviewers and ecologists in general to recognize that different opinions currently exist on these matters, and a lack of recognition that in these cases any of several procedures are acceptable at this time.

An apparent lack of agreed-upon standards on how ecology (and marine algal ecology) should be done

"You get a guy with four Ph.D.'s saying no fish were hurt, then you get a guy with four Ph.D.'s saying, yeah, a lot of fish were hurt. ... They just kind of delete each other out" [Barker (1994: 74), cited in Paine et al. (1996)].

As a result of disagreements about definitions of ecological terms and concepts, and apparent disagreements among statisticians regarding appropriate techniques of data analysis, as marine ecologists we often find ourselves at odds with each other about these matters. This manifests itself in our frequent inability to choose an appropriate statistical test (at least in the eyes of another ecologist), in our inability to accept that this disagreement exists and that thus more than one type of data analyses may be correct and, more generally, in the absence of an internally agreed-upon set of standards for what constitutes an acceptable paper for publication.

Peters (1991) argued similarly when he wrote that ecologists submitting papers for review work by different standards from those of the reviewers of those papers. He cited the unusually large proportion of submitted papers that are rejected by ecological journals (62%), compared to other science journals (27%), as evidence.

The infinite regress of causality

Peters (1991) also argued that ecologists are not as useful as they could be when it comes to proposing specific courses of action in the face of an ecological threat. He suggested that as ecologists are rarely convinced they fully understand an ecological system or process, the need for 'more studies' is more frequently stated than is advice given on a specific course of action.

Elizabeth Mann Borgese (1995) made the point that, since 1945, the information that science supplies has become a necessary element in the making of any political decision, e.g. in decisions about environmental policy. Yet she also asked "Is science able – or will it ever be able - to deliver the answers to questions which must be answered for sound political decisions to be made?". Her doubt about the utility of ecology to supply necessary advice is substantiated by the conclusions of many papers published in ecological journals which conclude that 'further research is required' or that 'more work must be done'.

To circumvent this problem, Peters (1991) proposed that we (ecologists) must avoid questions that ask 'Why'; instead he suggests we ask 'How much', 'How many', 'When' and 'Where' questions. He went even further and suggested that ecologists avoid questions that search for explanation, cause, mechanism, and understanding!

For example, in the context of the 1998 El Niño event, what is the likelihood that we can answer questions such as: Why was there a decrease in biomass of species X as a result of El Niño? A more answerable question might be "How much did the biomass of species X change during an El Niño year vs. other years?" In a sense this leaves begging the question of whether El Niño was the cause, but the likelihood of finding that answer is probably small. Nevertheless, an answer to the 'How Much' question is useful both to an ecologist and to policy makers attempting to judge the impact of another ENSO event.

Similarly one could ask "What is the mechanism by which *Katharina* (a herbivorous chiton) affects the abundance of *Hedophyllum*?" Instead, one might ask: "By how much must the density of *Katharina* be increased to produce a 50% decrease in juvenile kelp survivorship?"

The unwillingness of most funding agencies to support long term studies, ones that monitor the environment to establish a baseline of what constitutes 'normal'. This topic will be addressed below.

South-east Asia

All of the above problems apply to ecology in general, and thus to marine ecological phycology. Similarly, and again in the context of this conference, if this analysis is correct then these problems apply to S.E. Asia as well. Problems more specific to S.E. Asia are:

(A) The difficulty for local scientists critically to assess scientific research (e.g. peer review). This occurs when the predominant culture in the country is such that younger scientists do not argue with older ones because it is considered impolite to argue and question results produced by older colleagues. This observation is certainly not new with me; it was noted also by Mervis & Normile (1998) where an Asian biologist is quoted as saying "the culture has to be changed. . . by the researchers themselves. The culture manifests itself in deferring to superiors and avoiding risks in plotting out a course of research."

(B) Young scientists who did successfully attain their post-graduate degrees promptly disappeared into the administration of their home university, and their careers as scientists effectively came to a halt. This occurred despite the shortage of qualified working scientists in many of the S.E. Asian countries, e.g. perhaps 2000 in Malaysia and 4000 in Indonesia (Mervis & Normile, 1998).

(C) Choice of research topic – Basic or Applied? Within many of the countries of S.E. Asia the debate on this issue is an active one (Mervis & Normile, 1998; pers. obs.). Administrators and researchers are quoted (Mervis & Normile, 1998) with opinions from ". . . Research for its own sake is not something that we can afford." And ". . . Solving a practical problem is better than producing one publication that nobody reads.", to ". . . The pool of knowledge is drying up, and the 21st century will belong to those who are doing fundamental research.", and ". . . We need to do basic biology before we can apply it to biotechnology or genetic engineering."

When these opinions translate into policy, the ability of an individual scientist to make a choice of research direction may be limited by the policy dictates of the local government, and also by the lack of coherence within the larger scientific community as represented by foreign funding agencies. Again, to hark back to my own experience, while CIDA (Canadian International Development Agency) was working hand-in-hand with government agencies to promote basic science, another foreign-aid agency, funded by a different country, was actively promoting applied science (also in Mervis & Normile, 1998). Similarly, several marine oriented programs were operating at the same time in the same place, and frequently this resulted in a duplication of effort or in a competition for scarce resources (e.g. the university staff were being pulled in multiple directions by foreign projects, since each project required local resource personnel).

Given the limited local resources (money and faculty liaison) this was a counterproductive situation.

Moving forward

How do I see our field moving forward? Do we take Gertrude Stein's comment to heart, or do we proceed with Albert Schweitzer's comments in mind?

Generality of results

I think it important that as ecologists we do not accept easily the generality of the results of any given experiment. We need to accept the necessity of repeating experiments both in different habitats, and in the same habitat but at different times. As both Foster (1990) and Hurlbert (1984) suggested, we should not assume homogeneity and generality until we have done the requisite repetitions, and we should take care to not commit mensurative pseudoreplication. Furthermore, as pointed out by Csada et al. (1996), we must also become more aware of good experiments with results that indicate a lack of significance of some factor. Hopefully, if we recognize the validity of such results, we can make a case for their publication in one form or another. With the increased use of meta-analysis of data, the lack of publication of papers with negative results will seriously skew such analyses (Csada et al., 1996).

Vague and untestable hypotheses

This has been addressed by various ecologists and philosophers, e.g. Peters (1991) and Popper (1968), to name two, and in previous parts of this paper.

Disagreements re. statistical analyses

There has been progress here as well. Papers by Underwood (1981, 1997) on Use and Mis-use of AN-OVA, by Day & Quinn (1989) on appropriate choice of post-hoc tests, by Hurlbert (1984) on Pseudoreplication, and a group of papers on the appropriate analyses of single and multiple choice feeding experiments (Peterson & Renaud, 1989; Roa, 1992; Manley, 1993) have resulted in a raised common awareness among ecologists of potential problems.

However, there are still disagreements on even the procedures discussed by the authors mentioned above. Until we get the definitive word on such analytical problems, I propose that we attempt to maintain an awareness of the conflicting opinions regarding these issues. Furthermore, that we recommend to editors and reviewers that perhaps of the options available, any given one may be as correct as we can be at this time. In other words, we (ecologists) should be able to achieve some agreement on standards so that we can get on with the ecology and not get bogged down in statistical matters on which even the statisticians disagree.

Lack of agreed upon standards on how Ecology should be done

I do not mean that marine ecologists do not have standards regarding what constitutes an acceptable paper in our discipline, but rather that among ourselves there is much disagreement on these matters. Surely it behooves our science to discuss these issues and come to agree on some common set of standards. Once agreed, these standards should be promulgated. This does not in any way imply that discussion should be stifled, but rather that there should be a recognition of disagreement based on currently inadequate knowledge, and that this should not be the basis for rejecting an otherwise good piece of research.

Infinite chain of causality

This matter has already been referred to when discussing the ideas of Peters (1991).

The unwillingness of most funding agencies to support long term studies, ones that monitor the environment to establish a baseline of what constitutes 'normal'

The necessity for studies of this sort are proclaimed each time there is a an ecological disaster (real or imagined). Events such as the Exxon Valdez oil spill in Alaska (Paine et al., 1996), and the *Acanthaster* events in the tropical Pacific (Sapp, 1999), are just two instances of the absence of baseline knowledge, and a subsequent inability to judge the extent, and even the reality, of the suggested impact.

Paine et al. (1996) wrote that: "In our estimation, the research initiated after EVOS (Exxon Valdez Oil Spill) failed in three ways. First, much of it was carried out to assess injury in terms of changes in population size, using species lacking adequate baseline information.". Sapp (1999) stated: "Although the issues were similar (among *Acanthaster* events and coral reef bleaching), the response of the coral-reef science community differed dramatically from that in the early controversy surrounding the crown-of-thorns. Those differences highlight . . . the coral reef scientists' hard-won awareness of the general need for baseline data from which to distinguish between human-induced changes and long-term natural processes."

In conclusion, a greater common awareness among marine ecologists and phycologists of the issues raised above will, I believe, go a long way to establishing a more coherent set of standards for judging the merits of scientific papers in our field, will accelerate the ability of young marine ecologists to become succesfully published scientists, and may also help to establish a unity amongst ourselves that will strengthen our ability to influence the decisions of governments for funding directions, and increase our ability to provide coherent guidelines regarding environmentally sensitive actions to managers and politicians.

References

Barker, E., 1994. The Exxon trial: a do it yourself jury. Am. Lawyer, Nov.: 68–77.

Borgese, E. Mann, 1995. The challenge to marine biology in a changing world: future perspectives, responsibility, ethics. Helgolander wiss. Meeresunters. 48: 895–902.

Csada, R. D., P. C. James & R. H. M. Espie, 1996. The 'file drawer problem' of non-significant results: does it apply to biological research? Oikos 76: 591–593.

Day, R. W & G. P. Quinn, 1989. Comparisons of treatments after an analysis of variance in ecology. Ecol. Monogr. 59: 433–463.

Dennis, B., 1996. Should ecologists become Bayesians? Ecol. Appl. 6: 1095–1103.

Denny, M., 1995. Predicting physical disturbance: mechanistic approaches to the study of survivorship on wave-swept shores. Ecol. Monogr. 65: 371–418.

Dethier, M. N. & D. O. Duggins, 1988. Variation in strong interactions in the intertidal zone along a geographic gradient: a Washington-Alaska comparison. Am. Nat. 124: 205–219.

Duffy, J. E. & M. E. Hay, 1990. Seaweed adaptations to herbivory. BioScience 40: 368–375.

Duggins, D. O. & M. N. Dethier, 1985. Experimental studies of herbivory and algal competition in a low intertidal habitat. Oecologia 67: 183–191.

Ellison, A. M., 1996. An introduction to Bayesian inference for ecological research and environmental decision making. Ecol. Appl. 6: 1036–1046.

Elton, C. S., 1927. Animal Ecology. Sedgewick and Jackson, London.

Foster, M., 1990. Organization of macro-algal assemblages in the northeast Pacific: the assumption of homogeneity and the illusion of generality. Hydrobiologia 192: 21–33.

Foster, M. & D. R. Schiel, 1988. Kelp communities and sea otters: keystone species or just another brick in the wall? In Van Blaricom, G. R. & J. A. Estes (eds), The Community Ecology of Sea Otters. Springer-Verlag, Berlin, Germany: 92–108.

Fowler, J., L. Cohen & P. Jarvis, 1998. Practical Statistics for Field Biology. John Wiley & Sons, New York: 110 pp.

Fulghum, R., 1999. Words I Wish I Wrote. Harper Collins, New York. Globe & Mail, April 3/99: A1.

Hay, M. E. & W. Fenical, 1992. Chemical mediation of seaweed herbivore interactions. In John, D. M., S. S. Hawkins & J. H. Price (eds), Plant–Animal Interactions in the Marine Benthos. Systematics Association Special Volume, Clarendon Press, Oxford: 319–337.

Hinrichsen, D., 1998. Coastal Waters of the World: Trends, Threats, and Strategies. Island Press, Washington, D.C.

Holloway, M., 1996. Sounding out science. Scientific American, October: 106–112.

Hommersand, M. H., S. Fredericq, D. W. Freshwater & J. Hughes, 1999. Recent developments in the systematics of the Gigartinaceae (Gigartinales, Rhodopyta) based on rbcL sequence analysis and morphological evidence. Phycol. Res. 47: 139–151.

Hurlbert, S. H., 1984. Pseudoreplication and the design of ecological field experiments. Ecol. Monogr. 54: 187–211.

Krebs, C. J., 1985. Ecology, the Experimental Analysis of Distribution and Abundance, 3rd edn. Harper & Row, New York.

Kusumo, H. T., 1998. A Parallel Assessment of Morphological and Genetic Diversity of the Kelp *Alaria marginata*. Ph.D. Thesis, Department of Biology, Simon Fraser University, Burnaby, B.C. Canada: 93 pp.

Laur, D. R., A. W. Ebeling & D. A. Coon, 1988. Effects of sea otter foraging on subtidal reef communities off central California. In Van Blaricom, G. R. & J. A. Estes (eds), The Community Ecology of Sea Otters. Springer-Verlag, Berlin, Germany: 151–167.

Littler, M. M., P. R. Taylor & D. S. Littler, 1986. Plant defense associations in the marine environment. Coral Reefs 5: 63–71.

Manley, B. F. J., 1993. Comments on the design and analysis of multiple-choice feeding experiments. Oecologia 93: 149–152.

Maurer, B. A., 1999. Untangling Ecological Complexity: The Macroscopic Perspective. The University of Chicago Press, Chicago & London.

Mcintosh, R. P., 1976. Ecology since 1900. In Taylor, B. J. & T. J. White (eds), Issues and Ideas in America. University of Oklahoma Press, Norman: 353–372.

Menge, B. A., E. L. Berlow, C. A. Blanchette, S. A. Navarrete & S. B. Yamada, 1994. The keystone species concept: variation in interaction strength in a rocky intertidal habitat. Ecol. Monogr. 64: 249–286.

Mervis, J. D. & D. Normille, 1998. Science in Southeast Asia. Science 279: 1465–1482.

Paine, R. T, J. L. Ruesink, A. Sun, E. L. Soulanille, M. J. Wonham, C. D. G. Harley, D. R. Brumbaugh & D. L. Secord, 1996. Trouble on oiled waters: Lessons from the Exxon Valdez oil spill. Ann. Rev. Ecol. Syst. 27: 197–235.

Pavia, H., G. Cervin, A. Lindgren & P. Aberg, 1997. Effects of UV-B radiation and simulated herbivory on phlorotannins in the brown algal *Ascophyllum nodosum*. Mar. Ecol. Prog. Ser. 157: 139–146.

Peters, R. H., 1991. A Critique for Ecology. Cambridge University Press, New York.

Peterson, C. H. & P. E. Renaud, 1989. Analysis of feeding preference experiments. Oecologia 80: 82–86.

Peres-Neto, P. R., 1999. How many statistical tests are too many? The problem of conducting multiple ecological inferences revisited. Mar. Ecol. Prog. Ser. 176: 303–306.

Pfister, C. A., 1992. Cost of reproduction in an intertidal kelp: patterns of allocation and life history consequences. Ecology 73: 1586–1596.

Phillips, J. A., 1998. Marine conservation initiatives in Australia: their relevance to the conservation of macroalgae. Bot. mar. 41: 95–103.

Popper, K., 1968. Conjectures and Refutations: the Growth of Scientific Knowledge. Harper Torchbooks, Harper & Row, New York.

Power, M. E. & L. S. Mills, 1995. The keystone cops meet in Hilo. TREE 10: 182–184.

Roa, R., 1992. Design and analysis of multiple-choice feeding-preference experiments. Oecologia 89: 505–515.

Sapp, J., 1999. What is Natural? Coral Reef Crisis. Oxford University Press, Oxford.

Steinberg, P. D., 1994. Lack of short-term induction of phlorotannins in the Australian brown algae *Sargassum vestitum* and *Ecklonia radiata*. Mar. Ecol. Prog. Ser. 121: 129–133.

Steinberg, P. D., 1995. Interaction between the canopy dwelling echinoid *Holopneustes purpurescens* and its host kelp *Ecklonia radiata*. Mar. Ecol. Prog. Ser. 127: 169–181.

Stille, A., 1999. The man who remembers. The New Yorker: Feb. 15: 50–63.

Sussmann, A., B. K. Mable, R. E. DeWreede & M. Berbee, 1999. Identification of green algal endophytes as the alternate phase of *Acrosiphonia* (Codiolales, Chlorophyta) using ITS1 and ITS2 ribosomal DNA sequence data. J. Phycol. 35: 607–614.

Underwood, A. J., 1981. Techniques of analysis of variance in experimental marine biology and ecology. Oceanogr. mar. biol. Ann. Rev. 19: 513–605.

Underwood, A. J., 1985. Physical factors and biological interactions: the necessity and nature of ecological experiments. In Moore, P. G. & R. Seed (eds), The Ecology of Rocky Coasts. Hodder & Stoughton, London: 372–390.

Underwood, A. J., 1991. The logic of ecological experiments: a case history from studies of the distribution of macro-algae on rocky intertidal shores. J. mar. biol. Ass., U.K. 71: 841–866.

Underwood, A. J., 1997. Experiments in Ecology. Cambridge University Press, New York.

van Alstyne, K. L., 1988. Herbivore grazing increases polyphenolic defenses in the intertidal brown algae *Fucus distichus*. Ecology 96: 655–663.

Van Blaricom, G. R. & J. A. Estes (eds), 1988. The Community Ecology of Sea Otters. Springer-Verlag, Berlin, Germany.

Yates, J. & P. Peckol, 1993. Effects of nutrient availability and herbivory on polyphenolics in the seaweed *Fucus vesiculosus*. Ecology 74: 1757–1766.

Zar, J. H., 1996. Biostatistical Analysis. Prentice Hall, New York.

Hydrobiologia **512:** 11–20, 2004.
P.O. Ang, Jr. (ed.), Asian Pacific Phycology in the 21ˢᵗ Century: Prospects and Challenges.
© 2004 *Kluwer Academic Publishers.*

The past, present and future of phycology in China

C. K. Tseng

Institute of Oceanology, Chinese Academy of Sciences, Qingdao 266071, China
E-mail: cktseng@qd-public.sd.cninfo.net

Key words: Chinese phycology, algal taxonomy, ecology, algal biotechnology, cultivation, science history

Abstract

Algae have been part of Chinese life for thousands of years. They are widely used as food and have been cited in Chinese literature as early as 2500 years ago. However, formal taxonomic studies on Chinese algae were initiated by foreign scientists only about 200 years ago, and by Chinese phycologists only about 90 years ago. This paper summarizes the history of modern phycological studies on Chinese algae and provides an overview of the achievements of phycological studies by Chinese scientists, especially on algal taxonomy, morphology, genetics, ecology and environmental research, physiology, biotechnology, algal culture, applied phycology and space phycology, in the last century. Recent development in phycological research focuses on algal floristic and molecular systematics, algal molecular biotechnology, applied phycology including micro and macroalgal cultivation and algal product development, and the roles of algae in environmental pollution control. These areas will also be the main focuses of Chinese phycological research in the foreseeable future.

Introduction

Although phycology, the study of algae, was started in the west, the earliest bibliographic record of the algae appeared in an old Chinese book *Shijing* (Poem Classic), at least 2500 years ago. This paper reviews only contributions to Chinese phycology.

Reports on taxonomy of Chinese algae by foreign scientists

Modern study of the algae started about 200 years ago by Dawson Turner (1809) on a Chinese alga, *Fucus tenax* Turner (= *Gloiopeltis tenax* (Turner) Decaisne), a silk sizing seaweed. In the last century six other scientists had dealt with Chinese seaweeds, including C. Agardh (1820), Montagne (1840), J. Agardh (1848, 1879), Kuetzing (1849), Martens (1866) and Debeaux (1875). In the present century, there were at least 14 foreign scientists interested in Chinese seaweeds, including Gepp (1904), Cotton (1915), Grunow (1915, 1916), Ariga (1919), Collins (1919), Reinbold (1919), Cowdry (1922), Howe (1924, 1934),

Tilden (1929), Yamada (1925, 1950), Okamura (1931, 1936), Setchell (1931a, b; 1933, 1935, 1936), Grubb (1932) and Noda (1971).

The first report on Chinese fresh water algae was undoubtedly that by the French scientist P. Petit who published a paper on Ningbo algae in 1880. Borge published in 1899 a collection of freshwater algae by the Russian military officer N.-M. Przewalski from North and Northwest China and determined by K. Maximoviz. The German scientists, E. Lemmerman and H. Schauinland, reported their collection of Asian algae and algae from Jinjiang to Zhenjiang in 1899, 1901, 1907. The Russian P.K. Kozov collected algal specimens from Mongolia and Tibet in 1899-1901, and the lists were published in 1907 by C.S. Mereshkowsky. The American scientist N.G. Gee published a series of papers on Zhejiang and Jiangsu algae in 1919 and 1926. The French L. Vaillant and C. Gaudichaud collected some fresh water algae at Macau in 1836–1837; these algae were reported in 1844–1846 by C. Montagne. The famous Swedish explorer A. Sven Hedin visited China six times and collected some algae which were later published by F. Hustedh in 1922, W. Wille in 1930 and Borge in 1934. Another famous explorer, the Austrian H. Handel-Mazzetie came to China in 1913 and had also collected some algae which were studied and published by H.Skuja in his Symbolla Sinica in 1937. Other investigators were J.E. Tilden in 1920, Okada in 1932 and 1936, K. Negora in 1940, 1941, 1943 and 1953, Mashiko in 1951 and M. Noda in 1963 and 1971. This brief discussion on the reports of foreign scientists on Chinese fresh-water algae is extracted from a report on the history of systematic classification of Chinese freshwater algae by Professors Bi, Hu and Liu (Bi et al., 2001) containing the necessary references.

Modern study of phycology by Chinese scientists

Modern study of phycology by Chinese scientists was undoubtedly initiated by C.S. Chien (Professor Qian Chong Shu) who published a physiology paper in 1917 on 'Peculiar Effects of Barium, Strontium and Cesium on *Spirogyra*' (Chien, 1917). Professor H.H. Chung came back to China after some years of postgraduate study at Harvard University in U.S. Apparently, he took a course on algae at the Woods Hole Marine Biology Laboratory in the early twenties and became interested in phycology. In his trip to the vicinity of Xiamen and other places in Fujian, Guizhou and Hubei Provinces, he collected not only vascular plant specimens but also algal specimens including seaweeds and freshwater algae. He did not study the algal specimens himself but sent them to Dr N.L. Gardner of U.S. He even taught a course on algology at Xiamen University in 1929 in which I had the opportunity to attend.

Taxonomic phycology

The first Chinese to study taxonomic phycology in China is Wang Chu-Chia (Prof. Wang Zhi-Jia) who published his first paper in 1930 (Wang, 1930), and altogether 12 papers. The next one is Li Liang-Ching (Dr L.C. Li) who published in 1932 an abstract of his dissertation of Doctor's degree (Li, 1932), and altogether 29 papers. The third phycologist is C.K. Tseng (Dr Zeng Cheng-Kui) who published his first phycological paper in 1933 (Tseng, 1933), and altogether, to date, 344 papers, of which 97 concerned taxonomy and resources, and edited 12 books. The fourth phycologist is Jao Chin-Chih (Dr Rao Qin-Zhi) who published his first paper on Chinese phycology in 1935 (Jao, 1934), and altogether 76 phycological

papers and edited two books. The fifth phycologist is Nie Dashu (Prof. Ni Da-Shu), a protozoologist turning to a Dinophyta phycologist and published his first paper in 1934 (Wang & Nie, 1934), and altogether 12 papers. Professor Nie eventually became the first Chinese specialist on aquatic animal diseases. The sixth phycologist is Chin Teh-Chiang (Prof. Jin De-Xiang) who was originally a zoologist and after obtaining M.Sc. at Lingnan University in 1935 became interested in phytoplankton; he published his first phycological paper in 1937 (Chin, 1937) and studied diatoms ever since, publishing 65 papers and four books. The seventh phycologist is Chu Hao-Ran (Zhu Hao-Ran), a former student of Dr Jao who published his first paper in 1944 (Chu, 1944), and altogether 40 papers and one book. The eighth phycologist is Ley Shang-Hao (Prof. Li Shang-Hao), also a former student of Dr Jao, published his first phycological paper in 1944 (Ley, 1944), and altogether 9 papers and one book.

The above eight phycologists, including three marine and five freshwater phycologists, were practically the few scientists devoted to the study of algae before 1949 who continued their studies of taxonomic phycology after 1949. After the establishment of the People's Republic of China, the Chinese people especially people of the coastal provinces, pay more attention to the algae and a few universities even offer courses in phycology. Many scientists are involved in the study of different phases of algae. In taxonomic study of the algae, we may mention just a few, such as Profs Chiang Young-Ming of Taiwan and Fan Kung-Chu of mainland China, both students of the late Dr George Papenfuss of U.S., Profs Zheng Bo-Lin, Zhang Jun-Fu, Xia Bangmei and Lu Baoren, students of Dr C.K. Tseng, Prof. Bi Lie-Jiao, a student of Dr C.C. Jao.

As is in the other botanical and zoological sciences, taxonomy always preceeds other sciences. A total of 40 000 specimens of marine algae and 30 000 specimens of freshwater algae have been collected respectively by the Institute of Oceanology and Institute of Hydrobiology of the Chinese Academy of Sciences (CAS) alone. The publication of the Cryptogamic Floras of China started in the seventies, consisting of five floras (1) Marine algal flora, (2) Fresh water algal flora, (3) Fungal flora, (4) Lichen flora and (5) Mosses and Liverworts flora. The algal flora was split into two because persons involved and collecting methods in these studies are different from one another. In the marine algal flora (Flora Algarum Marinarum Sinicarum) two volumes have been published: Vol. 2, Rhodophyta, No. 5 Ahnfeltiales, Gigartinales and Rhodymeniales by Xia & Zhang (1999) describing 17 families, 40 genera and 104 species; and Vol. 3, Phaeophyta No. 2 Fucales by Tseng & Lu (1999) describing 3 families, 6 genera and 141 species. Two more volumes on Ceramiales (Rhodophyta) by Zheng Bo-Lin et al. and on Centricae (Bacillarophyta) by Guo Yu-Jie are in the process of editing and will be ready for printing soon. In the fresh water algal flora (Flora Algarum Sinicarum Aqua Dulcis) six volumes have already been published: Vol. 1, Zygnemataceae by C.C. Jao (1986); Vol. 2, Chroococcophyceae by H.R. Chu (1991); Vol. 3, Charophyta by Han Fushan & Li Yaoying (1994); Vol. 4, Centric diatoms by Qi Yuzao (1995); Vol. 5, Ulothricales, Ulvales, Chaetophorales, Trentepohliales and Sphaeropleales by Li Shang-Hao & Bi Lie-Jiao (1998); and Vol. 6, Euglenophyta by Shi Zhixin (1999). Two more volumes are in the process of being edited, and will be ready for printing soon.

Basic study on morphology, ecology and physiology of the algae

Algal morphology
Tseng & Chang (1954, 1955) started their investigation on the life history of *Porphyra tenera* Kjellman in early 1952, independent of Kurogi of Japan and solved the problem of the 'seed' of *Porphyra* in its cultivation. This they called 'conchospore' which has been followed by most phycologists. In the eighties, Tseng & Sun (1989) studied the chromosome numbers of the *Conchocelis* stage of *Porphyra* and the alternation of the nuclear phases and chromosome numbers. They revealed the astonishing phenomenon that meiosis occurs in the germinating conchospores (Tseng & Sun, 1989).

Algal ecology
The most difficult job in algal ecology takes place in marine expeditions. One has to collect the specimens of phytoplankton by special devices, to analyze the collected specimens, to identify them and to write out reports. Algal ecological work actually started in 1954 when we initiated fishery expedition under the ichthyologist Professor Zhang Xiao-Wei and the ecological group was led by Dr Zhu Shup-Ping with Guo Yu-Jie as his assistant. The population was analyzed and the ecological characteristics of the dominated species decided. This was followed by national ocean-

ological expeditions (1958–1962), coastal zones and beaches investigation (1981–1986), islands investigation (1990–1992), and Xisha and Nansha Islands expeditions (1980–1999). In these studies, Guo was in charge of the phytoplankton study. She contributed a paper on the primary productivity and phytoplankton of the Kuroshio (Y.J. Guo, 1991). She believed that the Kuroshio area was one of the unproductive regions in the world oceans, although it was a little more productive than the regions near the Equator. In a study of the characteristics of phytoplankton distribution in the Yellow Sea, Guo & Zhang (1996) pointed out that the horizontal and vertical distribution patterns of the phytoplankton in the Yellow Sea in May and September in 1992 were similar but the average abundance and species number of phytoplankton in September were higher than those in May. Diatoms dominated the 142 species of phytoplankton identified.

Since the seventies, our scientists paid special attention to marine pollution ecology, especially the ecology of red tides. On the ecological dynamics of the red tide, it was found that between *Noctiluca scintillans* (Macartney) Ehrenberg and *Skeletonema costatum* (Grev.) Cleve, *Prorocentrum minimum* (Pavilard) Schiller and *Skeletonema*, there was in existence a definite interspecific competition and that the appearance of different red tides was related to the N/P ratio (Zou, 1999). Investigation on the poisonous substance present in shellfish may indicate the distribution of the poisonous algae (Zhou et al., 1999).

Gymnodinium mikimotoi Miyake *et* Kominami *ex* Oda was one of the principal red tide organisms in the red tide occurring in Hong Kong and Guangdong Province, resulting in the death of many aquaculture fish in 1998 (Qi et al., 2004). An investigation was conducted by Jinan University on the mechanism of this Dinophyte on the fish and found out that this Dinophyte can cause swelling on the upper skin tissue of the fish gills. This resulted in the complete obstruction of the gills, leading to the death of the fish.

Algal physiology
In the early forties when C.K. Tseng worked on a special agar project, he initiated a study of the photosynthesis of *Gelidium cartilagineum* (L.) Gaillon and published a paper in 1946 (Tseng & Sweeney, 1946). In the sixties a study was made on the pigment system and photosynthesis of *Porphyra yezoensis* Ueda (Zhou et al., 1966). Unfortunately the researches had to stop in late 1966 because of the so-called 'Cultural Revolution'. The work was resumed in 1973, and a

comparative study of the spectrum absorption of some species of green, brown and red seaweeds was made and published in 1974 (Zhou et al., 1974). A series of papers on comparative photosynthesis of benthic seaweeds was published by Tseng et al. (1980, 1981a, b). The discovery of *Prochloron* in the Xisha (Paracel) Islands (Zeng et al., 1982), led us to discussion on the evolution of photosynthetic organisms (Tseng & Zhou, 1983a, b; 1984).

Algal genetics

Talking about algal genetics, we must pay due respect to the late Prof. T.C. Fang (Fang Zong-Xi), the founder of this science in China. Fang was trained as a human geneticist in England and returned to China in 1950. In 1953, he accepted the invitation of Shandong University, then in Qingdao, and joined the staff. Later in 1959 when the Shandong University moved to Jinan, he stayed and joined the staff of Shandong College of Oceanography. He was invited by the Institute of Oceanology, CAS, to be a part-time research fellow and in charge of genetic research. He used the cultivated *Laminaria* as the main subject for genetic research. In 1959, the principal problems of the cultivation of *Laminaria japonica* Aresch. had been solved and large-scale cultivation had just started. The problems concerning genetics of the *Laminaria* became significant. The *Laminaria* employed is the wild natural population, rather than selected strains. However, *Laminaria* has sexual reproduction. In the spore collecting process, two or more plants were used. The resulting sporophytes were a mix of those from hybridization and those from a single plant self-fertilization. Fang made an experiment with both and showed that hybrid sporophytes exhibited a two times faster in growth in area than those obtained from self fertilization (Fang & Jiang, 1962).

From 1959–1984, Fang and his students at the Institute of Oceanology, CAS and the Shandong College of Oceanology conducted 25 years of genetical research on *Laminaria* and later also on *Undaria*. On the conviction that the cultivated *Laminaria* is a hybrid, they employed self-fertilization to produce a few new strains of *Laminaria*, such as the broad leaf strain 'Haiqing No. 1', the long leaf strain 'Haiqing No. 2', and the thick leaf strain 'Haiqing No. 3'. They obtained clones of male and female gametophytes and discovered that parthenogenetically developed female gametophytes gave rise to female sporophytes which produced zoospores that grew to become female sporophytes (Fang & Dai, 1984).

The Shandong College of Oceanography, in co-operation with Rongcheng *Laminaria* Seedling Station and Rongcheng Aquaculture Station, used a selected female gametophyte to cross with the male gametophyte from a thick frond sporophyte and obtained a new strain called 'Danhai No. 1' which gave a higher yield and better quality than the best variety used in cultivation at that time (Fang et al., 1983). In 1981, Fang and his group employed the hybrid between the female haploid clone No. 10 and the male haploid cell introduced from Hokkaido, Japan. A new strain was obtained, the 'Danza No. 10', which was estimated to increase the yield by 30% (Fang et al., 1985). The bountiful *Laminaria* harvest became a reality and this must be partially credited to Prof. Fang.

Studies on microalgae, biotechnology and space phycology

Microalgae

In the fifties, the investigation on feeding marine juvenile aquatic animals with some microalgae such as *Tetraselmis*, *Phaeodactylum*, *Pavlova* was carried out. The animal growers were provided with the seeds of these algae and instructions for their cultivation. In the early sixties, cultivation of the freshwater alga *Chlorella* sp. in large scale was tried but failed. In the eighties, the brine alga *Dunaliella* was started to be cultivated in large scale and *Dunaliella* cultivation is now an established industry (Guo, 1991). In the seventh five-year plan, one of the problems was to decrease the dependence of aquaculture on imported fish protein. It was decided to study three microalgae because of their high protein contents, namely, *Dunaliella*, *Anabaena* and *Spirulina*. It was found that *Spirulina* gave the best and most proteinaceous products. It was further found that *Spirulina* is an excellent health food for humans (Wu et al., 1993). During the following years, there had been a *Spirulina* 'fever' in Chinese society and more than 100 enterprises were involved in *Spirulina* production. There are now still quite a few enterprises producing tablets for human consumption (Liang et al., 2004).

Biotechnology

Biotechnology has been effected in marine algae in the following six categories: (1) In hybridization technique, hybrids of *Laminaria japonica* have been obtained. (2) In cellular and protoplast technique, cultivation of *Porphyra haitanensis* Chang *et* Zheng has been effected with vegetative cell; growth of isolated seaweed protoplast of *Ulva, Enteromorpha, Monostroma, Porphyra* and *Chondrus* has been obtained. (3) Algal immobilization technique has been attempted for seedstock and seedling production. (4) In tissue culture technique, tissue cultures of *Laminaria japonica* and *Undaria pinnatifida* (Harv.) Sur. have been effected. (5) In processing of natural products, including bioactive substances, some work on anti-tumor and anti-radiation effect of marine algae has been done and the hypolipodermis medicine PSS from *Laminaria japonica* has been awarded golden medal in the 15th International Fair of Invention in Yugoslavia. (6) In biotechnology industry, we have promoted mass cultivation of *Dunaliella* for the β-carotene and that of *Spirulina* for its high protein content (Tseng & Qin, 1991).

Space phycology

Space biology is quite a new thing in China, especially space phycology for which we have to thank the Institute of Hydrobiology, Chinese Academy of Sciences. A good discussion of this research is found in Liu et al. (2001). In the Chinese Journal of Space Science, Vol. 17, supplement 1997, several papers were published by Prof. Y.D. Liu and his colleagues. In there, Prof. Liu and his colleagues summarized the interesting points about microgravity biology, space physiology, cell culture and tissue engineering, space exploitation and some details concerning gravi-sensing (Hu & Liu, 1997a, b; Song et al., 1997). They found in the flight experiment that in *Chlorella pyrenoidosa* Chick the number of pyrenoides conspicuously diminished and in *Anabaena* its fat body significantly decreased in size. Their experiments on *Anabaena oryza* Fritsch strains retrieved from space flight and reflight indicated two types of biological responses, the recoverable phenotype responses and the heritable genotype responses (Hu et al., 1996; Hu & Liu, 1997c). Their study of a strain of the microalga *Anabaena* carried in the retrievable canister for 15 days showed that the growth rate of the alga was slower than that of the ground control (Hu et al., 1997). In another experiment, nine species of algae were flown in space for eight or fifteen days, then retrieved and analyzed in the laboratory (Hu & Liu, 1997b). The results indicated that the algae have a strong adaptability to space environment. Cytological observation on *Anabaena siamensis* Antarikanonda after space flight showed obvious difference between the space flight and the ground control samples (Chen et al.,

1997). The possible mechanisms of responses of *Dunaliella salina* (Dunal) Teodoresco to simulated microgravity showed the plasma membrane to be the most direct site of paraperception and a theoretical model for micro-gravi-sensing, transduction and responses of the organism was proposed (Hu & Liu, 1997d).

Applied studies on cultivation of seaweeds and the seaweed industry

Cultivation of algae
Cultivation of seaweeds by traditional methods, to the knowledge of the author, dealing with *Gloiopeltis* and *Porphyra* has been in existence in China for several hundred years. Modern methods of cultivation are, however, only about forty something years old. Rafts in various forms are generally applied. Three kinds of methods are practiced.

Zoospores method, the Laminaria *type of mariculture*
The zoospores of *Laminaria* or *Undaria* are collected and cultured. *Laminaria japonica* is a cold temperate alga and can survive in summer water temperature of 20–25 °C, but cannot survive in temperature above 25 °C for a long time. We have devised a summer sporeling method in which zoospores are collected in early June, the gametophyte and young sporophyte stages pass the hot summer in cooled rooms of 10 °C, and the young sporophytes are taken out to the sea in Autumn when water temperature gets down to 20 °C. The summer sporeling method is very suitable for cultivation of cold temperate plants in warm temperate or even subtropical regions (Tseng et al., 1955; Tseng, 1981).

Conchospore method, the Porphyra *type of mariculture*
This is especially applicable for the cultivation of purple laver, *Porphyra* spp. The leafy *Porphyra* is the object of cultivation. Carpospores from *Porphyra* are planted on mollusc shells in late Spring. The carpospores will penetrate into the shell and develop the filamentous sporophyte stage – the *conchocelis* stage. This will eventually give rise to conchospores which upon germination will grow to become the leafy *Porphyra* (Tseng & Chang, 1955; Tseng, 1981). The filamentous thalli of *Bangia fusco-purpurea* (Dillw.) Lyngb. are also cultivated by this method.

Vegetative multiplication method, the Betaphycus *type of mariculture*

This is applicable to quite a few species of seaweeds beside *Betaphycus*, such as *Gracilaria*, *Kappaphycus* and *Hizikia* (= *Sargassum fusiforme* (Harv.) Setch.). Carpospores of *Gracilaria lemaneiformis* (Bory) Daws. upon germination will give rise to a callous like structure which will have to take about half a year to grow up to become young sporelings and another half year to grow to become an adult plant. For practical purposes, the growth rate is too slow. In the vegetative multiplication method, cuttings of *Gracilaria* are planted on raft during the best season for their growth and they grow to 100–200 times their original thallus in weight in one season. In the case of *Gelidium* and *Hizikia*, very young plants are employed.

Fertilizer application and Laminaria *transplantation to the south*

In the cultivation of *Laminaria*, a method of applying fertilizer in the open sea was devised (Tseng et al., 1955b). Southward transplantation of *Laminaria* to Zhejiang and Fujian Provinces of the East China Sea, which is subtropical in nature has been conducted with success (Tseng, 1958). With the help of the summer sporeling method, the height of the hot summer temperature and the duration of hot summer are not of major concerns. Rather, it is the length of the winter and spring seasons that is important, i.e. whether the seasons are long enough to effect growth of the plant to market size. By this way, the cold temperate *Laminaria japonica* is now maricultured in warm temperate and subtropical regions such as Zhejiang and Fujian (Tseng, 1981).

Current algal cultivation industry

At present two species each of *Porphyra* and *Gracilaria*, and one species each of the following: *Bangia*, *Gelidium*, *Kappaphycus*, *Betaphycus*, *Laminaria*, *Undaria* and *Hizikia* (= *Sargassum fusiforme*) are under cultivation by the above methods of mariculture.

There is also a small microalgae cultivation industry, a *Dunaliella* and a *Spirulina* mariculture industry. The *Dunaliella* culture group is located in Tanggu, Tianjin, a part of the Salt Research Institute, which received funding from the United Nation Development Programme (UNDP). This produces β-carotene present in the cells in as much as 6% of the dry weight (Guo, 1991). The *Spirulina* culture group is located in Sanya, Hainan Province in south China. It belongs to the South China Sea Institute of Oceano-logy. *Spirulina* is well known for its protein contents, 60–70% of the plants in dry weight (Wu et al., 1993). There are about 50 *Spirulina* culture groups in China, mostly freshwater, producing altogether less than 1000 tons of *Spirulina* powder per year.

Marine algal industry

At present China has an alginate industry composed of a few factories and employing *Laminaria japonica* as the principal raw material. The alginate was used as a substitute for starch grains in sizing cotton fibers in Qingdao and a small factory was built for the production of sodium alginate. It is now popularly employed in textile industry and produced about 8000–10 000 tons of the alginate annually. The industry was initiated in the early fifties employing wild *Sargassum confusum* Ag. as the raw material (Tseng & Ji, 1962). In a few years resources of the raw material were practically depleted. So the industry had to turn to the cultivated *Laminaria*, which is more expensive but dependable as a raw material.

China has a small agar industry dating back to the thirties. There has been a small agar factory in Qingdao employing *Gelidium amansii* (Lamx.) Lamx. as the raw material. There are now a few small factories in Fujian Province employing over-mature thalli of *Porphyra haitanensis* as raw materials and a few small factories in Guangdong and Hainan Provinces employing *Gracilaria tenuistipetata* var. *liui* Zhang *et* Xia as raw material. Total annual production of agar at present is only a few hundred tons.

China has also a small carrageenan industry with a small factory in Hainan. For many years this factory employed locally produced *Betaphycus gelatinum* (Esper) Doty as the raw material and produced a product called 'agar' in the market. It is actually a carrageenan instead of an agar. The annual production is also a few hundred tons.

The future of Chinese phycology

Taxonomy and floristic study

Chinese phycology has started rather late in the 1930s. Compilation of Chinese algal floras, although planned in the early seventies, did not get the actual publication started until the eighties. The first volume, Tomus 1 of Floras Algarum Sinicarum Aqua Dulcis, Zygnemataceae is by Dr Chin-Chih Jao. At present, six volumes have been published in this series. In the Flora Algarum Marinarum Sinicarum two volumes have been published. Four volumes of these two series have just finished compilation and are in the process of being edited. Fourteen more volumes in these two series of algal floras will be compiled in the next few years. Preparation for the other algal groups, mostly microscopic and uncommon species, will follow. Studies on soil algae, desert algae, snow algae and hot spring algae should be encouraged.

Ecological study

Phytoplankton in terms of nannoplankton and the smaller picoplankton have received careful studies for the last forty something years. A few years ago, Dr Jiao Nian-Zhi found the picoplankton *Prochlorella* in the Pacific. This picoplankton was found in great quantity below 100 m of sea water. Recently it was also found in the South China Sea. We are actually unaware taxonomically just how many genera and species of Prochlorophyte there are. The only way to find out is to keep on investigating. But it will involve a lot of expenses for ship times and instrumentations. Survey of deep lakes should also be taken. I believe we shall find similar picoplankton. Since the seventies, algal ecological study in terms of the environment has become more and more important.

China's marine environment has been troubled more and more with red tides and ways and means to predict and prevent the occurrence of red tides are becoming more and more important. We must protect our seas from pollution. In the fifties when we promoted cultivation of seaweeds, we had to fertilize our seas because there was then a very low N and P content. But now with the advance of animal aquaculture, the N and P contents in seawater have risen greatly. It is therefore suggested that large perennial seaweeds should be planted to help absorb these excess nutrients and large scale cultivation of seaweeds could be carried out for the same purpose to protect the environment (Fei & Tseng, 2003). With the development of space technique, space phycology must be emphasized.

Biotechnology study

Emphasis should be laid on molecular biotechnology of seaweeds, referring to the biotechnology on identification, modification, production and utilization of seaweed molecules, not only manipulating macromolecules such as DNA, RNA and proteins, but also dealing with low molecular weight compounds such as

secondary metabolites. Studies on molecular genetic labeling techniques and genetic engineering should be made (see Qin et al., 2004).

Comparative photosynthesis and evolution

Photosynthetic pigments of the algae are much more complicated than those of the seed plants, especially the phycoerythrin and phycocyanin, which are not found in the seed plants. Comparative photosynthesis of the algae with different photosynthetic pigments will show many things, especially the course of evolution. It is therefore suggested that studies on comparative photosynthesis should be continued.

Cultivation of algae

It may be said that modern seaweed cultivation has started in China in the early fifties when raft cultivation and the summer sporeling method of *Laminaria japonica* cultivation were initiated, making the *Laminaria* cultivation an important successful industry. In some years, just the *Laminaria* industry alone reached an annual production of over three millions tons, fresh weight. The method currently employed is basically that of the fifties with some small changes. For instance, in the cultivation of the summer sporelings, big green houses are still in use, and the zoospores are collected by numerous *Laminaria* fronds of one or several strains dumped together. We firmly believe that biotchnology should be applied to the cultivation of *Laminaria* sporelings and changes of the method be applied.

The *Porphyra* cultivation, thanks to Professor Fei Xiu-Geng, who holds a few patents in *Porphyra* cultivation method and has collected about 20 species from different parts of the world and more than 100 strains in his culture room, our *Porphyra* cultivation has been modernized to some extent. We believe that application of modern biotechnology can further modernize our *Porphyra* cultivation method.

In the present method of cultivation of *Gracilaria lemaneiformis,* the favorable growth temperature of 12–22 °C necessitates the storage of the seaweed during unfavorable season in low temperature room. Its growth is, however, best in the southern provinces probably because of higher concentration of nutrients elements, although its home in China is up north in Qingdao. We believe that by mean of modern biotechnology, we shall be able to improve our current farming method.

Hizikia (=*Sargassum fusiforme*) is now cultivated by transplants of young plants. The method has to be modernized because the current method consumes too much raw materials, depleting the natural resources.

At present, we cultivate nine genera and 11 species of seaweeds. There are a few more algae that deserve cultivation. For instance, *Caloglossa leprieurii* (Mont.) J. Ag., the famous antihelmintic *Zhegucai*, is in need of cultivation. *Gloiopeltis furcata* Post. *et* Rupr. and *G. tenax* (Turn.) J. Ag. are excellent silk sizing materials and the first algae to e cultivated in China. Modern cultivation method should be applied. So far, cultivation of freshwater algae is limited to a few microalgae, especially *Spirulina*. There is a demand on the market to have more 'Facai', *Nostoc commune* var. *flagelliformis* (Berkeley *et* Curtis) Bornet & Flahault. I hope that our freshwater algologist will see to it that *Facai* will be cultivated in the near future.

Algal industry

There is at present an alginate industry with producing ability of 13 000 tons of alginate annually. Its quality is, however, inferior to that of the American product and some improvement is needed. Our agar industry is in need of good raw materials. Recently, *Gracilaria lemaneiformis*, which gives a good agar, has been subjected to cultivation. We expect to have a good agar industry. Our carrageenan industry depends on *Betaphycus gelatinum* which produces a product called 'agar' for a long time but is actually a beta-carrageenan. We just began to cultivate *Kappaphycus alvarezii* (Doty) Doty *ex* Silva and the production of Kappa-carrageenan will be expected.

PSS and FPS are the two medicines now made from extract of *Laminaria japonica*. We believe there are many more bioactive substances in this group of lower plants, especially the microalgae. A few health food products are in the process of production. There is a great number of interesting and valuable compounds in the microalgae, not only the marine but also the freshwater microalgae, waiting to be developed (see Liang et al., 2004).

References

Agardh, C. A., 1820. Icones Algarum Ineditae. Lunda (Lund.) Fascicle 1: 4, X Pls.

Agardh, J. G., 1848. Species, Genera et Ordines Algarum. I. Species, Genera et Ordines Fucoidearum. Gleerup, Lund, viii + 363 pp.

Agardh, J. G., 1889. Species Sargassorum Australiae Descriptae et Dispositate. Kgl. Svenska Vet.-Akad. Handl. 23(3): 1–133, Pls. 1-31.

Ariga, K., 1919. Marine algae of Amoy vicinity, China. Formosa J. Aquatic Products 43: 12–16 (in Japanese).

Bi, L. J., Z. Y. Hu & G. X. Liu, 2001. History of taxonomy of Chinese freshwater algae. In Liu Yongding, Xiao Fan & Zhengyu Hu (eds), Research on Phycology in China. Wuhan Press, Wuhan, Hubei: 15–30 (in Chinese).

Chen, H. F., Y. Fu, L. R. Song & Y. D. Liu, 1997. The cytological observation of *Anabaena siamensis* after space-flight by a retrieved satellite. Chin. J. Space Science 17 (Suppl.): 102–106 (in Chinese).

Chien, Chungshu, 1917. Peculiar effects of barium, strontium and cerium on *Spirogyra*. Bot. Gaz. 63: 406–409.

Chin, T. G., 1937. Notes on the marine planktonic diatoms from Amoy. Amoy Mar. Biol. Bull. 3: 37–86.

Chu, H. R., 1944. Some Chinese Myxophyceae from Omei, Western Szechun. Sinensia 15: 153–156.

Chu, H. R. (ed.), 1991. Flora Algarum Sinicarum Aquae Dulcis. Vol. 2. Chroococcohyceae. Science Press, Beijing (in Chinese).

Collins, F. S., 1919. Chinese marine algae. Rhodora (Journ. New Eng. Bot. Club) 21: 203–207.

Cotton, A. D., 1915. Some Chinese marine algae. Kew Bull. Misc. Inform. 3: 107–113.

Cowdry, N. H., 1922. Algae in plants of Peitaiho. Journ. N. China Branch, Roy. Asia Soc. 53: 180–181.

Debeaux, O., 1875. Algues marines recoltees on Chine pendant l'expedition-francaise de 1860-1862. Actes de la Soc. Linn. De Bordeaux 30: 41–56.

Fang, T. C. & B. Y. Jiang, 1962. Heredity of the natural population of *Laminaria japonica* and its utilization in the future. J. Shandong Coll. Oceanol. 1: 1–5 (in Chinese).

Fang, T. C., Y. L. Ou, J. X. Dai, M. L. Wang, Q. S. Liu & Q. M. Yang, 1983. Breeding of the new variety 'Danhai No. 1' of *Laminaria japonica* by using a female haploid clone of the kelp. J. Shandong Coll. Oceanol. 13: 63–70 (in Chinese with English abstract).

Fang, T. C. & J. X. Dai, 1984. 1959-1984, Genetic studies in multi-cellular marine algae. J. Shandong Coll. Oceanol. 14: 16–19 (in Chinese with English abstract).

Fang, T. C., Y. L. Ou & J. J. Cui, 1985. Breeding of hybrid 'Danza No. 10' – an application of the Laminarian haploid cell clone. J. Shandong Coll. Oceanol. 15: 64–72 (In Chinese with English abstract).

Fei, X. G. & C. K. Tseng, 2003. Large scale seaweed cultivation for nutrient removal – A possible solution to the problem of coastal eutrophication. In: Lee, C. S. (ed.), Proceedings of the AIP Status of Aquaculture in China Seminar Series. The Oceanic Institute, University of Hawaii, Hawaii (in press).

Gepp, E. S., 1904. Chinese marine algae. Jour. Bot. 42: 161–165, Pl. 460.

Grubb, V. M., 1932. Marine algae of Korea and China, with notes on the distribution of Chinese marine algae. Jour. Bot. 70: 245–251.

Grunow, A., 1915. Additamenta ad cognitionem Sargassorum. Verh. Zool. Bot. Bes. Wien 65: 329–448.

Grunow, A., 1916. Additamenta ad cognitionem Sargassorum. Verh. Zool. Bot. Ges. Wien 66: 1–48, 136–185.

Guo, L. C., 1991. Large-scale culture of *Dunaliella* and its applications. Proc. Int. Symp. Biotechn. Saltponds: 23–30.

Guo, Y. J., 1991. The Kuroshio. Part II. Primary productivity and phytoplankton. Oceanogr. Mar. Biol. Annu. Rev. 29: 155–189.

Guo, Y. J. & Y. S. Zhang, 1996. Characteristics of phytoplankton distribution in Yellow Sea. J. Yellow Sea 2: 90–103.

Han, F. S. & Y. Y. Li, 1994. Flora Algarum Sinicarum Aquae Dulcis, Vol. 3, Charophyta. Science Press, Beijing, vii + 267 pp. (in Chinese).

Howe, M. A., 1924. Chinese marine algae. Bull. Torrey Bot. Club 51: 133–144, Pls. 1, 2.

Howe, M. A., 1934. Some marine algae of the Shantung Peninsula. Lingn. Sci. J. 13: 667–670, f. 1.

Hu, Z. L. & Y. D. Liu, 1997a. Studies on the mechanism in biological responses of algae to space environment. Chin. J. Space Science 17 (Suppl.): 23–34 (in Chinese).

Hu, Z. L. & Y. D. Liu, 1997b. Biological responses of algae to space environment. Chin. J. Space Science 17 (Suppl.): 14–23 (in Chinese).

Hu, Z. L. & Y. D. Liu, 1997c. The comparative studies on the metabolic characteristics between *Anabaena oryza* HB23 and its monoclonal strains retrieved from space-flight and reflight. Chin. J. Space Science 17 (Suppl.): 56–61 (in Chinese).

Hu, Z. L. & Y. D. Liu, 1997d. The possible mechanism for responses of algal cell *Dunaliella salina* to simulated microgravity. Chin. J. Space Science 17 (Suppl.): 117–124. (in Chinese).

Hu, Z. L., Y. D. Liu & L. R. Song, 1997. Effect of space-flight on survivorship and adoption of microalgae. Chin. J. Space Science 17 (Suppl.): 95–101 (in Chinese).

Hu, Z. L., Y. D. Liu, L. F. Dai, Sh. Yang, H. M. Lin & L. R. Song, 1996. Space-reflight strains of *Anabaena oryza* (Blue-green alga) carried by retrievable satellite. Chinese Science Bulletin 41(8): 679–683 (in Chinese).

Jao, C. C., 1934. New Oedogonia collected in China. I. Papers Mich. Acad. Sci. 19: 83–92.

Jao, C. C., 1988. Flora Algarum Sinicarum Aquae Dulcis, Vol. 1., Zygnemataceae. Science Press, Beijing. v + 228 pp. (in Chinese, with keys in English).

Kuetzing, F. T., 1849. Species Algarum. Brockhaus, Leipzig, 922 pp.

Ley, S. H., 1944. The Vauchariaceae from northern Kwangtung, China. Sinensia 15: 97–100.

Li, L. C., 1932. The freshwater algae of China. Abstr. Doct. Diss. Ohio State Univ. 9: 166–175.

Li, S. H. (S. H. Ley) & L. J. Bi, 1996. Flora Algarum Sinicarum Aquae Dulcis. Vol. 5. Ulothricata, Ulvales, Chaetophorales, Trentepohliales and Sphaeropleales. Science Press, Beijing. viii + 136 pp., Pls. 1–54 (in Chinese, with keys in English).

Liang, S., X. Liu, F. Chen & Z. Chen, 2004. Current microalgal health food R & D activities in China. Hydrobiologia 512/Dev. Hydrobiol. 173: 45–48.

Liu, Y. D., K. P. Li, K. F. Wang & L. R. Song, 2001. Algal space biology. In Liu Yongding, Xiao Fan & Zhengyu Hu (eds), Research on Phycology in China. Wuhan Press, Wuhan, Hubei: 49–66 (in Chinese).

Martens, G. V., 1866. Die Tange. Die Preussische Expedition nach Ost-Asien. Bot.Theil Die Tange: 1–152.

Montagne, J. F. C., 1840. Seconde centurie de plantes cellulaires exotiques nouvelles. Decades I et II. Ann. Sc. Nat. Bot. Sér. 2. 13: 193–207.

Noda, M., 1966. Marine algae of the North-Easterm China and Korea. Sci. Rep. Niigata Univ. Ser. D. (Biology) 3: 19–85.

Okamura, K., 1931. On the marine algae from Kotosho (Botel Tobago). Bull. Biogeogr. Soc. Jap. 2: 95–122, Pls. 10–12.

Okamura, K., 1936. Nippon Kaiso-shie (Manual of Japanese seaweeds). Uchida Rokakuho, Tokyo, 9+964+11 pp. (in Japanese).

Qi, Y. Z., 1995. Flora Algarum Sinicarum Aquae Dulcis. Vol. 4. Bacillariophyta Centreae. Science Press, Beijing. v+104 pp., Pls. 1–9 (in Chinese, with keys in English).

Qi, Y. Z., J. Chen, Z. Wang, N. Xu, Y. Wang, P. Shen, S. Lu & I. J. Hodgkiss, 2004. Some observations on harmful algal bloom (HAB) events along the coast of Guangdong, Southern China in 1998. Hydrobiologia 512/Dev. Hydrobiol. 207–212.

Qin, S., J. Peng, & C.-K. Tseng, 2004. Molecular biotechnology of marine algae in China. Hydrobiologia 512/Dev. Hydrobiol. 21–26.

Reinbold, Th., 1919. Algae. In Loesener, Th. (ed.), Prodromus Florae Tsingtauensis. Die Pfanzenwelt der Kiautschou-Gebietes. Beih. Bot. Central 37: 76–77.

Setchell, W. A., 1931a. Hong Kong seaweeds I. Hong Kong Naturalists 11: 39–60.

Setchell, W. A. 1931b. Hong Kong seaweeds II. Hong Kong Naturalists 11: 237–253.

Setchell, W. A. 1933. Hong Kong seaweeds III. Sargassaceae. Hong Kong Naturalists Suppl. 2: 33–49, Pls. 1–20.

Setchell, W. A., 1935. Hong Kong seaweeds IV. Sargassaceae. Hong Kong Naturalists Suppl. 4: 1–24, Pls. 1–17.

Setchell, W. A., 1936. Hong Kong seaweeds, V. Sargassaceae. Hong Kong Naturalists Supp. 5: 1–20, Pls. 1–8.

Shi, Z. X., 1999. Flora Algarum Sincarus Aquae Dulcis, Vol. 6, Euglenophyta. Science Press, Beijing (in Chinese, with keys in English).

Song, L. R., X. M. Zhang, D. H. Li & Y. D. Liu. Chlorophyll fluorescence as photosynthetic competence in *Poterioochromonas malharnensis*. Chin. J. Space Science 17 (Suppl.): 90–94 (in Chinese).

Tilden, J. E., 1929. The marine and freshwater algae of China. Lingn. Sci. J. 7: 349–398, Pl. 13.

Tseng, C. K., 1933. *Gloiopeltis* and the other economic seaweeds of Amoy, China. Lingn. Sci. J. 12: 43–63.

Tseng, C. K., 1958. The southward transplantation of kelp on China coast. Kexue Tongbo 17: 531–533 (in Chinese).

Tseng, C. K., 1981. Chapter 20: Commercial Cultivation. In Lobban, C. S. & M. J. Wynne (eds), The Biology of Seaweeds. Botanical Monographs, Vol. 17. Blackwell Scientific Publications, London: 680–725.

Tseng, C. K. & T. J. Chang, 1954. Studies on the *Porphyra* I. Life history of *Porphyra tenera* Kjellm. Act. Bot. Sin. 3: 287–301, Pls. 1–5 (in Chinese with English abstract).

Tseng, C. K. & T. J. Chang, 1955. Studies on the life history of *Porphyra tenera* Kjellm. Sci. Sin. 4: 378–398, Pls. 1–7.

Tseng, C. K. & M. H. Ji, 1962. Studies on the algin from *Sargassum* I. Conditions for extraction of algin from *Sargassum pallidum*. Stud. Mar. Sinica 1: 140–158 (in Chinese with English abstract).

Tseng, C. K. & B. R. Lu, 1999. Flora Algarum Marinarum Sinicarum Vol. 3. Phaeophyta, No. 2. Fucales. Science Press, Beijing (in Chinese, with keys in English).

Tseng, C. K. & S. Qin, 1991. Marine algal biotechnology in China: Present conditions and prospect. Proc. Int. Symp. Biotechn. Saltponds: 61–68.

Tseng, C. K. & A. S. Sun, 1989. Studies on the alternation of the nuclear phases and chromosome numbers in the life history of some species of *Porphyra* from China. Bot. mar. 32: 1–8.

Tseng, C. K. & B. M. Sweeney, 1946. Physiological studies of *Gelidium cartilagineum*. I. Photosynthesis, with special reference to the carbon dioxide factor. Amer. J. Bot. 33: 706–715.

Tseng, C. K. & B. C. Zhou, 1983a. On *Prochloron* and its significance in algal phylogeny. In Tseng, C. K. (ed.), Proceedings of the Joint China – U.S. Phycology Symposium. Science Press, Beijing: 29–38.

Tseng, C. K. & B. C. Zhou, 1983b. On the evolution of photosynthetic organism. In Editorial Board of Selected Papers on

Evolution (ed.), Selected Papers on Evolution. Science Press, Beijing: 34–43 (in Chinese).

Tseng, C. K. & B. C. Zhou, 1984. Some problems on the evolution of algae. Collected Papers 1st Chin. Phyco. Symp.: 1–7 (in Chinese).

Tseng, C. K., K. Y. Sun & C. Y. Wu, 1955a. On the cultivation of *Haitai* (*Laminaria japonica* Aresch.) by summering young sporophytes at low temperature. Acta Bot. Sinica 4: 255–264 (in Chinese with English abstract).

Tseng, C. K., K. Y. Sun & C. Y. Wu, 1955b. Studies on fertilizer application in the cultivation of Haitai (*Laminaria japonica* Aresch.). Acta Bot. Sin. 4: 375–392 (in Chinese with English abstract).

Tseng, C. K., B. C. Zhou & Z. Z. Pan, 1980. Comparative photosynthetic studies on benthic seaweeds. I. Photosynthetic properties and pigment composition of intertidal green algae. Oceanol. Limnol. Sin. 11: 134–140 (in Chinese with English abstract).

Tseng, C. K., B. C. Zhou & Z. Z. Pan, 1981a. Comparative photosynthetic studies on benthic seaweeds II. The effect of light intensity on photosynthsis of intertidal brown algae. Oceanol. Limnol. Sin. 12: 254–258 (in Chinese with English abstract).

Tseng, C. K., B. C. Zhou & Z. Z. Pan, 1981b. Comparative photosynthetic studies of benthic seaweeds III. Effect of light intensity on photosynthesis of intertidal red algae. Proc. Xth Intl. Seaweed Sym.: 515–520.

Turner, D., 1809. Fuci, sive Plantarum Fucorum Generi a Botanicis Asscriptarum Icones Descriptiones et Historia, Vol. 2. Mc'Reery, London: 1–164+2, Pls. 1–71.

Wang, C. C. & D. S. Nie, 1934. Notes on *Trachelomonas* of Nanking. Sinensia 5: 122–146.

Wang, Chu Chia, 1930. Notes on some subaerial Myxophyceae in Nanking and Soochow. Nanking Nat. Centr. Univ. Sci. Rep. Ser. B 1(1): 59–65.

Wu, B. T., C. K. Tseng & W. Z. Xiang, 1993. Large-scale cultivation of *Spirulina* in seawater based culture medium. Bot. Mar. 36: 99–102.

Xia, B. M. & J. F. Zhang (C. F. Chang), 1999. Flora Algarum Marinarum Sinicarum. Vol. 2. Rhodophyta No.5. Ahnfeltiales, Gigartinales, Rhodomeriales. Science Press, Beijing, vii + 199 pp., Pls.1–9 (in Chinese, with keys in English).

Yamada, Y., 1925. Studien über die Meeresalgen von der Insel Formosa I. Chlorophyceae. II. Phaeophyceae. Bot. Mag. Tokyo 39: 77–95, 239–254.

Yamada, Y., 1950. A list of marine algae from Ryukyusho, Formosa I. Chlorophyceae and Phaeophyceae. Sci. Pap. Inst. Algol. Res. Hokkaido Univ. 3: 173–194.

Zeng Chengkui (C. K. Tseng), B. C. Zhou, A. S. Sun, Z. Z. Pan & R. B. Zang, 1982. A preliminary report on *Prochloron* from China. Kexue Tongbao 27: 778–781.

Zhou, B. C., B. G. Wu, C. K. Tseng & G. Y. Xiao, 1966. The differences of pigment system and photosynthesis in *Porphyra yezoensis*. Kexue Tongbao 17: 427–429 (in Chinese).

Zhou, B. C., S. Q. Zheng & C. K. Tseng, 1974. Comparative studies on the absorption spectra of some green, brown and red algae. Act. Bot. Sin. 16: 146–156 (in Chinese with English abstract).

Zhou, M. J., J. Li, B. Luckas, R. Yu, I. Yan, C. Hummort & S. Kastrup, 1999. A recent shellfish toxin investigation in China. Mar. Pollut. Bull. 39: 331–334.

Zou, J. Z., M. J. Zhou, Z. M. Yu & R. G. Lin, 1999. Harmful algal blooms and eutrophication of aquaculture waters. In Li, Y. Q., J. Z. Zou & D. S. Li (eds), Prevention and Improvement of Ecology and Environment in Aquaculture Waters. Shandong Science and Technology Press, Jinan, Shandong: 74–131 (in Chinese).

Hydrobiologia **512**: 21–26, 2004.
P.O. Ang, Jr. (ed.), Asian Pacific Phycology in the 21ˢᵗ Century: Prospects and Challenges.
© 2004 *Kluwer Academic Publishers.*

Molecular biotechnology of marine algae in China

Song Qin, Peng Jiang & Cheng-Kui Tseng
Institute of Oceanology, Chinese Academy of Sciences, Qingdao 266071, China
E-mail: sqin@ms.qdio.ac.cn

Key words: genetic engineering, marine algae, molecular biotechnology, molecular genetic marker

Abstract

Molecular biotechnology of marine algae is referred to as the biotechnology on the identification, modification, production and utilization of marine algal molecules. It involves not only the manipulation of macromolecules such as DNA, RNA and proteins, but also deals with low molecular weight compounds such as secondary metabolites.

In the last decade, molecular systematic researches to investigate the relationship and to examine the evolutionary divergence among Chinese marine algae have been carried out by Chinese scientists. For example, RAPD has been widely used in several laboratories to elucidate genetic variations of the reds, such as *Porphyra*, *Gracilaria*, *Grateloupia* and the greens such as *Ulva* and *Enteromorpha*. Some important data have been obtained. The study on molecular genetic markers for strain improvement is now in progress.

In 1990s, genetic engineering of economic seaweeds such as *Laminaria*, *Undaria*, *Porphyra*, *Gracilaria* and *Grateloupia* has been studied in China. For *Laminaria japonica,* the successfully cultivated kelp in China, a model transformation system has been set up based on the application of plant genetic techniques and knowledge of the algal life history. Progress has been made recently in incorporating a vaccine gene into kelp genome. Evidence has been provided showing the expression of gene products as detectable vaccines.

In the present paper, the progress of molecular biotechnological studies of marine algae in China, especially researches on elucidating and manipulating nucleic acids of marine algae, are reviewed.

Introduction

Algae form a broad and special group of living organisms. People from both the east and the west have long been interested in algae because of their morphological diversity, great biomass, abundance of metabolites and foreseeable economic potentiality (Critchley & Ohno, 1998). At present, the algal industry is worth several billion US dollars world-wide and the challenges facing the industry include stable supply of high quality raw materials, new algal product development and continuous search for new algal species with novel properties (Critchley, 2003). Large-scale cultivation of economically important algal species (Ask, 2003) and genetic engineering of desirable algal strains (Minocha, 2003) are approaches to ensure steady supply of quality raw materials to the algal industry.

China has been very successful in the cultivation of several commercially important algal species, such as *Laminaria*, *Undaria*, *Porphyra* and some agarophytes such as *Gracilaria*. It is now the world largest producer of kelps. The great success in Chinese kelp farming was partly due to the application of biotechnological approaches, such as seed selection and breeding techniques at the cytological level, in the process of cultivation. During the past ten years, more extensive biotechnological approaches at the molecular level, e.g. molecular genetic labeling techniques, have been adopted in marine algal research. The research goals were determination of taxonomic status of some species in dispute, establishment of evolutionary relationship and investigation of characters to be used to improve the quality of cultivated species. Genetic engineering of seaweeds was aimed at strain improvement and the utilization of cultivated seaweeds as bioreactors. This paper reviews and summarizes some of the most recent results of algal molecular biotechnological researches carried out in China.

Molecular genetic markers

Analyses of phenotype mutation, chromosome polymorphism and protein polymorphism have been used to select and determine genetic markers. DNA polymorphism analysis is nowadays a more and more frequently used method. Many new genetic labeling techniques such as AFLP, RFLP, DNA fingerprinting, minisatellite DNA, microsatellite DNA, mitochondrion DNA, RAPD and sequence analysis have been set up (e.g. van Oppen et al., 1994, 1996; Ho et al., 1995; Minocha, 1998; Kusumo & Druehl, 2000; Wright et al., 2000). In the last decade, Chinese researchers have tried to employ some of these methods in marine algal research.

Sequence analysis

Partial sequence of the subunit gene of phycoerythrin (PE) in a red alga *Gracilaria lemaneiformis* (Bory) Dawson was analyzed by Sui & Zhang (1999), and high conservation level was seen by comparison with sequences from other three red algae *Rhodella violacea* (Kornmann) Wehrmeyer, *Polysiphonia boldii* Wynne *et* Edwards, and *Aglaothamnion neglectum* Feldmann–Mazoyer. Both α and β subunit genes have been highly expressed in *E. coli* (Sui & Zhang, 2001).

More sequence analysis researches have been performed on red tide-microalgae. The sequences of the 5.8S rDNA and the flanking ITS (internal transcribed spacers) 1 and 2 from twelve strains (five species) of *Alexandrium*, a dinophyceous genus with many toxic species, were analyzed by Chen & Qu (1999). The calculated genetic distance among these species suggested that ITS regions are practical molecular genetic markers for *Alexandrium* (Chen et al., 1999a).

Similar work was performed on another red tide-related species *Ceratium furca* Ehrenberg (Claparède *et* Lachmann) collected from Liaoning Province in northern China (Zhuang et al., 2001). Sequence alignments on over 1700 nucleotides of 18S rDNA of *C. furca* with 15 other representative species of dinoflagellates from Genebank have been analyzed in order to investigate the phylogenetic relationships within this highly divergent and taxonomically controversial group. A coherent and convincing evolutionary tree

was obtained, using *Tetrahymena corlissi* Corliss as the outgroup. The results also showed that the ITS region of rDNA had a high level of sequence divergence, which could be a suitable target sequence for developing genus or species-specific oligonucleotide probes. Such probes would also be available in genus *Microcystis* based on the sequence analysis on its ISR region between 16S and 23S rDNA (Chen et al., 1999c). These results provide new perspectives on instant diagnosis of red tide marine algae and cloning of functional gene with known conservative sequences.

Restriction fragment length polymorphism (RFLP)

The small subunit ribosomal RNA genes (Ss-rDNA) of eight strains of the red tide toxic alga *Alexandrium tamarense* (Labour) Balech were amplified for a RFLP assay. PCR-RFLP analysis of the Ss-rDNA of samples from the South China Sea revealed that they only had gene A but lack gene B. This result suggested that gene B could be used as a molecular biogeographic marker for this species (Chen et al., 1999b). Chen et al. (1999a) conducted RFLP analysis in ITS region of two morphologically alike species *Alexandrium catenella* (Whedon *et* Kofoid) Balech and *A. tamarense* and found that all *A. catanella* samples (different collection sites) shared the same polymorphism pattern.

Random amplified polymorphic DNA (RAPD)

RAPD is regarded as a rapid, convenient and economic method (van Oppen et al., 1996). This method is now being used by Chinese phycologists and is widely applied in algal systematic research.

Under the optimized reaction condition for polymerase chain reaction (PCR) by using a single arbitrary primer, a high degree of reproducibility of the amplified bands could be obtained by gel electrophoresis. Such genetic markers are derived from priming sites randomly distributed throughout a genome, whose polymorphism makes the analysis of a complex genome without any prior knowledge of the DNA sequence possible. Application of RAPD to investigate phylogenetic relationship, to evaluate genetic variation and to clarify taxonomic ambiguities within several groups of seaweeds has been documented in China.

New ideas challenging traditional taxonomy were proposed when RAPD was carried out on *Ulva* and *Enteromorpha* (Chlorophyta) (Yang et al., 2000).

Based on the dendrogram of UPGMA and N-J analysis, convincing results showed that RAPD was effective in discriminating these two genera. The divergence between *Enteromorpha linza* (L.) J. Ag. and *Ulva* is shown to be even closer than that between *E. linza* and other *Enteromorpha* species. Similar results were also found in *Grateloupia* (Q. Wang et al., 2000). The deduced genetic distance between *G. filicina* f. *lomentaria* Howe and other *Grateloupia* species suggested that the former should be listed as a new genus, namely *Sinoyubimorpha*.

Genetic diversity of *Gracilaria* was also studied by using RAPD (Li et al., 1998). Different PCR patterns were derived in three mutants of *G. lemaneiformis*, and the relationship between different strains from different habitats was elucidated. In the same report, the phase and sex related markers were also discussed.

Most RAPD analysis was done to the important cultivated red alga *Porphyra*. Phylogenetic divergence was generated between geographic populations (Jia et al., 2000; Kuang et al., 1998a; Song et al., 1998; Y. Wang et al., 2000), between wild and cultivated populations (Xu et al., 2001), or among different cultivated populations (Mei et al., 2000; Shi et al., 2000). Some species trait related markers have been proposed which may be used further in strain improvement (Song et al., 1998).

Genetic engineering

Difficulties in genetic engineering of seaweeds were recognized in early 1990s (Saga, 1991). There was very poor knowledge on how to introduce foreign DNA into seaweed cells as well as on how to regenerate and select transformed plants. Moreover, nobody at that time knew where were the vectors for expressing foreign genes in seaweeds. From then on, efforts have been made in the world towards establishing a so-called transformation model for seaweeds.

In China, scientists have studied genetic transformation of economic seaweeds such as *Laminaria, Undaria, Porphyra, Gracilaria* and *Grateloupia* since 1991, and visible progress has been made in model research, especially in *Laminaria*.

The work in red algae

Wang et al. (1994) reported transient expression of the exogenous GUS gene in the protoplasts of *Porphyra haitanensis* Chang *et* Zheng by using electroporation.

Kuang et al. (1998b) used five species of economically important red algae, i.e., *Porphyra yezoensis* Ueda, *Gracilaria asiatica* Zhang *et* Xia, *G. lemaneiformis*, *Grateloupia filicina* (Wulf.) C. Ag. and *Ceramium tenuissimum* J. Ag. as materials for transformation studies. The gene donors were plasmids containing GUS gene and CaMV35S promoter. Four DNA introduction methods, i.e., electroporation, PEG treatment, PEG + electroporation and Biolistic bombardment (Minocha & Wallace, 2000) were employed on *P. yezoensis*. Meantime, Biolistic bombardment was employed on explants of other four reds. The results suggested that PEG is the best method for transformation of protoplasts, and that Biolistic bombardment is effective in the transformation of protoplasts, thallus tissues and free-living conchocelis. In the transformation of *Gracilaria asiatica* by using Biolistic bombardment, a positive PCR result was obtained. The workers on reds are now trying to find applicable selectable markers to get true transformants at a certain scale. The development of a transformation model for the red algae is still incomplete.

The work in brown algae

Research has been done on the kelps *Laminaria japonica* Aresch. and *Undaria pinnatifida* (Harv.) Sur. in China. People want to use these algae as a cheap source of high value products by gene transfer. Genetic engineering is being expected to make kelp a useful bioreactor to produce drugs such as edible vaccines and to absorb excess N, P and other eutrophication elements from the marine environment. So far a model transformation system for kelp has been set up. The model includes a series of methods of gene introduction, vector construction, transformants' regeneration and screening (Qin et al., 1998a–c). By using this model, progress has been made in strain improvement with genes that serve particular purposes.

Application of Biolistic bombardment

Due to incomplete understanding of genomes of seaweed associated bacteria and viruses, direct physical methods have to be tried first in the transformation of seaweeds. Equally or even more difficult, protoplasts from either the sporophytes or the gametophytes of *L. japonica* failed to regenerate. Therefore, the search for method which can get the target DNA through the cell wall directly, e.g. ultrasonic treatment or Biolistic bombardment, was given the first priority.

Ultrasonic treatment was tried since it was less costly and more time saving, but the result did not suggest that it was a promising method. It could break up the filamentous female gametophytes into shorter fragments and partially break down cell walls, but at the same time it killed the cells and decreased their parthenogenesis efficiency (Wang et al., 1998a).

It has been proved that Biolistic bombardment could effectively introduce foreign DNA through cell walls into intact kelp cells, with either haploid or diploid thallus as recipient. By using this method, activities of CAT gene (*cat*) and LacZ gene (*lacZ*) have been detected in mature sporophytes regenerated by parthenogenesis. This suggested that random integration of foreign genes could occur in this way (Qin et al., 1998a–c).

Promoter selection

Promoter availability and selection is a critical factor in genetic transformation. Promoters from seaweed or seaweed-infective viruses have seldom been isolated (Henry & Meints, 1994). It was therefore necessary to examine some effective promoters from other organisms including higher plants and unicellular algae.

CaMV35S promoter was first used in brown seaweeds (*Laminaria* and *Undaria*) and transient expression of GUS reporter gene was observed (Qin et al., 1994). Further study was performed to select better promoters, including two from land plant, i.e. CaMV35S promoter and Ubiquitin promoter (from maize), and two from unicellular algae, i.e. fcp promoter (from diatom fucoxanthin, chlorophyll a/c-binding protein gene,) and amt promoter (from adenine methyltransferase gene of *Chlorella* virus). With young parthenogenetic sporophytes used as gene recipients and Biolistic bombardment as method, the quantitative detection of GUS transient expression using fluorometric assay indicated that CaMV35S and fcp promoter were more efficient in kelp than in the other algae (Wu, 2001).

So far two promoters have been shown to have efficient power in driving stable expression of foreign genes in kelp. They are fcp promoter (Wu, 2001) and SV40 promoter (from simian virus) (Qin et al., 1998a; Jiang et al., 2003). Utilization of SV40 promoter even resulted in uniform expression of *lacZ* reporter gene in regenerated *Laminaria* sporophytes, suggesting its high transcription recognition efficiency without histospecificity (Jiang et al., 2003). This promoter also worked well in *Undaria* for both transient (Yu et al., 2002) and stable expression (Qin et al., 2003).

Introduction of foreign genes into gametophytes and generation of sporophytes

Protoplasts, single cells and tissues are routine hosts for gene introduction in land plants, while neither haploid and diploid protoplasts nor single cells of *L. japonica* can be regenerated into new plants. Tissue culture in Laminariales has been studied extensively, but as yet no efficient regeneration system has been obtained in *L. japonica* since the regeneration efficiency from calli to sporophytes should still be increased (Wang et al., 1998b).

Fang (1978) reported that female gametophytes could develop by parthenogenesis. Also, female gametophytes could grow vegetatively to form filamentous clones which could be maintained for a long time in the laboratory. Stimulated under certain conditions, these vegetative clones can be developed into parthenogenetic sporophytes. Transgenic kelp has been obtained by bombarding foreign genes into female gametophytes and inducing the development of parthenogenetic plants (Qin et al., 1998a, c). Recently, success has also been obtained by using male gametophytes of *L. japonica* as transformation targets to generate zygotic sporophytes after hybridization (Jiang et al., 2003). The reproducibility of this newly developed pathway has also been confirmed in *U. pinnatifida* (Qin et al., 2003).

Employment of antibiotics and herbicide to select positive transformant

Antibiotics and herbicides are widely used as screening agents in genetic engineering of higher plants. Sensitivities of *L. japonica* to nine antibiotics and one herbicide were tested, including lincomycin, ampicillin, streptomycin, kanamycin, neomycin, chloramphenical, hygromycin, zeocin, G-418 and basta (Qin et al., 1998c). Results showed that both *L. japonica* and *U. pinnatifida* were sensitive only to chloramphenical, hygromycin and basta. The LD_{50} of hygromycin to parthenogenetic sporophytes was much lower than that of chloramphenical, and was not correlated with thallus length of the kelp while that of chloramphenical was. So *cat*, *hpt* and *bar* are applicable selectable markers for screening kelp transformants (Li et al., 1999; Wu et al., 2000; Yu et al., 2003).

By using the transformation model described above, stable expression of reporter genes such as *cat* and *lacZ* have been detected in parthenogenetic sporophytes after cultivation in the sea (Qin et al., 1998a, c; 1999). Recently progress has been made by using hepatitis B surface antigen gene (*HBsAg*). Positive results of PCR and ELISA for *HBsAg* detection were obtained, which suggested the integration of *HBsAg* into the genome of *L. japonica* (Jiang et al., 2002).

Conclusion

Much remains to be done for successful application of molecular biotechnological techniques in marine algal research in China to obtain strain improvement, development of algae as bioreactors, and for clarification of phylogenetic and evolutionary relationships among algal species of interest. Some progress has been achieved and it is anticipated that more will be coming in the near future as more and more people and resources are being invested towards these efforts.

References

Ask, E. I., 2003. Creating a sustainable commercial *Eucheuma* cultivation industry: the importance and necessity of the human factor. Proc. Int't Seaweed Symposium 17: 13–18.

Chen, Y. Q. & L. H. Qu, 1999. Molecular criteria for the delimination of *Alexandrium* species based on the analyses of rDNA ITS. J. Zhongshan University 38: 7–11 (in Chinese with English abstract).

Chen, Y. Q., L. H. Qu, L. M. Zeng, Y. Z. Qi & L. Zheng, 1999a. Molecular determination of two toxic red tide dinoflagellates *Alexandrium catenella* and *A. tamarense* from the South China Sea. Acta Oceanol. Sin. 21: 106–111 (in Chinese with English abstract).

Chen, Y. Q., X. Z. Qiu, L. H. Qu, L. M. Zeng, Y. Z. Qi & L. Zheng, 1999b. Analysis of molecular biogeographic marker on red tide toxic *Alexandrium tamarense* in the South China Sea. Oceanol. Limnol. Sin. 30: 45–51 (in Chinese with English abstract).

Chen, Y. Q., L. Zhuang, L. H. Qu, T. L. Zheng, D. Z. Wang & Y. L. Wang, 1999c. Sequence analysis and comparison of rDNA ISR regions of *Microcystis* species from red tide and normal regions. Mar. Sci. 23: 48–50 (in Chinese with English abstract).

Critchley, A. T., 2003. World-wide carrageenophyte cultivation: recent successes and challenges. Third European Phycological Congress, Programme & Book of Abstracts: 20.

Critchley, A. T. & M. Ohno (eds), 1998. Seaweed Resources of the World. Kanagawa International Fisheries Training Centre, Japan International Cooperation Agency. Yokosuka, Japan.

Fang, Z. X., 1978. The first record of female sporophytes of *Laminaria japonica*. Chin. Sci. Bull. 23: 43–44 (in Chinese).

Henry, E. C. & R. H. Meints, 1994. Recombinant viruses as transformation vectors of marine macroalgae. J. appl. Phycol. 6: 247–253.

Ho, C. L., S. M. Phang & T. Pang, 1995. Molecular characterisation of *Sargassum polycystum* and *S. siliquosum* (Phaeophyta) by polymerase chain reaction (PCR) using random amplified polymorphic DNA (RAPD) primers. J. Appl. Phycol. 7: 33–41.

Jia, J. H., P. Wang, D. M. Jin, X. P. Qu, Q. Wang, C. Y. Li, M. L. Weng & B. Wang, 2000. The application of RAPD markers in diversity detection and variety identification of *Porphyra*. Acta Bot. Sinica 42: 403–407.

Jiang, P., S. Qin & C. K. Tseng, 2002. Expression of hepatitis B surface antigen gene (*HBsAg*) in *Laminaria japonica* (Laminariales, Phaeophyta). Chin. Sci. Bull. 47: 1438–1440.

Jiang, P., S. Qin & C. K. Tseng, 2003. Expression of the *lacZ* reporter gene in sporophytes of the seaweed *Laminaria japonica* (Phaeophyceae) by gametophyte-targeted transformation. Plant Cell Rep. 21: 1211–1216.

Kuang, M., S. J. Wang, Y. Li, D. L. Shen & C. K. Tseng, 1998a. RAPD study on some common species of *Porphyra* in China. Chin. J. Oceanol. Limnol. 16 (Suppl.): 140–146.

Kuang, M., S. J. Wang, Y. Li, D. L. Shen & C. K. Tseng, 1998b. Transient expression of exogenous GUS gene in *Porphyra yezoensis* (Rhodophyta). Chin. J. Oceanol. Limnol. 16 (Suppl.): 56–61.

Kusumo, H. T. & L. D. Druehl, 2000. Variability over space and time in the genetic structure of the winged kelp *Alaria marginata*. Mar. Biol. 136: 397–409.

Li, X. F., Z. H. Sui & X. C. Zhang, 1998. Application of RAPD in genetic diversity study on *Gracilaria lemaneiformis*. Chin. J. Oceanol. Limnol. 16 (Suppl.): 147–151.

Li, X. P., S. Qin & C. K. Tseng, 1999. Sensitivity of sporophytes of parthenogenetic *Laminaria japonica* to chloramphenicol and hygromycin. Oceanol. Limnol. Sin. 30: 186–190 (in Chinese with English abstract).

Mei, J. X., D. M. Jin, J. H. Jia & X. G. Fei, 2000. DNA polymorphism of *Porphyra yezoensis* and its application to cultivar discrimination. J. Shandong Uni. 35: 230–234 (in Chinese with English abstract).

Minocha, S. C., 1998. Genetic manipulation of marine macroalgae. World Aquacult. 29–30: 57.

Minocha, S. C., 2003. Genetic engineering of seaweeds: current status and perspectives. Proc. Int'l Seaweed Symposium 17: 19–26.

Minocha, S. C. & J. C. Wallace, 2000. Gene transfer techniques and their relevance to woody plants. In Jain, S. C. & S. C. Minocha (eds), Molecular Biology of Woody Plant, Vol. 2. Kluwer Academic Publishers, Dordrecht: 1–24.

Qin, S., P. Jiang, X. P. Li & C. K. Tseng, 1998a. The expression of *lacZ* in regenerated sporophytes of parhenogenetic *Laminaria japonica*. Proceedings of the 2nd Asia-Pacific Marine Biotechnology Conference and 3rd Asia-Pacific Conference on Algal Biotechnology, Phuket (Thailand): 205–208.

Qin, S., P. Jiang, X. P. Li, X. H. Wang & C. K. Tseng, 1998b. A transformation model for *Laminaria japonica* (Phaeophyta, Laminariales). Chin. J. Oceanol. Limnol. 16 (Suppl.): 50–55.

Qin, S., D. Z. Yu, P. Jiang, C. Y. Teng & C. K. Zeng, 2003. Stable expression of *lacZ* reporter gene in seaweed *Undaria pinnatifida*. High Technology Letters 13: 87–89 (in Chinese with English abstract).

Qin, S., G. Q. Sun, P. Jiang, L. H. Zou. Y. Wu & C. K. Tseng, 1999. Review of genetic engineering of *Laminaria japonica* (Laminariales, Phaeophyta) in China. Hydrobiologia 398/399: 469–472.

Qin, S., J. Q. Wu, X. H. Wang, X. P. Li, P. Jiang & C. K. Tseng, 1998c. The expression of foreign genes in *Laminaria japonica* (Laminariales: Phaeophyta). In Morton, B. (ed.), The Marine

Biology of the South China Sea. Hong Kong University Press, Hong Kong: 209–217.

Qin, S., J. Zhang, W. B. Li, X. H. Wang, S. Tong, Y. R. Sun & C. K. Tseng, 1994. Transient expression of GUS gene in phaeophytes using Biolistic Particle Delivery System. Oceanol. Limnol. Sin. 25: 353–356 (in Chinese with English abstract).

Saga, N., 1991. Protoplasts and somatic hybridization-controversal discussion 'No' side. Proceeding of a COST-48 Workshop, Spain: 25–30.

Shi, J. F., J. H. Jia, P. Wang, D. M. Jin, P. Xu, X, G. Fei, B, Wang & M. L. Weng, 2000. Development of specific molecular markers for *Porphyra* lines. High Technology Letters 10: 1–3 (in Chinese with English abstract).

Song, L. S., D. L. Duan, X. H. Li & C. X. Li, 1998. Use of RAPD for detecting and identifying *Porphyra* (Bangiales, Rhodophyta). Chin. J. Oceanol. Limnol. 16: 237–242.

Sui, Z. H. & X. C. Zhang, 1999. Cloning and analysis of partial sequence of phycoerythrin gene from *Gracilaria lemaneiformis*. J. Ocean Uni. Qingdao 29: 81–86 (in Chinese with English abstract).

Sui, Z. H. & X. C. Zhang, 2001. Cloning of subunit genes of phycoerythrin of *Gracilaria lemaneiformis* and its expression in *E. coli*. High Technology Letters 11: 14–19 (in Chinese with English abstract).

van Oppen, M. J. H., O. E. Diekmann, C. Wiencke, W. T. Stam & J. L. Olsen, 1994. Tracking dispersal routes: phylogeography of the Arctic-Antarctic disjunct seaweed *Acrosiphonia arcta* (Chlorophyta). J. Phycol. 30: 67–80.

van Oppen, M. J. H., H. Klerk, M.de Graaf, W. T. Stam & J. L. Olsen, 1996. Assessing the limits of random amplified polymorphic DNAs (RAPDs) in seaweed biogeography. J. Phycol. 32: 433–444.

Wang, Q., L. J. An, J. Yang, Q. Su, H. W. Wang & X. H. Kang, 2000. RAPD analysis of *Grateloupia filicina* and *Sinoyubimorpha porracea*. Oceanol. Limnol. Sin. 31: 506–510 (in Chinese with English abstract).

Wang, S. J., H. Li, Y. Li, 1994. Transient expression of GUS gene in *Porphyra haitanensis* by electroporation transformation. J. Shanghai Fisheries Univ. 3: 145–150 (in Chinese with English abstract).

Wang, X. H., S. Qin, X. P. Li, P. Jiang, C. K. Tseng & M. Qin, 1998a. Effects of ultrasonic treatment on female gametophytes of *Laminaria japonica* (Phaeophyta). Chin. J. Oceanol. Limnol. 16 (Suppl.): 62–66.

Wang, X. H., S. Qin, X. P. Li, P. Jiang, C. K. Tseng & M. Qin, 1998b. High efficient induction of callus and regeneration of sporophytes in *Laminaria japonica* (Phaeophyta). Chin. J. Oceanol. Limnol. 16 (Suppl.): 67–74.

Wang, Y., B. Q. Liu, Q. J. Luo, Z. Q. Fei, L. Q. Pei & Q. Z. Xue, 2000. RAPD analysis of genetic variation among *Porphyra haitanensis* strains. J. Ocean Uni. Qingdao 30: 225–229 (in Chinese with English abstract).

Wright, J. T., G. C. Zuccarello & P. D. Steinberg, 2000. Genetic structure of the subtidal red alga *Delisea pulchra*. Mar. Biol. 136: 439–448.

Wu, Y., 2001. Application of GUS gene to modify the expression system of *Laminaria japonica*. Master's Thesis, Institute of Oceanology, Chinese Academy of Sciences (in Chinese with English abstract).

Wu, Y., L. H. Zou, P. Jiang, G. Q. Sun, S. Qin & C. K. Tseng, 2000. Sensitivity of parthenogenetic *Laminaria japonica* to basta. Oceanol. Limnol. Sin. 32: 19–22 (in Chinese with English abstract).

Xu, D., L. S. Song, S. Qin, A. Pala, X. G. Fei & C. K. Zeng, 2001. RAPD analysis of genetic variation among cultivated *Porphyra*. High Technology Letters 11: 1–8 (in Chinese with English abstract).

Yang, J., L. J. An, Q. Wang, H. W. Wang, Q. Su & X. H. Kang, 2000. Application of RAPD in *Ulva* and *Enteromorpha* (Chlorophyta). Oceanol. Limnol. Sin. 31: 408–413 (in Chinese with English abstract).

Yu, D. Z., S. Qin & C. K. Zeng, 2002. Transient expression of *lacZ* reporter gene in economic seaweed *Undaria pinnatifida*. High Technology Letters 12: 93–95 (in Chinese with English abstract).

Yu, D. Z., L. L. Yang, P. Jiang, S. Qin & C. K. Zeng, 2003. Study on selectable markers in genetic engineering of *Undaria pinnatifida*. High Technology Letters 13: 74–77 (in Chinese with English abstract).

Zhuang, L., Y. Q. Chen, Q. L. Li & L. H. Qu, 2001. Sequence determination and analysis of 18S rDNA and internal transcribed spacer regions of red tide-related *Creratium furca*. Oceanol. Limnol. Sin. 32: 148–154 (in Chinese with English abstract).

Hydrobiologia **512**: 27–32, 2004.
P.O. Ang, Jr. (ed.), Asian Pacific Phycology in the 21st Century: Prospects and Challenges.
© 2004 *Kluwer Academic Publishers.*

Microalgal studies for the 21st Century

Norihide Kurano & Shigetoh Miyachi*
Marine Biotechnology Institute, Kamaishi Laboratories, Heita,
Kamaishi City, Iwate 026-0001, Japan
**Author for correspondence: E-mail: BXW02571@nifty.ne.jp*

Key words: microalgae, Marine Biotechonology Institute, Japan, CO_2 fixation, culture collection

Abstract

Microalgal photosynthesis is efficient enough to fix CO_2 in both atmosphere and industrially discharged gases, and is a possible future alternative for CO_2 reduction. This paper describes physiological responses of microalgal cells to extremely high CO_2 concentrations, capability of microalgal cells to fix CO_2 at both indoor and outdoor culture experiments, and efforts to establish a culture collection of marine microalgae. Recent researches indicate that microalgae are likely to play a key role in worldwide issues of the coming century.

Introduction

Since Marine Biotechnology Institute (MBI) started its work in April of 1990, one of the research targets at the Kamaishi Laboratories has been marine microalgae. Many microalgae have been collected from various parts of the Pacific Ocean and elsewhere. The features of these microalgae have been characterized and, with some, their capacity for CO_2 fixation has been extensively studied.

Since the continued increase in atmospheric CO_2 in parallel with the increase in fossil fuel consumption was first reported, a possible rise in the Earth's temperature due to greenhouse effect has become a matter of global concern. Photosynthesis seems to be the only feasible way to remove atmospheric CO_2. Reduction of atmospheric CO_2 by means of chemical or physical methods will be too expensive, since its concentration is as low as about 0.04%. The efficiency of photosynthesis is much higher in microalgae than in terrestrial plants including C4s (Miyachi, 1995), so the use of microalgae, especially marine microalgae, for reducing the CO_2 concentration in the atmosphere, attracted us.

Carbonic anhydrase of *Porphyridium*

The enzyme carbonic anhydrase (CA, EC 4.2.1.1), which enhances the equilibrium between CO_2 and bicarbonate in water, is involved in the mechanism underlying the high efficiency of photosynthesis in microalgae (Suzuki et al., 1994). Studies on this enzyme are being carried out at MBI. We will therefore briefly mention first our studies on CA of the red alga, *Porphyridium purpureum.*

This enzyme was purified to nearly a single band on SDS-PAGE, and its nucleotide and deduced amino acid sequences were elucidated (Mitsuhashi & Miyachi, 1996). The primary structure of this enzyme is characterized by two nearly identical domains, each homologous with CAs found in eukaryotes. It was, therefore, assumed that *Porphyridium* CA had evolved through the duplication of an ancestral and hypothetical CA gene with subsequent fusion of the duplicated genes. Mitsuhashi et al. (2000a) have determined the X-ray structure of this enzyme, and revealed a novel catalytic site for CO_2 hydration (Mitsuhashi et al., 2000b)

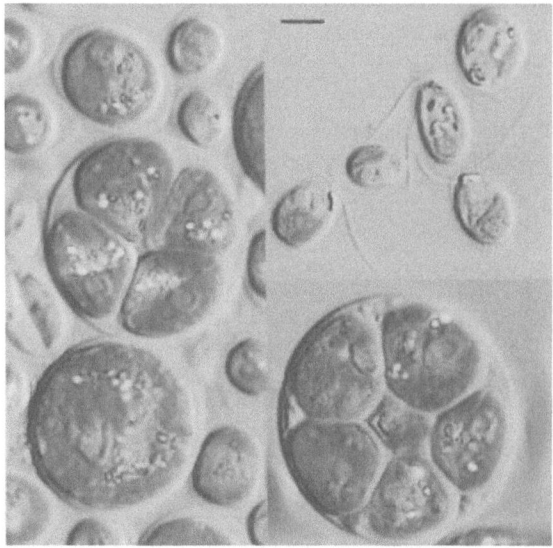

Figure 1. Photomicrographs of *Chlorococcum littorale*. Scale = 2 μm.

High CO$_2$-tolerant strain

It has been well documented that microalgae can play a very important role in the bioremediation of a low level of CO$_2$, CA being one of the essential tools. On the other hand, it is considered that microalgae are unsuitable for treating a high partial pressure of CO$_2$, since studies on some *Chlorella* species have revealed that CO$_2$ above 5% inhibited algal growth. Although this growth inhibition has been described as 'narcotic', the mechanism for the inhibition remains unclear.

Soon after MBI started its work, one scientist found a new species of a marine green microalga which could grow rapidly in an extremely high CO$_2$ concentration (Kodama et al., 1993). The new species, *Chlorococcum littorale* (Fig. 1), could grow under bubbled air enriched with 60% CO$_2$. At 10% CO$_2$, the growth rate was twice that in *Chlorella regularis* which had shown the highest growth rate in our culture collection. One of the surveys in 1988 indicated that about one third of CO$_2$ discharged from Japanese industries was from power stations. If we add CO$_2$ discharged from steelworks, the total would have covered about half of all CO$_2$ discharged by industry. These industrial flue gases usually contain 10–20% CO$_2$. The finding that *C. littorale* can grow rapidly under an extremely high CO$_2$ concentration raised the new possibility of absorbing CO$_2$ discharged from these industries by means of a biological method. Since the discovery of *C. littorale*, other green and blue-green

algae that can grow rapidly under extremely high CO$_2$ concentrations have been found. After the finding of *C. littorale* had been reported, a report on the red alga *Cyanidium caldarium* (Tilden) Geitler, which could grow under 100% CO$_2$ (Seckbach et al., 1970), was brought to our attention. Unfortunately, there has been no follow up on this important finding.

Physiology under high CO$_2$ pressure

In order to understand how *C. littorale* cells can grow under an extremely high CO$_2$ concentration, we studied the change in photosynthetic characteristics during the course of adaptation of air-grown cells to 40% CO$_2$.

When air-grown cells were inoculated into a fresh culture medium and aerated with ordinary air supplemented with 40% CO$_2$, they did not grow for the first 3–4 days (Pesheva et al., 1994). However, logarithmic growth started after that. During the induction period, the specific growth rate was nearly zero and photosynthesis was inhibited. During the initial phase of inhibition of photosynthesis by 40% CO$_2$, the activity of PS II in intact cells measured by the Hill reaction in the presence of benzoquinone was suppressed. In contrast, the PS I activity in cell homogenates measured by oxygen uptake as a result of the photoreduction of methylviologen was greatly enhanced. These results obtained in batch cultures were confirmed by other measurements from a continuous culture (Iwasaki et al., 1996).

The cell concentration of *C. littorale* was kept constant with constant in- and out-flow of air. When the bubbling gas was changed from air to 40% CO$_2$, the concentration of algal cells temporarily decreased, then recovered to the original level by the 4th day followed by a further increase. In a green alga, *Stichococcus bacillaris*, which is intolerant to extremely high CO$_2$, cell concentration was constant in air, progressively decreasing and finally the cells disappeared at 40% CO$_2$, indicating that the *Stichococcus* cells could not survive at this CO$_2$ concentration. A quenching analysis of chlorophyll fluorescence carried out by the saturation pulse method supported the suggestion that there was photoinhibition in PS II during the induction period after the cells were transferred to 40% CO$_2$ from air. The change in excitation energy distribution between PS I and PS II was measured by the ratio of fluorescence at 714 nm and 618 nm (F714/F678) at 77K. The transition from state 1 to state 2 was enhanced immediately after the *C. littorale* cells had been transferred from air to 40% CO_2 (Iwasaki et al., 1998). No such change in excitation energy distribution was apparent when *S. bacillaris* cells were transferred from air to 40% CO_2. During the transition period, the level of D1 protein in *C. littorale* cells was approximately constant, indicating that PS II (including D1 protein) in these cells was protected from photoinhibition by controlling the state transition. Such a protective mechanism is not likely to have functioned in *S. bacillaris* cells because the level of their D1 protein dropped during the transient period.

The next query was why PS II in *S. bacillaris* was damaged under extremely high CO_2 concentrations. One possibility is acidification of the algal cells due to the formation of protons during the course of CO_2 hydration. ^{31}P-NMR spectroscopy and the DMO method showed a drop in the cytoplasmic pH value when *S. bacillaris* cells were exposed to 40% CO_2, while the cytoplasmic pH value remained constant in *C. littorale* cells exposed to 40% CO_2. This indicated that the inhibition of photosynthesis by an extremely high CO_2 concentration was associated with intracellular acidification (Pronina et al., 1993).

In this connection, an increase in the number and size of vacuoles has been observed in *C. littorale* cells under extremely high CO_2 concentrations, but no such change was apparent in *S. bacillaris* cells (Kurano et al., 1998). With *C. littorale* cells, Sasaki et al. (1999) have further reported that the activity of vacuolar-type H^+-ATPase increased under 40% CO_2. These results suggest that vacuole development associated with enhanced vacuolar H^+-ATPase activity occurred during the acclimatization of *C. littorale* cells to an extremely high CO_2 concentration.

Acidification of the cytoplasm and possibly of the chloroplast would inhibit the Calvin-Benson cycle, decreasing the population of open-state PS II reaction centers. We, therefore, assume that during the early period after the transfer of *C. littorale* cells from air to 40% CO_2, a large increase in their PS I/PS II ratio will support the energy supply for ATP synthesis, which would be enhanced by cyclic electron transfer around PS I. Accordingly, the state 2 transition in photosynthesis will support the pumping activity of H^+-translocating ATPase across the cytoplasmic and chloroplastic membranes to maintain the pH values of the cytoplasm and chloroplast under extremely high CO_2 concentrations. In contrast, *S. bacillaris* cells, which did not show the state 2 transition, cannot maintain the cytoplasmic and chloroplastic pH values constant under extremely high CO_2 conditions.

CO_2 fixation

In parallel to these photosynthetic studies, studies on the growth capability of microalgae are also being carried out at MBI. To evaluate the maximum ability for growth and CO_2 uptake by microalgae, a novel photobioreactor of the flat-plate type has been developed. This reactor is mainly characterized by its short light path, which is less than 2 cm, curved baffles which stimulate liquid mixing, and intensive aeration which enhances gas–liquid mass transfer and the flashing light effect. The flat plate is made of transparent acrylic plastic and is illuminated from both sides by banks of fluorescent lamps. Ordinary air or air enriched with CO_2 is supplied from the bottom of the reactor through a fine perforated tube. The culture temperature is controlled in a water bath made of acrylic plastic. Under continuous illumination by an average light intensity of 2000 μmol m^{-2} s^{-1}, *C. littorale* cells exhibited a very high linear growth rate of 384 ± 30 mg of dry cells l^{-1} h^{-1} (Hu et al., 1998a) in a small-scale flat-plate photobioreactor with 1 cm light path and 1.5 l working volume. This growth rate corresponds to a CO_2 uptake rate of 16.7 g in 24 h, and is almost 4 times faster than the highest value so far reported for a marine cyanobacterium (Takano et al., 1992). By replacing the culture medium daily with a fresh medium, the maximum cell density reached as high as 84 g of dry cells l^{-1}. Frequent medium replacement has two positive effects: compensation of nutrient limitation and removal of autoinhibitory activity. This high cell concentration holds economic significance, because cell separation from the culture medium requires a large amount of power and a high cell density saves on this cost.

Larger-scale culture experiments have been conducted outdoors to evaluate the CO_2 fixation ability of microalgae under solar irradiation (Zhang et al., 1999). Three to five flat plates, each being 1.5 cm wide, 100 cm long and 80 cm high and giving a 9 l working volume, were stood vertically in parallel at a distance of 25 to 50 cm (Fig. 2). The thermophilic cyanobacterium, *Synechocystis aquatilis* SI-2, which had been isolated from a hot spring for the purpose of CO_2 fixation in a temperature range higher than that at which *C. littorale* could grow, was grown in this reactor system. The optimum temperature for the

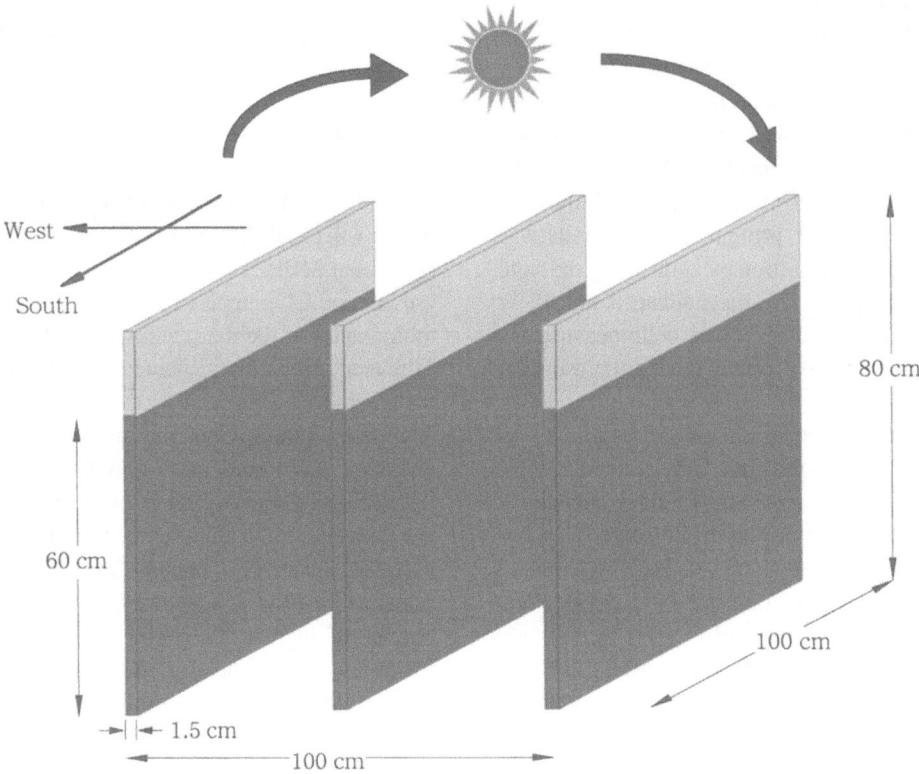

Figure 2. Flat-plate photobioreactor system for outdoor culture.

growth of this alga is 40 °C (being 25 °C for *C. littorale*), and 2 h of doubling time was observed at this temperature. We have chosen *S. aquatilis* SI-2 as a candidate organism for outdoor experiments, because *C. littorale* cannot survive above 30 °C and the temperature of outdoor cultures often exceeds this limit. The effects of the following three factors on cell productivity were studied during the summer of 1998 at Kamaishi (39° N, 142° E): direction of the flat-plate surface, distance between the plates and temperature control. Japan has a shortage of usable land and land is therefore expensive; the smaller the land area needed by the reactor system for CO_2 fixation, the better the economics. In these outdoor experiments, we therefore evaluated the CO_2 fixation rate on the basis of the land area occupied by individual plates. When five plates were placed in an east-west orientation separated by a distance of 25 cm and the culture temperature was regulated in the optimal range (37–43 °C), a CO_2 fixation rate of 53 g of CO_2 m^{-2} of ground area d^{-1} was achieved. This is an average value obtained in the middle three plates during 8 days, the average daily irradiation being 6.3 MJ m^{-2} d^{-1} in this period. This CO_2 fixation ability is about 10 times

greater than that of the average forestland in the temperate regions. This high capability implies that the reactor design provides effective utilization of solar irradiation. Unlike conventional open-pond culture facilities, the flat-plate system has a three-dimensional structure which can function to diffuse unnecessarily strong sunlight into a moderate intensity and distribute it uniformly on to the flat surface. The microalgal cells utilize this moderate irradiation for photosynthesis without any unfavorable effect of photoinhibition. The high growth potential of *S. aquatilis* SI-2 also contributes to the performance of this system.

Culture collection

The search for microalgae with high efficiency for CO_2 fixation has resulted in many strains collected and stored at MBI. Based on this culture stock, we are going to establish a specialized and open culture collection of marine microalgae. The purposes of this effort are as follows: (1) to maintain and to supply marine strains for scientific studies as well as for industrial application, (2) to develop a tech-

Table 1. Some characteristic microalgal strains mentioned in this paper.

Species	Unique properties	Reference
Acaryochloris marina Miyashita et al.	Possesses chlorophyll *d* as a major photosynthetic pigment	Miyashita et al. (1996, 1997) Hu et al. (1998b)
Chlorococcum littorale N. Chihara, T. Nakayama et I. Inouye	Grows at high CO_2	Kodama et al. (1993)
Porphyridium purpureum (Bory de Saint-Vincent) K. Drew et Ross	Possesses unique gene structure and novel catalytic site of CA	Mitsuhashi et al. (1996, 2000a, 2000b)
Stichococcus bacillaris Nägeli	Sensitive to high CO_2	Iwasaki et al. (1996, 1998)
Synechocystis aquatilis Sauvageau	Very short doubling time (2 h)	Zhang et al. (1999)

nique for effective preservation and (3) to construct a database containing taxonomic information, strain history, characteristics of each strain, growth conditions, photo- and electron-micrographs, 18S rDNA (16S rDNA for the Cyanophyta) gene sequences, etc.

With strong emphasis on the phylogenetic relationship, the 18S rDNA sequences of nearly all of our green algal strains have been determined. The results indicate that most of these strains form clades consisting of marine microalgae, while a few belong to the fresh-water group. This means some algae of fresh-water origin can be found in seawater, even in pelagic waters. Traditional methods for taxonomy, for example morphological observations by electron microscopy and pigment analysis, are also core activities of this project.

Cryopreservation is an important method to effectively maintain algae; however, many strains cannot survive after being frozen. Periodical subculturing is the only method at this moment, and extending the subculture interval by applying sub-optimum growth conditions for culturing helps to decrease the labor required and the risks accompanying such manipulation. To develop a universal medium for supporting the life of a broad range of marine strains is also one of our targets to achieve effective preservation. One result of this effort, the Daigo IMK medium, is now commercially available and is manufactured by Nihon-Seiyaku of Japan (contact e-mail: life-tc@nihon-pharm.co.jp, or labchem-tect@wako-chem.co.jp). This medium is applicable to Cyanophyta, Rhodophyta, Chlorophyta, Cryptophyta, Chlorarachniophyta, Dinophyta, Heterokontophyta and Haptophyta.

At present, we are privately maintaining more than a thousand strains, with ca. 870 original isolates comprising Cyanophyta, Rhodophyta, Cryptophyta, Dinophyta, Heterokontophyta, Haptophyta, Chlorarachniophyta, Chlorophyta, unidentified strains and protozoa. Some of them produce a large amount of intracellular starch, extracellular polysaccharides (Miyashita et al., 1993), and polyunsaturated fatty acids (Kawachi et al., 1996). It should be noted that the novel marine prokaryote, *Acaryochloris marina*, which has chlorophyll *d* as its major photosynthetic pigment, has been isolated and studied in detail at MBI (Miyashita et al., 1996; 1997; Hu et al., 1998b). Some of these microalgae with outstanding feature are listed in Table 1. New classes, orders, genera and species exist in this collection, showing that the marine environment still embraces a great number of unknowns.

Conclusion

The research activities that have been described indicate that microalgae are likely to play a key role in solving some environmental problems, in studies on photosynthesis, in the production of useful substances, and in understanding the marine ecosystem. Technology and knowledge on these small organisms will contribute to the welfare of humans in the coming century.

Acknowledgement

This work was partly supported by New Energy and Industrial Technology Development Organization (NEDO).

References

Hu, Q., N. Kurano, M. Kawachi, I. Iwasaki & S. Miyachi, 1998a. Ultrahigh-cell-density culture of a marine green alga, *Chlorococcum littorale*, in a flat-plate photobioreactor. Appl. Microbiol. Biotech. 49: 655–662.

Hu, Q., H. Miyashita, I. Iwasaki, N. Kurano, S. Miyachi, M. Iwaki & S. Itoh, 1998b. A photosystem I reaction center driven by chlorophyll *d* in oxygenic photosynthesis. Proc. Natl. Acad. Sci. USA 95: 13319–13323.

Iwasaki, I., N. Kurano & S. Miyachi, 1996. Effects of high-CO_2 stress on photosystem II in a green alga, *Chlorococcum littorale*, which has a tolerance to high CO_2. J. Photochem. Photobiol. B: Biology 36: 327–332.

Iwasaki, I., Q. Hu, N. Kurano & S. Miyachi, 1998. Effect of extremely high-CO_2 stress on energy distribution between PS I and PS II in a 'High-CO_2' tolerant green alga, *Chlorococcum littorale*, and the intolerant green alga, *Stichococcus bacillaris*. J. Photochem. Photobiol. B: Biology 44: 184–190.

Kawachi, M., M. Kato, H. Ikemoto & S. Miyachi, 1996. Fatty acid composition of a new marine picoplankton species of the Chromophyta. J. appl. Phycol. 8: 397–401.

Kodama, M., H. Ikemoto & S. Miyachi, 1993. A new species of highly CO_2-tolerant fast-growing marine microalga for high-density cultivation. J. Mar. Biotech. 1: 21–25.

Kurano, N., T. Sasaki & S. Miyachi, 1998. Carbon dioxide and microalgae. In Inui T. et al. (eds), Advances in Chemical Conversions for Mitigating Carbon Dioxide Studies in Surface Science and Catalysis, vol. 114. Elsevier Science B.V., Amsterdam: 55–63.

Mitsuhashi, S. & S. Miyachi, 1996. Amino acid sequence homology between N- and C-terminal halves of a carbonic anhydrase in *Porphyridium purpureum*, as deduced from the cloned cDNA. J. Biol. Chem. 271: 28703–28709.

Mitsuhashi, S., T. Mizushima, E. Yamashita, S. Miyachi & T. Tsukihara, 2000a. Crystallization and preliminary X-ray diffraction studies of a carbonic anhydrase from the red alga, *Porphyridium purpureum*. Acta Cryst. D56: 210–211.

Mitsuhashi, S., T. Mizushima, E. Yamashita, M. Yamamoto, T. Kumasaka, H. Moriyama, T. Ueki, S. Miyachi & T. Tsukihara, 2000b. X-ray structure of β-carbonic anhydrase from the red alga, *Porphyridium purpureum*, reveals a novel catalytic site for CO_2 hydration. J. Biol. Chem. 275: 5521–5526.

Miyachi, S., 1995. Diversity of microalgae and their possible application. OECD Documents, Environmental Impacts of Aquatic Biotechnology: 28–31.

Miyashita, H., H. Ikemoto, N. Kurano, S. Miyachi & M. Chihara, 1993. *Prasinococcus capsulatus* gen. et sp. nov., a new marine coccoid prasinophyte. J. Gen. appl. Microbiol. 39: 571–582.

Miyashita, H., K. Adachi, N. Kurano, H. Ikemoto, M. Chihara & S. Miyachi, 1997. Pigment composition of a novel oxygenic photosynthetic prokaryote containing chlorophyll *d* as the major chlorophyll. Plant Cell Physiol. 38: 274–281.

Miyashita, H., H. Ikemoto, N. Kurano, K. Adachi, M. Chihara & S. Miyachi, 1996. Chlorophyll *d* as a major pigment. Nature 383: 402.

Pesheva, I., M. Kodama, M. L. Dionisio-Sese & S. Miyachi, 1994. Changes in photosynthetic characteristics induced by transferring air-grown cells of *Chlorococcum littorale* to high-CO_2 conditions. Plant Cell Physiol. 35: 379–387.

Pronina, N. A., M. Kodama & S. Miyachi, 1993. Changes in intracellular pH values in various microalgae induced by raising CO_2 concentrations. XV Int. Botanical Cong., Yokohama, Japan: 419.

Sasaki, T., N. A. Pronina, M. Maeshima, I. Iwasaki, N. Kurano & S. Miyachi, 1999. Development of vacuoles and vacuolar H^+-ATPase activity under extremely high-CO_2 conditions in *Chlorococcum littorale* cells. Plant Biol. 1: 68–75.

Seckbach, J., A. F. Baker & P. M. Shugarman, 1970. Algae thrive under pure CO_2. Nature 227: 744–745.

Suzuki, E., Y. Shiraiwa & S. Miyachi, 1994. The cellular and molecular aspects of carbonic anhydrase in photosynthetic microorganisms. Progress in Phycological Research 10: 1–54.

Takano, H., H. Takeyama, H. Nakamura, H. Sode, J. G. Burges, E. Manabe, E. Hirono & T. Matsunaga, 1992. CO_2 removal by high-density culture of a marine cyanobacterium *Synechococcus* sp. using an improved photobioreactor employed light-diffusing optical fibers. Appl. Biochem. Bioeng. 34/35: 449–458.

Zhang, K., N. Kurano & S. Miyachi, 1999. Outdoor culture of a cyanobacterium with a vertical flat-plate photobioreactor: effects on productivity of the reactor orientation, distance setting between the plates, and culture temperature. Appl. Microbiol. Biotechnol. 52: 781–786.

Hydrobiologia **512**: 33–37, 2004.
P.O. Ang, Jr. (ed.), Asian Pacific Phycology in the 21st Century: Prospects and Challenges.
© 2004 *Kluwer Academic Publishers.*

Principles for attaining maximal microalgal productivity in photobioreactors: an overview

Amos Richmond
Microalgal Biotechnology Lab., Albert Katz Department of Dryland Biotechnologies, Blaustein Institute for Desert Research, Ben-Gurion University of the Negev, Sede-Boker Campus, 84990 Israel
E-mail: amosr@bgumail.bgu.ac.il

Key words: microalgae, photoreactors, productivity

Abstract

Efficient management of mass algal cultures requires appreciation of the most important factors governing the light regime of the average cell, i.e. the interrelationships between the intensity of the light source – never the sole factor involved in mass culture productivity – and the optimal cell density affected by the optical path. The latter is a dominant factor in photosynthetic productivity of ultra high cell density cultures (UHDC) cultured in flat plate reactors. Indeed, a very short optical path (5–10 mm) permits a most efficient use of strong light by facilitating ultra-high cell densities (ca. 10–20 g dry cell mass 1^{-1}), in which the condensed cells are exposed to very high frequency light/dark cycles. Another important feature of dense cultures concerns the very small but highly efficient light dose available to cells under extreme mutual shading. The low productivity of the single cell in the culture is well compensated, in terms of culture productivity, by the high culture cell mass exposed to very high frequency light/dark cycles. The combined effects of all these factors result in high efficiency of strong light-use for photosynthesis. UHDC are associated with growth inhibition which represents a severe production obstacle. Once this aspect is better understood and managed, UHDC in ultra short optical path reactors may become a useful production mode of photoautotrophic cell mass and secondary metabolites.

Introduction

Maximal culture productivity of phototrophic microorganisms is obtainable when light represents the sole limitation to productivity, i.e. the nutritional requirements are satisfied and the temperature not far from optimal. Therein rests the major challenge involved in microalgal biotechnology, both scientific and economic: devise efficient, cost-effective reactors and protocols by which to utilize best high irradiance such as solar energy for mass production of phototrophic algae.

Light zones in photobioreactors

Efficient utilization of light by the cells is associated with many constrains. One relates to the exponential

Table 1. Light penetration depth[a] (cm) into cultures of *Nannochloropsis* sp. as effected by the concentration of cell mass (Gitelson, unpublished).

DW (g 1^{-1})	2	10	50
Blue (410-450 nm)	0.960115	0.192023	0.038405
Green (580-600 nm)	9.433679	1.886736	0.377347
Red (670-678 nm)	1.252403	0.250481	0.050096

[a]To the depth in which light energy is 10% of incident light.

attenuation of light energy in passing through the culture column. As cell density increases, light penetration into the culture, expressed as a percent of total incident light impinging on culture surface, decreases exponentially (Table 1). Two light zones are thereby affected in the photobioreactor: The illuminated photic volume in which light supports photosynthesis, and the dark volume, in which light intensity is below the compensation point and net photosynthesis cannot take place. The higher the population density (and the longer the light path), the more complex it becomes to attain one basic requirement for efficient utilization of light, i.e. an even distribution of light to all cells in the reactor. Another difficulty in harnessing effectively solar, or any other strong light source for photosynthesis, rests upon the fact that the higher the light intensity, the lower, as a rule, the efficiency by which light energy impinging on the reactor surfaces is converted into chemical energy.

Adjusting the photon flux density to cell density

The difficulty in understanding the complex mode by which light effects mass cultures of phototrophic microorganisms has been augmented over the years by erroneous application of the so called 'light curve' (Goldman, 1979) to interpret the growth response of mass cultures to light. This model provides a generalized shape of the light response curve, relating the photosynthetic or growth rate of the culture to the intensity of the light source, i.e. the photon flux density (PFD) impinging on culture surface. The light source is thus placed as the sole rate-limiting factor in a light-limited system. This relationship, however, is only correct for optically thin cultures, of such low population density that mutual shading by the cells is essentially absent. In reality, however, mass cultures exposed to strong light cannot be maintained in optically thin concentrations with no mutual shading, since the photon flux density (PFD) outdoors is much higher, up to an order of magnitude or more, than the photosynthetic saturating light intensity. Excess light may cause photoinhibition followed by culture death and the most practical approach by which to cope with this is to increase cell density to the point in which mutual shading causes cells to receive strong light intermittently. The high PFD prevailing outdoors is thereby diminished or 'diluted' for the individual cells and the light energy each cell receives along time is thus not only a function of the intensity of the light source, but is also, and often more so, dependent on cell density. Any growth rate therefore may indeed be manifested in a culture in response to a given intensity of the light source, depending on the population density through its effect on modifying light availability in the culture. Second, the major parameter in mass culture is the output rate of cell mass or some specific product, per unit reactor-volume or illuminated

reactor-area. The output rate in continuous cultures at steady- state is a function of both the growth rate and cell density, and both optically very thin or extremely dense cultures yield significantly below maximal output rates. As is in any biological phenomenon related to optimal exploitation of resources per unit area, there is a certain optimal algal 'stand' which results in the highest areal output rate. This is the 'optimal cell density' (OCD, $g\,l^{-1}$) which is species-specific and which is mandatory to maintain in a culture in order to exploit light most efficiently.

Light–dark (L–D) cycles

A salient feature of OCD is that at this cell concentration, except in optically thin cultures, light penetrates to only a small fraction of the reactor volume. At any given instant, therefore, most of the cells at OCD are exposed to darkness. For a given light-path length, the fraction of the photic volume over the entire, mostly dark reactor volume is a function of both the intensity of the light source and the extent of the population density. The length of the light-path (corresponding to the width of a plate reactor, perpendicular to the direction of the light source), greatly affects the frequency (measured in ms) of the light/dark (L–D) cycle which originates from the movement of cells in and out of the photic volume, in optimally stirred cultures. The shorter the light path, the higher the frequency of the L–D cycle which, as a rule, accelerates the photosynthetic rate. At optimal cell density (i.e. which results in highest productivity per irradiated area), cells are exposed to relatively short flashes of light followed by a relatively long period of darkness; the higher the frequency of this L–D cycle therefore, the more efficiently light (particularly high PFD) may be utilized for photosynthesis (Grobbelaar et al., 1996; Hu et al., 1998b; Janssen 2002; Richmond et al., 2003). Indeed, under specific laboratory conditions, reducing the light-path thereby increasing L–D cycle frequency results in a significant rise in areal productivity. Since the OPD is inversely related to the LP, using reactors with a very narrow LP (e.g. 1–2 cm) exposed to high PFD permits maintenance of cultures of extremely high cell concentrations, e.g. over 100 and up to 1000 mg chlorophyll l^{-1} (Zou et al., 2000).

Mixing the culture

This is yet another factor which affects the light regime to which the average cell in the culture is exposed in UHCD. Mixing represents the most practical means by which to attempt to distribute radiation evenly to all cells in the culture, as well as accelerate growth by reducing the diffusion barriers around the cells. Mixing affects a higher L–D cycle frequency, mandatory for efficient utilization of light for photosynthetic productivity. This may be readily observed in open raceways, in which insufficient stirring reduces the efficiency of solar energy utilization as the pattern of flow becomes increasingly laminar. In addition, dissolved oxygen (DO) builds up in cultures in which mixing is inadequate, inhibiting photosynthesis. The positive effect optimal stirring exerts on the output-rate of biomass is accentuated as the population density and light limitation in the culture increase. Inadequate mixing resulting in high O_2 tensions and in laminar – instead of turbulent – flows resulting in cell precipitation and wall growth, is at the root of colossal industrial failures well described recently by Tredici (1999).

The interrelationships existing between the intensity of the light source, the OCD, the mixing rate and the output rate are clearly evident in *Spirulina* cultures (Table 2). Varying the rate of mixing in cultures exposed to 'low' and 'high' incident light showed that at 'low' light and low OCD, mixing rate had no effect on productivity. Increasing light intensity (all other conditions kept constant), resulted in a significant shift-up of OCD, accompanied by an increase in the output rate. The magnitude of the mixing effect on productivity was strictly dependent on the strength of the light source through its effect on culture density. At 'high' PFD, the output rates of cell mass reflected strong dependence on the rate of mixing, and as the mixing rate was increased to optimal, the output rate doubled (Hu & Richmond, 1996). It should be stressed however, that the dominant effect of the mixing rate shown in Table 2, took place in a culture of *Spirulina platensis* ((Norst.) Gietler, a very large filamentous cell-aggregate requiring therefore high stirring energy for induction of optimal turbulent streaming. Small, single-cell species, e.g. *Nannochloropsis salina* D.J. Hibberd, require much less energy for optimal stirring. In such species, the effect of the mixing rate on the output rate would be considerably smaller than in *Spirulina* sp , and would become clearly manifested only at relatively high cell densities.

Growth inhibition

Mass production of ultra high cell-density cultures (abbreviated UHCD, Hu et al., 1996, 1998b) however, is presently unattainable due to inhibition of cell growth which unfolds in the culture as soon as cell concentration (of at least some species) increases above a certain, species specific, threshold (Javanmardian & Palsson, 1991; Imada et al., 1992; Zou et al., 2000). The economic advantages of growing phototrophic cultures at very high cell concentrations in terms of reduced production and capital costs, are presently curtailed by the necessity to continuously alleviate growth inhibition in such cultures.

Preliminary experiments conducted recently with *Nannochloropsis* sp. by Zhang Cheng-Wu in the author's laboratory showed that when the inhibitory activity is removed from UHDC, a decrease in the LP is associated with a marked increase in areal yield. If growth inhibition is not removed, which would practically be the case in large scale cultivation, a short 1.0 cm LP resulted in a sharp reduction in areal yield ($g\,m^{-2}\,h^{-1}$) compared with the areal yield of a 9.0 cm LP reactor (Richmond et al., 2003). The increase in volumetric yield ($g\,l^{-1}\,h^{-1}$) taking place when the LP is shortened does not represent a real increase in productivity, being simply proportional to the reduction in reactor volume. The contrasting response to reduction in optical path is depicted in Fig. 1 in which the effect of the light-path on areal yield of soluble polysaccharides in *Porphyridium* is compared to its effect on productivity of cell mass in *Spirulina*. In the latter, growth inhibition was routinely removed whereas it was not removed in the former. An increase in areal yield was associated with a decrease in LP in *Spirulina* (interpreted as due to a higher L–D cycle frequency), an effect which in *Porphyridium* was not manifested due apparently to growth inhibition (Fig. 1).

Considerations of cost-effectiveness

Efficient, cost-effective photobioreactors would be characterised by high areal- as well as volumetric-productivity. This may be obtained by establishing in the culture an optimal light regime, i.e. in which all parameters effecting the illumination of the average cell in the culture are optimised. Reactor efficiency in terms of cost effectiveness involves high volume- and areal yields of cell mass or products, coupled with a low annual investment cost and, very impor-

Table 2. Interrelationships between intensity of the light source, the optimal cell density, the rate of mixing and the output rate of cell mass in *Spirulina platensis* (after Hu & Richmond, 1996)

Light μmol photon m^{-2} s^{-1}	Mixing rate[a] L(air) l^{-1} min^{-1}	Optimal cell density g dw l^{-1}	Maximal output rate mg dw l^{-1} hr^{-1}
	0.6	2.0	70
500	2.0	5.0	100
	4.0	5.0	100
	0.6	6.0	200
1800	2.0	9.0	300
	4.0	17.0	400

[a] A linear relationship was found to exist between the rate of aeration and the rate of stirring.

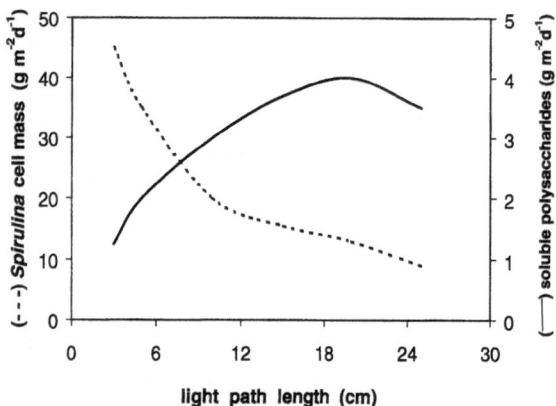

Figure 1. Area-output rates of *Spirulina* cell mass and *Porphyridium* polysaccharides as a function of light-path length (after Hu et al., 1998a; Singh et al., 2000).

tant, simple and easy procedures for reactor operation and maintenance. Some practical points in this context must be addressed in designing efficient reactors: Is the reactor illuminated on all its surfaces, displaying a high area to volume ratio? Is the reactor likely to sustain continuous, mono-algal cultures? I.e. would bio-fouling and wall-growth be readily checked and controlled, reducing the likelihood of culture deterioration? Is the oversaturated oxygen in the culture prevented from reaching prohibitive tensions? Is the cooling system optimised for the local weather conditions? Is turbulent streaming readily introduced and maintained? Is the cost of a unit reactor volume reasonable in terms of the amortisation as well as maintenance costs? Cost-effective photobioreactors may significantly augment the economic potential involved in microalga culture, representing therefore an important research goal.

Summary

Our recent researches (Hu et al., 1996, 1998b; Richmond, 2000; Richmond et al., 2003) permit an attempt to offer a unified concept with which to elucidate light-mediated growth limitations in photoautotrophic mass cultures related to reactor design and culture maintenance protocols, as follows: assuming all other growth conditions are optimal, the sole limitation to photosynthetic activity and cell growth in optically thin cultures is the intensity of the light source, as depicted by the 'light curve'. In high cell density mass cultures, in which cells receive light intermittently, cell density becomes ever more dominant in determining culture productivity, and the quantity light cell^{-1}, i.e. the light energy received by the average cell in the culture, becomes the significant growth parameter. Further up the cell concentration scale, the rate of mixing becomes ever more effective in controlling cell growth and culture productivity. Intensifying the PFD and shortening the length of the optical path facilitates maintenance of ultra high cell concentrations. The combined effects of short optical paths and high L–D cycle frequency which shortens the photosynthetically wasteful dark periods to which the cells are exposed as mutual shading is mounting, combined with high PFD and high cell concentrations result in efficient use of strong light, thus producing maximal areal yields of cell mass. Inhibitory substances or conditions however, arrest cell growth and development at high cell densities, and if not removed, bring about gradual deterioration of the culture. Limitation to growth at that stage is shifted from light, the major growth limitation in mass cultures, to an altogether different mode of growth limitation, which is not yet sufficiently understood.

References

Goldman, J. C., 1979. Outdoor algal mass cultures. 2; Photosynthetic yield limitations. Water Res. 13: 119–136.

Grobbelaar, J. U., L. Nedbal, & V. Tichy, 1996. Influence of high frequency light/dark fluctuations on photosynthetic characteristics of microalgal photoacclimated to different light intensities and implications for mass algal cultivation. J. Appl. Phycol. 8: 335–343.

Hu, Q. & A. Richmond, 1996. Productivity and photosynthetic efficiency of *Spirulina platensis* affected by light intensity, cell density and rate of mixing in a flat plate photobioreactor. J. Appl. Phycol. 8: 139–145.

Hu, Q., H. Guterman & A. Richmond, 1996. Physiological characteristics of *Spirulina platensis* (cyanobacteria) cultured at ultrahigh cell densities. J. Phycol. 32: 1066–1073.

Hu, Q., Y. Zarmi & A. Richmond, 1998a. Combined effects of light intensity, light-path and culture density on output rate of *Spirulina platensis* (cyanobacteria). Eur. J. Phycol. 33: 165–171.

Hu, Q., N. Kurano, M. Kawachi, I. Iwasaki & S. Miyachi, 1998b. Ultrahigh-cell-density culture of a marine green alga *Chlorococcum littorale* in a flat-plate photobioreactor. Appl. Microbiol. Biotechnol. 49: 655–662.

Imada, N., K. Kobayashi, K. Isomura, H. Saito, S. Kimura, K. Tahara & Y. Oshima, 1992. Isolation and identification of an autoinhibitor produced by *Skeletonema costatum*. Nippon Suisan Gakkaishi 58: 1687–1692.

Janssen, M., 2002. Cultivation of Microalgae: Effect of Light/Dark Cycles on Biomass Yield. Thesis, Wageningen University, Wageningen, The Netherlands.

Javanmardian, M. & B. O. Palsson, 1991. High-density photoautotrophic algal cultures: design, construction, and operation of a novel photobioreactor system. Biotechnol. & Bioengin. 38: 1182–1189.

Richmond, A., 2000. Microalgal biotechnology at the turn of the millennium: A personal view. J. Appl. Phycol. 12: 441–451.

Richmond, A., Zhang C.-W. & Y. Zarmi, 2003. Interrelationships between the optical path, the optimal population density and cell growth inhibition with reference to efficient use of strong light for photosynthetic productivity in high density algal cultures. Biomolecular Engineering (in press).

Singh, S., S. M. Arad & A. Richmond, 2000. Extracellular polysacchride production in outdoor mass cultures of *Porphyridium* sp. in flat plate glass reactors. J. Appl. Phycol. 12: 269–275.

Tredici, M. R., 1999. Bioreactors, photo. In Flickinger, M. C. & S. W. Drew (eds), Encyclopedia of Bioprocess Technology: Fermentation, Biocatalysis and Bioseparation. John Wiley & Sons, New York.

Zou, N., C. W. Zhang, Z. Cohen & A. Richmond, 2000. Production of cell mass and eicosapentaenoic acid (EPA) in ultrahigh cell density cultures of *Nannochloropsis* sp. (Eustigmatophyceae). Eur. J. Phycol. 35:127–133.

Hydrobiologia **512**: 39–44, 2004.
P.O. Ang, Jr. (ed.), Asian Pacific Phycology in the 21ˢᵗ Century: Prospects and Challenges.
© 2004 *Kluwer Academic Publishers.*

Mass production of *Spirulina*, an edible microalga

Hidenori Shimamatsu
5-1-72 Mitsuwadai, Wakaba-ku, Chiba 264-0032, Japan
E-mail: splshima@mxg.mesh.ne.jp

Key words: Spirulina, Arthrospira, microalgae, mass production, open pond culture safety

Abstract

Spirulina (*Arthrospira*) is a filamentous cyanobacterium that is grown commercially for food and feed and as a food coloring and additive. Currently there are many companies producing *Spirulina* in different countries to the tune of 3000 tons a year. This paper attempts to describe the problems of mass culture of *Spirulina*, deriving information from two commercial facilities: Siam Algae Company (Thailand) and Earthrise Farms (U.S.A.).

Introduction

Spirulina (*Arthrospira*) grows naturally in alkaline lakes, and has a long history of being used as a human food. The rediscovery of *Spirulina* by J. Leonard and Compere in the 1960s (Leonard & Compere, 1967) led to the beginning of mass production of *Spirulina* for commercial purposes in the late 1970s. The first attempt of mass production was carried out by Sosa Texcoco Co. in Mexico, supported by the technology of Institute Francais du Petrole (IFP) (Duran-Chastel, 1980; Ciferi, 1983; Ciferi & Tiboni, 1985). Since then, many producers have become successful in the outdoor open pond production of *Spirulina*. The production in the world may now exceed 3000 tons on a dry weight basis. Table 1 shows some of the past and present producers. Most of the market for *Spirulina* is as a food supplement, namely health food; although it is also marketed as feed, especially for the aquaculture industry, and as a blue color extract for food.

In the past, many producers were confronted with various problems on how to achieve the high productivity and cost performance that they enjoy today. As long as *Spirulina* is used for food purpose, its safety and quality must be assured. Mass production of quality *Spirulina* requires special techniques and control practices to solve the problems which are associated with outdoor open pond culture under recycling condition of culture water. Problems related

Table 1. Representative producers of *Spirulina* in the past and present.

Companies	Country/territory
Spirulina Mexicana (Sosa Texcoco) SA	Mexico
Siam Algae Co., Ltd.	Thailand
Nippon Spirulina Co., Ltd.	Japan
Koor Foods Co., Ltd	Israel
Earthrise Farms	U.S.A.
Cyanotech Corporation	U.S.A.
Nan Pao Resins Chemical Co., Ltd.	Taiwan
Blue Continent Co., Ltd.	Taiwan
Far East Microalgae Co., Ltd.	Taiwan
Tung Hai Chlorella Co., Ltd.	Taiwan
Parry Agro Industries Ltd.	India
Yunnan Spirin Co., Ltd.	Mainland China
Hainan DIC Microalgae Co., Ltd.	Mainland China

to harvesting, drying and packaging must also be addressed.

This paper will touch upon some of the problems encountered in mass production of *Spirulina* at two *Spirulina* farms, Siam Algae Co. (SAC) in Bangkok, Thailand and Earthrise Farms (EF) in California, U.S.A. Based on the author's experience in DIC (Dainippon Ink & Chemicals Inc., Japan), which is the parent company of these two *Spirulina* farms

Table 2. A typical composition of basal medium for mass production of *Spirulina*

Ingredients	Amount ($g\ l^{-1}$)
$NaHCO_3$	16.8
K_2HPO_4	0.5
$NaNO_3$	2.5
K_2SO_4	1.0
NaCl	1.0
$MgSO_4 \cdot 7H_2O$	0.2
$CaCl_2 \cdot 2H_2O$	0.04
$FeSO_4 \cdot 7H_2O$	0.01
EDTA	0.08

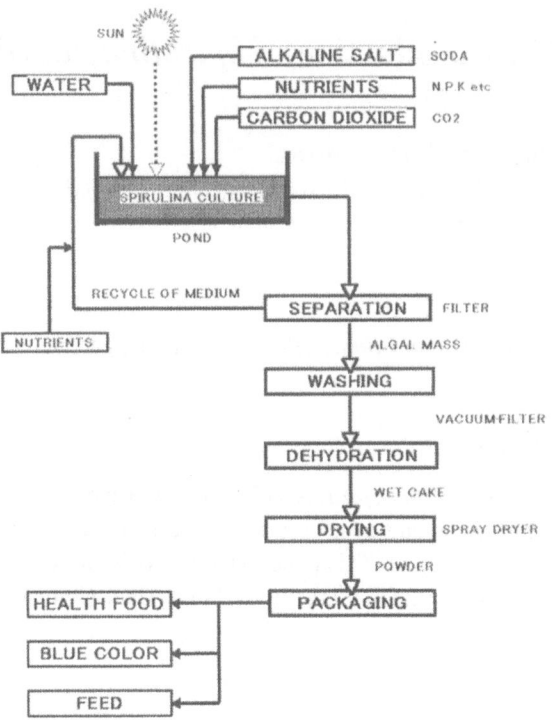

Figure 1. A schematic diagram of production system of *Spirulina*.

mentioned above, some solutions to these problems from an industrial perspective will be discussed.

Description of sites studied

Most of the studies were done at two *Spirulina* farms owned by DIC, namely SAC and EF. The facility of SAC was constructed in 1977 in the suburbs of Bangkok, Thailand and started mass production in 1978. Mean ambient temperature at the site ranges from 25 to 29 °C, which enables outdoor production of *Spirulina* throughout the year. Mean annual rainfall in this area is around 1400 mm. Most of the rainfall happens during the rainy season from May to October. EF is located in Imperial Valley, California, USA and has been in operation since 1983. Mean ambient temperature in this place is from zero to 40 °C, sometimes reaching up to 50 °C and down to the freezing point, restricting the production season within April to October. During the production season, there is almost no rain. The mean annual rainfall in this production site is a mere 10 mm.

Mass production system of *Spirulina*

A schematic diagram of the typical production system of *Spirulina* is presented in Fig. 1. Only a brief description of the production process will be given here as detailed information can be found in the reviews by Belay et al. (1994) and Belay (1997).

The culture pond

The basic shape and design of the culture pond developed by the author (Shimamatsu, 1987) are shown in Fig. 2. The pond is a rectangular open channel with a paddle wheel generating a circulation flow. Special provision is made at every corner of the pond as shown in Fig. 3. This system gives an uniform flow speed and uniform mixing effect at every point in the pond, free from any stagnation spot. SAC has 13 concrete-made ponds in total, each having an area of 2000 m^2 and EF has 30 PVC-lined ponds of 5000 m^2 each for mass production of food grade *Spirulina*.

The harvest system

Algal mass is recovered from the culture through a series of filtration steps utilizing filtration equipment such as inclined gravity screen and vibrating screen and is further concentrated into a wet algal cake by a horizontal vacuum filter. The final cake will have about 15% solids.

Drying and packaging

The wet algal cake is dried into fine powder by the spray dryer, and packed in an oxygen barrier bag

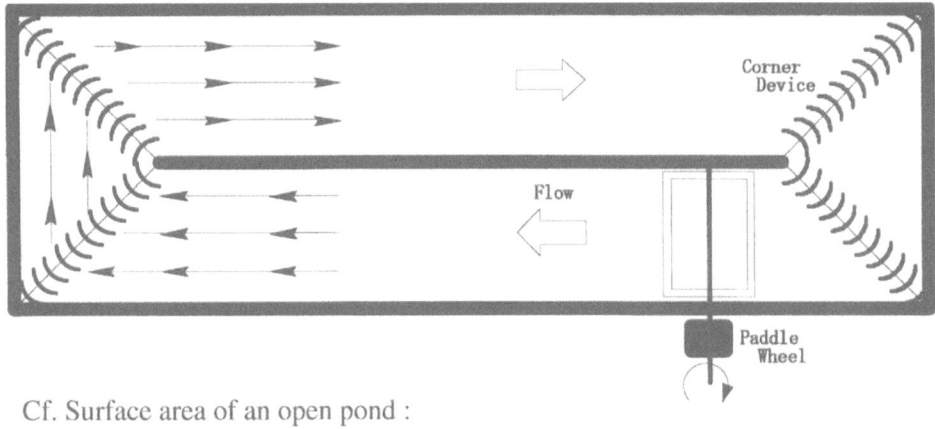

Cf. Surface area of an open pond :

    Siam Algae   : 2,000 m²

    Earthrise Farms : 5,000 m²

Figure 2. A pond design for mass culture of *Spirulina*.

in order to keep the quality of the product for prolonged period. This process is particularly important in order to maintain the content of less stable components, such as beta-carotene. Both facilities in Thailand and U.S.A. have accumulated enough experience in this area and have developed their own drying and packaging proprietary technology.

Problem of outdoor commercial biomass production

In the past, SAC and EF were confronted with various problems in the mass production of *Spirulina* under the conditions of recycling of culture water and continuous cultivation. How to keep an unialgal culture throughout the production season without contamination and to maintain consistent quality of the product were the major problems that had to be solved. The problems encountered are discussed below in relation to other important factors.

Strain selection

The strain for mass production is selected carefully among the extensive collections of *Spirulina* from around the world which are periodically subcultured in the laboratory in order to maintain actively growing cells. Major criteria in the selection of strains are growth rate, biochemical composition and resistance to environmental stress at each production site. It must be emphasized, however, that a strain which

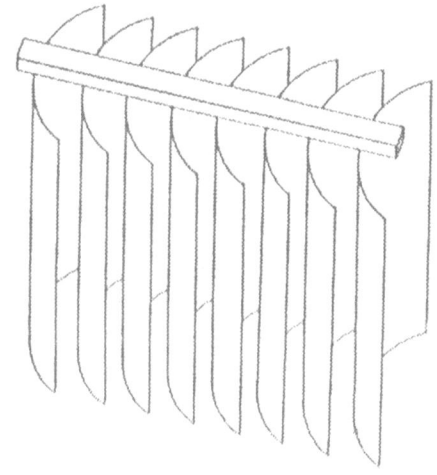

Figure 3. Corner device of the pond (shown in Fig. 2).

shows a good performance in a laboratory does not always display the same behavior in outdoor open pond operation.

The culture medium

The typical medium used in growing *Spirulina* is shown in Table 2 as a simplified Zarrouk medium. Trace elements are adjusted depending on the quality of the culture water. The pH of the medium is kept between 9.5 to 10.5 using carbon dioxide. The desired growth rate and the cost of the nutrients may dictate the choice of the nutrients used. For example, while nitrate is usually used as a nitrogen source, ammonium hydroxide (ammonia water) can also be used.

Table 3. A typical analysis of *Spirulina* product. Based on sample of dried *Spirulina* powder of Siam Algae Company (SAC) analyzed by Japan Food Research Laboratories.

Composition	Quantity (per 100 g dry wt)
General composition	
Moisture	3.00 g
Protein	61.40 g
Fat (Lipids)	8.50 g
Fibre	3.00 g
Ash	7.70 g
N-free extract	16.40 g
Colorants	
Phycocyanin	16.20 g
Carotenoids	477.00 mg
Chlorophyll-*a*	1.20 g
Vitamins	
Provitamin A	214.00 mg
Thiamin (V.B$_1$)	1.98 mg
Riboflavin (V.B$_2$)	3.63 mg
Vitamin B$_6$	0.59 mg
Vitamin B$_{12}$	0.11 mg
Vitamin E	11.80 mg
Niacin	13.20 mg
Folic acid	42.00 μg
Panthothenic acid	0.88 mg
Inositol	74.00 mg
Minerals	
Phosphorus	914.00 mg
Iron	57.40 mg
Calcium	171.00 mg
Potassium	1.77 g
Sodium	1.05 g
Magnesium	257.00 mg

Table 4. Quality standard of *Spirulina* in Japan based on the criteria of the Japan Health Food Association (JHFA) announced on August, 1986.

1.	**Appearance and Characteristics**	
	Dark green in color,	
	Slight seaweed smell,	
	No nasty smell and bad taste.	
2.	**Essential Components**	
	Crude Protein	>50 %
	Total Carotenoids	>100 mg %
	Chlorophyll-*a*	>500 mg %
	c-Phycocyanin	>2000 mg %
3.	**Pheophorbide**[a]	
	Existing Pheophorbide	<50 mg %
	Total Pheophorbide	<100 mg %
4.	**Arsenic**	< 2 ppm as As
5.	**Heavy metals**	< 20 ppm as Pb
6.	**Standard plate count**	$<5 \times 10^4$ (per g)
7.	**Coli form**	Negative
8.	**Moisture**	<7 %

[a]Revised in July 1990.

Table 5. Food and Drug Administration (FDA)'s requirement for *Spirulina* in USA.

1.	*Spirulina*	No foreign algae
		No contaminants
2.	Insect fragment:	<30 pcs per 10 g
3.	Rodent hair:	< 1.5 pcs per 150 g

However, *Spirulina* cells are easily damaged by higher ammonia concentrations. More than 3 ppm of ammonia would be toxic to *Spirulina*. Urea might be toxic also because *Spirulina* has a high urease activity. The higher calcium content of the make-up water results in the precipitation of calcium salt which causes the loss of alkalinity and some minerals such as iron and phosphorus in the culture.

Operational practices in pond culture

It is said that higher flow speed of the culture in the ponds is recommendable for effective photosynthesis (Richmond, 1988; Richmond et al., 1990). However, in our experience, increase of flow speed above 30 cm s^{-1} sometimes results in fragmentation of the algal trichomes which leads to decreases in harvest efficiency. This increases the risk of contamination and foaming of the culture. Such accumulation of organic matter can actually inhibit photosynthesis and growth (Belay, 1997). Algal density in the culture is generally kept between 400 to 600 mg dry weight l^{-1}. This range of density is mainly dictated by harvest efficiency. At density lower than 100 mg dry weight l^{-1}, some photoinhibition or even photooxidation due to high light intensity may occur (Vonshak & Richmond, 1988; Vonshak et al., 1988). Thus, during the scaling-up stage of the culture from a seed pond to a production pond, special consideration must be paid on the density of algal biomass. The culture depth of the ponds is usually kept between 15–30 cm. Depths greater than 30 cm will result in severe reduction in photosynthesis due to light limitation.

Contamination problems

In mass production of *Spirulina* with open ponds, how to maintain the unialgal culture for a long time without any contamination by other organisms is a major challenge. In Thailand, SAC suffered from the problems caused by rainfall. Dilution of the culture medium by heavy rainfall facilitates the growth of unfavorable organisms including bacteria, green algae, protozoa and insects. The challenge under such circumstance is how quickly could the medium be restored to its original concentration, preferably within one day. Occasionally, phage-like phenomena were experienced that destroyed all the culture at once. It is not known why such phenomenon occurs but we believe it is probably associated with some sort of viral infection. We have not observed permanent bleaching of the algae although there was some pigment loss. Contamination by green algae and bacteria are the common and major problems in mass production of *Spirulina* with outdoor open ponds, particularly under the condition of continuous recycling of the culture medium. This recycling operation can easily result in the accumulation of organic matter because of decomposition and death of algae. This encourages the growth of various contaminants. Accordingly, special practices are required so as not to induce much stress or damage to the algae mechanically and operationally. How to minimize the accumulation of organic matters in the culture is the key issue to avoid or control contamination. On the other hand, ecological technique can also be useful occasionally to control the single cell green algae such as *Chlorella* or *Oocystis*. Some examples of this type of ecological control can be observed from the unialgal culture of *Spirulina* in natural lakes (e.g. Lake Ankorongo, Madagascar; Lake Texcoco, Mexico) where some predators weed out all of the green algae but not *Spirulina*. The problem of contamination by aquatic insects and ground insects is unavoidable. Removal of these insects is done commonly by netting. Knowledge of the ecological interaction of these aquatic insects is also useful for their control.

Harvesting and drying

Problems encountered in harvesting are usually related to the efficiency of harvesting techniques. The latter are affected by factors such as amount of algal biomass, contamination, suspended matter and the nature of the harvesting machinery used. Both farms in Thailand and U.S.A. have developed their own proprietary harvesting and drying methods to maximize harvest efficiency and minimize loss in quality.

Production cost

This is an area where little information is available because companies are not willing to disclose their cost of production. Success in reducing production cost is dependent upon (1) growth rate of the algae, (2) control of contamination and hence reduction in culture renewal time, (3) increased harvest efficiency and (4) the overall operational efficiency of the farm.

Safety and quality assurance

There are established national and international quality standards for *Spirulina* products. Tables 3–5 show the typical analysis of the contents of *Spirulina* product, the quality standard in Japan and United States Food and Drug Administration's (FDA) requirement for *Spirulina* product respectively. Recently, cyanobacterial toxins have become a major issue in public health due to the increased occurrence of toxic cyanobacterial blooms. These toxic blooms contain algae that produce hepatotoxins called microcystins (Carmichael, 1994). *Spirulina* companies like Earthrise Farms have already developed methods for the determination of these toxins and actually certify each lot of their product to be toxin free. *Spirulina* does not normally contain microcystins but contamination of outdoor culture by other cyanobacteria is a possibility.

Conclusion

In the past 20 years, mass production of *Spirulina* has developed rapidly. However, most producers have been confronted with several problems before they achieved economical success in the *Spirulina* market. These problems relate mainly to outdoor production and harvest efficiency. Another major problem is the maintenance of stable and high quality in an outdoor-grown *Spirulina*. Several companies have overcome these problems and are now the major players in the domestic and international *Spirulina* market.

The future calls for a concerted effort to maximize output from such outdoor pond production facilities. There are lots of challenging opportunities for research in academia and private enterprises. Indeed, cooperation between researchers in the academia and those with rich experience in actual outdoor mass culture can bring significant improvement in this area and bring down the cost of production so that many more people can benefit from the health food qualities of *Spirulina*.

References

Belay, A., 1997. Mass culture of *Spirulina* outdoor – The earthrise farms experience. In Vonshak, A. (ed.), *Spirulina platensis (Arthrospira)*, Physiology, Cell-biology and Biotechnology. Taylor & Francis, New York: 131–158.

Belay, A., Y. Ota, K. Miyakawa & H. Shimamatsu, 1994. Production of high quality *Spirulina* at Earthrise Farms. In Phang, S. W., Y. K. Lee, M. A. Borowitzka & B. A. Whitton (eds), Algal Biotechnology in the Asia Pacific Region. University of Malaya. Kuala Lumpur: 92–102.

Carmichael, W., 1994. The toxins of cyanobacteria. Scientific American 279: 78–86.

Ciferi, O., 1983. *Spirulina*, the edible microorganism. Microbiol. Rev. 47: 551–578.

Ciferi, O. & O. Tiboni, 1985. The biochemistry and industrial potential of *Spirulina*. Ann. Rev. Microbiol. 39: 503–526.

Durand-Chastel, H., 1980. Production and use of *Spirulina* in Mexico. In Shelef, G. & C. J. Soeder (eds), Algal Biomass. Elsevier/North-Holland Biochemical Press, Amsterdam: 39.

Leonard, J. & P. Compere, 1967. *Spirulina platensis* Geitler, algue bleue de grande valeur alimentaire par sa richesse en proteines, Bull. Jard. Bot. Nat. Belg. 37: 1.

Richmond, A., 1988. *Spirulina*. Microalgal Biotechnology. Cambridge University Press, New York.

Richmond, A., E. Lichtenberg, B. Stahl & A. Vonshak, 1990. Quantitative assessment of the major limitations on productivity of *Spirulina platensis* in open raceways. J. Appl. Phycol. 2: 195–206.

Shimamatsu, H., 1987. A pond for edible *Spirulina* production and its hydraulic studies. Hydrobiologia 151/152: 83–89.

Vonshak, A. & A. Richmond, 1988. Mass production of the blue-green alga *Spirulina*: an overview. Biomass 15: 233–247.

Vonshak, A., R. Guy & M. Guy, 1988. Photoinhibition and its recovery in two different strains of *Spirulina*. Pl. Cell Physiol. 29: 721–726.

Hydrobiologia **512**: 45–48, 2004.
P.O. Ang, Jr. (ed.), Asian Pacific Phycology in the 21st Century: Prospects and Challenges.
© 2004 *Kluwer Academic Publishers.*

Current microalgal health food R & D activities in China

Shizhong Liang[1], Xueming Liu[1], Feng Chen[2] & Zijian Chen[3]
[1]*The College of Food Engineering & Biotechnology, South China University of Technology, Guangzhou 510640, China*
[2]*Department of Botany, The University of Hong Kong, Pokfulam Road, Hong Kong, China*
[3]*Jiangmen Biotechnology Center, Jiangmen, Guangdong, China*
Author for correspondence; E-mail: fesliang@scut.edu.cn

Key words: microalgae, *Spirulina*, *Chlorella*, microalgal foods, China

Abstract

The major microalgal genera presently cultivated in China are *Spirulina* and *Chlorella*. They are initially manufactured in the form of algal biomass or extracts by the food industry. The biomass is then used for producing a variety of health products such as tablets, capsules, powder or for extracting bioactive ingredients such as beta-carotene, and phycocyanin. The algal biomass is supplemented to noodles, breads, biscuits, candies, ice cream, bean curd and other common foods as food additives so as to enhance their nutritive and health values. The extracts are mainly used to enrich liquid foods such as health drink, soft drink, tea, beer or spirits.

Introduction

Edible microalgae such as *Chlorella* and *Spirulina* are rich in protein, lipid, polysaccharide, edible fiber, microelements and bioactive substances (Li & Li, 1997). They have such health and pharmacological properties that can help to prevent and cure peptic ulcer and anemia, enhance immunity, anti-tumor, anti-radiation, anti-pathogenic activities against microorganisms, decrease blood lipid and some as anti-arteroclerosis agent (Sonoda, 1972; Sano & Tanaka, 1987; Hasegawa et al., 1995; Singh et al., 1995; Hayashi & Hayashi, 1996; Tanaka et al., 1997, 1998). They can be used to develop nutritional and health foods (Liu & Liang, 1999; Becker, 1988).

In fact, microalgae have been used for food long time ago. For example, *Spirulina* naturally grew in Chad Lake was used as food by Kanembu who lived around the lake since hundreds, or even thousands of years ago. In 1964, health food was produced with microalgae cultivated in artificial media in Japan. *Chlorella* tablets made from dry powder were sold in the markets. In 1975, *Spirulina* tablets were marketed (Yamaguchi, 1997).

Microalgal biotechnology in China has a good developmental momentum, though it started late. The first pilot cultivation base was constructed at Chenghai Lake in Yunnan, China in 1989 (Li & Qi, 1997). In 1991, the first microalgae corporation, Shenzhen Lanzao Biotechnology Corporation was founded in Shenzhen. Now there are more than one hundred research institutes and manufacturing enterprises concerned with the study and development of microalgae as food. The output of *Spirulina* powder has reached 1000 t in 1996. China has now the most extensive cultivation of *Spirulina* in the world.

The College of Food Engineering and Biotechnology of South China University of Technology (SCUT), in cooperation with the Research Center of Biological Fermentation and Food Technology of Hong Kong University and Jangmen Biotech. Center, has successfully conducted pilot research on suspension cultivation of *Chlorella* in fermentors. The production of *Chlorella* has reached a dry weight of 20 kg m^{-3}. It is expected to realize the industrial production of edible microalgae by high-density heterotrophic culture in reactors and to provide cheap raw material for the food industry.

Along with the development of microalgal biotechnology in China, the utilization of microalgae has also developed very rapidly. The major microalgal genera presently cultivated in China are *Spirulina* and *Chlorella*. They are initially manufactured in the form of algal biomass or extract by the food industry. Up to now, a series of nutritional and health microalgal food has been developed.

Nutritional and health microalgal food

Microalgal tablet or capsule

Microalgal tablet (or capsule) is made from dry microalgal powder directly. For example, Qizheng *Spirulina* Tablet and Mingxing Hukangbao Tablets are made from *Spirulina* powder and sticky material. The former is produced by Shenzhen Lanzao Biotechnology Corporation and Guangzhou Guanghua Pharmaceutical Company Ltd, and sold by Guangzhou Maoyuan Imp. & Exp. Corporation. Guangzhou Mingxing Pharmaceutical Factory produced the latter. The only microalgal health drug is *Spirulina* capsule, manufactured by Yunan Green-A Biotechnology Co., Ltd.

Microalgal tablets find favors in consumers' eyes because they are rich in protein, vitamins, polysaccharide, polyunsaturated fatty acids (PUFAs), microelements and edible fiber. For example, Green-A *Spirulina* Tablets have the following nutritional compositions (per 100 g): Protein 55–70 g, carbohydrates 15–25 g, fat 2–6 g, dietary fiber 2–4 g, moisture 4–9 g; chlorophyll 0.6–1 g, carotenoids 0.1–0.4 g, phycocyanin 3.6 g; Mineral nutrition (calcium 0.5–1 g, magnesium 0.2–0.6 g, iron 30–100 mg, zinc 2–4 mg selenium 10–30 μg, germanium 40–80 μg), Vitamins (A 100–200 mg, B_1 1.5–4 mg, B_2 3–5 mg, B_3 10–30 mg, B_6 0.5–0.8 mg, B_{12} 0.05–1.5 mg, E 5–10 mg, PP 200–400 mg, folic acid 0.05–0.1 mg, pantothenic acid 0.5–1.5 mg). They are believed to improve health, prevent disease and reduce weight (Liu & Liang, 1999).

Microalgal nutrition liquid

Up to now, although microalgal tablets are still the most popular algal products, diversification of other microalgal products must be encouraged to ensure the development of microalgal biotechnology. Microalgal extract is one of the important microalgal products. It is convenient to drink, easy to digest, and suitable for all ages. It has now occupied some share of the market. Microalgal extracts, produced from edible microalgae as the main material, will have good development and market prospects.

Some reports showed that *Chlorella* growth factors (CGF) could activate the growth of other organisms and enhance their immunity (Konishi et al., 1985). The main bioactive component of *Chlorella* is glycoprotein with D-galactose, MW 63 000. Similarly, *Spirulina* and its hot water extract can also achieve the function of activating the growth of target organism, reducing weight, preventing disease infection and curing sickness (Belay et al., 1993). In order to preserve the health function and to reduce the unfavorable algal flavor, extracts from traditional Chinese medicinal food can be added to the microalgal extracts.

The development of microalgal extract is in an experimental or small production scale. Wuhan Plant Research Institute, in cooperation with Wuhan Pharmaceutical Factory, studied and produced in lot size the *Spirulina* Oral Liquid. On the other hand, in collaboration with Guangzhou Maoyuan Imp. & Exp. Corporation, SCUT developed the health and nutritional *Spirulina* Liquid with *Spirulina* extract and honey (Liang et al., 2001). The liquid has the nutrition and health function of *Spirulina* and the flavor of softdrink. We also made some progress in the study of *Chlorella* growth factor.

Other microalgal health and functional foods

There are other microalgal health and functional foods produced with microalgal powder as the main ingredient besides microalgal tablet (or capsule) and microalgal extract (or nutrition liquid). Examples of these include Yingkang *Spirulina* Cosmetic Cake and Show & Me Complete Nutrient. The former was researched and developed by the College of Food Engineering & Biotechnology of SCUT and Guangdong Maoyuan Imp. & Exp. Corporation and produced by Guangdong Guanghua Pharmaceutical Factory. It was made from *Spirulina* powder (food grade) and mixed with traditional Chinese medicinal food such as the seed of Job's tears and malt, bean dregs and milk powder (Liang et al., 1998). The latter, a powder food with more than 22% protein, has abundant and balanced nutrients. It is produced by Tianjin Meilin Health Products Co., Ltd. with *Spirulina* powder and extracts of more than ten natural plants and microorganisms as ingredients.

Health foods such as *Spirulina* Cosmetic Cake not only have abundant and balanced nutrients, they also have the flavor of traditional food as well. This is because these foods are produced with microalgae as the main ingredient and mixed with other nutritional foods such as milk powder or other food additives with natural flavor as supplements. So these products find favors in consumers' eyes, and have good economic benefits and development prospects.

Common microalgal foods

Although microalgae such as *Spirulina* and *Chlorella* are mainly used in the production of microalgal tablets (or capsule) and extract, they are also utilized in the processing of common foods in recent years. To date, there are various general foods produced with the addition of edible *Spirulina* and *Chlorella* powder, extracts, pigment of microalgae (e.g. phycocyanin and carotene) and vitamins. They are new and possess the nutritional and health properties of microalgae and the flavor of common foods.

Microalgal noodles

Among the application of microalgae in common food processing, the largest quantity of microalgae is being used in instant noodles or general noodles. Microalgal noodles, produced with flour, *Chlorella* and *Spirulina*, will have abundant nutrients, good color, smell and taste and strong ductility with the color and flavor of microalgae. Examples of microalgal noodles include *Spirulina* noodle cakes, produced by Guangzhou Nanfang Flour Factory, and are marketed in China. Chinese Educational Ministry recommended it as the ideal lunch for the primary school pupils and secondary school students.

To produce microalgal noodles with light color, 0.1–1.0% of microalgal powder is added to flour. Xu studied the manufacturing of instant *Spirulina* noodles and applied a patent for it in China (Xu, Chinese Patent 1993). Similar result was reported by Chen & Li (1999).

Microalgal bread

To improve the nutritional value of bread, the microalgal powder can be added to it. Microalgal bread has the color and flavor of microalgae, contains more vitamins, microelements, and special bioactive substances present in microalgae. With the help of water

retention function of microalgae, the valid duration time of the product will be longer (Liang, 1999).

Microalgal biscuits

Biscuits are always made of the main ingredients like flour, sugar, cream, edible oil, essence, leavening agent and phospholipids, so their nutrients are always insufficient for human body. In recent years, biscuits with vegetables or fiber have received favorable acceptance by consumers. If green algae or blue-bacteria are added to the biscuit, they can increase the nutritional value of the food. There are some microalgal (*Spirulina*) nutrition biscuits for children and microalgal (*Spirulina*) soda biscuits in the market of China produced by the Shenzhen Haiwang Food Factory or are being developed (Lu & Wang, 1998).

Microalgal drink

Theoretically, microalgae can be added to many drinks to strengthen their nutritional value. In fact, a series of microalgal drinks such as microalgal health drink, microalgal sour milk, microalgal green tea has been developed. In order to keep the clear color or homogeneity of these drinks, microalgae must be used in the form of microalgal extract (green algae extract, *Spirulina* extract or phycocyanin). Because of the nutritional value of microalgae, they are frequently utilized in the development of drinks in recent ten years. Feng and Zeng reported *Spirulina* drink (Feng & Peng, Chinese Patent 1991; Zeng & Liang, 1995). *Spirulina* drink powder was also produced (Bai et al., 1996). We are researching on and developing *Chlorella* and *Spirulina* nutrition drink. Some *Spirulina* drink, for example, *Spirulina* drink produced by Guangzhou Jingxiutang Pharmaceutical Factory, is sold in the markets.

Microalgal green tea

Tea, especially green tea, the leading health drink, is rich in vitamin C, while microalgae such as *Spirulina* and *Chlorella* contain less vitamin C but abundant protein, chlorophyll, carotenoids, PUFA, polysaccharide and micro-elements. If microalgal biomass is mixed with green tea, the nutritional components could be complementary. Foods containing more and balanced nutrients can be acquired. Thanks to low molecular protein, lipid and polysaccharide, microalgae can change the dispersion feature of green tea micropowder in water, and bring up good foam (Zhou et al.,

48

1999). Microalgae are rich in chlorophyll and phycocyanin. These pigments can change the color of tea into green. In addition, microalgal protein and polysaccharide can decrease the irritation brought about by caffeine in green tea. Hence, microalgal green tea can serve the function of health protection and help to alleviate fatigue.

Microalgal beer

Shandong Shuyi Group Corporation, in cooperation with China Aquaculture Science Research Institute, produces *Spirulina* beer. This health *Spirulina* beer keeps the nutritional components and flavor of traditional beer, and also increases its nutrient content derived from *Spirulina* (Chen & Ying, 1998).

Microalgal candy

Guangzhou Dongguan Zhenshanmei Food Factory produces *Spirulina* candy and iodine-rich crunchy candy with granulated sugar, pectin, *Spirulina* powder, peanut and agar-agar. In addition, South China Ocean Research Institute produces phycocyanin from *Spirulina* as edible pigment on a trial basis. EPA, DHA and polysaccharide from microalgae can also be used as functional food additives.

With the increase in the demand for health food products and the acceptance of more and more consumes for microalgal food, it is expected that the research and development of microalgal health foods, flavor foods and additive will speed up in China. The utilization of microalgae in food industry will become widespread in this country. *Spirulina*, *Chlorella* and other edible microalgae will play an important role in the health and nutrition of human beings.

Acknowledgement

This research was supported by the Scientific Reasearch Funds of Guangdong Province, China.

References

Bai W. D., Y. S. Liang & W. H. Zhao, 1996. Production of solid *Spirulina* drink. Food Science: 17 (7): 32–34 (in Chinese).

Becker, E. W., 1998. Micro-algae for human and animal consumption. In Borowitzka, M. A. & L. J. Borowitzka (eds), Micro-algal Biotechnology. Cambridge University Press, Cambridge, MA.

Belay A., Y. Ota, K. Miyakawa & H. Shimamatsu, 1993. Current knowledge on potential health benefits of *Spirulina*. J. appl. Phycol. 5: 235–241.

Chen S. M. & F. Y. Ying, 1998. A super nutritious and health beer- *Spirulina* beer. Niang Jiu (4): 58–59 (in Chinese).

Chen, Y. Z., & Y. M. Li, 1999. Development of nutritious *Spirulina* noodle. J. Chinese Cereals and oils Association 14(4): 13–15 (in Chinese).

Feng, C. F. & S. P. Peng, 1991. Production method of blue-bacteria – *Spirulina* drink. Chinese Patent CN1035425A.

Hasegawa T., M. Okuda , M. Makino, K. Hiromatsu, K. Nomoto & Y. Yoshikai, 1995. Hot water extracts of *Chlorella vulgaris* reduce opportunistic infection with *Listeria* monocytogenes in C57BL/6 mice infected with LP-BM5 murine leukemia viruses. Int. J. Immunopharmacol. 17: 505–512.

Hayashi, T. & K. Hayashi, 1996. Calcium Spirulan, an inhibitor of enveloped virus replication, from a blue-green alga *Spirulina*. Journal of Natural Products 59: 83–87.

Konishi F., K. Tanaka & K. Himeno, 1985. Antitumor effect induced by a hot water extract of *Chlorella vulgaris* (CE): resistance to meth A tumor growth mediated by CE-induced polymorphonuclear leukocytes. Cancer Immunol. Immunother. 19(2): 73–78.

Li, D. M. & Y. Z. Qi, 1997. *Spirulina* industry in China: present status and future prospects. J. Appl. Phycol. 9: 25–28.

Li, S. W. & H. Q. Li, 1997. Nutriological and toxicological analysis of the *Chlorella* dry powder. Food Science 18(7): 48–51 (in Chinese).

Liang, S. Z., 1999. Microalgal use in food production. In Chen F & Y. Jiang (eds), Microalgal Biotechnology. China Light Industries Press, Beijing: 254–259 (in Chinese).

Liang, S. Z., W. P. Liang, Z. Q. Wu & R. Q. Yu, 2001. Production of health *Spirulina* liquid. J. Wuhan Polytechnic Univ.1: 14–16 (in Chinese).

Liang, W. P., S. Z. Liang, R. Q. Yu & Z. Q. Wu, 1998. Research and development of *Spirulina* diet cake. Food and Fermentation Industries 24(6): 34–35 (in Chinese).

Liu, X. M. & S. Z.Liang, 1999. Pharmacological effects of *Chlorella* and its use as a health care supplement. Chinese Traditional and Herbal Drugs 30: 383–386 (in Chinese).

Lu, Z. X. & L. M. Wang, 1998. *Spirulina* biscuits development. Food Science and Technology (4): 1 (in Chinese).

Sano, T., & Y. Tanaka, 1987. Effect of dried, powdered *Chlorella vulgaris* on experimental arterosclerosis and alimentary hypercholesterolemia in cholesterol-fed rabbits. Artery 14(2): 76–84.

Singh, S. P., A. B. Tiku, & P. C. Kesavan, 1995. Post-exposure radioprotection by *Chlorella vulgaris* (E-25) in mice. Indian J. Exp. Biol. 33: 612–615.

Sonoda, M., 1972. Effect of *Chlorella* extract on pregnancy anemia. Jpn. J. Nutrition 30: 218–225 (in Japanese).

Tanaka, K., A. Yamada, K. Noda, Y. Shoyama, C. Kubo & K. Nomoto, 1997. Oral administration of a unicellular green algae, *Chlorella vulgaris*, prevents stress-induced ulcer. Planta Med. 63: 405–406.

Tanaka K., A.Yamada, K. Noda, T. Hasegawa, M. Okuda, Y. Shoyama & K. Nomoto, 1998. A novel glycoprotein from *Chlorella vulgaris* strain CK22 shows antimetastatic immunopotentiation. Cancer Immunol. Immunother. 45: 313–320.

Xu, C. W., 1993. An instant algal noodle and its production method. Chinese Patent CN1077857A.

Yamaguchi, K., 1997. Recent advances in microalgal bio-science in Japan, with special reference to utilization of biomass and metabolites: a review. J. appl. Phycol. 8: 487–502.

Zeng, Y. & M. S. Liang, 1995. Production of *Spirulina* drink. Food Science. 16 (7): 39–41 (in Chinese).

Zhou, Y. C., Y. X. Hu, F. Qiu, & Y. Nie, 1999. Health function of an algal green tea. Practical Preventive Medicine 16 (1): 78 (in Chinese).

Hydrobiologia **512**: 49–56, 2004.
P.O. Ang, Jr. (ed.), Asian Pacific Phycology in the 21ˢᵗ Century: Prospects and Challenges.
© 2004 *Kluwer Academic Publishers.*

Spatial pattern of intertidal macroalgal assemblages associated with tidal levels

Tae Seob Choi & Kwang Young Kim*
Department of Oceanography, Chonnam National University, Kwangju 500-757, Korea
E-mail: kykim@chonnam.ac.kr
*(*Author for correspondence; E-mail: kykim@chonnam.ac.kr*

Key words: macroalgal assemblages, spatial pattern, tidal level

Abstract

The zonation pattern of macroalgal assemblages was investigated from December 1995 to October 1996 on a semi-exposed, rocky intertidal shore at Chungdori (southwestern Korea) based on quantitative and qualitative estimates of species occurrences in 31 permanent quadrats. Variation in cover associated with tidal levels was described for 30 species (that could be discerned with the unaided eyes) including three green, five brown and 22 red algae. Macroalgae inhabiting intertidal zone exhibited distinct zonation patterns. The number of species increased with decreasing intertidal height and was independent of season. The community was dominated by five species (*Gloiopeltis furcata, Gelidium divaricatum, Ulva pertusa, Sargassum horneri, Hizikia fusiformis*). The intertidal assemblage at the study site can be divided into two groups based on the number of species and the population structure with the division occurring at the critical level of 34 cm above MLW (mean low water). *Gloiopeltis furcata, Gelidium divaricatum, Sargassum thunbergii, Monostroma grevillei,* and *Myelophycus simplex* were more abundant in the upper shore zone and rapidly declined in abundance with depth, relative to all other species. *Gelidium amansii, Pachymeniopsis elliptica, Hizikia fusiformis, Gigartina intermedia, Laurencia* sp., *Chondrus ocellatus, Corallina* spp. and *Gigartina tenella* became more dominant in the lower shore zone.

Introduction

Studies of macroalgal zonation patterns in southwestern Korea (e.g., Lee & Boo, 1982; Boo & Choi, 1989; Choi et al., 1989; Lee et al., 1991; Choi et al., 1994) have previously been restricted to qualitative descriptions of species distribution and seasonal variation with little understanding of the roles of environmental factors.

Zonation patterns of macroalgal assemblage in general are recognized to be the result of the effects of biological factors such as competition and grazing as well as physical factors such as wave action, aerial exposure, irradiance, temperature range and time available for nutrient exchange (see Lobban & Harrison, 1994). Physical parameters are important in the determination of upper reaches where conditions become increasingly harsh for macroalgae (Mathieson et al., 1981; Dring, 1982; Underwood & Jernakoff, 1984;

Seapy & Littler, 1993; Chapman, 1995). Numerous studies have also shown that biotic factors such as competition and predation set the lower limits to the zonation of rocky intertidal macroalgae (Santelices et al., 1981; Dethier, 1982; Underwood & Jernakoff, 1984; Wolfe & Harlin, 1988). However, some investigators reported that grazing could have a significant role in the determination of upper limits of macroalgae (Lubchenco, 1980; Santelices et al., 1981).

Here we describe the results of a detailed study of intertidal macroalgal distribution at Chungdori, on the southwestern limit of the Korean Peninsula. No comparable spectrum of rocky intertidal systems has previously been examined with the level of sampling effort and resolution of data undertaken in this study in Korea. The scope of this work is such that spatial variation of the macroalgal populations has been assessed in terms of tidal levels and species assemblages. The aims of our study are to describe the zonation patterns of macroalgae commonly found in this intertidal community and to collect information on the distribution and zonation pattern of the algae by studying the percent cover of the macroalgae in relation to intertidal levels.

The rocky intertidal shore of Chungdori is subject to exceptionally high harvesting pressure by local fishers. The excessive usage (e.g., collecting, cultivation, trampling) may stress this interface between land and sea, making it particularly sensitive to disturbance. This creates some unique conservation and management problems. Hence, an understanding of the zonation pattern of intertidal macroalgae in Chungdori is important, not only in terms of an assessment of Korean shores, but also in the context of general resource management in southwestern Korea.

Study area

The study area is located on the southwestern coast of the Korean Peninsula (Fig. 1), on the south side of Wando Island (34° 17′ 59″ N, 126° 42′ 23″ E). The site is protected from oceanic wave surge but experiences strong tidal currents. The rocky intertidal zone consists of a platform, ca. 50 m in width, that extends from the base of a cliff to an abrupt drop-off into the subtidal zone.

The coast of Chungdori has a semi-diurnal tidal regime with a maximum tidal amplitude (at spring tide) of 3.2 m and a minimum (at neap tide) of less than 1.5 m (OHA ROK, 1995, 1996). Air temperature ranges from 2.4 °C in winter (February) to 26.3 °C in summer (August). There are large seasonal variations in rainfall with distinct dry and wet periods. From September to February average monthly rainfall normally does not exceed 100 mm. Monthly rainfalls are high from March to August (usually more than 80% of the total annual precipitation) with monthly rainfall frequently exceeding 150 mm. Monthly total hours of sunshine are higher during April and May (ca. 220 h per month) than during the rest of the year (80–200 h per month). Data on air temperature, rainfall and sunshine were obtained from the local meteorological institute (KMA, 1995, 1996). Seawater temperature varies from 7.5 °C (February) to 24.6 °C (August). Salinity is more or less stable throughout the year with extremes of 34.7‰ (May) and 32.7‰ (September).

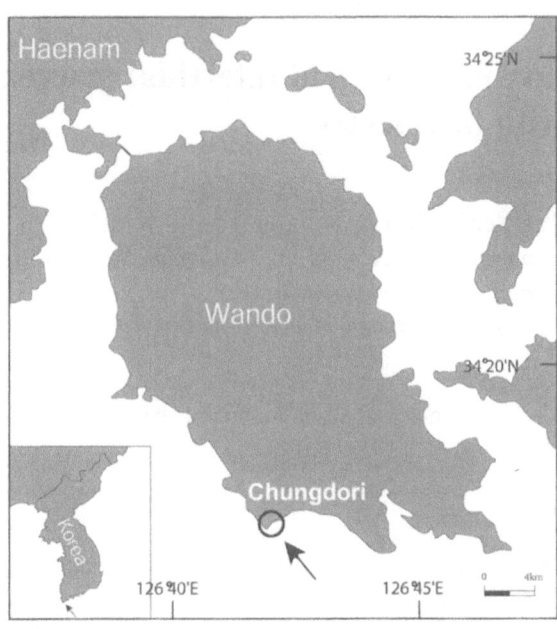

Figure 1. Map of Korea (insert) showing the general location of the study area (arrow) and Chungdori intertidal shores sampled in the present study (circle).

Materials and methods

Qualitative and quantitative sampling of macroalgae was carried out from December 1995 to October 1996 using non-destructive permanent quadrat sampling. Seven censuses were performed at intervals ranging from 1 to 2 months. Thirty-one 50 × 50 cm randomly-scattered quadrats were sampled in the intertidal zone, spanning ca. 50 m of shore. The quadrat locations ranged from immediately above the highest intertidal level to just below the waterline at low water level.

To locate permanent quadrats, holes were drilled and metal studs cemented into the substratum at the upper and lower ends of the quadrats; this enabled the precise relocation of the quadrats during seasonal studies. Vertical height and emersion duration of each quadrat with respect to the mean low water (MLW) were determined (Table 1) using stadia rod, inclinometer and predictions of local tidal heights (OHA ROK, 1996). The difference in tidal height between the uppermost and lowermost quadrats was 126 cm.

Percent cover was used to quantify abundance of macroalgae. For every sample we identified species that could be discerned with the unaided eyes and visually estimated the cover of each species in a 0.25 m² quadrat. To aid estimate of % cover, each quadrat was subdivided into 25, 10 × 10 cm subquadrats us-

Table 1. Tidal height of the permanent quadrats above mean lower water (MLW) and emersion duration per day as measured in February 1996

Quadrat no.	Tidal height above MLW (cm)	Emersion duration hours/day (h:m)
1	101.0	12:04
2	100.6	12:00
3	94.0	11:45
4	80.2	11:14
5	80.2	11:14
6	79.8	11:10
7	77.5	11:05
8	76.3	11:01
9	67.9	10:41
10	66.4	10:37
11	65.8	10:35
12	61.8	10:25
13	57.4	10:12
14	54.8	10:05
15	43.8	9:41
16	39.9	9:30
17	36.9	9:21
18	34.0	9:12
19	30.3	9:01
20	30.3	9:01
21	18.4	8:30
22	17.5	8:26
23	15.2	8:18
24	12.3	8:10
25	10.1	8:03
26	7.3	7:55
27	6.0	7:51
28	3.1	7:41
29	0.5	7:32
30	−7.8	7:07
31	−25.1	6:11

rat was taken to quantify each stratum after upper strata had successively been moved aside. The only organisms removed from the permanent undisturbed quadrats were very small samples taken occasionally for purposes of identification.

Mean covers of each taxon were computed over the entire sampling period for use in plotting kite diagrams of taxon abundance as a function of tidal height. Significance of the vertical distribution pattern of the macroalgal assemblage was determined by a skew statistic using % cover data obtained from permanent quadrats. The Monte Carlo method was used for determining significances of skews (Sheppard, 1995a). The Monte Carlo method computes 1000 simulations, after which it is noted whether the real value falls within the most extreme 5% of the range of simulations (see Sheppard, 1995a, b). In the skew diagrams each horizontal line represents the skew of a species. Species were arranged (top to bottom) in the order of greatest skew towards the upper zone to greatest skew at the lower intertidal zone.

Results

Our observations were restricted to 30 macroalgae that could be discerned with the unaided eyes. Thirty species were present in the permanent quadrats, of which 22 were red, five brown and three green algae (Table 2). Species number showed only minor seasonal variation over the study period and ranged from 19 (May) to 24 (February) (Table 2). One green alga (*Ulva pertusa*), two brown algae (*Hizikia fusiformis* and *Sargassum thunbergii*) and nine red algae (*Gelidium amansii*, *G. divaricatum*, *Corallina* spp., *Carpopeltis affinis*, *Pachymeniopsis elliptica*, *Gloiopeltis furcata*, *Chondrus ocellatus*, *Gigartina tenella* and *Laurencia* sp.) were found throughout the entire sampling period.

The number of species varied in relation to tidal height (Fig. 2). The intertidal zone can be divided into two zones based on species number with the division occurring at quadrat No. 18, 34 cm above MLW. The lower zone (between 30 cm above MLW and 25 cm below MLW) showed much higher species number than the upper zone (between 34 cm and 101 cm above MLW). Accordingly, over the entire sampling period means of 4 and 12 species were present in the upper and lower intertidal zones. Macroalgal abundances were also unevenly distributed over the vertical range of the sampled intertidal, cover being much higher in

ing nylon thread. Percent cover for each taxon was determined by dividing the number of subquadrats occupied by the total number of subquadrats. Species which were observed within a subquadrat but did not (cover) half the space were arbitrarily assigned a cover value of 0.1. The present study was based on undisturbed, repetitively sampled quadrat data collected over a year at a single site. In addition, a photographimetric method modified from Littler & Littler (1985) had the advantage of being rapid and simple to use, thus enabling a greater number of samples to be taken per unit of time. When macroalgal assemblages had multiple layers, more than one photograph per quad-

Table 2. Seasonality of 30 macroalgae at Chungdori, southwestern coast of Korea, between December 1995 and October 1996

Taxa	Dec	Jan	Feb	Apr	May	Jul	Oct
Green algae							
Monostroma grevillei Wittrock		+	+	+	+	+	+
Ulva pertusa Kjellman	+	+	+	+	+	+	+
Codium adhaerens (Cabrera) C. Agardh	+	+	+			+	
Brown algae							
Leathesia difformis (L.) Areschoug			+				
Ishige okamurae Yendo	+					+	+
Myelophycus simplex (Harvey) Papenfuss		+	+	+	+	+	+
Hizikia fusiformis (Harvey) Okamura	+	+	+	+	+	+	+
Sargassum thunbergii (Roth) Kuntze	+	+	+	+	+	+	+
Red algae							
Porphyra sp.				+			
Gelidium amansii (Lamouroux) Lamouroux	+	+	+	+	+	+	+
Gelidium divaricatum Martens	+	+	+	+	+	+	+
Gelidium sp.							+
Pterocladia capillacea (Gmelin) Bornet	+	+	+	+	+		
Corallina spp.	+	+	+	+	+	+	+
Carpopeltis affinis (Harvey) Okamura	+	+	+	+	+	+	+
Carpopeltis cornea (Okamura) Okamura			+	+		+	+
Grateloupia turuturu Yamada					+		
Pachymeniopsis elliptica (Holmes) Yamada	+	+	+	+	+	+	+
Prionitis patens Okamura	+						
Gloiopeltis furcata (Postels et Ruprecht) J. Agardh	+	+	+	+	+	+	+
Callophyllis adhaerens Yamada			+	+			
Gracilaria verrucosa (Hudson) Papenfuss	+	+					
Gymnogongrus flabellifomis Harvey	+	+	+	+	+		
Chondrus ocellatus Holmes	+	+	+	+	+	+	+
Gigartina intermedia Suringar	+	+	+	+		+	+
Gigartina tenella Harvey	+	+	+	+	+	+	+
Campylaephora hypnaeoides J. Agardh					+		
Acrosorium uncinatum (Turner) Kylin	+	+	+			+	+
Laurencia sp.	+	+	+	+	+	+	+
Symphyocladia latiuscula (Harvey) Yamada	+	+	+		+	+	+

the lower region (Fig. 3). The mean cover of macroalgae was 39.1% and 69.6% in the upper and lower zones, respectively, over the whole assessment period (Fig. 3).

The highest cover was contributed by the brown alga, *Hizikia fusiformis* (16.7%). Only four other species contributed mean cover values over the study period greater than 2.0%, two turf-forming red algae, *Gloiopeltis furcata* (10.7%) and *Gelidium divaricatum* (5.6%), the brown alga *Sargassum thunbergii* (5.7%) and the sheet-like green alga, *Ulva pertusa* (5.6%).

Macroalgal cover was averaged over the study period as a function of intertidal height to depict prevailing patterns of zonation (Fig. 4). The upper zone was dominated by *Gloiopeltis furcata*, *Gelidium divaricatum* and *Ulva pertusa*. Other upper zone species, *Sargassum thunbergii* and *Hizikia fusiformis*, generally occurred in the upper intertidal, although they were present throughout the entire vertical range. *Pachymeniopsis elliptica*, *Chondrus ocellatus*, *Gigartina tenella* and *Laurencia* sp. were restricted to the lower region, especially the lowermost part which was dominated by *Gelidium amansii*.

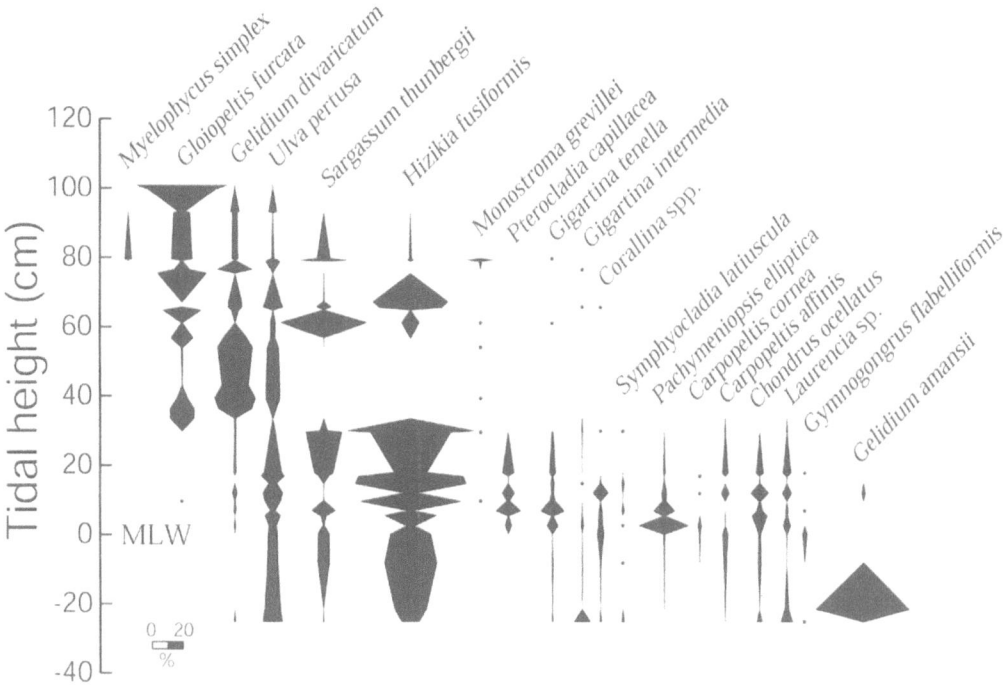

Figure 4. Kite diagram illustrating the abundance (% cover) and distribution of 19 macroalgae on the Chungdori intertidal zone. Cover values below 0.1% are designated with *.

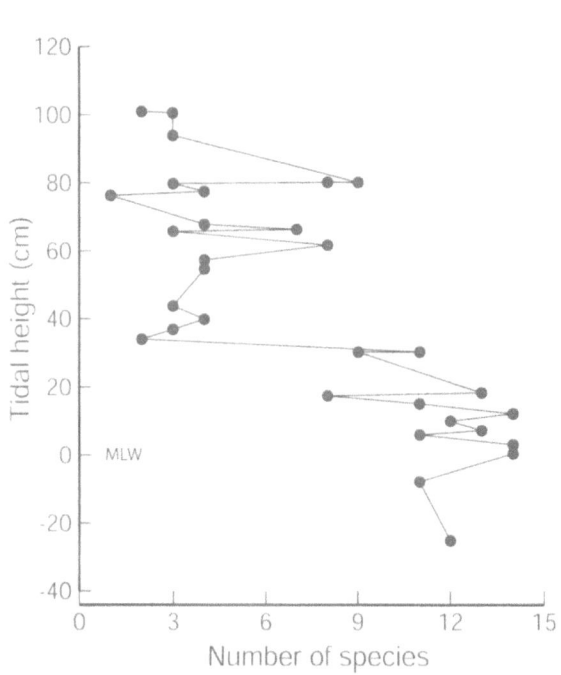

Figure 2. The number of species for each permanent quadrat along the tidal levels.

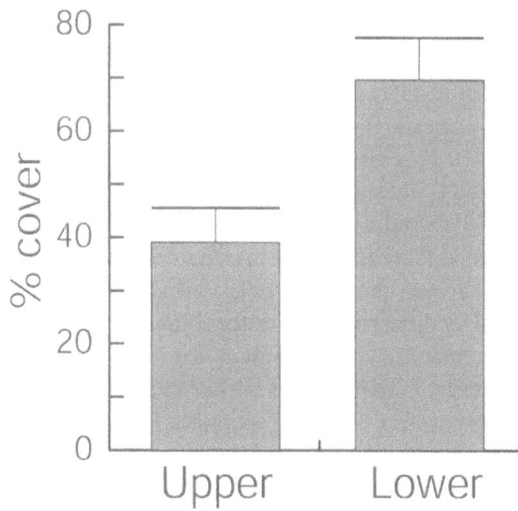

Figure 3. Mean differences in macroalgal cover ($\pm$ S.D; $n = 7$) associated with different tidal levels.

Patterns of zonation were determined objectively by skew analysis of the permanent quadrat samples using the % cover data for the macroalgal assemblages (Fig. 5). The red alga *Gloiopeltis furcata* showed a strong skew ($p < 0.05$) towards upper zone while *Gelidium amansii* showed an equally strong and significant skew to the lower intertidal zone. Other upper shore species such as *Gelidium divaricatum*, *Sargassum thunbergii*, *Monostroma grevillei* and *Myelophycus simplex*, and seven other lower species also showed significant skews to the upper and lower zones, respectively.

Discussion

Previous observations at several localities on the southwestern coast of Korea suggest that *Gloiopeltis furcata*, *Gelidium divaricatum*, *Ulva pertusa*, *Caulacanthus okamurae* and *Enteromorpha* spp. are dominant in the high intertidal zone, and *Ulva pertusa*, *Chondria crassicaulis* and *Sargassum thunbergii* in the mid intertidal zone. The lower intertidal zone is dominated by *Sargassum thunbergii*, *Ulva pertusa*, *Corallina* spp. *Sargassum* spp., *Ecklonia cava* and *Undaria pinnatifida*, and most of these species appear to be able to grow in the subtidal zone (Kang et al., 1980; Lee & Boo, 1982; Lee et al., 1983, 1991; Boo & Choi, 1989; Choi et al., 1989, 1994; Kim & Lee, 1995; Kim, 1999).

Past studies on the macroalgae at the present site have only consisted of qualitative observations (e.g., Choi et al., 1989), and recorded a total of 119 macroalgal taxa in the intertidal zone. Although we considered 30 taxa as representative of the macroalgal assemblage in this study, there were no previously unreported species. In addition, species occurring in this study were also common in the intertidal zone along the southwestern coast of Korea. Kim (1999) suggested that differences in intertidal macroalgal assemblages on the southwestern coast of Korea were caused by differences in substrate stability and suspended sediments. The substrata of the study site consist primarily of massive bed rocks with relatively low turbidity waters. This may permit a high diversity of macroalgae in this study site.

There was a large spatial difference in species number from the macroalgal assemblage of Chungdori, the number being higher in the low tidal zone. The subdivision of the intertidal zone into upper and lower zones can be partly correlated with the gradient of physical factors such as emersion: immersion duration, and possibly tolerance of desiccation. The algae in the upper zone experience more than 9 h per day of aerial exposure suggesting harsh physical factors prevail in this zone during low tides, which may explain why this zone is composed of only a few algae.

In contrast to the relatively poor macroalgal cover on the southwestern coast of Korea (see Kim, 1999), the cover at this study site was relatively high and, as discussed above, it was likely to be effected by substrate condition supporting the macroalgal assemblage. There was also a dramatic decrease in the percent cover of macroalgae with increasing tidal heights. Especially, the cover value abruptly increased in the lower zone (between 30 cm above MLW and 25 cm below MLW). Spatial changes of the abundance may also be related to a critical time for emersion duration.

Most of the macroalgal cover in the upper zone consisted of turf-forming growths of *Gloiopeltis furcata* and *Gelidium divaricatum*, and the thick leathery brown algae *Sargassum thunbergii* and *Hizikia fusiformis*. *Gelidium amansii* as well as *Hizikia fusiformis* were generally found in the lower zone. The overall zonation pattern described above matches well with previous reports in other locations, e.g. Uido Island (Choi et al., 1994), Neobdo Island (Kim & Lee, 1995) and Hawon-Pando (Kim, 1999) on the southwestern coast of Korea. However, *H. fusiformis*, *Ulva pertusa* and *S. thunbergii* can extend their vertical distribution substantially, modifying the limits of intertidal algal zonation.

Kite diagrams are historically perhaps the most common way to show zonation patterns along rocky shores. However, these methods do not by themselves show whether or not the distributions are statistically significant. Skew analysis reflects the zonation pattern observed in the present study site. The vertical distribution of some intertidal macroalgae at Chungdori can be assigned significantly into the upper zone or lower zone. *Gloiopeltis furcata*, *Gelidium divaricatum*, *Sargassum thunbergii*, *Monostroma grevillei* and *Myelophycus simplex* can be assigned to the upper zone, with *Gelidium amansii* and seven other species to the lower zone.

Other studies of macroalgal community on the southwest and south coasts of Korea showed fairly dramatic seasonal fluctuations in the abundance of macroalgal species inhabiting rocky shore (Lee et al., 1991; Choi et al., 1994; Kim et al., 1998; Kim, 1999). The greater seasonal variation observed by all of these authors may be explained by greater changes of air and water temperatures in their study areas. The rocky intertidal zone of Chungdori received little direct solar radiation due to shading by the cliff on the western side of the site. The presence of the cliff may explain why the abundance of macroalgae showed only minor

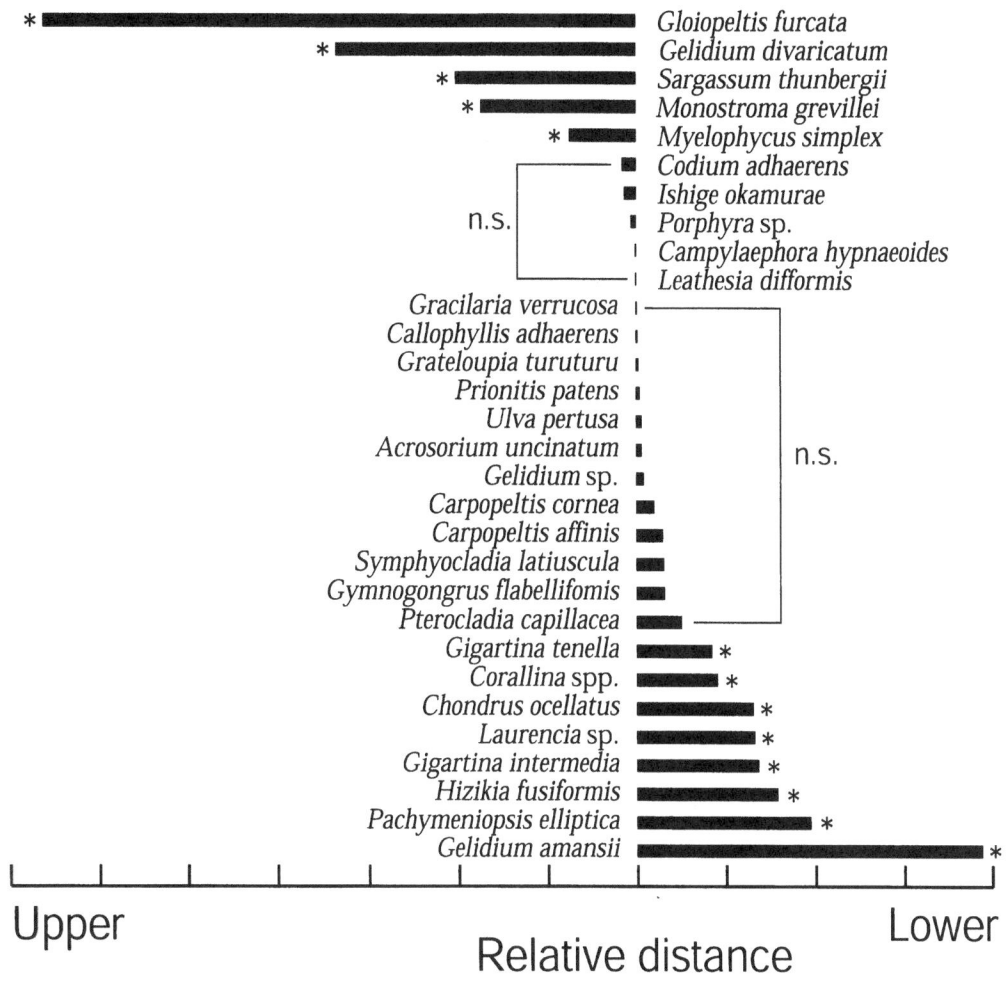

Figure 5. Skew diagram of quantitative distribution of 30 macroalgae along the tidal levels on the Chungdori intertidal zone. The results of Monte Carlo statistics for species distribution along the tidal levels are shown. Level of significance: $p < 0.05$, n.s.: not significant.

seasonal variation within permanent quadrats over the study period.

Based on the review of intertidal community structure in Korea, this study represents the most thorough investigation to date of ecological descriptions in understanding the structure of a rocky intertidal community. Although the roles of competition, predation and recruitment in determining algal zonation patterns in this study were not investigated, these factors probably accounted for the spatial and temporal distribution patterns at this site.

Acknowledgements

We are grateful to Dr C.R.C. Sheppard (University of Warwick), for providing the software for skew analysis and Dr D.J. Garbary (University of St. Francis Xavier) for the benefit of his perceptive comments. Two anonymous referees provided useful comments to improve the paper. This work was supported by the Korea Research Foundation (KRF-2002-070-C00088) and RRC program (SERC) of KOSEF.

References

Boo, S. M. & D. S. Choi, 1989. A summer marine flora of Anma Islands. Rep. Surv. Natur. Environ. Korea 9: 207–218.

Chapman, A. R. O., 1995. Functional ecology of fucoid algae: twenty-three years of progress. Phycologia 34: 1–32.

Choi, D. S., T. S. Yoon & I. K. Lee, 1989. A marine algal flora of Jungdori, Wando. Bull. Inst. Litt. Biota 6: 97–109.

Choi, D. S., K. Y. Kim, W. J. Lee & J. H. Kim, 1994. Marine algal flora and community structure of Uido Island west-southern coast of Korea. Korean J. Environ. Biol. 12: 65–75.

Dethier, M. N., 1982. Pattern and process in tide pool algae: Factors influencing seasonality and distribution. Bot. mar. 25: 55–66.

Dring, M. J., 1982. The Biology of Marine Plants. Edward Arnold, London.

Kang, J. W., C. H. Sohn & C. W. Lee, 1980. The summer marine algal flora of Uido and Maeseom, southwestern coast of Korea. Rep. KACN 16: 95–107.

Kim, K. Y., 1999. Vertical distribution and seasonality of intertidal macroalgae on the coast of Hawon-Pando, southwestern Korea. J. Korean Soc. Oceanogr. 34: 172–178.

Kim, K. Y. & I. K. Lee, 1995. Community and structure of subtidal macroalgae around Neobdo Island on the west-southern coast of Korea. J. Plant Biol. 38: 153–158.

Kim, K. Y., T. S. Choi, S. H. Huh & D. J. Garbary, 1998. Seasonality and community structure of subtidal benthic algae from Daedo Island, southern Korea. Bot. mar. 41: 357–365.

KMA (Korea Meteorological Administration), 1995. Annual Climatological Report: 119 pp.

KMA (Korea Meteorological Administration), 1996. Annual Climatological Report: 119 pp.

Lee, I. K. & S. M. Boo, 1982. A summer marine algal flora of Island in Wando-kun. Rep. Surv. Natur. Environ. Korea 2: 209–227.

Lee, I. K., H. B. Lee & S. M. Boo, 1983. A summer marine algal flora of Island in Jindo-gun. Rep. Surv. Natur. Environ. Korea 3: 291–312.

Lee, I. K., D. S. Choi, Y. S. Oh, G. H. Kim, J. W. Lee, K. Y. Kim & J. S. Yoo, 1991. Marine algal flora and community structure of Chongsando Island on the south sea of Korea. Korean J. Phycol. 6: 131–143.

Littler, M. M. & D. S. Littler, 1985. Nondestructive sampling. In Littler, M. M. & D. S. Littler (eds), Handbook of Phycological Method: Ecological Field Methods: Macroalgae. Cambridge Univ. Press, New York: 161–175.

Lobban, C. & P. J. Harrison, 1994. Seaweed Ecology and Physiology. Cambridge Univ. Press, New York.

Lubchenco, J., 1980. Algal zonation in the New England rocky intertidal community: an experimental analysis. Ecology 61: 333–344.

Mathieson, A. C., N. B. Reynolds & E. J. Hehre, 1981. Investigations of New England marine algae II: the species composition, distribution and zonation of seaweeds in the Great Bay estuary system and the adjacent open coast of New Hampshire. Bot. mar. 24: 533–545.

OHA ROK (Office of Hydrographic Affairs, Republic of Korea), 1995. Tide table. Hydrographic Office Annual Report: 250 pp.

OHA ROK (Office of Hydrographic Affairs, Republic of Korea), 1996. Tide table. Hydrographic Office Annual Report: 250 pp.

Santelices, B., S. Montalva & P. Oliger, 1981. Competitive algal community organization in exposed intertidal habitats from Central Chile. Mar. Ecol. Prog. Ser. 6: 267–276.

Seapy, R. R. & M. M. Littler, 1993. Rocky intertidal community structure on Santa Barbara Island and the effects of wave surge on vertical zonation. In Hochberg, F. G. (ed.), Third California Island Symposium: Recent Advances in Research on the California Island. Santa Barbara Museum of Natural History: Santa Barbara, CA: 273–292.

Sheppard, C. R. C., 1995a. Transect analysis: interactive ecology and social science programs with a novel approach. Scientific Software, U.K.

Sheppard, C. R. C., 1995b. Species and community changes along environmental and pollution gradients. Mar. Pollut. Bull. 30: 504–514.

Underwood, A. J. & P. Jernakoff, 1984. The effects of tidal height, wave-exposure, seasonality and rock-pools on grazing and the distribution of intertidal macroalgae in New South Wales. J. exp. mar. Biol. Ecol. 75: 71–96.

Wolfe, J. M. & M. M. Harlin, 1988. Tidepools in Southern Rhode Island, U.S.A. I. distribution and seasonality of macroalgae. Bot. mar. 31: 525–536.

Hydrobiologia **512**: 57–62, 2004.
P.O. Ang, Jr. (ed.), Asian Pacific Phycology in the 21ˢᵗ Century: Prospects and Challenges.
© 2004 *Kluwer Academic Publishers.*

A short-term response of macroalgae to potential competitor removal in a mid-intertidal habitat in Korea

Baek Jun Kim, Hyuk Je Lee, Seungshic Yum, Hyun Ah Lee, Yong Ju Bhang, Sang Rul Park, Hyun Jin Kim & Jeong Ha Kim*
Department of Biological Science, Basic Science Institute, Sungkyunkwan University, Suwon 440-746, Korea
Author for correspondence; Tel: 82-31-290-7009. Fax: 82-31-290-7015. E-mail: jhkimbio@skku.ac.kr

Key words: rocky intertidal, reciprocal test, positive interaction, competition, turf-forming algae

Abstract

Interspecific interactions among three dominant macroalgae, *Pterocladia capillacea* (Rhodophyta), *Hizikia fusiformis* (Heterokontophyta) and *Chondracanthus intermedius* (Rhodophyta), were experimentally investigated on the rocky mid-intertidal zone of Sungsan, Jeju Island, Korea from March 1998 to June 1999. Each of the potentially competing species was removed in permanent plots (20 × 20 cm), and percent covers of non-manipulated species were measured by an image analyzing method using a digital camera. *Pterocladia capillacea* was the most abundant during all seasons, except for winter. Its abundance was lowered by the removal of the turf-forming alga *C. intermedius*, indicating that turf had a positive effect on *P. capillacea*. Conversely, there was a negative effect of *P. capillacea* on the abundance of *C. intermedius*. Interactions between *C. intermedius* and *P. capillacea* can probably be explained as a consequence of the water-trapping ability of the former and the canopy-forming ability of the latter. There was, however, no apparent effect related to *H. fusiformis* since the abundance of this alga remained low. This study supports that both negative and positive effects between same pair of species could be common depending on the morphological differences of algae and particular habitat conditions.

Introduction

Interspecific interactions among marine benthic algae have been considered as a fundamental force to understanding dynamics in marine macroalgal communities. Thus, in natural communities, patterns of algal interaction have often turned out more complex due to other biotic factors, such as the presence of herbivores, algal life history, morphology, and seasonality (Olson & Lubchenco, 1990; Dudgeon et al., 1999). The majority of previous researches has, however, shown that the outcome of interaction is often one-directional, which means species A outcompetes species B, or A has a positive effect on B, even though the necessity of reciprocal test to detect an interactive effect in both directions has been suggested (Denley & Dayton, 1985; Olson & Lubchenco, 1990; Kim, 1995). The presence of both negative (i.e., competition) and positive (i.e., facilitation) effects between the same pair of species has rarely been reported. If this is the case, the

result can theoretically cause the exclusion of one species. This extreme outcome of interaction, however, is not common in natural communities, because the coexistence of species is maintained by other factors, such as consumers, other algal species, disturbance, or the changes in interaction strength with season or life history stages. In this study we report an example of both negative and positive effects occurred between the same pair of species of intertidal macroalgae and discuss its implication for the community structure.

Although positive interactions have been investigated less frequently than competition in marine benthic organisms, their possibility and relevance have been discussed theoretically by some ecologists (Connell & Slatyer, 1977; Vandermeer, 1980; Connell, 1983; Dethier & Duggins, 1984; Bertness & Callaway, 1994). Some positive interactions occur as a consequence of the direct beneficial influence of one species on another (Dayton, 1975; Brawley & Johnson, 1993; Kim, 2002a); others occur as a consequence of negative interactions that are manifested indirectly through other species (Kastendiek, 1982; Dethier & Duggins, 1984; Hay, 1986; Kim, 1997). In physically stressful habitats (e.g., intertidal zones under desiccation stress), positive interactions (e.g., neighbor habitat-amelioration; Bertness & Callaway, 1994) rather than competition may be a particularly important factor leading to species coexistence in a community.

The present work is the first report regarding the interspecific interactions of marine benthic macroalgae on Korean shores. The aim of this study is to investigate a short-term response of macroalgae to the press effect of potential competitor removal in a three species-dominated intertidal community. The three dominants used in the experiment differed in their morphology, upright and turf-forming, and co-existed in an apparently space-limited community in the mid-intertidal zone of Jeju Island, Korea.

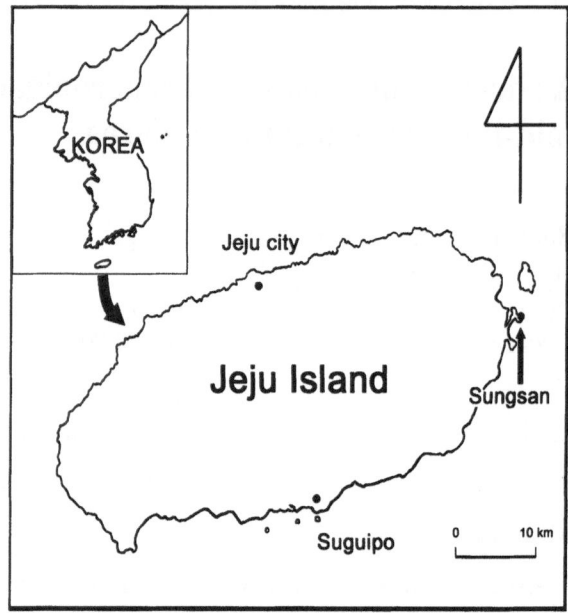

Figure 1. Map of the study site in Jeju Island, Korea.

Study site and organisms

The study site is located at Sungsan on Jeju Island, South Korea (33° 27′ N, 126° 56′ E; Fig. 1). The rocky shore of Jeju Island, mostly consisting of volcanic rocks, supports the most diverse marine intertidal flora in Korea, probably due to exposure to low levels of pollutants and the influence of a mixture of the warm Kuroshio current with the relatively cold Yellow Sea current (Lee & Lee, 1982; Kim, 1991; Park et al., 1994). Along the rocky coast at Sungsan, the intertidal zone stretches over 60 m at low tides, and has a gentle slope with many shallow tide pools. The shoreline is exposed to moderate wave action due to a barrier effect by nearby Woo Island. The tidal regime usually exposes the habitat for prolonged period (about 4–6 hours a day) during daylight hours except for winter months when low tides mostly occur in the evening.

The mid-intertidal zone at Sungsan is dominated by three macroalgae, *Pterocladia capillacea* (Gmelin) Bornet (Gelidiales, Rhodophyta), *Chondracanthus intermedius* (Suringar) Hommersand comb. nov. (Gigartinales, Rhodophyta) and *Hizikia fusiformis* (Harvey) Okamura (Fucales, Heterokontophyta). Both *P. capillacea* and *H. fusiformis* are upright, fleshy algae and their sizes at maturity in this site are 5–7 cm and 6–10 cm, respectively. However, thallus size of *H. fusiformis* at the site was variable depending on tidal height and slope, and plants >20 cm are common. *Chondracanthus intermedius* is a turf-forming

alga (< 1 cm in turf thickness) and its branched flat thalli grow horizontally on the rock surface. This alga sometimes grows together with *Gigartina teedii* (Roth) Lamouroux and *Caulacanthus okamurae* Yamada in this habitat. Other algae occurred less frequently or seasonally in the experimental plots, including *Sargassum thunbergii* (Mertens *ex* Roth) Kuntze, *Gloiopeltis complanata* (Harvey) Yamada, *Gelidium amansii* (Lamouroux) Lamouroux, *Chondrus ocellatus* Holmes, *Ulva conglobata* Kjellman, crustose coralline algae and benthic diatoms. Percent covers of these algae in the plots hardly reached >7% for the study period, except for *S. thunbergii* appearing in a substantial amount from a few plots. We recorded their abundance, and then removed them at every sampling time to focus on the interactive effects among the three dominants.

Materials and methods

A permanent transect line (6 m in length) was established along a gently sloped portion of the shore where the three species of macroalgae occurred in a mixed stand. Anchor bolts were placed into the substratum to mark both ends of the line. The line was marked at every 30 cm to position 20 permanent plots (20 × 20 cm) alternately placed on the upper and the lower sides of the line. We assumed that effects of experimental treatment on the adjacent plots were not significant, because plots were placed alternately along the line with a 10 cm gap between the marked points and the edge of the next plot. By using the same transect line hooked to the anchor bolts, plots could be relocated at each sampling time.

Plots were randomly assigned to one of 4 treatments with 5 replicates per treatment. The treatments included: (1) *Hizikia fusiformis*-removal, (2) *Pterocladia capillacea*-removal, (3) *Chondracanthus intermedius*-removal and (4) undisturbed controls. For the treatment effects, each species was continuously removed on sampling dates which were distributed over intervals of 2–3 months. Other species, which usually occurred in low abundance, were also removed from all the plots. Plants were removed by scraping using a small flat screw driver, and care was taken not to affect the remaining species. Percent coverage of remaining species was visually estimated on a computer monitor using Adobe Photoshop (version 5.0) after photographing all plots in the field using a digital camera (Kodak DC260 zoom). For accurate assessment of percent covers of understory species, such as *C. intermedius*, extra photos were taken after laying the canopy aside. Visual estimation of cover in the lab rather than in the field (Dethier et al., 1993; Kim & DeWreede, 1996a) allowed us more time for the assessment of percent coverage with a greater accuracy and repeatability.

Percent cover data were analyzed for treatment effects using an one-way ANOVA on each sampling date, followed by Tukey's HSD tests if the results of ANOVA were significant. Data were arcsine-transformed prior to analysis to provide variance homogeneity. We did not perform a test for normality since the sample size was low ($n = 5$) and because ANOVA is generally robust against violation of the assumption of normality in the case of equal sample size (Keppel, 1991). All data analyses were done using SPSS (version 6.1).

Results

Effects of one species removal on the abundance of remaining species under each experimental condition are shown in Figure 2. Effects of the treatments on the percent covers of each seaweed are presented in Figure 3. *Pterocladia capillacea* was the most abundant alga in control plots and comprised 30–60% cover during the study period except for February, 1999 when cover dropped to ~20% (Fig. 2D). *Chondracanthus*

intermedius cover decreased to 10% from June to October, 1998; however, cover of this alga slightly increased in February, 1999, as *P. capillacea* cover declined. Mean percent covers of *Hizikia fusiformis* were low (< 10%) except for the first sampling date (Fig. 2D). Although the abundance of the three species was similar at the beginning of the experiment (March, 1998), both *C. intermedius* and *H. fusiformis* became less abundant than *P. capillacea*, as shown in control plots.

Among the three macroalgae, the interaction between *P. capillacea* and *C. intermedius* showed an interesting result. In the *Chondracanthus*-removal plots, percent covers of *P. capillacea* were relatively lower than those in the control plots, except for February, 1999 (ANOVA, $F_{(2,12)} = 3.954$, $p = 0.048$ for October, 1998; Fig. 3B). This result supports the hypothesis that *C. intermedius* has a positive effect on the abundance of *P. capillacea*. Conversely, in the *Pterocladia*-removal plots, percent covers of *C. intermedius* were significantly higher than those in the controls (ANOVA, $F_{(2,12)} = 4.543$, $p = 0.034$ for October, 1998; $F_{(2,12)} = 6.042$, $p = 0.015$ for April, 1999; $F_{(2,12)} = 18.768$, $p < 0.001$ for June, 1999; Fig. 3C). This result indicated that *P. capillacea* had a negative effect on *C. intermedius*. Therefore, both positive and negative effects occurred between the two species. Percent covers of *H. fusiformis* were higher in the *Pterocladia*-removal plots than in the control and the *Chondracanthus*-removal plots in October, 1998 (ANOVA, $F_{(2,12)} = 5.663$, $p = 0.019$; Fig. 3A).

Discussion

The present study represented short-term reponses of neighbor species to the effects of potential competitor removal in a rocky mid-intertidal habitat. The interaction pattern between *Pterocladia capillacea* and *Chondracanthus intermedius* showed an opposite effect on each other, which could be explained by the morphological characteristics of the algae. Turf-forming algae with fast vegetative propagation have been shown to inhibit erect algae by space preemption (Sousa et al., 1981; Airoldi, 1998; Kim, 2002a). However, in this study the positive effect of the turf-forming *C. intermedius* on the upright *P. capillacea* might result from habitat-amelioration (Bertness & Callaway, 1994). The potential advantages of being a neighbor of turf algae can be explained as follows: less desiccation pressure because of the moisture-trapping ability of

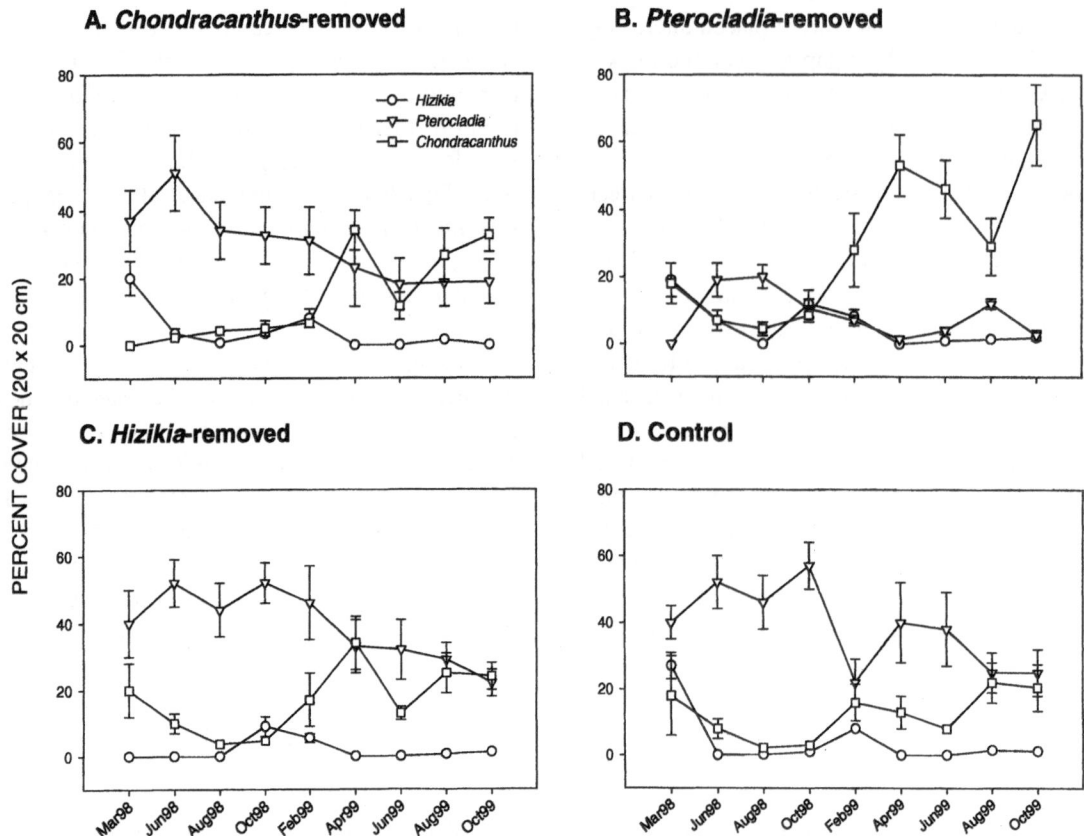

A. *Chondracanthus*-removed

B. *Pterocladia*-removed

C. *Hizikia*-removed

D. Control

PERCENT COVER (20 x 20 cm)

—O— Hizikia
—▽— Pterocladia
—□— Chondracanthus

Figure 2. Changes in the percent cover of each alga under each treatment. Data are means ± SE of 5 replicates.

turf (Hay, 1981; Padilla, 1984) and a buffering effect of the turf against wave impact (Kim, 2002a). It seems obvious that dense turf forms are particularly resistant when exposed to desiccating conditions since gaps between the compact thalli trap a substantial amount of water. This characteristic of turf may lead to better photosynthetic performance compared with non-turf morphology (Hay, 1981). However, the ability of turf algae to ameliorate desiccating habitat conditions by enhancing the growth rates and survivorship of neighboring species has not been well documented. In a previous study on the interspecific interaction between *Fucus gardneri* P.C. Silva and *Mazzaella cornucopiae* (Postels *et* Ruprecht) Hommersand in the upper intertidal zone of western Canada, the survivorship of *F. gardneri* growing close (about 2 cm) to the turf-edge of *M. cornucopiae* was enhanced in comparison with that of other microhabitat conditions (Kim, 2002a). Kim (2002a) claimed that this could be the result of a potential benefit of *Mazzaella* turf ameliorating microhabitat conditions through lessening desiccation stress or buffering wave impact, or both. In this study, thalli of *C. intermedius* form a dense turf, unlike the com-

pact upright fronds of *M. cornucopiae,* by growing close to the substratum as creeping fronds. Therefore, it seems unlikely that the thin (usually <1 cm in thickness) *C. intermedius* turf can provide a buffer against wave impact. Rather, this alga may improve microhabitat conditions for other seaweeds by retaining moisture and reducing desiccation stress. This explanation is supported by the fact that the positive effect of *C. intermedius* on *P. capillacea* became weak during winter when desiccation was less severe.

The negative effects of *Pterocladia capillacea* on *Chondracanthus intermedius* might be due to shading since highly branched thalli of *P. capillacea* form an effective canopy that reduces the amount of irradiance that reaches the understory *C. intermedius* turf. Although there is no information available on the irradiance requirement for *C. intermedius*, it is apparent that when this species co-occurred with *P. capillacea* it showed inhibited growth even during its active growing season, February–June, 1999 (Fig. 3C). Also, it was observed that the abundance of *C. intermedius* under the canopy was often much less than that occurring in nearby open areas (B. J. Kim, pers. obser.).

Other possibilities for the mechanism of the negative effect of *P. capillacea* on *C. intermedius* might be whiplash and allelopathy. A few reports have claimed that some small sized intertidal algae, such as *Fucus gardneri* or *Pelvetiopsis limitata* (Setchell) Gardner (about 7–8 cm in height at maturity), did not affect neighboring species by whiplash (Grant, 1977; Farrell, 1989; Kim, 2002b). *Pterocladia capillacea* in this habitat grows up to a similar size as these fucoids, but this alga is lighter because of its thinner thallus; therefore, the whiplashing effect by *P. capillacea* would not be strong enough to affect *C. intermedius*. Allelopathic effects might be possible since thalli of the two species were usually in contact when emergent, however there is no evidence that either species produces allelochemical.

No influence of *Hizikia fusiformis* on either *P. capillacea* or *C. intermedius* and *vice versa* was detected in this study because *H. fusiformis* abundance remained low regardless of the treatment throughout the study period. Except for the first sampling date, we found most of *H. fusiformis* thalli were < 1 cm. This small thallus size might be in part due to the influence of commercial harvesting by local people who use this alga as a foodstuff (Lee & Kamura, 1997).

The occurrence of both positive and negative effects between the two macroalgae provides an interesting ecological implication. Denley & Dayton (1985) suggested that an experimental approach for interspecific competition should accommodate an experimental design for detecting reciprocal effects, rather than simply the effect of species A on species B. According to the results in this study, it is clear that *Chondracanthus intermedius* enhances the abundance of *Pterocladia capillacea*, which then competitively inhibits the former. If so, will the positive and negative effects of these species on each other result in *P. capillacea* increasingly outcompeting *C. intermedius* and excluding the turf-forming alga from the local community? However, our observations suggest that exclusion does not occur since both species appear to coexist in the intertidal habitat. There are many factors that may potentially lead to coexistence; unpredictable perturbation, herbivory, seasonality of the species, life history traits. Although herbivores may switch the competitive dominance of macroalgae through their differential food preferences (Lubchenco, 1983; Kim & DeWreede, 1996b), in this experiment we found few major herbivores, such as limpets and snails, in our plots except for a few crabs and polychaetes. Moreover, *P. capillacea* is not collected by people un-

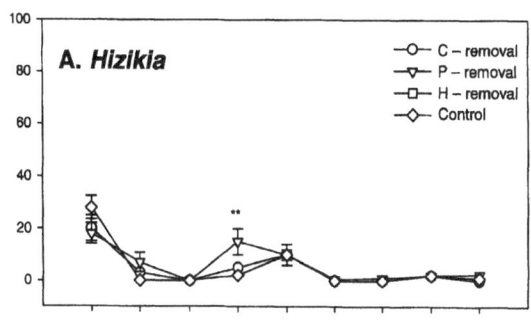

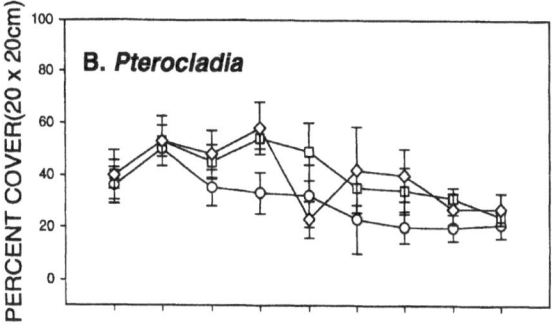

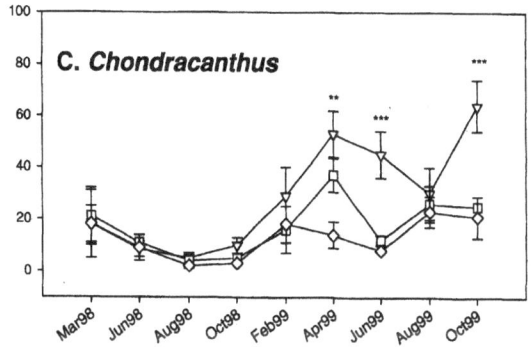

Figure 3. Effects of the treatments on the percent covers of each alga. C-removal: *Chondracanthus intermedius* removed; P-removal: *Pterocladia capillacea* removed; H-removal: *Hizikia fusiformis* removed; Control: no alga removed. Data are means ± SE of 5 replicates. * for <0.05, ** for <0.025, *** for <0.01.

like *H. fusiformis*. The most likely explanation for the coexistence of these species at Sungsan would be the difference in their seasonality. The abundance of *P. capillacea* decreased in March, 1998 and February, 1999 when *C. intermedius* showed peaks in its abundance (Fig. 2D).

This study presents a possible example of a reciprocal interaction between two morphologically distinct macroalgae, and may be extended to other sets of macroalgae and to other habitats. Together with experimental evidence for the mechanisms underlying these interactions, the information regarding the two species will be useful for understanding the dynamics of large-scale algal assemblage in this habitat.

62

Acknowledgements

This research was supported by a grant (981-0513-066-2) from the Korea Science and Engineering Foundation to J.H.K. The authors thank Drs S. M. Boo and Y. S. Oh for their comments on taxonomic problems of seaweeds, and two anonymous reviewers provided helpful comments on the manuscript.

References

Airoldi, L., 1998. Roles of disturbance, sediment stress, and substratum retention on spatial dominance in algal turf. Ecology 79: 2759–2770.

Bertness, M. D. & R. Callaway, 1994. Positive interactions in communities. TREE 9: 191–193.

Brawley, S. H. & L. E. Johnson, 1993. Predicting desiccation stress in microscopic organisms: the use of agarose beads to determine evaporation within and between intertidal microhabitats. J. Phycol. 29: 528–535.

Connell, J. H., 1983. On the prevalence and relative importance of interspecific competition: evidence from field experiments. Am. Nat. 122: 661–696.

Connell, J. H. & R. O. Slatyer, 1977. Mechanisms of succession in natural communities and their role in community stability and organization. Am. Nat. 111: 1119–1144.

Dayton, P. K., 1975. Experimental evaluation of ecological dominance in a rocky intertidal algal community. Ecol. Monogr. 45: 137–159.

Dethier, M. N. & D. O. Duggins, 1984. An 'indirect commensalism' between marine herbivores and the importance of competitive hierarchies. Am. Nat. 124: 205–219.

Dethier, M. N., E. S. Graham, S. Cohen & L. M. Tear, 1993. Visual versus random-point percent cover estimation: 'objective' is not always better. Mar. Ecol. Prog. Ser. 96: 93–100.

Denley, E. J. & P. K. Dayton, 1985. Competition among macroalgae. In Littler, M. M. & D. S. Littler (eds), Ecological Field Methods: Macroalgae. Handbook of Phycological Methods. Cambridge University Press. New York: 511–530.

Dudgeon, S. R., R. S. Steneck, I. R. Davison & R. L. Vadas, 1999. Coexistence of similar species in a space-limited intertidal zone. Ecol. Monogr. 69: 331–352.

Farrell, T. M., 1989. Succession in a rocky intertidal community: the importance of disturbance size and position within a disturbed patch. J. exp. mar. Biol. Ecol. 128: 57–73.

Grant, W. S., 1977. High intertidal community organization on a rocky headland in Maine, U.S.A. Mar. Biol. 44: 15–25.

Hay, M. E., 1981. The functional morphology of turf-forming seaweeds: persistence in stressful marine habitats. Ecology 62: 739–750.

Hay, M. E., 1986. Associational plant defenses and the maintenance of species diversity: turning competitors into accomplices. Am. Nat. 128: 617–641.

Kastendiek, J., 1982. Competitor-mediated coexistence: interactions among three species of benthic macroalgae. J. exp. mar. Biol. Ecol. 62: 201–210.

Keppel, G., 1991. Design and Analysis: A Researcher's Handbook. Prentice Hall. New Jersey.

Kim, J. H., 1995. Intertidal Community Structure, Dynamics and Models: Mechanisms and the Role of Biotic and Abiotic Interaction. Ph.D. Thesis. The University of British Columbia, Vancouver, BC, Canada.

Kim, J. H., 1997. The role of herbivory, and direct and indirect interactions, in algal succession. J. exp. mar. Biol. Ecol. 217: 119–135.

Kim, J. H., 2002a. Patterns of interactions among neighbor species in a high intertidal algal community. Algae 17: 41–51.

Kim, J. H., 2002b. Mechanisms of competition between canopy-forming and turf-forming intertidal algae. Algae 17: 33–39.

Kim, J. H. & R. E. DeWreede, 1996a. Effects of size and season of disturbance on algal patch recovery in a rocky intertidal community. Mar. Ecol. Prog. Ser. 133: 217–228.

Kim, J. H. & R. E. DeWreede, 1996b. Distribution and feeding preference of a high intertidal littorinid. Bot. mar. 39: 561–569.

Kim, Y. H., 1991. Marine algal resources in Cheju Island. J. Cheju Studies 8: 137–156.

Lee, Y. P. & S. Kamura, 1997. Morphological variations of *Hizikia fusiformis* (Harvey) Okamura (Sargassaceae, Pheophyta) from the western coast of the north Pacific. Algae 12: 57–72.

Lee, Y. P. & I. K. Lee, 1982. Vegetation analysis of marine algae in Jeju Island. Proc. Coll. Natur. Sci., SNU 7: 73–91.

Lubchenco, J., 1983. *Littorina* and *Fucus*: effects of herbivores, substratum heterogeneity and plant escapes during succession. Ecology 64: 1116–1123.

Olson, A. M. & J. Lubchenco, 1990. Competition in seaweeds: linking plant traits to competitive outcomes. J. Phycol. 26: 1–6.

Padilla, D. K., 1984. The importance of form: differences in competitive ability, resistance to consumers and environmental stress in an assemblage of coralline algae. J. exp. mar. Biol. Ecol. 79: 105–127.

Park, S. H., Y. P. Lee, Y. H. Kim & I. K. Lee, 1994. Qualitative and quantitative analyses of intertidal benthic algal community in Cheju Island: 1. Species composition and distributional patterns. Korean J. Phycol. 2: 193–203.

Sousa, W. P., S. C. Schroeter & S. D. Gaines, 1981. Latitudinal variation in intertidal community structure: the influence of grazing and vegetative propagation. Oecologia 48: 297–307.

Vandermeer, J., 1980. Indirect mutualisms: variations on a theme by Stephen Levine. Am. Nat. 116: 441–448.

Note added in proof

No Korean material has been examined for the basis of recently suggested taxonomic revisions (Santelices & Hommersand, 1997), hence we have retained current usage.

Hydrobiologia **512**: 63–78, 2004.
P.O. Ang, Jr. (ed.), Asian Pacific Phycology in the 21ˢᵗ Century: Prospects and Challenges.
© 2004 *Kluwer Academic Publishers.*

Seasonal occurrence and reproduction of *Hypnea charoides* (Rhodophyta) in Tung Ping Chau, N.T., Hong Kong SAR, China

Corsica S. L. Kong & Put O. Ang, Jr.*
Department of Biology, The Chinese University of Hong Kong, Shatin, N.T., Hong Kong SAR, China
E-mail: put-ang@cuhk.edu.hk
(*Author for correspondence)

Key words: Hong Kong, Tung Ping Chau, *Hypnea charoides*, reproductive seasonality, algal population dynamics

Abstract

Seasonal occurrence and reproduction in populations of *Hypnea charoides* Lamouroux were investigated along two coastal shores, A Ma Wan and Lung Lok Shui, in Tung Ping Chau, N. T., Hong Kong SAR, China, from 1996 to 1999. Annual growth of these populations was initiated in early winter (November–December) and ended in late spring (April–May). Mean length of *H. charoides* in A Ma Wan was significantly negatively correlated with photoperiod ($r = -0.359$, $n = 38$, $p < 0.05$) and seawater temperature ($r = -0.669$, $n = 38$, $p < 0.05$). Vegetative plants were dominant but relatively high abundance in tetrasporic plants was observed at the end of each growing season. Percentage occurrence of tetrasporic plants was significantly positively correlated with seawater temperature in samples collected at -1 m CD ($r = 0.635$, $n = 19$, $p < 0.05$), -2 m CD ($r = 0.690$, $n = 13$, $p < 0.05$) and from drifted populations ($r = 0.705$, $n = 17$, $p < 0.05$), suggesting that increase in seawater temperature might induce tetrasporogenesis of *H. charoides* in A Ma Wan. Plants in Lung Lok Shui were mostly vegetative but 100% tetrasporic samples were collected at -1 m CD during an unusual growth period in October 1998. High abundance of tetrasporic plants was also observed at a depth of -10 m CD on 9 April 1998 (97.5%) and 22-April 1999 (90%). Significantly negative correlation was found in percentage occurrence of vegetative plants at -10 m CD with photoperiod ($r = -0.553$, $n = 14$, $p < 0.05$) and seawater temperature ($r = -0.855$, $n = 8$, $p < 0.05$). Dominance of vegetative and tetrasporic plants and rarity of cystocarpic plants in both A Ma Wan and Lung Lok Shui suggested that the life span of *H. charoides* might be very short and/or majority of the plants underwent apomeiosis to complete their life cycles in Tung Ping Chau.

Introduction

Algae and seagrasses play very important roles in structuring the marine community as they are the dominant primary producers. Their population dynamics greatly affect those of all the other organisms. However, little attention has been paid in this aspect in algal vegetation studies. For population dynamic analysis, seasonal growth and reproduction have been of considerable interest to many phycologists. They are known to be closely tied to changes in different environmental conditions and the availability of resources. Among these parameters, light and temperature are the physical factors which have been most widely investigated with respect to the growth and reproductive responses of marine algae (Burns & Mathieson, 1972a,b; Adey, 1973; Mathieson & Burns, 1975; De Wreede, 1976; Kapraun, 1978; Prince & O'Neal, 1979; Schoschina et al., 1996; Voskoboinikov et al., 1996; Zamorano & Westermeier, 1996; Vásquez et al., 1998). This is particularly so in many red algal species where irradiance and photoperiod have received the greatest attention as factors that control their reproductive processes. Out of the 46 red algae examined by various authors, 38 species showed that the development or release of their reproductive structures was a type of photoperiodic responses (see review by Murray & Dixon, 1992). Many of these responses were also associated with changes in temperature and some, in nutrient levels. This indicates that temperature or nutrient concentrations could modify or block the inductive effects of photoperiod. Ranges of temperature, irradiance, or photoperiod can therefore serve as thresholds that trigger growth or reproduction in many algae.

Many marine red algae have a triphasic life cycle characterized by a diploid generation of tetrasporophyte, a haploid generation of gametophyte and a diploid generation of carposporophyte that develops on the female gametophyte. Natural populations of these marine algae, however, do not always have equal proportions of these generations at any one time. The dominance of one generation (phase) over the other is commonly reported. For example, gametophyte was found to be the dominant generation in populations of *Gigartina skottsbergii* Setchell et Gardner (Piriz, 1996; Zamorano & Westermeier, 1996), whereas tetrasporophyte was dominant in populations of *Rhodymenia howeana* Dawson (Vásquez et al., 1998). Several hypotheses have been advanced to explain the significance of these differences in phase dominance as part of the life history strategy of these algae. Environmental factors, such as temperature, photoperiod, irradiance, have also been shown to affect phase dominance in different seasons within a year (Dyck & De Wreede, 1995; Piriz, 1996; see also review by Murray & Dixon, 1992).

Hong Kong (22° 12′ N; 114° 6′ E) is located in the southeastern part of China. It experiences a subtropical, highly seasonal monsoonal climate with annual die-off of the intertidal epibiota (Williams, 1993). The weather in summer (June–August) is hot (mean air temperature = 30 °C) and wet while that in winter (December–February) is cool (mean air temperature = 17 °C) and dry (Kennish et al., 1996; see also Hodgkiss, 1984). This characteristic monsoonal climate also brings about marked differences in the seawater conditions (more particularly in temperature). Being an important component of many coastal communities, most intertidal marine algae in Hong Kong display clear seasonal patterns in which they start to grow in winter and disappear in summer (Hodgkiss & Lee, 1983). Their diversity, abundance or zonation patterns were strongly influenced by the monsoonal nature of Hong Kong's climate (Kennish et al., 1996; Kaehler & Williams, 1996). Strong sunlight was suggested to be the major factor responsible for their seasonal pattern of distribution (Hodgkiss, 1984). Nonetheless, distribution patterns of subtidal algal populations have never been studied in Hong Kong except for some anecdotal observations. In Tung Ping Chau, an island in the northeastern part of Hong Kong, for

example, subtidal populations dominated by *Hypnea charoides* Lamouroux also show a seasonal pattern in both their occurrence and reproduction. However, details of this seasonal pattern are not known, especially with respect to the occurrence and abundance of the different reproductive stages (phases).

This study investigated the seasonal growth and reproduction of *H. charoides* in Tung Ping Chau with an attempt to relate these seasonal patterns with the variations in photoperiod and seawater conditions, e.g. temperature and nutrient concentrations. This is the first study of its kind in Hong Kong waters.

Materials and methods

Study site

Field experiments were carried out along the rocky shores of A Ma Wan and Lung Lok Shui in Tung Ping Chau, Hong Kong (Fig. 1). A Ma Wan, which is facing the northeast, is a more sheltered area, while Lung Lok Shui, located on the southwest side of the island, is subjected to strong waves generated by the SE monsoon. In winter, a number of algal species grow densely on the siltstone substratum, which forms a series of slightly inclined terraces (15° to 20°) lying perpendicularly to the shore, at a depth of −1 to −3 m CD (Chart Datum) in A Ma Wan. These marine algae include *Colpomenia sinuosa* (Roth) Derbes & Solier, *Dictyota* spp., *Enteromorpha* spp., *Galaxaura fruticulosa* Kjellman, *Hypnea* spp., *Padina arborescens* Holmes, *P. australis* Hauck, *Sargassum* spp. and *Ulva* spp., with *Hypnea charoides* being one of the most dominant red algal species observed. Areas beyond −3 m CD are dominated by coral communities. In Lung Lok Shui, similar inclined terraces (15° to 20°) are also found, but in contrast to A Ma Wan, *Sargassum* spp. are the most dominant, canopy-forming species growing on these terraces. Compared with A Ma Wan, marine algae in Lung Lok Shui are far more dominant than corals, growing from −1 to −10 m CD.

Populations of Hypnea charoides

Individuals of *Hypnea charoides* are either loosely attached on the rocky substratum forming clumps or are entangled with other marine algae (e.g. *Sargassum* spp.). Many of them, however, are easily swept away by waves and thus become drifted in the water column. In A Ma Wan, comparatively, more attached and drifted individuals are observed. An attached population

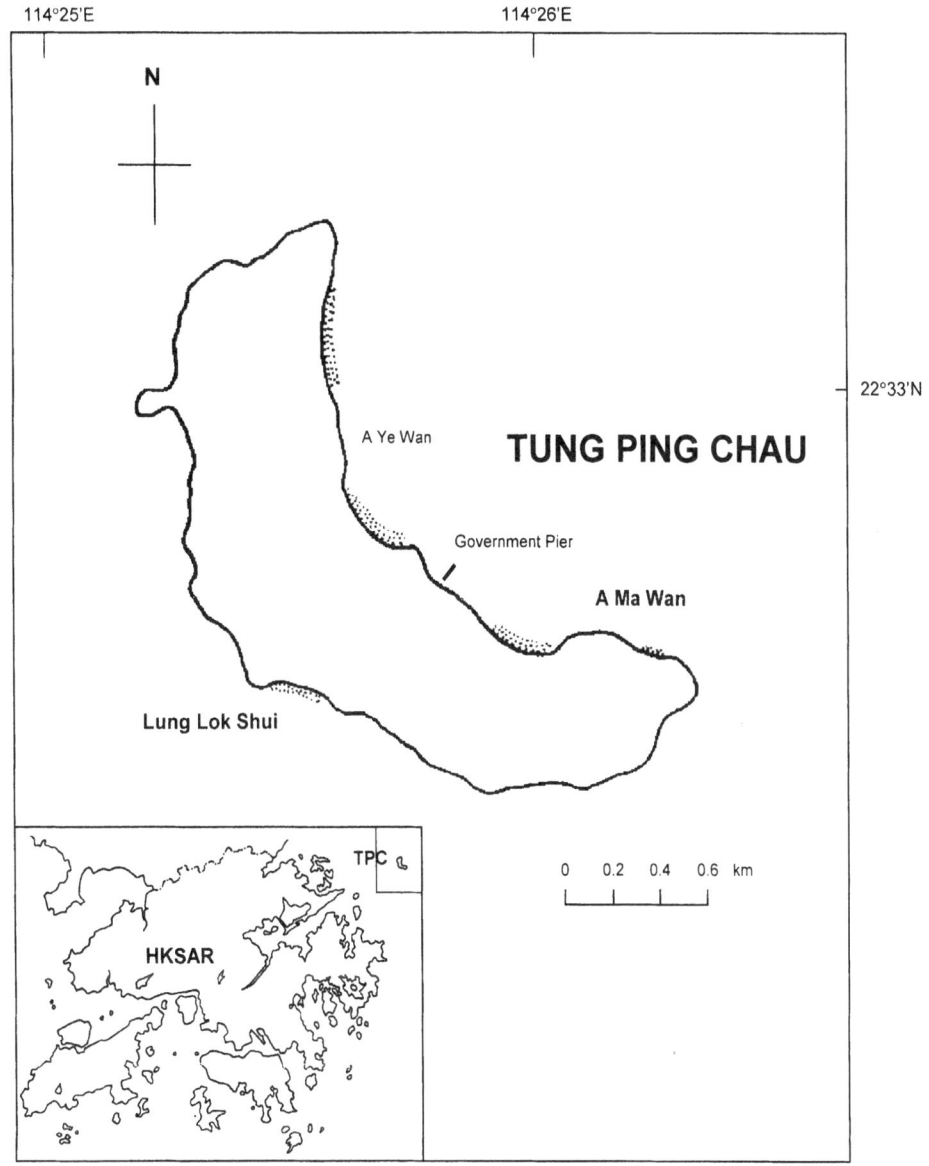

Figure 1. Map of Tung Ping Chau, showing the locations of A Ma Wan and Lung Lok Shui. The location of Tung Ping Chau (TPC) is shown in the insert map of Hong Kong Special Administrative Region (HKSAR).

of *H. charoides* is found on the inclined terraces at −1 m CD while another grows on the corals (alive or dead) at −2 to −3 m CD. Drifting individuals are mostly found at a depth of −2 to −3 m CD. In Lung Lok Shui, most of the clumps are found entangled with the holdfast of *Sargassum* spp. on the terraces at −1 to −3 m CD while a much deeper population was observed at a depth of −10 m CD.

Measurement of plant length

Measurement of plant length and sample collection of *Hypnea charoides* were carried out by SCUBA or by snorkeling in both A Ma Wan and Lung Lok Shui. One hundred individuals (clumps) were haphazardly selected and measured at a depth of −1 to −2 m CD every two weeks from January 1997 to June 1999 in each site. Plant length was measured from the base to the tip of each clump of *H. charoides*. Due to typhoons or strong monsoon, field trips were sometimes interrupted (especially to Lung Lok Shui). Therefore, *in situ* length measurement could not be carried out regularly at exactly biweekly intervals and at times, had to be delayed for one or two weeks. All the plants measured in A Ma Wan were attached individuals while those in Lung Lok Shui were individuals entangled with the holdfasts of *Sargassum* spp.

Examination of reproductive structures

To examine the reproductive structures of *Hypnea charoides*, samples were collected at approximately biweekly intervals during the growing seasons from January 1996 to June 1999. At each sampling in A Ma Wan, 30 individuals (clumps) were haphazardly collected from the population attached on the siltstone substratum at a depth of −1 m CD and another 30 were from the population attached on corals at depths of −2 to −3 m CD. Additional drifting samples (30 individuals at each sampling) within the water column of −2 to −3 m CD were also collected. In 1999, an extra 30 attached samples were collected approximately biweekly (February–April) at a depth of −5 m CD for further examination. In Lung Lok Shui, 30 drifting individuals entangled with *Sargassum* spp. (fronds or holdfasts) were collected at depths of −1 to −2 m CD. Starting from April 1998, additional samples (30 individuals at each sampling) were collected from a depth of −10 m CD. All the plants collected were taken back to the laboratory and examined under the microscope for the presence or absence of reproductive structures (cystocarps vs. tetrasporangia). Search for male plants was also attempted. The occurrence of different reproductive stages (phases) was expressed as a percentage of the number of cystocarpic, tetrasporic or vegetative plants over the total number of plants observed.

Environmental parameters

Information on photoperiod over the sampling period was obtained based on the sunrise and sunset data from the Hong Kong Observatory. Seawater temperature was recorded by temperature probes (Minilog TP, Vemco Inc., Halifax, Canada) in both A Ma Wan and Lung Lok Shui over the study period. These probes were set to record the bottom temperature every 30 min and were retrieved for replacement approximately every other month. In A Ma Wan, the temperature probe was placed at a depth (−1 to −2 m CD) in between the −1 m CD and −2 to −3 m CD populations. There was no significant difference in the seawater temperature between −1 m and −3 m CD.

Hence, only one set of temperature data was collected. In Lung Lok Shui, three temperature probes were placed, respectively, at depths of −1 m CD, −2 to −3 m CD and −10 m CD. However, due to strong waves and possible human interferences, temperature probes at −1 m CD were frequently lost. So, only data recorded at −2 to −3 m CD and −10 m CD were available. For nutrient analysis, three water samples (mostly surface water) were collected from A Ma Wan and Lung Lok Shui at approximately biweekly intervals. Concentrations of ammonium ions, nitrites, nitrates and phosphates were determined by an autoanalyzer in the laboratory using the standard methods prescribed by the American Public Health Association (APHA, 1995).

Statistical analysis

Pearson Product Moment Correlation (SigmaStat, Jandel Scientific Software) was used to correlate the patterns of change in mean plant length and percentage occurrence of reproductive phases with variations in the environmental parameters investigated.

Results

Seasonal occurrence and growth of Hypnea charoides

Populations of *Hypnea charoides* in A Ma Wan were observed from January to May in 1997, from January to April in 1998 and from December 1998 to May in 1999 (Fig. 2). Mean plant length (±SD) was first measured in January 1997. It reached its maxima on 16 February 1997 (17.0 ± 9.5 cm), 11 April 1998 (11.7 ± 6.8 cm) and 21 March 1999 (14.0 ± 4.0 cm) and minima on 8 March 1997 (10.6 ± 4.5 cm), 24 January 1998 (4.8 ± 2.0 cm) and 11 May 1999 (4.1 ± 1.3 cm) respectively in the three growing seasons from 1996 to 1999 (Fig. 2). The population usually appeared as creeping thalli or as very short branches at the start or the end of each growing season.

In Lung Lok Shui, *H. charoides* normally appeared within the same period as in A Ma Wan, from January to May in 1997, January to April in 1998 and from January to April in 1999 (Fig. 2). However, many individuals were also observed from July to December in 1998 during an unusual bloom. Plant length measurement started in January 1997. The maximum values were 12.8 ± 6.1 cm on 3 May 1997, 10.0 ± 5.7 cm on 22 April 1998 and 20.1 ± 6.1 cm on 22 April 1999

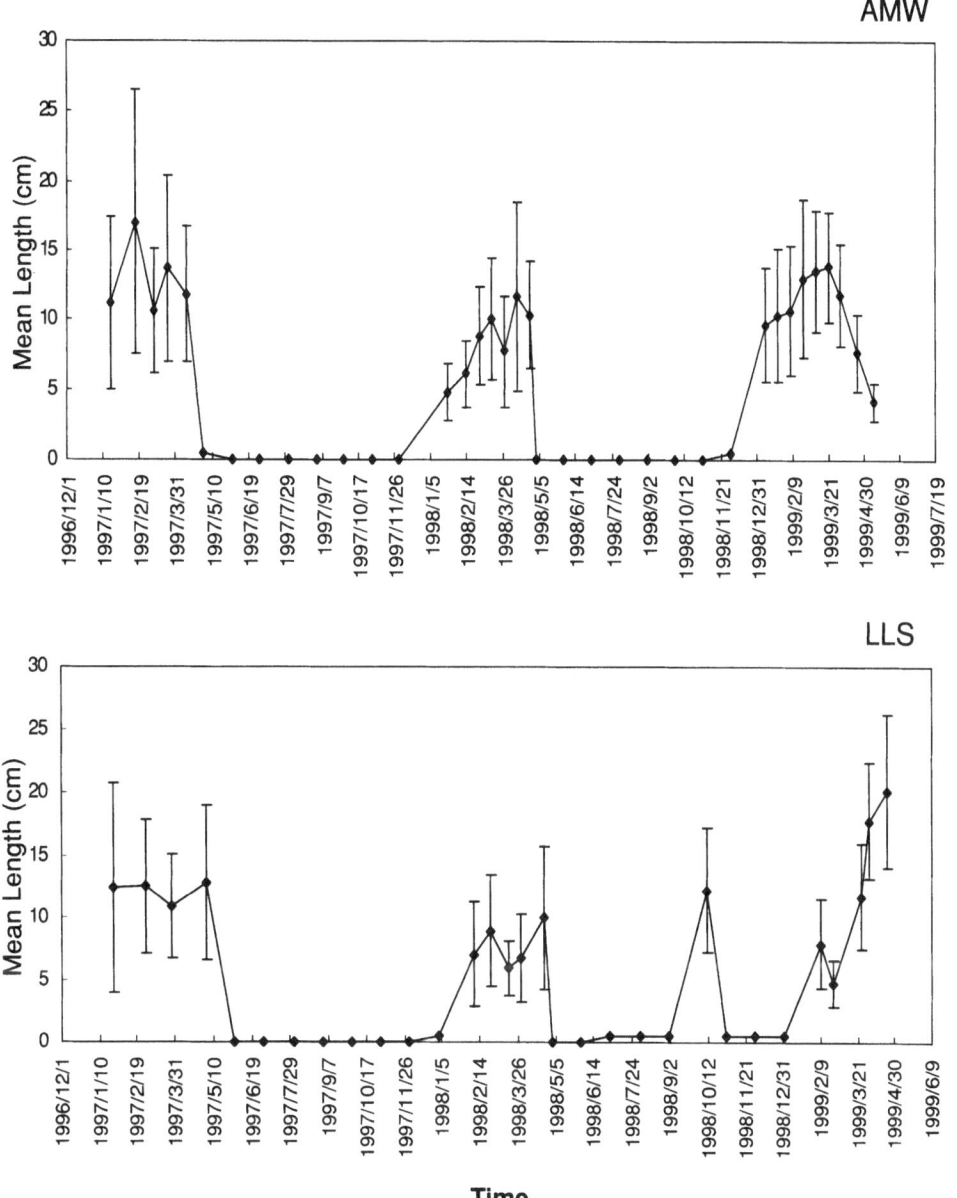

Figure 2. Variation in mean length (cm ± SD) of *Hypnea charoides* (*n* = 100) at −1 to −2 m CD in A Ma Wan (AMW) and Lung Lok Shui (LLS) from January 1997 to June 1999.

while the minimum values were 10.9 ± 4.1 cm on 28 March 1997, 6.0 ± 2.2 cm on 16 March 1998 and 4.7 ± 1.9 cm on 23 February 1999 in the three growing seasons from 1997 to 1999 (Fig. 2).

In July 1998, the population of *H. charoides* in Lung Lok Shui appeared as short branches entangled around the bases of *Sargassum* spp. This growth period was much earlier than what was normally observed. At the end of September, plants were observed to have two growth forms: prostrate and erect. The

prostrate parts appeared like a vegetative mat adhering onto the rocky substratum whereas the erect parts were long and bushy. Both of them exhibited healthy greenish red in coloration. They grew so rapidly and covered almost the whole area that was normally dominated by *Sargassum* spp. at that period of time. Measurement was done on the erect plants only once on 11 October 1998 (12.2 ± 4.9 cm) as all these plants disappeared during the following visit to Lung Lok Shui in November 1998. Only short creeping branches were observed thereafter (November 1998 to January 1999).

Reproductive seasonality

During the period from 1996 to 1999, the −1 m CD attached population of *Hypnea charoides* in A Ma Wan appeared in winter and early spring and disappeared completely in the summer (Fig. 3). All the plants observed from December 1996 to February 1997 were vegetative. Tetrasporic plants were first observed on 23 March 1997 and 15.9% of the plants were tetrasporic. This percentage occurrence increased to 50% on 13 April 1997. The next growing season (starting from January 1998) also showed dominance of vegetative plants and 13.3% of the plants were tetrasporic on 11 April 1998. This percentage increased to 60% on 25 April 1998. Cystocarpic plants were rarely seen and found only on 13 December 1998 (6.7%). When compared with the previous two growing seasons, the percentage occurrence of tetrasporic plants was relatively low during the growth period in 1999, with a maximum of 30% recorded on 3 April 1999.

Drifted and attached samples at −2 to −3 m CD were collected for examination starting from 1997 and 1998 respectively (Fig. 4). Drifted samples were first collected in March 1997 and majority of them were vegetative. Percentage occurrence of tetrasporic plants increased from 3.3% (23 March) to 56.7% (13 April) in 1997 and cystocarpic plants were only recorded on 23 March (3.3%). In 1998, no cystocarpic plants were observed and tetrasporic plants were only found in April, with an increase from 60% (11 April) to 82.8% (25 April). However, cystocarpic plants occurred quite frequently in the growing season from December 1998 to March 1999. In spite of this, the percentage occurrence of cystocarpic plants remained very low, ranging from 3.3% on 21 February 1999 to 8.3% on 13 December 1998. Similarly, tetrasporic plants were found in all the samples collected in the second growing season, ranging from 3.3% on 21 February 1999 to 58.3% on 13 December 1998.

For attached samples at −2 to −3 m CD, majority of the plants were vegetative (Fig. 4). A few cystocarpic plants were observed on 3 February (3.3%) and 25 April (6.5%) in 1998. Tetrasporic plants were only found in April 1998 with a maximum percentage of 90% on 25 April. In the second growing season from December 1998 to April 1999, however, they were present in all the samples collected, ranging from 3.3%

on both 7 February and 7 March 1999 to 50% on 3 April 1999.

All the plants (100%) collected at a depth of −5 m CD in 1999 from A Ma Wan were vegetative in February. This percentage occurrence dropped to 6.7% on 3 April when majority (83.3%) of the plants became tetrasporic. Only few cystocarpic plants (10%) were observed on 3 April 1999.

In Lung Lok Shui, the population of *H. charoides* at −1 to −2 m CD appeared in January and disappeared in May 1997 (Fig. 5). Majority of the plants were vegetative and 23.3% of them were tetrasporic on 29 March 1997. This percentage increased to 46.7% on 3 May 1997. In 1998, all the plants (100%) observed in February and March were vegetative but 80% of them were tetrasporic on 22 April 1998. In July 1998, when some individuals reappeared apparently earlier than normal as small clumps, all of them (100%) were vegetative. In September, however, they grew very quickly and 41.7% of the plants became tetrasporic. By October, all of them (100%) were tetrasporic. After this period, there were only creeping parts left behind. Tetrasporic plants were observed again in the following January 1999 (23.3% on 3 January) and this percentage occurrence rose to 66.7% on 22 April 1999. No cystocarpic plants were found in Lung Lok Shui during the whole study period.

Samples of *H. charoides* were first collected at a depth of −10 m CD in April 1998 and most of the plants were tetrasporic (97.5% on 9 April). In the following growing season in December 1998, 30% of the plants were tetrasporic and this percentage increased to 90% on 22 April 1999. Majority of the plants remained vegetative in January and February in 1999.

Other observations

Among all the individuals of *Hypnea charoides* examined, the presence of cystocarps and tetrasporangia on the same branch was occasionally observed. The first one was found in Lung Lok Shui on 3 May 1997 and another one among the drifted samples in A Ma Wan on 25 April 1998. This phenomenon was not common during the whole study period. Of hundreds of plants observed, only two individuals were found to bear both of these structures.

Although attempts were made to look for male plants, they were not successful.

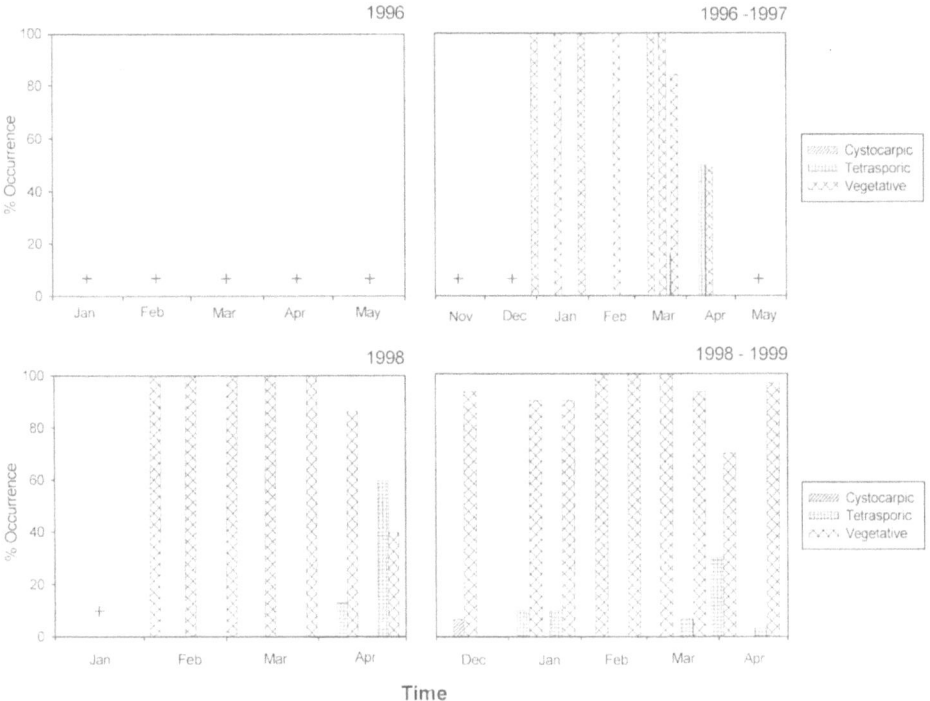

Figure 3. Seasonal occurrence and percentage of reproductive phases in attached population of *Hypnea charoides* (*n* = 30) at −1 m CD in A Ma Wan from January 1996 to June 1999. Only those months when *H. charoides* was present are shown. "+" = months when *H. charoides* was present, but percentage of reproductive phases was not estimated.

Environmental parameters

From January 1996 to June 1999, the mean monthly photoperiod was recorded to be the longest in July (13.5:10.5h L:D), while the shortest was in December (10.75:13.25h L:D) in 1996, 1997 and 1998 (Fig. 6).

Mean monthly water temperature was calculated as an average of all the temperatures recorded at 30 min interval within a month by the temperature probe. In A Ma Wan, the mean temperature of seawater at −1 to −2 m CD varied from a maximum of 29.5 °C in August 1998 to a minimum of 16.0 °C in February 1996 (Fig. 6). Slight fluctuations were observed during 1997 and the overall mean values were relatively lower than those of the other study years. In Lung Lok Shui, mean temperature at −2 to −3 m CD varied from a maximum of 29.7 °C in August 1998 to a minimum of 16.7 °C in February 1998 during the period from January 1998 to June 1999 (Fig. 6). While at −10 m CD (measured only from December 1998 to June 1999), mean seawater temperature decreased from 19.5 °C in December 1998 to a minimum of 17.9 °C in January and February in 1999 and rose to 26.7 °C in June 1999.

Variations in the concentrations of ammonium ions, nitrites, nitrates and phosphates in seawater samples collected from A Ma Wan and Lung Lok Shui during the study period (August 1996 to June 1999) were also recorded. In A Ma Wan, concentrations of ammonium ions, nitrates and phosphates did not fluctuate (mostly remained at a level below 1 μM l^{-1}) except a very sharp peak (21.5 μM l^{-1}) was recorded on 10-May 1999 for ammonium ions. Similarly, nitrate concentration remained steady, with values around or below 1 μM l^{-1}, but a sharp peak (22.3 μM l^{-1}) was also recorded on 10-May 1999. While for phosphates, concentration rose from 0.4 μM l^{-1} (24-September 1998) to a level higher than 3 μM l^{-1} (from November 1998 to April 1999) and a maximum of 5.1 μM l^{-1} was reached on 16 June in 1999.

Similar results were recorded in Lung Lok Shui, except that a sharp peak (14.5 μM l^{-1}) was observed in the concentration of nitrites on 6 October 1996. For all the other periods of sampling, nitrite concentration remained around or below 1 μM l^{-1}. For the other three nutrients, fluctuations occurred starting from November 1998. Concentrations of both ammonium

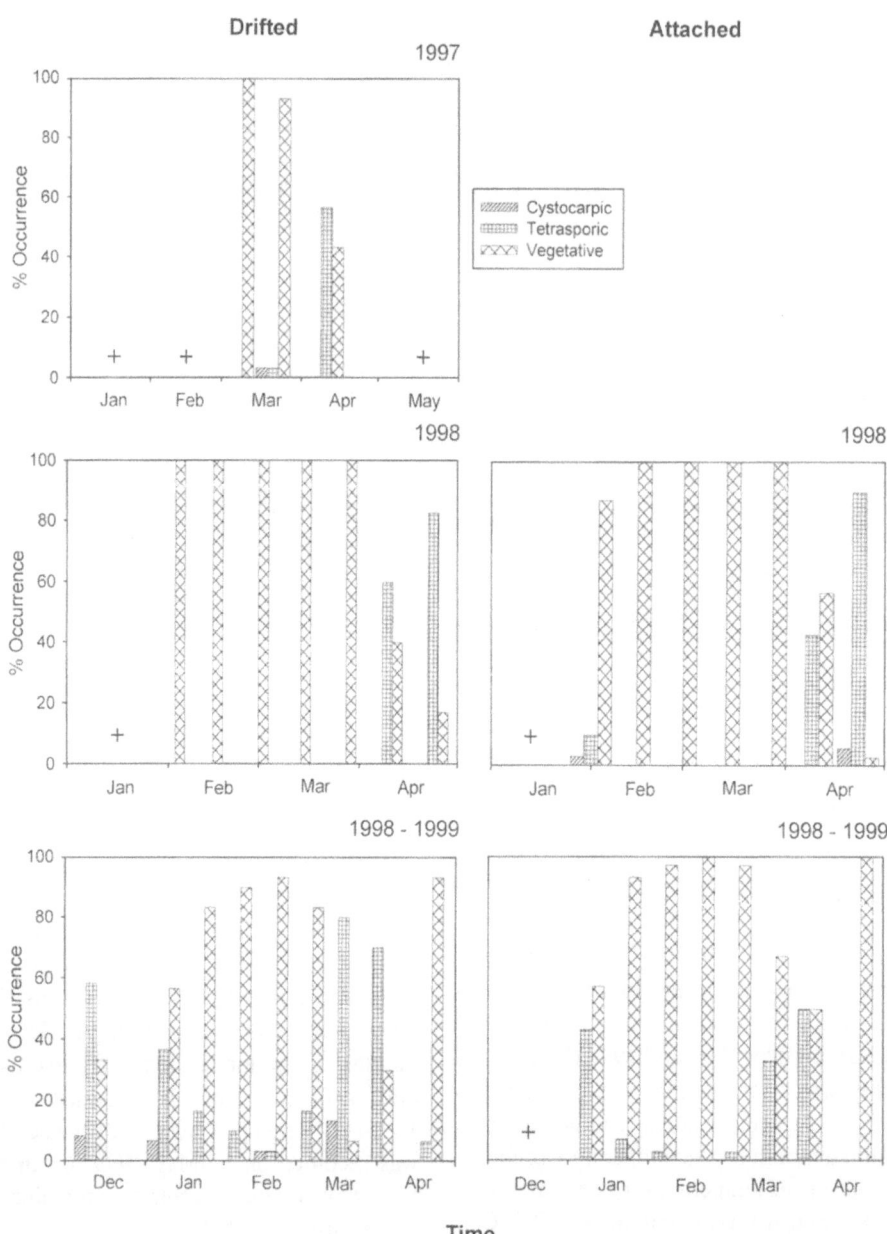

Figure 4. Seasonal occurrence and percentage of reproductive phases in drifted and attached populations of *Hypnea charoides* ($n = 30$) at -2 to -3 m CD in A Ma Wan from January 1997 to June 1999. Only those months when *H. charoides* was present are shown. "+" = months when *H. charoides* was present, but percentage of reproductive phases was not estimated.

ions and phosphates varied from 1 μM l^{-1} to 3 μM l^{-1} in 1999. While for nitrates, three peaks were recorded on 10 January (7.9 μM l^{-1}), 7 February (8.7 μM l^{-1}) and 30 May (10.8 μM l^{-1}) in 1999.

Statistical analysis

In A Ma Wan, the mean length of *Hypnea charoides* was significantly negatively correlated with photoperiod ($r = -0.359$, $n = 38$, $p < 0.05$) and seawater temperature ($r = -0.669$, $n = 38$, $p < 0.05$) but was not significantly correlated with any nutrient concentration in seawater. No correlation analysis

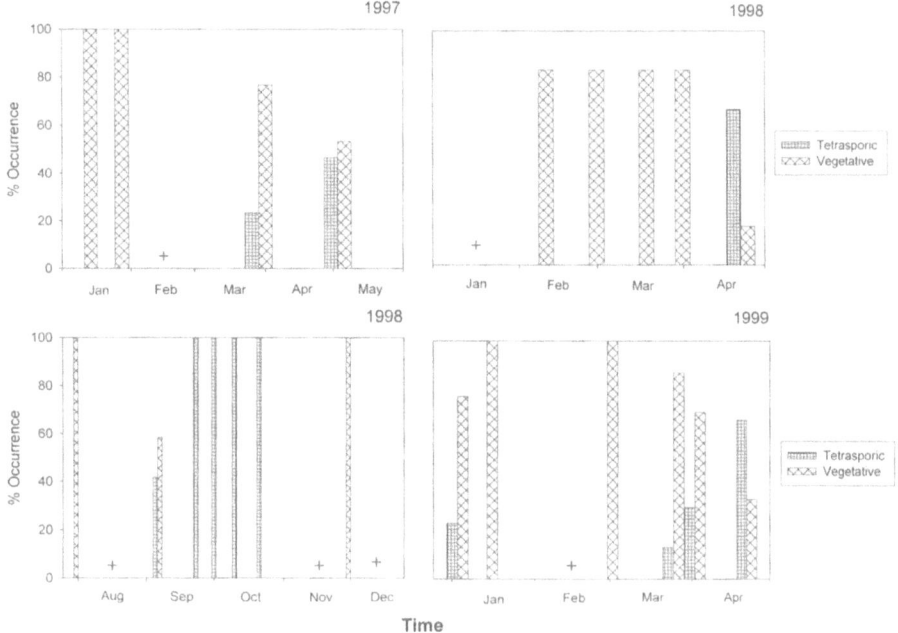

Figure 5. Seasonal occurrence and percentage of reproductive phases in attached population of *Hypnea charoides* ($n = 30$) at -1 to -2 m CD in Lung Lok Shui from January 1997 to June 1999. Only those months when *H. charoides* was present are shown. "+" = months when *H. charoides* was present, but percentage of reproductive phases was not estimated.

was carried out to relate the percentage occurrence of cystocarpic plants in -1 m CD attached samples with any physical parameters due to the small sample size of cystocarpic plants. Percentage occurrence of tetrasporic plants in -1 m CD attached samples was significantly negatively correlated with photoperiod ($r = -0.533$, $n = 24$, $p < 0.05$) but positively correlated with seawater temperature ($r = 0.635$, $n = 19$, $p < 0.05$), while that of vegetative plants was negatively correlated with both photoperiod ($r = -0.510$, $n = 24$, $p < 0.05$) and seawater temperature ($r = -0.635$, $n = 19$, $p < 0.05$). Percentage occurrence of cystocarpic plants in attached and drifted samples at depths of -2 to -3 m CD was not significantly correlated with any physical parameters. In attached samples, the percentage occurrence of tetrasporic plants showed a significant positive correlation with seawater temperature ($r = 0.690$, $n = 13$, $p < 0.05$) but that of vegetative plants showed a negative correlation ($r = -0.685$, $n = 13$, $p < 0.05$). Similarly, a significant positive correlation was found in the percentage occurrence of tetrasporic plants of drifted samples with seawater temperature ($r = 0.705$, $n = 17$, $p < 0.05$) whereas a negative correlation was observed in that of vegetative plants ($r = -0.689$, $n = 17$, $p < 0.05$). No significant correlations were

found in percentage occurrence of tetrasporic and vegetative plants in both attached and drifted samples at -2 to -3 m CD with photoperiod and nutrient concentrations.

In Lung Lok Shui, significant correlations with environmental parameters were only found in samples collected from the depth of -10 m CD. For these samples, the percentage occurrence of vegetative plants was significantly negatively correlated with both photoperiod ($r = -0.553$, $n = 14$, $p < 0.05$) and seawater temperature ($r = -0.855$, $n = 8$, $p < 0.05$).

Discussion

Seasonal occurrence and growth of Hypnea charoides

Annual growth of populations of *Hypnea charoides* in Tung Ping Chau was initiated in early winter (November–December) and ended in late spring (April–May). These periods were when photoperiod and seawater temperature showed marked changes. Since photoperiod and temperature are environmental signals for seasonal change (Lüning, 1990), it might be inferred that they are among the most important factors responsible for the seasonal growth of *H. char-*

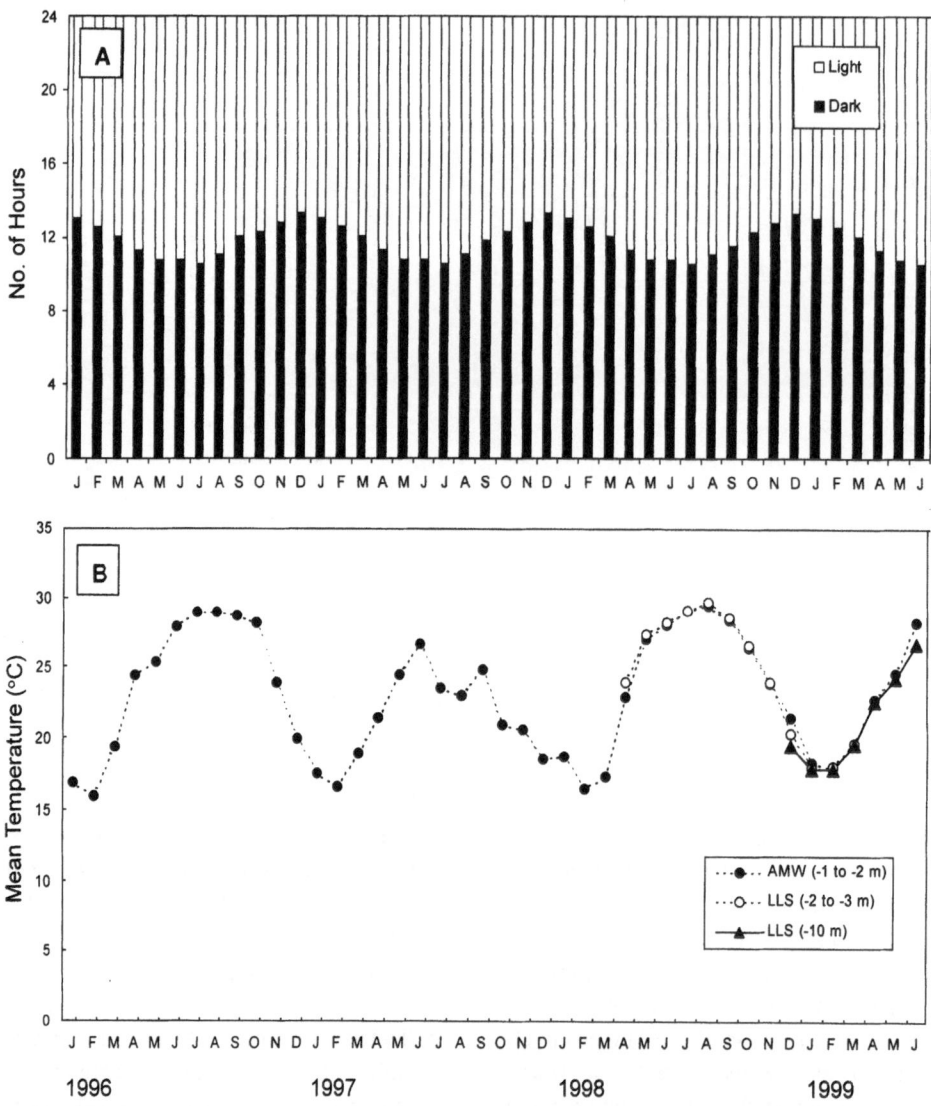

Figure 6. Mean monthly variations in (A) photoperiod and (B) seawater temperature (°C), recorded at depth of −1 to −2 m CD at A Ma Wan (AMW), −2 to −3 m CD and −10 m CD at Lung Lok Shui (LLS), from January 1996 to June 1999. Standard deviations are not shown.

oides in Tung Ping Chau. Besides these two parameters, high nutrient content of seawater is also an ecological factor that favours algal growth, particularly in winter as marine algae start to consume their reserve materials for growth during this period (Lüning, 1990). The effects of different nutrients (e.g. nitrates and phosphates) on algal growth have been documented in many studies in addition to the influences of light and temperature (e.g. Kilar & Mathieson, 1978; Zinoun & Cosson, 1996; Hurtado-Ponce & Pondevida, 1997; Reani et al., 1998; see also Murray & Dixon, 1992). However, algal growth is not always favoured by high nutrient content. In *Gracilariopsis bailinae* Zhang et Xia, for example, plants grew well with a

low concentration of nutrients (Rabanal et al., 1997). While in some economic marine algae, like *Chondrus crispus* Stackhouse (Neish et al., 1977) and *Hypnea musciformis* (Wulfen) J. V. Lamouroux (Guist et al., 1982), the carrageenan production of thalli increased under reduced nitrogen availability. Nitrogen in the form of ammonium ions was even toxic to *Solieria chordalis* (C. Agardh) J. Agardh at high concentrations (Brown, 1995).

In the present study, annual growth of *H. charoides* was correlated with shorter photoperiod and lower seawater temperature, implying that decrease in photoperiod and seawater temperature could initiate its growth in A Ma Wan. The results from the shallow water samples further indicate that change in seawater temperature might be more critical in such initiation than the photoperiod. Photoperiod is more predictable year after year, but change in seawater temperature is not. The timing of the initiation of growth of *H. charoides* populations is not consistent over the years. Populations of *H. charoides* in shallow water of A Ma Wan were first recorded in November 1996 when seawater temperature dropped from 28.2 °C (October 1996) to 24 °C (November 1996) and disappeared after May 1997 when seawater temperature rose from 24.5 °C (May 1997) to 26.7 °C (June 1997). In 1998, *H. charoides* first appeared in December 1998 when seawater temperature dropped from 23.9 °C (November 1998) to 21.5 °C (December 1998) and disappeared after April 1999 when seawater temperature rose from 22.8 °C (April 1999) to 24.6 °C (May 1999). This pattern was less clear in 1997 when the temperature fluctuated more irregularly. Mean temperature lower than 24 °C was recorded in October, but population of *H. charoides* did not appear until January 1998. However, the population disappeared in May 1998, at a time when temperature rose from 23 °C in April to 27.1 °C in May. The lowest seawater temperature throughout the whole study period was 16 °C recorded in February 1996. It thus appears that population of *H. charoides* in A Ma Wan grows best within a mean temperature range from 16 °C to 24 °C but culture experiment may be needed to further investigate such relationship.

In contrast, none of the physical parameters investigated was significantly correlated with the growth of *H. charoides* at −1 to −2 m CD in Lung Lok Shui, suggesting that factors other than photoperiod, seawater temperature and nutrient concentrations were also important in structuring its growth pattern. The apparent irregular growth pattern of *H. charoides* populations in Lung Lok Shui may be due to the much stronger wave action in the study site. This caused fragmentation of the algal thalli and resulted in various sizes of individuals in the population. Strong wave action may also explain why plants of *H. charoides* are mostly found entangled with *Sargassum* spp., a plant with a strong holdfast, rather than being attached on the substratum directly.

The unusual sudden appearance of *H. charoides* from July to November in 1998 also contributed to the irregular growth pattern of *H. charoides* observed in Lung Lok Shui. Although the reason for the appearance of this growth period is unknown, the ability of *H. charoides* to survive and grow during summer

months (i.e. longer photoperiod and warmer temperature) could not be ruled out. In some algal species, seasonal growth was not entirely controlled by physical factors alone but may be due to endogenous regulations (Molenaar & Breeman, 1994, 1997; Voskoboinikov et al., 1996; Makarov et al., 1999), allowing survival of these species under adverse conditions. Some tropical algae also exhibited no evident responses to environmental parameters (e.g. irradiance, temperature and salinity), indicating a natural ability to adapt to different marine conditions (Dawes et al., 1999). In a review of life history, reproduction and phenology of *Gracilaria*, Kain & Destombe (1995) found that there was no evidence of *Gracilaria* species anticipating the seasons by being triggered by a repeatable seasonal condition (e.g. photoperiod or water temperature) for optimal exploitation of predictable conditions. It seems that such species are responders (*sensu* Kain, 1989) and they grow when they can, when the environment allows. For the populations of *H. charoides* in Lung Lok Shui, there might also be some endogenous regulations involved in their growth. Perhaps they are similar to some *Gracilaria* spp. that they grow whenever they can.

During this unusual, sudden growth period, *H. charoides* was observed to exhibit a heterotrichous organization, i.e. its thallus was differentiated into a prostrate and an erect form. Members of the red algal order Gigartinales, to which *H. charoides* belongs, possess this categorization of thalli into two types of growth forms. An initial basal layer is first formed which then gives rise to filaments that form erect axes (Murray & Dixon, 1992). In *Cryptopleura ramose* (Hudson) Kylin ex Lily Newton or *Plocamium cartilagineum* (Linnaeus) P. S. Dixon, the realized developmental pathway for their particular mode of erect or prostrate growth appeared to be correlated with irradiance (Dixon, 1973). Though factors that induce *H. charoides* to exhibit heterotrichy are not known, it appears that heterotrichy could be a means of increasing the opportunities to successfully perennate its thallus throughout any open space (Murray & Dixon, 1992).

Reproductive seasonality

Effects of seawater temperature on phenology of marine algae have been well documented in the past (e.g. Adey, 1973; De Wreede, 1976; Luxoro & Santelices, 1989; Narasimha Rao, 1995; Voskoboinikov et al., 1996). Although in some cases, water temperature *per se* does not directly affect algal reproduction, it may initiate the reproductive events in conjunction with other environmental parameters like irradiance/photoperiod (Kapraun, 1978; Prince & O'Neal, 1979) and/or nutrient concentrations (Reed et al., 1996; Stimson et al., 1996; Rabanal et al., 1997). In some red algal species, photoperiod and seawater temperature were found to be particularly important to reproductive events (Molenaar & Breeman, 1994; Molenaar et al., 1996; Zamorano & Westermeier, 1996; Hall & Murray, 1998). In the present study, variations in photoperiod and seawater temperature seemed to play a more important role in controlling the occurrence of reproductive plants in the population of *H. charoides* in A Ma Wan than that in Lung Lok Shui. It was more obvious in the attached samples collected at −1 m CD in A Ma Wan, in which shorter photoperiod and higher temperature were more favourable for the growth of tetrasporic plants. Besides, percentage occurrence of tetrasporic plants in attached and drifted samples at depths of −2 to −3 m CD was also enhanced by higher temperature, indicating that increasing seawater temperature may favour tetrasporogenesis in the populations of *H. charoides* in A Ma Wan. Induction of tetrasporogenesis by change in water temperature was also reported in other red algal species. In a population of *Gracilaria heteroclada* (Montagne) Feldmann & G. Feldmann in Central Philippines, a positive correlation ($r = 0.6586$) was obtained between the percentage occurrence of tetrasporophytes and seawater temperature (Luhan, 1996). A sudden rise in water temperature (+ 5 °C) together with a decrease of incident light could even trigger a massive tetraspore release in *Gracilaria bursa-pastoris* (S.G. Gmelin) P.C. Silva (Marinho-Soriano et al., 1998). The onset and end of the growth and reproductive seasons may also be triggered by many factors other than temperature (Dring, 1974; Lobban & Harrison, 1994). Nevertheless, none of the other environmental parameters investigated appeared to induce development of reproductive plants in both A Ma Wan and Lung Lok Shui.

Throughout the whole study period, cystocarpic plants were very rare and even absent in Lung Lok Shui. The populations of *H. charoides* in Tung Ping Chau were dominated by vegetative and tetrasporic plants. Natural populations do not always have equal proportions of isomorphic generations. Although it was evident that the gametophytic:sporophytic ratio of fronds in populations of *Chondrus crispus* in sublittoral Prince Edward Island of Canada was a result of stochastic events, with both generations having equal

chances of becoming established (Lazo et al., 1989), cystocarpic plants of the same species were found to be relatively more common and abundant than tetrasporic plants in other places (Mathieson & Burns, 1975). Piriz (1996) observed clear predominance of cystocarpic plants in the population of *Gigartina skottsbergii* Setchell & N.L. Gardner in Argentina, suggesting that there were differential adaptive strategies for each growth phase in its life cycle. Predominance of gametophytes was also reported in other red algae, like *Iridaea cordata* (Turner) Bory de Saint-Vincent (May, 1986), *Iridaea laminarioides* Bory de Saint-Vincent (Luxoro & Santelices, 1989) and *Mazzaella cornucopiae* (Postels & Ruprecht) Hommersand (Scrosati, 1998). On the contrary, the rarity of gametophytes and/or the predominance of tetrasporophytes were more commonly documented in many other red algal species (e.g. Hansen & Doyle, 1976; Kaliaperumal & Umamaheswara Rao, 1982; Vásquez et al., 1998; Cecere et al., 2000; Reis & Yoneshigue-Valentin, 2000).

In addition to rarity of cystocarpic plants, male plants were also not examined in the present study. This could underestimate the present reproductive analysis. Absence of male plants and rare occurrence of female ones in *Hypnea* species were also reported in other places like India (Rama Rao, 1977) and Brazil (Schenkman, 1989; Reis & Yoneshigue-Valentin, 2000). Two questions were raised by Schenkman (1989): (1) Why are fertile gametophytes so rare even if there is high production of tetraspores in several genera of Rhodophyta? (2) Could the reason be the apomeiotic division of the sporangia or a greater ability of diploid plants to propagate vegetatively in comparison with the haploid phase (see also Dixon, 1973; Searles, 1980), or neither?

There are several explanations for the first question. Cecere et al. (2000) stated that gametophytes, which were ephemeral, might die back soon after reproduction or might become reproductive when they were still so small to be underestimated (see also Breeman et al., 1988). Similarly, Reis & Yoneshigue-Valentin (2000) thought that the reason why male plants of most red algae were rarely recorded was likely to be that the male plants manifested themselves during a very short period of time or it was a bit difficult to recognize them. While Mathieson (1989) explained that relatively longer lifespan of tetrasporophytes in wild populations allowed them to produce more upright thalli per plant and thus they became more robust under stressful conditions. In a culture study of *Stictosiphonia hookeri* (Harvey) Harvey from different localities, West et al. (1996) suggested two possibilities to explain the low frequency of gametophytes: (1) tetraspores do not survive germination due to their low viability, thus survival and dispersal are probably achieved by vegetative growth and fragmentation (see also Hansen & Doyle, 1976 and Cecere et al., 2000), (2) gametophytes reproduce and die more quickly than tetrasporophytes do. All these would lead to rare occurrence of gametophytes and a prevalence of tetrasporophytes (see also Rama Rao, 1977).

For the second question, both suggestions are possible. Variations in gametophyte-to-sporophyte ratios could be attributable to greater fecundity and survivorship (relative fitness) of one phase or to asexual phenomena, such as vegetative propagation and/or apomixis (see Hawkes, 1990). In certain algal species, apomixis can occur by means of apogamy/parthenogenesis or apomeiosis. Since in red algae, vegetative reproduction is an important and widespread strategy to maintain their populations in one place, many of them are able to develop apomeiotic tetrasporangia, thus giving rise to a life history in which new tetrasporophytes develop directly from tetraspores (Hansen & Doyle, 1976; Rama Rao, 1977; Hoyle, 1978; Magne, 1987; see also Hawkes, 1990; Murray & Dixon, 1992). Hence, the phenomenon showing high frequency of tetrasporophytes in certain species of the family Hypneaceae suggested that the principal reproductive mode could be resulted from apomeiosis (West & Hommersand, 1981). Nevertheless, Schenkman (1989) believed that the *Hypnea* population studied in Brazil maintained itself mainly by vegetative reproduction rather than apomeiosis as normal meiosis was observed in the tetrasporangia in a previous study (Schenkman, 1986).

Since *Hypnea* species exhibit isomorphic life cycles, the many vegetative plants of *H. charoides* observed in Tung Ping Chau during the growing seasons could be either infertile gametophytes or tetrasporophytes. If the suggestions mentioned above, i.e. that gametophytes might die soon after reproduction (Cecere et al., 2000) or that they manifested themselves only during a very short period of time (Reis & Yoneshigue-Valentin, 2000), are applicable to populations of *H. charoides* in Tung Ping Chau, the processes from fertilization of gametes to the development of carposporophytes and from carpospores to tetrasporophytes should be relatively short. This leads to the rare occurrence of cystocarpic plants. The tetrasporophytes became fertile and released tetraspores at the end of the growing season. These tetraspores persisted in summer months and then gave rise to gametophytes in the next growing season, completing the life cycles. However, the viability of free-floating spores is exceedingly low and this is true for tetraspores (Santelices 1990; Lobban & Harrison, 1994). Then, how could they survive in summer? Hoffmann & Santelices (1991) suggested that this role could be fulfilled by developing into microscopic forms that remained in a state of suspended growth. Some algal species existed as a microthallus phase (e.g. prostrate disclike phases) in order to survive adverse conditions like intense high summer temperatures and ice scour (Schoschina et al., 1996). In *Acanthophora najadiformis* (Delile) Papenfuss populations in the Ionian Sea, Cecere et al. (2000) explained that due to their inviability, tetraspores were substituted by propagules, which could act either as perennating organs or resting organs to overcome unfavourable environmental conditions. Such kinds of (asexual) propagules was found to have a substantially greater amount of nutrient reserves and greater photosynthetic potential than the spores (Cruz Adames & Ballantine, 1996). Hence, it is possible that tetraspores of *H. charoides* in Tung Ping Chau are able to develop and exist in a form (e.g. sporelings or propagules) that can survive during summer and then grow into adult plants when the conditions are optimal (i.e. late fall to winter).

Despite the low percentage occurrence of reproductive plants in populations of *H. charoides* in A Ma Wan during the whole study period, the frequency of observing cystocarpic (drifted samples) and tetrasporic (−1 m CD, −2 to −3 m CD and drifted samples, also in Lung Lok Shui populations) plants was relatively higher in 1999. There are several possibilities: (1) the process from fertilization of gametes to the development of tetrasporophytes was prolonged, increasing the chances of observing cystocarpic plants; (2) the tetraspores produced in the previous growing season in 1998 might undergo apomeiosis and develop into a tetrasporophyte; (3) the ripening of tetrasporangia was premature. These observed phenomena might be a response to stressful environment conditions (e.g. severe fluctuations of nutrient concentrations in seawater in 1999). The real reasons, however, could not be determined at present.

Population recruitment of *H. charoides* during the growing season was thought to be provided by vegetative fragmentation as the timing of occurrence and reproduction were similar among different groups of samples in A Ma Wan. Many drifted samples were found at the beginning of the growing season, suggesting that the loosely attached individuals could easily disperse themselves (or branches) in order to populate the whole area. The stability of this free-floating population ensured the vegetative fragmentation as an effective means of population recruitment of *H. charoides* in A Ma Wan (see Perrone & Cecere, 1997). Without considering the unusual, sudden growth of *H. charoides* in 1998, similar pattern of vegetative fragmentation and population recruitment would likely be occurring in Lung Lok Shui. This process of population recruitment is more likely to happen in Lung Lok Shui as indicated by the fact that thalli of *H. charoides* were mostly found entangled with *Sargassum* spp. rather than loosely attached on the substratum. Furthermore, this might be an adaptation to populate a place with strong wave action. The relatively high abundance of tetrasporic plants in Lung Lok Shui could also be another adaptation of *H. charoides*. Tetrasporophytes could be more resistant to hydrodynamic forces (Scrosati, 1998). In addition, the high abundance of tetrasporic plants (100%) observed in October 1998 during the unusual growth period might be the result of selective forces acting in different ways on the different phases, with one phase better adapted than the other (Zamorano & Westmeier, 1996; Reis & Yoneshigue-Valentin, 2000). It could also be a response of *H. charoides* populations to some undetected changes induced by the plants themselves to the surroundings. Selection will favor rapid growth, early reproduction, and short life spans in unstable areas (Lobban & Harrison, 1994). Among algae, different types of stress induce different physiological responses, resulting in different ecological outcomes (Davison & Pearson, 1996).

The onset of reproduction coinciding with the end of the growth phase has been recorded in several red algal species (Burns & Mathieson, 1972b; Kilar & Mathieson, 1978; Kain & Norton, 1990; Voskoboinikov et al., 1996; Molenaar & Breeman, 1997). In this study, populations of *H. charoides* in A Ma Wan became tetrasporic when the growing season came to the end. This could be advantageous as there would be more free substrata due to the mortality of many other marine algae during this period (Kim & De Wreede, 1996; Scrosati, 1998).

Occurrence of cystocarps and tetrasporangia on the same thallus in Hypnea charoides

During the examination of reproductive structures in *Hypnea charoides*, presence of cystocarps and tetrasporangia on the same branch was observed twice (25 April 1998 in A Ma Wan and 3 May 1997 in Lung Lok Shui). Abnormal pairing of these two reproductive organs had been recorded in certain red algae, like *Cystoclonium purpureum* (Hudson) Batters (Gigartinales), *Chondria baileyana* (Montagne) Harvey (Ceramiales) (Edelstein et al., 1974), *Polysiphonia harlandii* Harvey (Cheung et al., 1984) and many *Gracilaria* species (see Kain & Destombe, 1995). Plants can bear both structures either on separate parts of the thallus or on the same. Explanations, suggested by Kain & Destombe (1995), include the *in situ* germination of tetraspores (enabling gametophytes to develop as epiphytes on parental tetrasporophytes), the coalescence of spores or basal discs arising from spores to form a chimaera, mitotic recombination during cell division (resulting in occurrence of diploid male and female cells on the same tetrasporophyte), a natural mutation during which female expression is repressed, allowing the formation of carpogonia on male plants which thus became bisexual, and finally, incomplete formation of cross-walls in tetrasporangia. Though it is not clear which explanation(s) would be more applicable to the populations of *H. charoides* in Tung Ping Chau at the moment, this could be a strategy for the tetraspores to survive and germinate. The percentage occurrence of cystocarpic plants in both A Ma Wan and Lung Lok Shui could also be underestimated. This phenomenon of having both gametangial and non-gametangial sexual reproductive structures on a single thallus was not commonly recorded in Hong Kong (see Cheung et al., 1984). The present finding is thus a first record for *H. charoides* in Hong Kong.

Acknowledgements

This paper benefited from valuable comments from Professors H. M. Lam, C. K. Wong (Chinese University of Hong Kong (CUHK)) and S. W. Phang (University of Malaya). Special thanks are given to staff of Marine Science Laboratory for using the facilities and the following people for their kind assistance in the field: Louis Y. Cheung, Theo C. H. Hui, Donna C. Lee, May S. H. Liu, Henry T. W. Tam, Eva K. Y. Tso, Dickson C. C. Wong, Icy W. S. Yip and Y. H.

Yung. This project was supported by various CUHK University Grants Council Direct grants to POA.

References

Adey, W. H., 1973. Temperature control of reproduction and productivity in a subarctic coralline alga. Phycologia 12: 111–118.

APHA (American Public Health Association), 1995. Standard Methods for the Examination of Water and Waste Water (19th edn). American Public Health Association, Washington, DC.

Breeman, A. M., E. J. S. Meulenhoff & M. D. Guiry, 1988. Life history regulation and phenology of the red alga *Bonnemaisonia hamifera*. Helgoländer wiss. Meeresunters. 42: 535–551.

Brown, M. T., 1995. Interactions between environmental variables on growth rate and carrageenan content of *Solieria chordalis* (Solieriaceae, Rhodophyceae) in culture. J. appl. Phycol. 7: 427–432.

Burns, R. L. & A. C. Mathieson, 1972a. Ecological studies of economic red algae. II. Culture studies of *Chondrus crispus* Stackhouse and *Gigartina stellata* (Stackhouse) Batters. J. exp. mar. Biol. Ecol. 8: 1–6.

Burns, R. L. & A. C. Mathieson, 1972b. Ecological studies of economic red algae. III. Growth and reproduction of natural and harvested populations of *Gigartina stellata* (Stackhouse) Batters in New Hampshire. J. exp. mar. Biol. Ecol. 9: 77–95.

Cecere, E., O. D. Saracino, M. Fanelli & A. Petrocelli, 2000. Phenology of two *Acanthophora najadiformis* (Rhodophyta, Ceramiales) populations in the Ionian Sea (Mediterranean Sea). Bot. mar. 43: 109–117.

Cheung, K. W., K. Y. Lee & I. J. Hodgkiss, 1984. The occurrence of tetrasporangia and cystocarps on the same thalli in *Polysiphonia harlandii* (Rhodophyta: Ceramiales) (Note). Bot. mar. 27: 571–572.

Cruz Adames, V. M. & D. L. Ballantine, 1996. Asexual reproduction in *Laurencia poiteaui* (Rhodomelaceae, Rhodophyta). Bot. mar. 39: 75–77.

Davison, I. R. & G. A. Pearson, 1996. Stress tolerance in intertidal seaweeds. J. Phycol. 32: 197–211.

Dawes, C. J., J. Orduña-Rojas & D. Robledo, 1999. Response of the tropical red seaweed *Gracilaria cornea* to temperature, salinity and irradiance. J. appl. Phycol. 10: 419–425.

De Wreede, R. E., 1976. The phenology of three species of *Sargassum* (Sargassaceae, Phaeophyta) in Hawaii. Phycologia 15: 175–183.

Dixon, P. S., 1973. Biology of the Rhodophyta. Hafner Press, New York: 285 pp.

Dring, M. J., 1974. Reproduction. In Stewart, W. D. P. (ed.), Algal Physiology and Biochemistry. Blackwell Scientific Publications: 814–833.

Dyck, L. J. & R. E. De Wreede, 1995. Patterns of seasonal demographic change in the alternate isomorphic stages of *Mazzaella splendens* (Gigartinales, Rhodophyta). Phycologia 34: 390–395.

Edelstein, T., C. J. Bird & J. McLachlan, 1974. Tetrasporangia and gametangia on the same thallus in the red algae *Cystoclonium purpureum* (Huds.) Batt. and *Chondria baileyana* (Mont.) Harv. Br. Phycol. J. 9: 247–250.

Guist, G. G., Jr., C. J. Dawes & J. R. Castle, 1982. Mariculture of the red seaweed, *Hypnea musciformis*. Aquaculture 28: 375–384.

Hall, J. D. & S. N. Murray, 1998. The life history of a Santa Catalina Island population of *Liagora californica* (Nemaliales, Rhodophyta) in the field and in laboratory culture. Phycologia 37: 184–194.

Hansen, J. E. & W. T. Doyle, 1976. Ecology and natural history of *Iridaea cordata* (Rhodophyta; Gigartinaceae): population structure. J. Phycol. 12: 273–278.

Hawkes, M. W., 1990. Reproductive strategies. In Cole, K. M. & R. G. Sheath (eds), Biology of the Red Algae. Cambridge University Press, Cambridge: 455–476.

Hodgkiss, I. J., 1984. Seasonal patterns of intertidal algal distribution in Hong Kong. Asian mar. Biol. 1: 49–57.

Hodgkiss, I. J. & K. Y. Lee, 1983. Hong Kong Seaweeds. Urban Council, Hong Kong.

Hoffmann, A. J. & B. Santelices, 1991. Banks of algal microscopic forms: hypotheses on their functioning and comparisons with seed banks. Mar. Ecol. Prog. Ser. 79: 185–194.

Hoyle, M. D., 1978. Reproductive phenology and growth rates in two species of *Gracilaria* from Hawaii. J. exp. mar. Biol. Ecol. 35: 273–283.

Hurtado-Ponce, A. Q. & H. B. Pondevida, 1997. The interactive effect of some environmental factors on the growth, agar yield and quality of *Gracilariopsis bailinae* (Zhang et Xia) cultured in tanks. Bot. mar. 40: 217–223.

Kaehler, S. & G. A. Williams, 1996. Distribution of algae on tropical rocky shores: spatial and temporal patterns of non-coralline encrusting algae in Hong Kong. Mar. Biol. 125: 177–187.

Kain (Jones), J. M., 1989. The seasons in the subtidal. Br. Phycol. J. 24: 203–215.

Kain (Jones), J. M. & C. Destombe, 1995. A review of the life history, reproduction and phenology of *Gracilaria*. J. appl. Phycol. 7: 269–281.

Kain (Jones), J. M. & T. A. Norton, 1990. Marine ecology. In Cole, K. M. & R. G. Sheath (eds), Biology of the Red Algae. Cambridge University Press, Cambridge: 377–422.

Kaliaperumal, N. & M. Umamaheswara Rao, 1982. Seasonal growth and reproduction of *Gelidiopsis variabilis* (Greville) Schmitz. J. exp. mar. Biol. Ecol. 61: 265–270.

Kapraun, D. F., 1978. Field and culture studies on growth and reproduction of *Callithamnion byssoides* (Rhodophyta, Ceramiales) in North Carolina. J. Phycol. 14: 21–24.

Kennish, R., G. A. Williams & S. Y. Lee, 1996. Algal seasonality on an exposed rocky shore in Hong Kong and the dietary implications for the herbivorous crab *Grapsus albolineatus*. Mar. Biol. 125: 55–64.

Kilar, J. A. & A. C. Mathieson, 1978. Ecological studies of the annual red alga *Dumontia incrassata* (O. F. Müller) Lamouroux. Bot. mar. 21: 423–437.

Kim, J. H. & R. E. DeWreede, 1996. Effects of size and season of disturbance on algal patch recovery in a rocky intertidal community. Mar. Ecol. Prog. Ser. 133: 217–228.

Lazo, M. L., M. Greenwell & J. McLachlan, 1989. Population structure of *Chondrus crispus* Stackhouse (Gigartinaceae, Rhodophyta) along the coast of Prince Edward Island, Canada: distribution of gametaphytic and sporophytic fronds. J. exp. mar. Biol. Ecol. 126: 45–48.

Lobban, C. S. & P. J. Harrison, 1994. Seaweed Ecology and Physiology. Cambridge University Press, Cambridge.

Luhan, M. R. J., 1996. Biomass and reproductive states of *Gracilaria heteroclada* Zhang et Xia collected from Jaro, Central Philippines. Bot. mar. 39: 207–211.

Lüning, K., 1990. Seaweeds: Their Environment, Biogeography and Ecophysiology. John Wiley & Sons, Inc., New York.

Luxoro, C. & B. Santelices, 1989. Additional evidence for ecological differences among isomorphic reproductive phases of *Iridaea laminarioides* (Rhodophyta: Gigartinales). J. Phycol. 25: 206–212.

Magne, F., 1987. Is the frequency of apomeiosis in the Rhodophyta a genetic character? Hydrobiologia 151/152: 221–232.

Makarov, V. N., M. V. Makarov & E. V. Schoschina, 1999. Seasonal dynamics of growth in the Barents Sea seaweeds: endogenous and exogenous regulation. Bot. mar. 42: 43–49.

Marinho-Soriano, E., T. Laugier & M. L. De Casabianca, 1998. Reproductive strategy of two *Gracilaria* species, *G. bursa-pastoris* and *G. gracilis*, in a Mediterranean Lagoon (Thau, France). Bot. mar. 41: 559–564.

Mathieson, A. C., 1989. Phenological patterns of New England seaweeds. Bot. mar. 32: 419–438.

Mathieson, A. C. & R. L. Burns, 1975. Ecological studies of economic red algae. V. Growth and reproduction of natural and harvested populations of *Chondrus crispus* Stackhouse in New Hampshire. J. exp. mar. Biol. Ecol. 17: 137–156.

May, G., 1986. Life history variations in a predominantly gametophytic population of *Iridaea cordata* (Gigartinaceae, Rhodophyta). J. Phycol. 22: 448–455.

Molenaar, F. J. & A. M. Breeman, 1994. Ecotypic variation in *Phyllophora pseudoceranoides* (Rhodophyta) ensures winter reproduction throughout its geographic regions. J. Phycol. 30: 392–402.

Molenaar, F. J. & A. M. Breeman, 1997. Latitudinal trends in the growth and reproductive seasonality of *Delesseria sanguinea*, *Membranoptera alata* and *Phycodrys rubens* (Rhodophyta). J. Phycol. 33: 330–343.

Molenaar, F. J., A. M. Breeman & L. A. H. Venekamp, 1996. Ecotypic variation in *Cystoclonium purpureum* (Rhodophyta) synchronizes life history events in different regions. J. Phycol. 32: 516–525.

Murray, S. N. & P. S. Dixon, 1992. The Rhodophyta: some aspects of their biology. III. Oceanogr. mar. biol. Ann. Rev. 30: 1–148.

Narasimha Rao, G., 1995. Seasonal growth, biomass, and reproductive behavior of three species of red algae in Godavari estuary, India. J. Phycol. 31: 209–214.

Neish, A. C., P. F. Shacklock, C. H. Fox & F. J. Simpson, 1977. The cultivation of *Chondrus crispus*. Factors affecting growth under greenhouse conditions. Can. J. Bot. 55: 2263–2271.

Perrone, C. & E. Cecere, 1997. Regeneration and mechanisms of secondary attachment in *Solieria filiformis* (Gigartinales, Rhodophyta). Phycologia 36: 120–127.

Piriz, M. L., 1996. Phenology of a *Gigartina skottsbergii* Setchell et Gardner population in Chubut Province (Argentina). Bot. mar. 39: 311–316.

Prince, J. S. & S. W. O'Neal, 1979. The ecology of *Sargassum pteropleuron* Grunow (Phaeophyceae, Fucales) in the waters off South Florida I. Growth, reproduction and population structure. Phycologia 18: 109–114.

Rabanal, S. F., R. Azanza & A. Hurtado-Ponce, 1997. Laboratory manipulation of *Gracilariopsis bailinae* Zhang et Xia (Gracilariales, Rhodophyta). Bot. mar. 40: 547–556.

Rama Rao, K., 1977. Studies on Indian Hypneaceae. II. reproductive capacity in the two *Hypnea* over the different seasons. Bot. mar. 20: 33–39.

Reani, A., J. Cosson, A. Parker & D. Zaoui, 1998. Seasonal variation of growth, carrageenan content and rheological properties of *Cystoclonium purpureum* (Huds.) Batters (Rhodophyta, Cystocloniaceae) from the Calvados Coast (France). Bot. mar. 41: 383–387.

Reed, D. C., A. W. Ebeling, T. W. Anderson & M. Anghera, 1996. Differential reproductive responses to fluctuating resources in two seaweeds with different reproductive strategies. Ecology 77: 300–316.

Reis, R. P. & Y. Yoneshigue-Valentin, 2000. Phenology of *Hypnea musciformis* (Wulfen) Lamouroux (Rhodophyta, Gigartinales) in three populations from Rio de Janeiro State, Brazil. Bot. mar. 43: 299–304.

Santelices, B., 1990. Patterns of reproduction, dispersal and recruitment in seaweeds. Oceanogr. mar. biol. Ann. Rev. 28: 177–276.

Schenkman, R. P. F., 1986. Cultura de *Hypnea* (Rhodophyta) in vitro como subsidio para estudos morfologicos, reprodutivos e taxonomicos. Ph. D. thesis, Inst. Biociências, Universidade de São Paulo, Sao Paulo, Brazil.

Schenkman, R. P. F., 1989. *Hypnea musciformis* (Rhodophyta): ecological influence on growth. J. Phycol. 25: 192–196.

Schoschina, E. V., V. N. Makarov, G. M. Voskoboinikov & C. van den Hoek, 1996. Growth and reproductive phenology of nine intertidal algae on the Murman Coast of the Barents Sea. Bot. mar. 39: 83–93.

Scrosati, R., 1998. Population structure and dynamics of the clonal alga *Mazzaella cornucopiae* (Rhodophyta, Gigartinaceae) from Barkley Sound, Pacific Coast of Canada. Bot. mar. 41: 483–493.

Searles, R. B., 1980. The strategy of the red algal life history. Am. Nat. 115: 113–120.

Stimson, J., S. Larned & K. McDermid, 1996. Seasonal growth of the coral reef macroalga *Dictyosphaeria cavernosa* (Forskål) Børgesen and the effects of nutrient availability, temperature and herbivory on growth rate. J. exp. mar. Biol. Ecol. 196: 53–77.

Vásquez, J. A., A. Vega, B. Matsuhiro & C. Faúndez, 1998. Biomass, reproductive phenology and chemical characterization of soluble polysaccharides from *Rhodymenia howeana* Dawson, (Rhodymeniaceae, Rhodymeniales) in Northern Chile. Bot. mar. 41: 235–242.

Voskoboinikov, G. M., A. M. Breeman, C. van den Hoek, V. N. Makarov & E. V. Shoschina, 1996. Influence of temperature and photoperiod on survival and growth of North East Atlantic isolates of *Phycodrys rubens* (Rhodophyta) from different latitudes. Bot. mar. 39: 341–346.

West, J. A. & M. H. Hommersand, 1981. Rhodophyta: life histories. In Lobban, C. S. & M. J. Wynne (eds), The Biology of Seaweeds. Blackwell Scientific Publications: 133–193.

West, J. A., G. C. Zuccarello & U. Karsten, 1996. Reproductive biology of *Stictosiphonia hookeri* (Rhodomelaceae, Rhodophyta) from Argentina, Chile, South Africa and Australia in laboratory culture. Hydrobiologia 326/327: 277–282.

Williams, G. A., 1993. Seasonal variation in algal species richness and abundance in the presence of molluscan herbivores on a tropical rocky shore. J. exp. mar. Biol. Ecol. 167: 261–275.

Zamorano, J. & R. Westermeier, 1996. Phenology of *Gigartina skottsbergii* (Gigartinaceae, Rhodophyta) in Ancud Bay, southern Chile. Hydrobiologia 326/327: 253–258.

Zinoun, M. & J. Cosson, 1996. Seasonal variation in growth and carrageenan content of *Calliblepharis jubata* (Rhodophyceae, Gigartinales) from the Normandy coast, France. J. appl. Phycol. 8: 29–34.

Hydrobiologia **512**: 79–88, 2004.
P.O. Ang, Jr. (ed.), Asian Pacific Phycology in the 21ˢᵗ Century: Prospects and Challenges.
© *2004 Kluwer Academic Publishers.*

Biomass production of two *Sargassum* species at Cape Rachado, Malaysia

Ching-Lee Wong & Siew-Moi Phang*
Institute of Biological Sciences, Faculty of Science, University of Malaya, 50603 Kuala Lumpur, Malaysia
**Author for correspondence; E-mail: phang@um.edu.my*

Key words: biomass, standing crop, seasonality, *Sargassum*, Malaysia

Abstract

Seasonality in biomass production of *Sargassum baccularia* (Mertens) C. Agardh and *Sargassum binderi* Sonder ex J. G. Agardh was analysed based on quarterly destructive sampling using a line transect-quadrat method from January 1995 to April 1996. Biomass for both *Sargassum* species showed an unimodal pattern. *S. baccularia* attained high biomass during January 1995 (520.23 g wet weight m^{-2}, 47.88 g dry weight m^{-2}, 13.35 g ash-free dry weight m^{-2}) and July 1995 (501.98 g wet weight m^{-2}, 64.92 g dry weight m^{-2}, 14.00 g ash-free dry weight m^{-2}). *S. binderi* attained the highest biomass in April 1996 (656.13 g wet weight m^{-2}, 80.81 g dry weight m^{-2}, 15.25 g ash-free dry weight m^{-2}) with another high value recorded in July 1995 (429.28 g wet weight m^{-2}, 54.30 g dry weight m^{-2}, 11.20 g ash-free dry weight m^{-2}). Both *Sargassum* species recorded the lowest biomass in January 1996 (*S. baccularia*: 76.14 g wet weight m^{-2}, 9.97 g dry weight m^{-2}, 1.97 g ash-free dry weight m^{-2}; *S. binderi*: 68.21 g wet weight m^{-2}, 8.36 g dry weight m^{-2}, 1.68 g ash-free dry weight m^{-2}). Biomass for both *Sargassum* species was strongly correlated to the thallus length. The population of both *Sargassum* species consisted mainly of young plants with 96% of the *S. baccularia* population and 89% of the *S. binderi* population being shorter than 199 mm. Both *Sargassum* populations recorded low percentage of fertility. The most important factor that controlled the biomass production and reproduction for both *Sargassum* species was rainfall.

Introduction

Malaysia has a high diversity of marine algae, with 261 taxa recorded by Phang (1998). However, only a few studies on the distribution and ecology of marine algae have been reported in Peninsular Malaysia (Sivalingam, 1977, 1978, Arumugam, 1981; Crane, 1981; Phang, 1984, 1986, 1988, 1995) and Sabah (Khew, 1978). *Sargassum* is common in Malaysian waters and only the subgenus *Sargassum* is found in Malaysia. There are a total of 21 species of *Sargassum* recorded in Malaysia (Phang, 1998).

Information on the ecological and phenological aspects of *Sargassum* species in Malaysia is scarce as compared to other neighbouring countries where this genus is gaining more attention because of its economic value. The aim of this research is to establish some baseline data on the *Sargassum* species of Malaysia for its conservation and management for commercial utilisation. This is achieved through:

1. observation of seasonal variation in the standing crop of two *Sargassum* species; that is *Sargassum baccularia* (Mertens) C. Agardh and *Sargassum binderi* Sonder ex J. G. Agardh,

2. determination of the fertility periods and relating them to the mean thallus length and biomass of the species, and

3. correlation of the seasonal variation of biomass with environmental parameters.

In terms of the relationship between phenology and commercial utilisation, observations such as the time of appearance of the reproductive organs, seasonal variation in growth, biomass and development of plants, are necessary for identifying optimal harvest periods and this, in turn, ensures sustainable utilisation of a species.

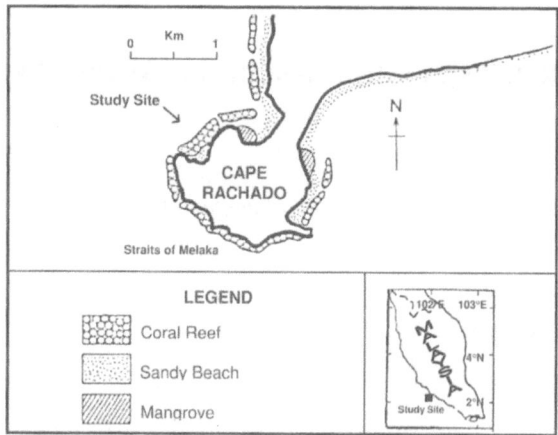

Figure 1. Map showing the location of the study site.

Materials and methods

Study site

Cape Rachado is a fringing coral reef located at Port Dickson, Negeri Sembilan, West Coast of Peninsular Malaysia (Fig. 1). The reef flats extend out to sea for about 110 m and gently slope towards the reef edge to a depth of about 8 m (Goh & Sasekumar, 1980). The reef flats are dominated by *Sargassum, Turbinaria* and *Padina* species. Cape Rachado is a famous tourist destination. The study site chosen is a bay surrounded by rocky shore that is less disturbed by tourists as compared with adjacent bays.

The reef flats are completely exposed when the tide level is 0.3 m or less above chart datum (CD). It is completely exposed for more than three hours when the tide level is 0.1 m or less above CD. The lowest tide level (0.1 m or less above CD) during this study period was from February to May 1995 and January to May 1996. The reef flats were exposed 145 times (daytime low tides of 0.3 m or less above CD), for about 1–3 h during the monitoring period (January 1995 to April 1996). Calculations were based on the Tide Tables Malaysia (1995, 1996) for Port Dickson, Negeri Sembilan Darul Khusus (Latitude $2°31'$ N, Longitude $101°47'$ E).

Sampling methods

Destructive sampling for *S. baccularia* and *S. binderi* was conducted quarterly from January 1995 to April 1996. The plants were collected from 0.25 m × 0.25 m quadrats along six line transects every three months during the lowest daytime spring tide. These line transects were laid perpendicular to the shoreline, each 100 m apart. Ten quadrat samples were placed at ten meter intervals along each line transect. Quadrats without both or either one of the *Sargassum* species were also recorded in order to provide information on the spatial distribution pattern of the species.

Twenty-two to 58 quadrat samples were processed each time to determine the seasonal variation in the biomass of both *Sargassum* species. Only *S. baccularia* and *S. binderi* plants within each quadrat were collected and placed in individually labelled plastic bags. The samples were stored in the ice-chest when brought back from the sampling site to the laboratory and were kept in the freezer before processing. All plants were defrosted and rinsed in fresh water to remove sand, silt, epiphytes and other debris before weighing.

Wet weight (WW), dry weight (DW) and ash-free dry weight (AFDW) of the samples were obtained after washing and cleaning. Wet weight of the samples was measured using a top pan balance (Ohaus portable advanced balance) to one decimal point after blotting the samples dry with paper towel. Wet weight values in terms of g m^{-2} were then calculated by multiplying the results obtained by 16, since each quadrat covered an area of 0.0625 m^2. Dry weight of the samples was obtained by drying the samples in the oven at 105 °C for 48 h. The samples were then weighed using the analytical balance (Mettler AJ100) to two decimal points. The values were converted to gram dry weight per m^2 (g DW m^{-2}). The dried samples were placed in dry and clean ceramic crucibles and combusted in the muffle furnace at 550 °C for 6 h. After combustion, the samples were weighed using the analytical balance (Mettler AJ100) to two decimal points and subtracted from the initial dry weight to give the ash-free dry weight. Subsampling was performed whenever combustion of an entire sample was impossible. These values were converted to gram ash-free dry weight per m^2 (g AFDW m^{-2}).

Thallus length of all the plants was measured before obtaining the wet weight. Thallus length was measured as the distance from the end of the holdfast to the apex of the longest branch. The measured length of all plants was averaged to give the mean plant length. Reproductive state was recorded by the presence of receptacles on plants in each quadrat for each sampling period. The receptacles were sectioned to determine the sexuality.

Physical parameters were recorded at the time of each destructive sampling. Salinity was measured using an Atago hand refractometer, water temperature and DO were measured using a salinity compensated dissolved oxygen meter (YSI Model 57), pH was measured using an ATC pH meter (Piccolo 2). Ambient temperature (minimum and maximum air temperatures), mean number of hours of sunshine per day, mean solar radiation and total rainfall for Malacca ($2°16'$ N, $102°15'$ E), were obtained from the Malaysian Meteorological Department, Petaling Jaya, Selangor. The protocol for the analysis of nutrients followed the standard method given for ammonia (Solorzano, 1969), nitrate (Isa et al., 1980) and phosphate (Strickland & Parsons, 1968).

Statistical analysis

One way ANOVA was employed to test for any significant difference among the monthly abiotic and biotic data. Data from destructive sampling (biomass, mean thallus length and number of plants bearing receptacles) were correlated with the environmental parameters using Pearson Product Moment correlation analysis. All statistical analyses were conducted using the computer package 'Statgraphic Version 5.0'.

Results

Mean quarterly biomass

The peak biomass of *S. baccularia* population was obtained in January 1995 (520.23 ± 724.37 g WW m^{-2}, 47.88 ± 65.47 g DW m^{-2}, 13.35 ± 16.69 g AFDW m^{-2}) and July 1995 (501.98 ± 341.05 g WW m^{-2}, 64.92 ± 40.99 g DW m^{-2}, 14.00 ± 9.65 g AFDW m^{-2}) (Fig. 2A). The lowest biomass occurred in January 1996 at 76.14 ± 37.97 g WW m^{-2}, 9.97 ± 4.62 g DW m^{-2}, 1.97 ± 1.00 g AFDW m^{-2}. One way ANOVA followed by LSD multiple range test (95% confidence level) denoted a significant difference among the monthly biomass values (WW: df = 30, $F = 16.66$, $p < 0.0001$; DW: df = 30, $F = 7.185$, $p < 0.003$; AFDW: df = 29, $F = 12.060$, $p < 0.0001$).

Biomass of *S. binderi* showed a more distinct unimodal pattern of change within a year (Fig. 2B). The highest value was first obtained in July 1995 (429.28 ± 385.64 g WW m^{-2}, 54.30 ± 50.25 g DW m^{-2}, 11.20 ± 11.43 g AFDW m^{-2}), followed by another peak in the following year in April 1996 (656.13 ± 735.27 g WW m^{-2}, 80.81 ± 76.02 g DW m^{-2}, 15.25

± 16.72 g AFDW m^{-2}). Biomass was the lowest in January 1996 (68.21 ± 33.08 g WW m^{-2}, 8.36 ± 4.07 g DW m^{-2}, 1.68 ± 0.89 g AFDW m^{-2}). Results of one way ANOVA, followed by LSD multiple range test (95% confidence level), also indicated a significant difference among the monthly biomass values (WW: df = 30, $F = 5.255$, $p < 0.002$; DW: df = 30, $F = 3.228$, $p < 0.022$; AFDW: df = 29, $F = 3.336$, $p < 0.019$).

Comparison between biomass, mean thallus length and percentage fertility

The peak period of the thallus length coincided with the peak period of biomass for both *Sargassum* species (Fig. 2). Correlation analysis indicated *S. baccularia* population to have a strong positive correlation between the mean thallus length and the WW ($r = 0.9826$), DW ($r = 0.8783$) and AFDW ($r = 0.9777$). Similar results were obtained for *S. binderi* population ($r = 0.9275$ for WW, $r = 0.9496$ for DW, $r = 0.8593$ for AFDW).

S. baccularia population recorded extremely low percentage fertility in July 1995 (0.54%) and April 1996 (0.44%), with zero fertility in other sampling periods (Fig. 3A). The fertility of *S. baccularia* population showed a significant ($p \leq 0.05$) positive correlation with WW ($r = 0.6690$), DW ($r = 0.7505$), AFDW ($r = 0.7301$) and mean thallus length ($r = 0.5952$).

S. binderi population was fertile throughout the sampling period with the highest percentage fertility recorded in April 1996 (59.70%) and the lowest in January 1996 (4.17%) (Fig. 3B). The percentage fertility of *S. binderi* population was significantly (p $\leq$0.05) correlated with WW ($r = 0.7770$), DW ($r = 0.8673$), AFDW ($r = 0.6267$) and mean thallus length ($r = 0.8959$).

Seasonal variation in length classes

The length class distribution of the *S. baccularia* plants was different for every sampling period from January 1995 to April 1996 (Fig. 4). For individuals in this population, those in 50–99 mm length class showed the highest percentage frequency (54%) in January 1995. This gradually decreased to about 20% in July and October 1995 before starting to increase in April 1996 (24%). Within this same period, those in length classes >100 mm increased from around 37% in January 1995 to about 60% in July 1995 before ex-

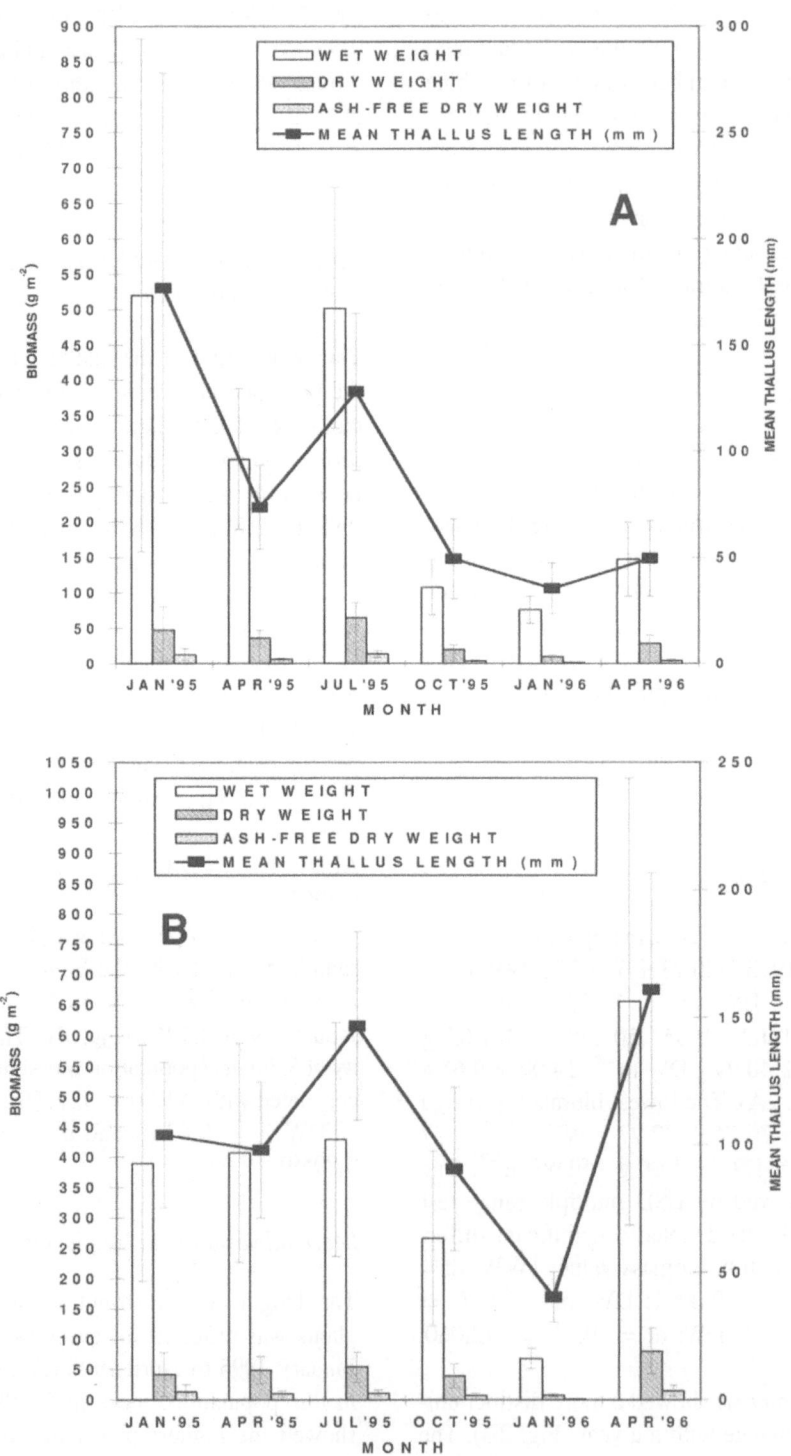

Figure 2. Mean (± SD) thallus length (mm) and biomass (g m^{-2}) of (A) *Sargassum baccularia* and (B) *S. binderi.*

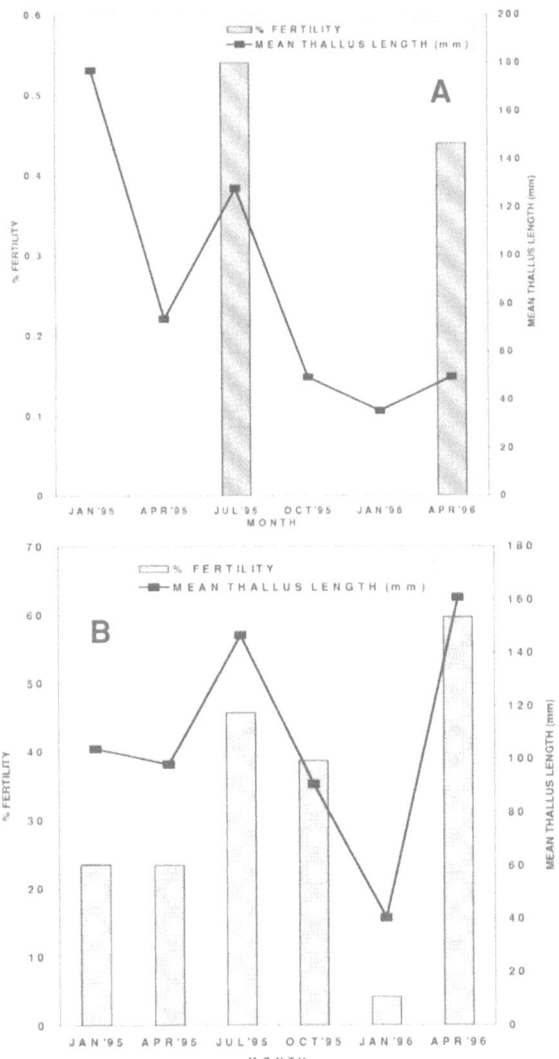

Figure 3. Comparison between percentage fertility and mean thallus length (mm) for (A) *Sargassum baccularia* and (B) *S. binderi.*

periencing a sudden drop to less than 12% in October 1995. Those in the smallest length class (>0–49 mm) showed a reverse trend. In January 1995, percentage frequency for >0–49 mm length class was less than 10% but it gradually increased to about 70% in October 1995 and reached the maximum (80%) in January 1996 before starting to decline again in April 1996. The overall average length of individuals in the population was 65.99 ± 60.32 mm with 54% of these being <49 mm in length.

The length class distribution of the *S. binderi* plants was also different for every sampling period from January 1995 to April 1996 (Fig. 5) although the general trend was similar to that observed for *S. bac-*

cularia. Almost half of the plants were found in the length class of 50–99 mm in January 1995 (47%) and April 1995 (43%). This gradually decreased to 19% in July 1995 and then slightly increased to 25–27% in October 1995 and January 1996 before dropping to 18% in April 1996. The frequency of those in length classes >100 mm increased from 38% in January 1995 to 72% in July 1995 before dropping to 39% in October 1995 and <2% in January 1996. It reached another peak of 71% again in April 1996. Individuals in length class >0–49 mm constituted <16% of the total population from January to July 1995. In October 1995 and January 1996 however, 35% and 71%, respectively, of the plants were found in this smallest length class before dropping to around 10% in April 1996. The overall average length of individuals of *S. binderi* population was 107.24 ± 74.17 mm with 89% of the population <199 mm and 54% <99 mm in length.

Correlation between environmental parameters, biomass, thallus length and fertility

Increase in biomass of *S. baccularia* population was significantly correlated with the increase in rainfall and phosphate level (except for DW) and the decrease in pH and radiation (Table 1). Whereas, increase in biomass of the *S. binderi* population was significantly correlated to the increase in salinity (except for DW and AFDW), water temperature (except for AFDW), ambient temperature, sunshine and rainfall, and decrease in ammonia (except for AFDW) and nitrate levels (Table 1).

Discussion

Both *Sargassum* populations attained two peaks and one low in standing crop over the 15-month quarterly monitored period. The values of the highest standing crop obtained for both species are relatively low (approximately ten times lower) as compared to the standing crop of *Sargassum* species reported in the Philippines (Ang, 1984; Trono & Lluisma, 1990; Largo et al., 1994), but are comparable to those from the earlier studies by Phang (1995) at Cape Rachado, Port Dickson in 1987–1988. The number of plants collected during the peak biomass period for *S. baccularia* in July 1995) and for *S. binderi* in April 1996 were 373 and 67, respectively (Wong, 1997). Although the amounts of standing crop attained for both species were almost the same, the number of

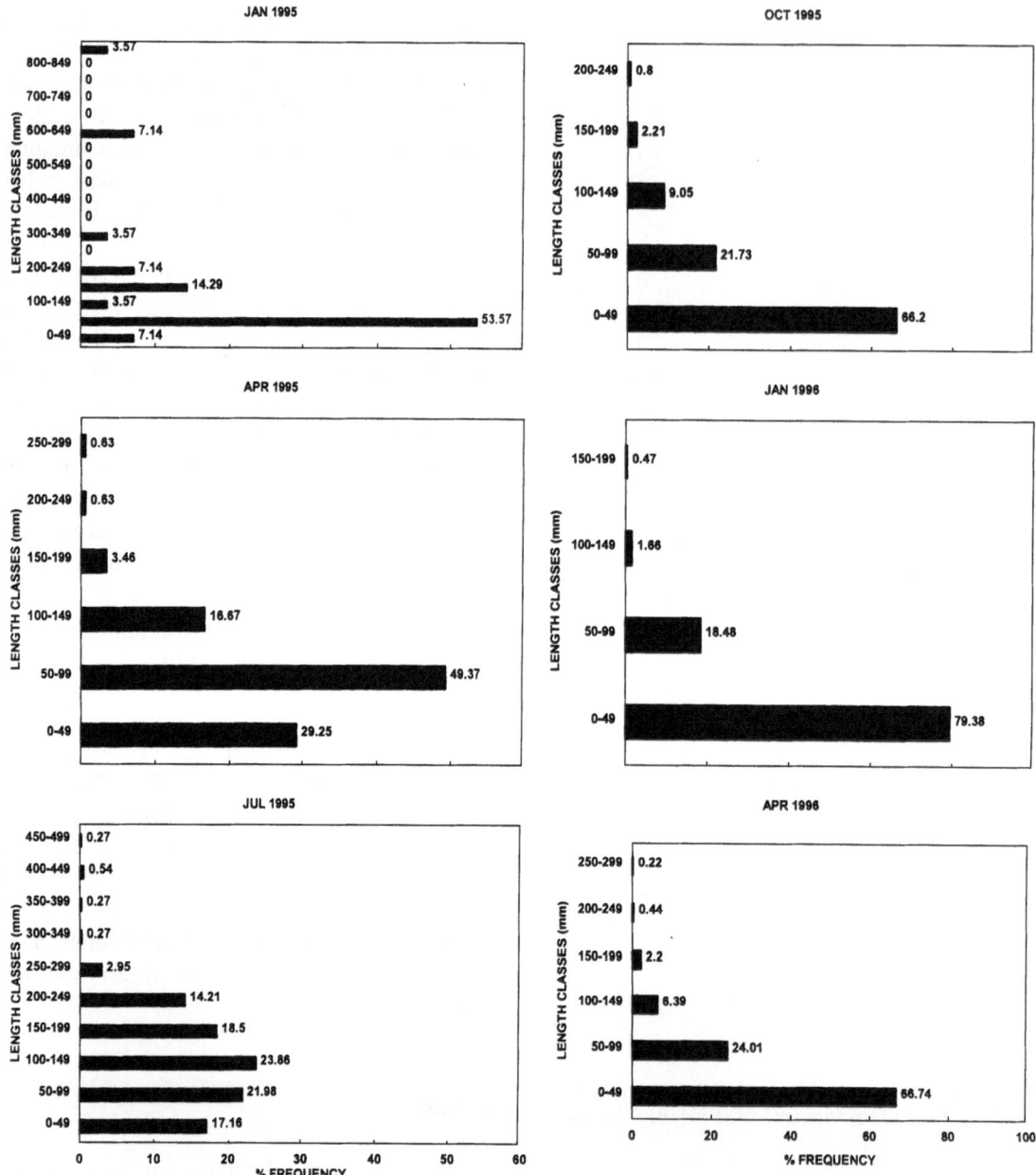

Figure 4. Percentage frequency of length classes of *Sargassum baccularia* in different sampling periods.

plants involved for *S. binderi* was almost six times lower than that for *S. baccularia*. *S. binderi* plants were usually larger and with more lateral branches. They were mostly found farther away from shore and hence were usually submerged in the water even during very low tide. This prevented them from being exposed to severe desiccation stress even when the reef was exposed. Furthermore, the presence of receptacle-bearing lateral branches increased their standing crop. This was also observed by Largo & Ohno (1992), who reported that the standing crop of *S. myriocystum* and

S. siliquosum in the Philippines increased as the plants became fertile. In contrast to *S. binderi*, *S. baccularia* plants tended to grow near shore where they were usually exposed to the full impact of desiccation.

A strong correlation was found between the standing crop and the mean thallus length as well as the number of plants bearing receptacles for both *Sargassum* species in this present study. The peak thallus length and reproductive period coincided with the peak biomass period. This is a phenomenon typical to *Sargassum* species as reported by various authors

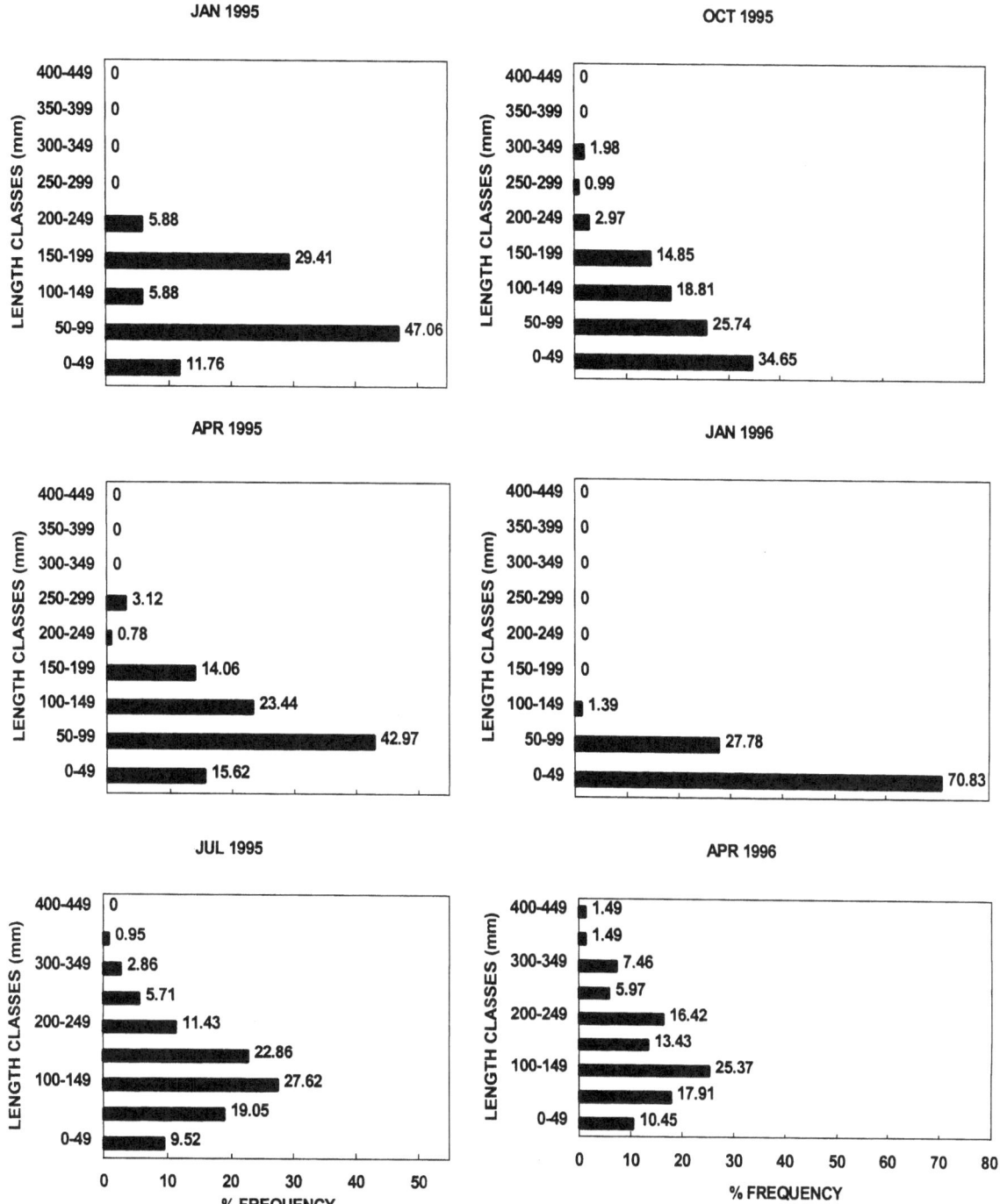

Figure 5. Percentage frequency of length classes of *Sargassum binderi* in different sampling periods.

(Tsuda, 1972; De Wreede, 1976; De Ruyter Van Steveninck & Breeman, 1987).

Mean thallus length for both *Sargassum* species (*S. baccularia* = 66 mm, *S. binderi* = 107 mm) in this study is smaller than what is reported for other species in other countries. Populations of both *Sargassum* species in the present study were exposed to air at low tide. This is especially so for individuals of *S. baccularia* as they were found higher up on the intertidal. For most of the other species studied elsewhere, they are either found in the subtidal environment or in low intertidal, such that exposure to air during low tides happens only occasionally. Exposure to air, especially during daytime low tides, results in loss of water from

Table 1. Correlation between *Sargassum* biomass [wet weight (WW), dry weight (DW), ash free dry weight (AFDW)], mean thallus length (TL) and number of plants bearing receptacles (*F*) with environmental parameters

Environmental parameters	WW	DW	AFDW	TL	*F*
I. *S. baccularia*					
Salinity	0.3679	0.4550	0.2704	0.2044	0.4457
Temperature	−0.2059	0.0959	−0.2010	−0.3179	0.1596
Min Temp.	0.0008	0.1670	−0.0266	−0.0696	−0.0374
Max Temp.	−0.0995	0.0568	−0.1657	−0.2201	0.0364
D.O.	−0.1186	0.0260	−0.0764	−0.0588	−0.4030
pH	−0.6864*	−0.6101*	−0.7391*	−0.7074*	−0.5806*
Radiation	−0.6658*	−0.5962*	−0.7301*	−0.7554*	−0.1822
Sunshine	−0.3553	−0.2434	−0.4174	−0.4766*	0.0588
Rainfall	0.5370*	0.7987*	0.5917*	0.4313	0.7619*
Ammonia	−0.1856	−0.4103	−0.1329	−0.0080	−0.5200*
Nitrate	−0.4543	−0.4821	−0.3931	−0.3482	−0.3722
Phosphate	0.5673*	0.2777	0.5156*	0.6254*	−0.0492
II. *S. binderi*					
Salinity	0.4997*	0.4198	0.4139	0.4236	0.1087
Temperature	0.5798*	0.6775*	0.3632	0.5914*	0.8040*
Min Temp.	0.5669*	0.6229*	0.4857	0.4622	0.6401*
Max Temp.	0.6638*	0.6858*	0.5314*	0.4767	0.4865
D.O.	−0.4027	−0.2797	−0.4782	−0.1778	0.4074
pH	0.0072	0.0831	−0.1643	−0.1666	0.1194
Radiation	0.3234	0.3424	0.1288	0.0718	0.0017
Sunshine	0.6390*	0.6457*	0.4823	0.3888	0.2559
Rainfall	0.7529*	0.7854*	0.7069*	0.9104*	0.8686*
Ammonia	−0.6228*	−0.6105*	−0.4452	−0.6232*	−0.4587
Nitrate	−0.8343*	−0.7700*	−0.8886*	−0.6241*	−0.3370
Phosphate	0.0560	−0.0712	0.3469	−0.1419	−0.4482

*$p < 0.05$.

the thallus. Furthermore, daytime low tides are much more damaging during hot summer. This situation causes desiccation stress to the intertidal seaweeds especially those seaweeds found higher above zero tide level, like S. *baccularia*, where the plants are exposed to desiccation stress for prolonged period of time. This could have restricted the growth of these plants and thus may explain why individuals of *S. baccularia* plants had much shorter mean thallus length than those of *S. binderi*, and both were generally shorter than those of other species reported in other countries.

The highest growth and reproductive activity of *S. binderi* occurred during periods of higher water temperature, but not for *S. baccularia*. The *S. binderi* population behaved more like the *Sargassum* species reported by Tsuda (1972) in Guam, Prince & O'Neal

(1979) in Florida, Ang (1985) in the Philippines and De Ruyter Van Stevenick & Breeman (1987) in Curacao, Netherlands Antilles. This is in contrast to the generalisation made by De Wreede (1976) and McCourt (1984), who stated that most tropical *Sargassum* species reached maximum growth and fertility in the cooler months of the year.

Analysis of length class frequency distribution of the two *Sargassum* populations revealed that the populations were mainly made up of small plants during most of the year. Shift in length class distribution was a result of three main processes: growth, die back (degeneration) and recruitment of new individuals. Similar patterns were observed for individuals in both the S. *baccularia* and *S. binderi* populations although the timing may not be the same. The initial shift in dominance from smaller to larger sizes was due to growth of the plants as the maximum growth rate was obtained in June 1995 (Wong, 1997). On the other hand, the highest degenerative rate (die back) was obtained after July 1995 (Wong, 1997). This coincided with the die back of the plants after peak reproduction and contributed to the shift in maximum size frequency of the population to the smallest length class from July 1995 to October 1995. Koh et al. (1993) recorded small size classes of *S. thunbergii* over the whole year in Korea. They pointed out that the occurrence of higher frequency of smaller size class individuals indicated a size reduction of the *S. thunbergii* plant after the growing period. The rapid increase in the dominance of smallest length class after July 1995 must be contributed in part by the recruitment of new and smaller plants. This contribution to shift in the population size structure should be more important in *S. baccularia* than in *S. binderi* as reproductive plants were always present in the latter. Continuous recruitment must be taking plant in the population of *S. binderi* so that new recruits would always contribute to a certain proportion of the individuals in the smallest length class of this population. In contrast, reproduction took place only within a narrow period in *S. baccularia*. Recruits would thus be available only within a few months after reproduction. Hence, for *S. baccularia* population, it is likely that recruits would only contribute mainly to the dominance of the smallest length class 3–6 months after reproduction, i.e. in Oct 1995 and January 1996.

Environmental factors play an important role in affecting the growth of algal populations. In this study, the most important among these factors in affecting the growth of both *Sargassum* populations appears to be rainfall. November is the monsoon month with heavy rain, strong waves and high turbidity (field observation). All of these could be detrimental to the growth of *Sargassum*. Hence, biomass and percentage fertility for both *Sargassum* species were the lowest in January 1996. Most plants would have died, or their annual parts died back in the previous months. Other than rainfall, there may be other factors which contributed to the seasonality of the *Sargassum* populations. Owing to the spatial distribution of *S. baccularia* population, being found mainly in the shallower, shoreward location when compared to the *S. binderi* population, factors like radiation may be more important to *S. baccularia* than to *S. binderi*. This is evidenced by the significant correlation found between radiation and the biomass and thallus length of *S. baccularia* but not with those of *S. binderi*. On the other hand, water temperature appears to be more important to *S. binderi* than to *S. baccularia*.

This study represents the first detailed study of *Sargassum* ecology in Malaysia. Data generated from this study are therefore very important in providing an insight into the growth and production of the two dominant *Sargassum* species found in the west coast of Peninsular Malaysia. Rainfall appears to be the most important factor in influencing the growth of *Sargassum*. Phang & Maheswary (1989) identified some Malaysian *Sargassum* species as potential sources of alginic acid. However, the results from the present study show that the biomass yield is too low to sustain harvesting from the wild populations for an alginate industry.

Acknowledgements

We thank the Ministry of Science and Technology (R&D 09-02-03-0222) and the University of Malaya for financial support, the Department of Fisheries for permitting research at the study site, Prof. Ang Put for the invaluable editorial advice, and KY Chew for assistance with the fieldwork.

References

Ang, P. O., 1984. Preliminary study of the alginate contents of *Sargassum* spp. in Balibago, Calatagan, Philippines. Hydrobiologia 116/117: 547–550.

Ang, P. O., 1985. Phenology of *Sargassum siliquosum* J. Ag. and *S. paniculatum* J. Ag. (Sargassaceae, Phaeophyta) in the reef flat of Balibago (Calatagan, Philippines). Proc. Int. Coral Reef Cong., Tahiti 5: 51–57.

Arumugam, P., 1981. Algal distribution in a Malaysia coral reef at Pulau Bidong Laut. Pertanika 44: 99–102.

Crane, P., 1981. The marine Chlorophyceae and Phaeophyceae of Penang Island. Malay. Nat. J. 34: 143–169.

De Ruyter Van Steveninck, E. D. & A. M. Breeman, 1987. Population dynamics of a tropical intertidal and deep-water population of *Sargassum polyceratium* (Phaeophyceae). Aquat. Bot. 29: 139–156.

De Wreede, R. E., 1976. The phenology of three species of *Sargassum* (Sargassaceae, Phaeophyta) in Hawaii. Phycologia 15: 491–500.

Goh, A. H. & A. Sasekumar, 1980. The community structure of the fringing coral reef, Cape Rachado, Malaysia. Malay. Nat. J. 34: 25–37.

Isa, Z., W. M. Yong & M. M. Singh, 1980. Manual of Laboratory Methods for Chemical Analysis of Rubber Effluent. Rubber Research Institute Malaysia, Kuala Lumpur. 35–39.

Khew, K. L., 1978. Marine algae. In Chua, T. H. & J. A. Mathias (eds), Coastal Resources of West Sabah. An Investigation into the Impact of Oil Spill. University Sains Malaysia Publication: 109–114.

88

Koh, C. H., Y. Kim & S. G. Kang, 1993. Size distribution, growth and production of *Sargassum thunbergii* in an intertidal zone of Padori, west coast of Korea. Hydrobiologia 260/261: 207–214.

Largo, D. B. & M. Ohno, 1992. Phenology of two species of brown seaweeds, *S. myriocystum* J. Agardh and *Sargassum siliquosum* J. Agardh (Sargassaceae, Fucales) in Liloan, Cebu, in Central Philippines. Bull. Mar. Sci. Fish, Kochi Univ. 12: 17–27.

Largo, D. B., M. Ohno & A. T. Critchley, 1994. Seasonal changes in the growth and reproduction of *Sargassum polycystum* C. Ag. and *Sargassum siliquosum* J. Ag. (Sargassaceae, Fucales) from Liloan, Cebu, in Central Philippines. Jap. J. Phycol. (sorui) 42: 53–61.

McCourt, R. M., 1984. Seasonal patterns of abundance, distribution and phenology in relation to growth strategies of three *Sargassum* species. J. exp. mar. Biol. Ecol. 74: 141–156.

Phang, S. M., 1984. Seaweed resources of Malaysia. Walaceana 33: 3–8.

Phang, S. M., 1986. Malaysian seaweed flora. Proc. Ninth Annu. Sem. Malaysian Soc. Mar. Sci. 17–45.

Phang, S. M., 1988. The effect of siltation on algal biomass production at a fringing coral reef flat, Port Dickson, Peninsular Malaysia. Wallaceana 51: 3–5.

Phang, S. M., 1995. Distribution and abundance of marine algae on the coral reef flats at Cape Rachado, Port Dickson, Peninsular Malaysia. Malaysia J. Sci. 16A: 23–32.

Phang, S. M., 1998. The seaweed resources of Malaysia. In Critchley, A. T. & M. Ohno (eds), Seaweed Resources of the World. JICA, Japan: 79–91.

Phang, S. M. & V. Maheswary, 1989. Phycocolloid content of some Malaysian seaweeds. Proceedings 12th Ann. Sem. Malaysian Mar. Sci. Soc. 65–76.

Prince, J. S. & S. W. O'Neal, 1979. The ecology of *Sargassum pteropleuron* Grunow (Phaeophyceae, Fucales) in the waters off South Florida. I. Growth, reproduction and population structure. Phycologia 19: 109–114.

Sivalingam, P. M., 1977. Algal zonation pattern at Batu Ferringhi. I. seaward side. Sains Malaysiana 6: 153–163.

Sivalingam, P. M., 1978. Algal succession pattern on the rock shores of Batu Ferringhi in Penang Island I. Jap. J. Phycol. 26: 161–164.

Solorzano, L., 1969. Determination of ammonia in natural waters by the phenolhypochlorite method. Limnol. Oceanogr. 14: 799–801.

Strickland, J. D. H. & T. R. Parsons, 1968. A practical handbook of seawater analysis. Bull. Fish. Res. Bd Can. 167: Section II.2.1.

Tide Tables Malaysia, 1995. Hydrographic Directorate and Royal Malaysian Navy. Vol. I.

Tide Tables Malaysia, 1996. Hydrographic Directorate and Royal Malaysian Navy. Vol. I.

Trono, G. C. & A. O. Lluisma, 1990. Seasonality of standing crop of a *Sargassum* (Fucales, Phaeophyta) bed in Balinao, Pangasinan, Philippines. Hydrobiologia 204/205: 331–338.

Tsuda, R. T., 1972. Morphological, zonation and seasonal studies of two species of *Sargassum* on the reef flats of Guam. Proc. Int. Seaweed Symp. 7: 40–44.

Wong, C. L., 1997. Phenological Studies of Two Species of *Sargassum* (Sargassaceae, Phaeophyta) on the Coral Reef Flats at Cape Rachado, Peninsular Malaysia. Master of Philosophy thesis, University of Malaya: 188 pp.

Hydrobiologia **512**: 89–95, 2004.
P.O. Ang, Jr. (ed.), Asian Pacific Phycology in the 21st Century: Prospects and Challenges.
© 2004 *Kluwer Academic Publishers.*

Cyanobacteria-dominated biofilms: a high quality food resource for intertidal grazers

Sanjay Nagarkar[1], Gray A. Williams[1,*], G. Subramanian[2] & S. K. Saha[2]
[1]*Department of Ecology & Biodiversity and The Swire Institute of Marine Science, The University of Hong Kong, Hong Kong, SAR, China*
[2]*National Facility for Marine Cyanobacteria, Bharathidasan University, Tiruchirapalli, India*
Author for correspondence; E-mail: snagarka@hkusua.hku.hk

Key words: cyanobacteria, biofilm, nutritional value, protein, carbohydrate, calorific value, tropical rocky shore

Abstract

Hong Kong rocky shores are dominated by cyanobacterial biofilms composed of a diversity of species. Thirteen common species, belonging to seven genera, were isolated in pure culture in MN+ and MN− media under defined growth conditions from a semi-exposed shore in Hong Kong. The nutritional values (i.e., protein, carbohydrate and calorific value) of these 13 species were determined. All species showed high nutritional quality in terms of protein, carbohydrate and calorific value, however, overall nutritional value varied between the species. Species of *Spirulina* and *Phormidium* were most nutritious (highest nutritional values) whereas species of *Calothrix* and *Lyngbya* were the least nutritious. Microphagous molluscan grazer density and diversity were relatively high at the study site, despite the seemingly low biomass (as assessed by chlorophyll *a* concentration) of the biofilm. It is suggested that the high nutritional quality of cyanobacteria, together with their fast turnover rates can support high levels of secondary production (biomass of grazers). The high nutritional quality of cyanobacteria on tropical, cyanobacteria-dominated, rocky shores is therefore of great importance in the benthic food web.

Introduction

Intertidal, epilithic biofilms are 3-dimensional structures, mainly composed of bacteria, cyanobacteria, diatoms, microalgae, protozoa and spores and sporelings of macroalgae embedded in a mucopolysaccharide matrix (Anderson, 1995). Rocky shores around the world are typically covered with these biofilms throughout the year. Temperate rocky shore biofilms are dominated by diatoms or cyanobacteria (MacLulich, 1987; Hill & Hawkins, 1991; Thompson et al., 1996). On tropical rocky shores, in contrast, the biofilms are mainly composed of cyanobacteria with only sporadic patches of bacteria, diatoms, protozoa and juvenile stages of macroalgae (Potts, 1980; Whitton & Potts, 1982; Nagarkar & Williams, 1999). Cyanobacteria are important primary producers, many species of which are able to fix atmospheric nitrogen (Stewart, 1973; Whitton & Potts, 1982). Cyanobacterial biofilms, therefore, form the energy base of the benthic food web and are very important in terms of overall productivity and community organization on tropical rocky shores (Nagarkar, 1996).

Although the importance of cyanobacteria as a food source for microphagous grazers such as intertidal molluscs (Foster, 1964; Raffaelli, 1985; Quinn, 1988) and zooplankton (Schmidt & Jónasdóttir, 1997; Repka, 1998) has long been realized, little attention has been paid to the nutritional quality of cyanobacteria (but see Ahlgren et al., 1992; Kaehler & Kennish, 1996). Many studies have found planktonic and benthic cyanobacteria to be a poor food source due to their nutritional inadequacy and toxicity (Lampert, 1987; Ahlgren et al., 1990; Thacker et al., 1997), but no such information is available for intertidal, epilithic cyanobacteria (but see Nicotri, 1977). The main reason for this nutritional inadequacy of cyanobacteria is the absence of long-chain, polyunsaturated, fatty acids (PUFA), which are known to be essential components of zooplankton diets and are an indicator of high nutritional quality (Jónasdóttir & Kiørboe, 1996). In contrast, no information is available on the nutritional requirements of intertidal molluscs' diet. The importance of cyanobacteria as a high quality food source, however, is mainly based on their protein content and the presence of essential amino acids, which has become a focus of biotechnological exploitation of cyanobacteria (Venkataraman, 1993). Chemical screening of many laboratory grown, commercially viable, marine cyanobacteria has revealed that they have a high nutritional value, especially in terms of protein (Venkataraman, 1993). The nutritional value of ecologically significant cyanobacteria has, however, received little attention.

Hong Kong rocky shores are dominated by cyanobacterial biofilms throughout the year (Nagarkar & Williams, 1999). During winter, macroalgal growth is common on Hong Kong shores (Kaehler & Williams, 1996). Several rocky shores, however, remain free of macroalgae and are only covered with a cyanobacterial biofilm (Nagarkar, pers. obs.). On visual observation, these shores appear bare, but chlorophyll *a* and microscopic analyses reveal the presence of these cyanobacteria-dominated biofilms (Williams, 1994; Nagarkar & Williams, 1997). Cyanobacteria are the only available food source on these shores which support a wide variety of microphagous grazers (Williams, 1993; Nagarkar, 1996). No information is available, however, on the nutritional quality of these cyanobacteria (except for one species, *Kyrtuthrix maculans*, Umezaki, 1996; Kaehler & Kennish, 1996) to their molluscan grazers. This paper, therefore, presents preliminary work on nutritional quality (i.e., protein, carbohydrate and calorific value) of cyanobacteria present naturally in the intertidal, epilithic biofilm on rocky shores in Hong Kong.

Materials and methods

Study site

The study site was an ∼100 m long, gently slopping granodiorite heterogeneous rock platform, with few crevices, on a semi-exposed rocky shore at Mo Tat Wan (22° 13′ N, 114° 10′ E), Lamma Island, Hong Kong. This site was located in a remote area far from anthropogenic activities and was relatively clean compared to other polluted shores. Absence of visually conspicuous macroalgae and thick biofilm at this site resulted in a bare appearance of the shore. This site, however, supports a wide variety and high density of intertidal grazers such as limpets, *Cellana grata* Gould, *Cellana toreuma* Reeve, the chiton *Acanthopleura japonica* Lischke and topshell *Monodonta labio* Linné (Nagarkar, pers. obs.).

Isolation and purification of cyanobacteria

Samples were collected during March 1999, from the intertidal zone between 1.25 m and 2.00 m above Chart Datum. Rock chips (∼2 cm²) with firmly attached cyanobacterial growth were removed using a hammer and chisel and thin biofilms were scraped from the rock with a single-sided razor blade. All the samples were transferred into plastic vials with a few drops of seawater. In the laboratory, within 6 h, each sample was divided into four parts which were inoculated into liquid MN+, MN−, ASN III+ and ASN III− media (Rippka et al., 1979). To support the growth of nitrogen fixing cyanobacteria, $NaNO_3$ was omitted from the MN− and ASN III− media. In all the media, 50 μg ml^{-1} cycloheximide (antibiotic) was added to inhibit eukaryotic growth especially macroalgal spores and microalgae. Bacterial growth was reduced by adding 25 μg ml^{-1} ampicillin and 15 μg ml^{-1} tetracycline. Samples were incubated at 25 °C in continuous low white fluorescent, diffuse light (<1000 Lux) for about seven days. Later, light intensity was gradually increased to 1200–1500 Lux. After 3 weeks, cyanobacterial growth was observed in the respective media. Individual filaments or cells were isolated using a long neck glass capillary pipette under a dissecting microscope (Leitz Diaplan, U.S.A.) and inoculated into respective fresh liquid MN and ASN III (+ or −) media; incubated again for another 3–4 weeks. Pure cultures were obtained by repeating this procedure several times.

Nutritional quality

Thirteen ecologically common cyanobacterial species which were isolated in pure culture were selected for nutritional analysis (i.e., protein, carbohydrate and calorific value). Qualitative microscopic observations, based on the number of filaments or cells under each microscopic field of view at a fixed magnification (×40), revealed that these 13 cyanobacterial species were abundant at the study site. To investigate the nutritional quality of laboratory cultured cyanobacteria, all the non-heterocystous species (absence of heterocyst, a special cell which performs nitrogen fixation) were maintained in MN+ media and heterocystous species (presence of heterocyst) in MN– media. Cyanobacterial cells were harvested at the log growth phase (rapid cell multiplication stage) by centrifuging at 10 000 rpm for 10 min. The resultant cyanobacterial pellet was rinsed with distilled water several times and blotted on blotting paper, air dried over night, ground into a powder and then extracted for analysis.

Protein estimation

Protein was estimated by the Coomassie Brilliant Blue (Bio-Rad Laboratories) dye-binding method of Bradford (1976). One millilitre of 0.5 M NaOH was added to 20 mg of dried powder of each species, incubated for 10 min in a water bath at 80 °C and then centrifuged at 3000 rpm for 10 min. The supernatant was removed, to which 4 ml of distilled water was added and then 1 ml of the reagent Coomassie Brilliant Blue and finally mixed with a vortex mixer. The absorbence of the sample was measured immediately using a Pye Unicam spectrophotometer at 595 nm. A similar procedure was repeated for the solid residue and the absorbance of the supernatant was again measured at 595 nm. Protein was estimated by comparison with a standard bovine serum-albumin curve (Kochert, 1978).

Carbohydrate estimation

Carbohydrate was estimated using the phenol sulphuric acid method (Dubois et al., 1956). Ten milligrams of sample were mixed with 3 ml of distilled water, added to 50 μl of 90% phenol, mixed and then finally 5 ml of concentrated H_2SO_4 was added onto the surface of the liquid. Treated samples were placed in a water bath at 60 °C for 30 min and then centrifuged at 3000 rpm for 10 min. The absorbance of the supernatant was measured spectrophotometrically at 485 nm. Carbohydrate was measured by comparison with a standard curve prepared using glycogen (Sigma Chemicals).

Calorific value

To estimate calorific value of each cyanobacterial species, dry powder (800 mg) of each species was combusted in a Parr 1261 semi-micro isoperibol bomb calorimeter, calibrated using benzoic acid.

Data analysis

To investigate trends in the nutritional value of the 13 cyanobacterial species, Principal Component Analysis (PCA) was performed on the appropriate correlation matrix (MVSP, Ver. 3.01, 1998, Kovach Computing Services, Wales).

Results

Thriteen cyanobacterial species (*Calothrix contarenii* (Zanard.) Born. *ex* Flah., *Calothrix crustacea* Thuret, *Gloeocapsa crepidinum* Thuret, *Lyngbya martensiana* Menegh. *ex* Gomont, *Lyngbya semiplena* (C. Ag.) J. Ag. *ex* Gomont, *Oscillatoria formosa* Bory ex Gomont, *Oscillatoria salina* Biswas, *Oscillatoria subbrevis* Schmidle, *Phormidium corium* (Ag.) Gomont, *Phormidium tenue* (Menegh.) Gomont, *Spirulina labyrinthiformis* (Menegh.) Gomont, *Spirulina subsalsa* Oerst. *ex* Gomont and *Synechococcus* sp.) were isolated in pure culture. Of the 13 species, only two were unicellular forms (spherical or ellipsoidal shape, i.e., *G. crepidinum* and *Synechococcus* sp.) and of the remaining 11 filamentous forms (unbranched filament shape), two were heterocystous (i.e., *C. crustacea* and *C. contarenii*) and the other nine were non-heterocystous species (Table 1).

Protein

Most of the cyanobacteria species contained a high percentage of protein (total dry weight). Protein values ranged from 18.9% for *L. martensiana* to a maximum of 70.8% for *S. subsalsa* (Table 1). Cyanobacteria belonging to *Lyngbya* and *Calothrix* genera contained the lowest protein concentrations (<28%) as compared to other species which contained >40% protein (Table 1).

Carbohydrate

Carbohydrate values ranged from a minimum of 5.4% for *L. martensiana* to a maximum of 16.6% of the total dry weight for *S. subsalsa* (Table 1). Cyanobacterial species belonging to *Phormidium* and *Spirulina* genera contained the highest carbohydrate concentrations (>14%) as compared to species belonging to other genera (<11.5%; Table 1).

Table 1. Mean (± S.D) protein, carbohydrate and calorific values of 13 cyanobacterial species isolated from a Hong Kong rocky shore ($n = 3$). (HF) Heterocystous filamentous cyanobacteria; (F) Non-heterocystous filamentous cyanobacteria; (C) Unicellular cyanobacteria

Species	Protein (% DW)	Carbohydrate (% DW)	Calorific value (kJ 10 g^{-1} DW)
Calothrix crustacea (HF)	21.50 ± 0.40	7.60 ± 0.50	25.16 ± 0.30
Calothrix contarenii (HF)	27.43 ± 0.47	8.23 ± 0.65	29.00 ± 0.45
Gloeocapsa crepidinum (C)	56.46 ± 0.25	7.63 ± 0.55	20.53 ± 0.55
Lyngbya martensiana (F)	18.86 ± 0.65	5.43 ± 0.41	21.73 ± 0.56
Lyngbya semiplena (F)	27.50 ± 0.45	8.93 ± 0.15	23.30 ± 0.36
Phormidium corium (F)	49.56 ± 0.55	16.46 ± 0.45	32.56 ± 0.41
Phormidium tenue (F)	62.96 ± 0.55	15.46 ± 0.40	31.33 ± 0.20
Spirulina subsalsa (F)	70.76 ± 0.90	16.63 ± 0.56	34.83 ± 0.20
Spirulina labyrinthiformis (F)	68.03 ± 0.85	14.73 ± 0.66	34.16 ± 0.37
Synechococcus sp. (C)	63.56 ± 0.60	8.56 ± 0.56	27.60 ± 0.45
Oscillatoria formosa (F)	50.85 ± 0.79	9.46 ± 0.45	15.30 ± 0.36
Oscillatoria salina (F)	41.80 ± 0.81	11.20 ± 0.36	19.30 ± 0.20
Oscillatoria subbrevis (F)	45.16 ± 0.41	11.53 ± 0.68	21.43 ± 0.30

Calorific values

Calorific values for cyanobacteria ranged from 15.30 kJ 10 g^{-1} DW for *O. formosa* to a maximum of 34.8 kJ 10 g^{-1} DW for *S. subsalsa* (Table 1). Species belonging to *Phormidium* and *Spirulina* genera had the highest calorific values (>31.3 kJ 10 g^{-1} DW) as compared to species belonging to other genera (<29 kJ 10 g^{-1} DW).

Overall nutritional value

Principal Component Analysis revealed a clear separation of three groups of species on axis 1 (Principal Component 1) which accounted for 90.11% of the total variance and axis 2 accounted only for 8.67% of the variance (Fig. 1). The first Principal Component has a high positive loading for protein and low positive loading for carbohydrate and calorific value (Fig. 1). Species on the positive side of axis one (a group containing *S. labyrinthiformis*, *S. subsalsa* and *P. tenue*), have the highest nutritional values, whereas species (*C. contarenii*, *C. crustacea*, *L. martensiana* and *L. semiplena*) on the negative side have the lowest nutritional values whilst the remaining species (*G. crepidinum*, *O. formosa*, *O. salina*, *O. subbrevis* and *P. corium*) fall in between these extremes (Fig. 1).

Discussion

A great diversity of cyanobacterial species, belonging to various morphological and functional groups, have been reported from various rocky shores around Hong Kong (Nagarkar, 1998a,b). In the present study, 13 species belonging to seven genera were isolated in pure culture from one shore, Mo Tat Wan. These 13 species were the most common cyanobacteria found in the biofilm on the study site. Most of these species are abundant components of the intertidal, epilithic biofilm in Hong Kong (Nagarkar & Williams, 1999). Similar species have also been reported on rocky shores from other geographical locations (Japan, Umezaki, 1961; Red Sea, Sinai Peninsula, Potts, 1980; India, Thajuddin & Subramanian, 1992; Red Sea, Saudi Arabia, Hussain & Khoja, 1993).

The nutritional value of a species is known to be influenced by culture media (Ben-Amotz et al., 1985), cell harvesting stage (Whyte, 1987), temperature (James et al., 1989), light intensity (Thompson et al., 1990) and pH (James et al., 1989). To compare the nutritional value of these cyanobacteria, all 13 species were grown under similar laboratory conditions and harvested at the same growth phase. Two different culture media were, however, used to support nitrogen fixing (i.e., *Calothrix* spp.) and non-nitrogen fixing cyanobacteria. Differences in protein, carbohydrate and calorific values between *Calothrix* and the

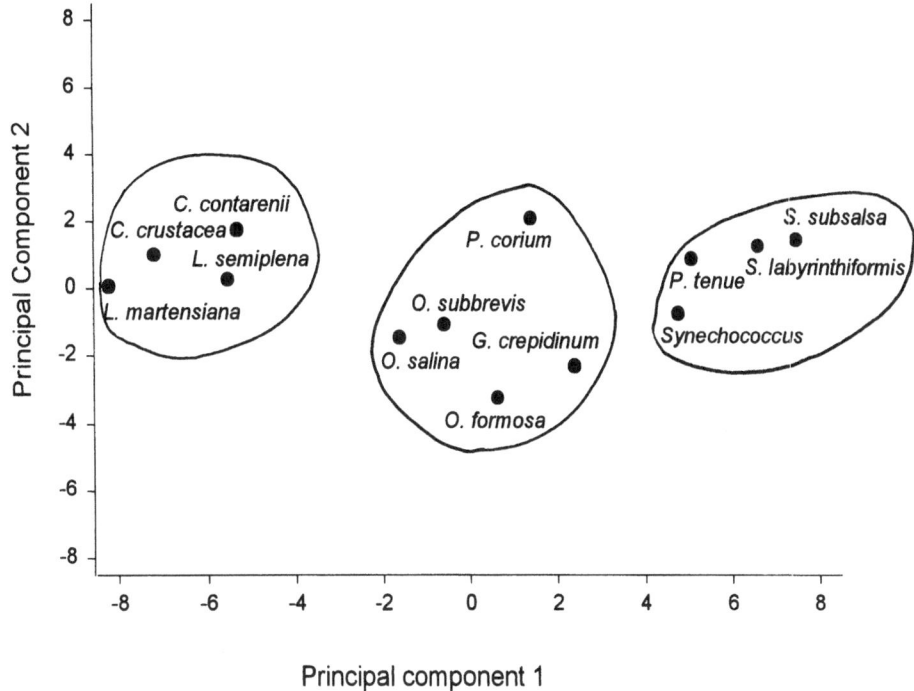

Figure 1. Principal Component Analysis (PCA) based on nutritional values (i.e., protein, carbohydrate and calorific value) of 13 isolated cyanobacterial species from a Hong Kong rocky shore.

other species may, therefore, be partially influenced by the culture media and hence this comparison should be interpreted with caution.

All the cyanobacterial species isolated from Hong Kong rocky shores showed high nutritional quality. Protein content was usually high and contributed up to 70.8% of the total dry weight. *Calothrix crustacea* had the lowest percentage of protein whereas *S. subsalsa* had the highest value. The protein content of Hong Kong species was similar to values for planktonic cyanobacteria (30–40%, Ahlgren et al., 1992) and those isolated in pure culture for biotechnology purposes from various sources such as freshwater, terrestrial or marine environments (65–71%, Venkataraman, 1993; Subramanian, 1998; Sujatha & Kaushik, 1998). Previously, *K. maculans* (one of the dominant encrusting cyanobacteria on tropical rocky shores) was, however, reported as a poor source of protein (7.0%, Kaehler & Kennish, 1996). The protein content of most of the cyanobacteria (>41%) in the present study was in contrast, much higher than planktonic microalgae (12–35%) used in mariculture except for species of *Calothrix* and *Lyngbya* which contained protein (21.5–27.5%) within the range of protein values of microalgae (Brown, 1991).

Carbohydrate values ranged from 5.4 to 16.6% of the total dry weight. Most of the isolates showed <11.5% carbohydrate, except species belonging to *Phormidium* and *Spirulina* genera. The maximum carbohydrate recorded in the present study was 16.6% for *S. subsalsa*. Carbohydrate values previously reported for *K. maculans* (>30%, Kaehler & Kennish, 1996) were much higher than the values recorded for all the 13 cyanobacterial species in the present study. Cyanobacterial species cultured for biotechnological purposes (Venkataraman, 1993) and microalgal species used in mariculture (Brown, 1991), however, contain a similar range of carbohydrate levels (6.0–16.0%) as reported in the present study.

The calorific values of the 13 cyanobacteria species ranged from 15.3 to 34.8 kJ 10 g^{-1} DW which fall within the range of values previously recorded (Venkataraman, 1993; Kaehler & Kennish, 1996). The calorific value of *K. maculans*, for example, was 17.5 kJ 10 g^{-1} DW (Kaehler & Kennish, 1996). In the present study, *G. crepidinum*, *O. formosa* and *O. salina* had calorific values (15.3–20.5 kJ 10 g^{-1} DW) close to that of *K. maculans*. The calorific values of many commercially viable species such as *Spirulina*, however, fall towards the higher side of calorific values recorded in the present study. Most of the *Spirulina* species, for example, have calorific values between 30 and 36 kJ 10 g^{-1} DW (Venkataraman, 1993). In the present study, only *P. corium*, *P. tenue*, *S. subsalsa* and *S. labyrinthiformis* had calorific values between 31.3 and 34.8 kJ 10 g^{-1} DW.

When considering protein, carbohydrate and calorific values together, *P. tenue*, *S. labyrinthiformis* and *S. subsalsa* contained the greatest concentration as compared to the other species. These results suggest that due to their high protein content and calorific values, and in some cases high carbohydrate content, cyanobacterial species are nutritionally superior to other micro- and macroalgal food resources available on rocky shores (Dawes et al., 1974; McQuaid, 1985; Kaehler & Kennish, 1996). The importance of intertidal, epilithic cyanobacteria as a high quality food source, however, should be interpreted with caution because the present comparison is made on the basis of nutritional values of laboratory cultured cyanobacteria which may vary from the field situation. Although cyanobacteria contained high nutritional values, nutritional adequacy of cyanobacteria for intertidal grazers needs further investigation based on their PUFA and toxicity.

During winter, Hong Kong has favourable conditions for macroalgal growth (Kaehler & Williams, 1996) and shores can support a dense cover of filamentous cyanobacteria, macroalgae and diatoms (Kaehler & Williams, 1996; Nagarkar & Williams, 1999) and chlorophyll *a* values can reach 40 μg cm^{-2} (Nagarkar & Williams, 1999). The study site at Mo Tat Wan did not support macroalgae in March 1999 and had very low, patchy, chlorophyll *a* values (2.6 μg cm^{-2} ± 1.9 S.D., $n = 25$) resulting in the shore being visually bare (Nagarkar, pers. obs.). This patchy, sparse biofilm might appear insufficient as a food supply to support the wide variety and high density of intertidal grazers which are present at this site. The present results, however, show that all the cyanobacterial species were highly nutritious, especially in terms of protein, although these species were not tested for their nutritional adequacy and toxicity. Since cyanobacteria were the only food source available on the study site, this suggests the low biomass of the biofilm together with the known high production rate of cyanobacteria was probably able to supply enough nutrition to support the large population of grazers at this site (Nagarkar, 1996). Cyanobacteria are important primary producers and often form the energy base of the benthic food web on tropical rocky shores, therefore, the nutritional value of these species could play an important role in the energetics of intertidal coastal ecosystems.

Acknowledgements

We would like to thank Dr Richard Corlett and Mr A. D. Weerasooriya (HKU) for their advice on statistics. This work was supported by a RGC grant (RGC project no. HKU 7231/98M) to Dr G. A. Williams. Mr S. K. Saha was supported by a Department of Biotechnology, Government of India, fellowship.

References

Ahlgren, G., I. B. Gustafsson & M. Boberg, 1992. Fatty acid content and chemical composition of freshwater microalgae. J. Phycol. 28: 37–50.

Ahlgren, G., L. Lundstedt, M. Brett & C. Forsberg, 1990. Lipid composition and food quality of some freshwater phytoplankton for cladoceran zooplankters. J. Plankton Res. 12: 809–818.

Anderson, M. J., 1995. Variations in biofilms colonizing artificial surfaces: seasonal effects and effects of grazers. J. mar. biol. Ass. U.K. 75: 705–714.

Ben-Amotz, A., T. G. Tornabene & W. H. Thomas, 1985. Chemical profile of selected species of microalgae with emphasis on lipids. J. Phycol. 21: 72–81.

Bradford, M., 1976. A rapid and sensitive method for the quantitation of microgram quantities of protein utilizing the principle of protein-dye binding. Analyt. Biochem. 72: 248–254.

Brown, M., 1991. The amino-acid and sugar composition of 16 species of microalgae used in mariculture. J. exp. mar. Biol. Ecol. 145: 79–99.

Dawes, C. J., J. M. Lawrence, D. P. Cheney & A. C. Mathieson, 1974. Ecological studies of floridian *Eucheuma* (Rhodophyta, Gigartinales). III. Seasonal variation of carrageenan, total carbohydrate, protein and lipid. Bull. mar. Sci. 24: 286–299.

Dubois, M., K. A. Gilles, J. K. Hamilton, P. A. Rebers & F. Smith, 1956. Colorimetric method for determination of sugars and related substances. Analyt. Chem. 28: 350–356.

Foster, M. S., 1964. Microscopic algal food of *Littorina planaxis* Philippi and *Littorina scutulata* Gould (Gastropoda: Prosobranchia). Veliger 7: 149–152.

Hill, A. S. & S. J. Hawkins, 1991. Seasonal and spatial variation of epilithic microalgal distribution and abundance and its ingestion by *Patella vulgata* on a moderately exposed rocky shore. J. mar. biol. Ass. U.K. 71: 403–423.

Hussain, M. I. & T. M. Khoja, 1993. Intertidal and subtidal blue-green algal mats of open and mangrove areas in the Farasan Archipelago (Saudi Arabia), Red Sea. Bot. mar. 36: 377–388.

James, C. M., S. Al-Hinty & A. E. Salman, 1989. Growth and ω3 fatty acid and amino acid composition of microalgae under different temperature regimes. Aquaculture 77: 337–357.

Jónasdóttir, S. H. & T. Kiørboe, 1996. Copepod recruitment and food composition: do diatoms affect hatching success? Mar. Biol. 125: 743–750.

Kaehler, S. & G. A. Williams, 1996. Distribution of algae on tropical rocky shores: spatial and temporal patterns of non-coralline encrusting algae in Hong Kong. Mar. Biol. 125: 177–187.

Kaehler, S. & R. Kennish, 1996. Summer and winter comparisons in the nutritional value of marine macroalgae from Hong Kong. Bot. mar. 39: 11–17.

Kochert, G., 1978. Protein determination by dye binding. In Hellebust, J. A. & J. S. Craigie (eds), Handbook of Phycological Methods. Physiological and Biochemical Methods. Cambridge University Press, Cambridge: 91–93.

Lampert, W., 1987. Laboratory studies on zooplankton-cyanobacteria interactions. Nz. J mar. Freshwat. Res. 21: 483–490.

MacLulich, J. H., 1987. Variation in the density and variety of intertidal epilithic microflora. Mar. Ecol. Prog. Ser. 40: 285–293.

McQuaid, C. D., 1985. Seasonal variation in ash-free calorific value of nine intertidal macroalgae. Bot. mar. 28: 545–548.

Nagarkar, S., 1996. The Ecology of Intertidal, Epilithic Biofilms with Special Reference to Cyanobacteria. Unpublished Ph.D. thesis. The University of Hong Kong, Hong Kong.

Nagarkar, S., 1998a. New records of marine cyanobacteria from rocky shores of Hong Kong. Bot. mar. 41: 527–542.

Nagarkar, S., 1998b. New records of coccoid cyanobacteria from Hong Kong rocky shores. Asian mar. Biol. 15: 119–125.

Nagarkar, S. & G. A. Williams, 1997. Comparative techniques to quantify cyanobacteria dominated epilithic biofilms on tropical rocky shores. Mar. Ecol. Prog. Ser. 154: 281–291.

Nagarkar, S. & G. A. Williams, 1999. Spatial and temporal variation of cyanobacteria-dominated epilithic communities on a tropical rocky shore in Hong Kong. Phycologia 38: 385–393.

Nicotri, M. E., 1977. Grazing effects of four marine intertidal herbivores on the microflora. Ecology 58: 1020–1032.

Potts, M., 1980. Blue-green algae (Cyanophyta) in marine coastal environments of the Sinai Peninsula; distribution, zonation, stratification and taxonomic diversity. Phycologia 19: 60–73.

Quinn, G. P., 1988. Ecology of the intertidal pulmonate limpet *Siphonaria diemenensis* Quoy et Gaimard. I. Population dynamics and availability of food. J. exp. mar. Biol. Ecol. 117: 115–136.

Raffaelli, D., 1985. Functional feeding groups of some intertidal molluscs defined by gut content analysis. J. moll. Stud. 51: 233–239.

Repka, S., 1998. Effects of food type on the life history of *Daphnia* clones from lakes differing in trophic state. II. *Daphnia cucullata* feeding on mixed diets. Freshwat. Biol. 38: 685–692.

Rippka, R., J. Deruelles, J. B. Waterbury, M. Herdman & R. Y. Stanier, 1979. Generic assignments, strain histories and properties of pure cultures of cyanobacteria. J. gen. Microbiol. 111: 1–61.

Schmidt, K. & S. H. Jónasdóttir, 1997. Nutritional quality of two cyanobacteria: how rich is 'poor' food? Mar. Ecol. Prog. Ser. 151: 1–10.

Stewart, W. D. P., 1973. Nitrogen fixation. In Carr, N. G. & B. A. Whitton (eds), The Biology of Blue-green Algae. Botanical Monographs 9, Blackwell Scientific Publications, Oxford: 260–278.

Subramanian, G., 1998. Marine cyanobacteria for feed, fine chemicals and pharmaceuticals. In Subramanian, G., B. D. Kaushik & G. S. Venkataraman (eds), Cyanobacterial Biotechnology. Science Publishers, Inc., U.S.A: 281–285.

Sujatha, V. S. & B. D. Kaushik, 1998. Cellular characteristics of a marine *Nostoc calcicola* BDU 40302. In Subramanian, G., B. D. Kaushik & G. S. Venkataraman (eds), Cyanobacterial Biotechnology. Science Publishers, Inc., U.S.A: 91–97.

Thacker, R. W., D. G. Nagle & V. J. Paul, 1997. Effects of repeated exposures to marine cyanobacterial secondary metabolites on feeding by juvenile rabbitfish and parrotfish. Mar. Ecol. Prog. Ser. 147: 21–29.

Thajuddin, N. & G. Subramanian, 1992. Survey of cyanobacterial flora of the southern east coast of India. Bot. mar. 35: 305–314.

Thompson, P. A., P. J. Harrison & J. N. C. Whyte, 1990. Influence of irradiance on the fatty acid composition of phytoplankton. J. Phycol. 26: 278–288.

Thompson, R. C., B. J. Wilson, M. L. Tobin, A. S. Hill & S. J. Hawkins, 1996. Biologically generated habitat provision and diversity of rocky shore organisms at a hierarchy of spatial scales. J. exp. mar. Biol. Ecol. 202: 73–84.

Umezaki, I., 1961. The Marine Blue-green Algae of Japan. In Memoirs of the College of Agriculture Kyoto University, Kyoto University, Kyoto, Japan 83: 1–149.

Venkataraman, L. V., 1993. *Spirulina* in India. In Subramanian, G. (ed.), Proceedings of the National Seminar on Cyanobacterial Research – Indian Scene. NFMC, BARD, Tiruchirapalli, India: 92–116.

Whitton, B. A. & M. Potts, 1982. Marine Littorals. In Carr, N. G. & B. A. Whitton (eds), The Biology of Cyanobacteria. Blackwell Scientific Publications, Oxford: 515–542.

Whyte, J. N. C., 1987. Biochemical composition and energy content of six species of phytoplankton used in mariculture of bivalves. Aquaculture 60: 231–241.

Williams, G. A., 1993. Seasonal variation in algal species richness and abundance in the presence of molluscan herbivores on a tropical rocky shore. J. exp. mar. Biol. Ecol. 167: 261–275.

Williams, G. A., 1994. Grazing by high-shore littorinids on a moderately exposed tropical rocky shore. In Morton, B. S. (ed.), Proceedings of the Third International Workshop on the Malacofauna of Hong Kong and Southern China. Hong Kong University Press, Hong Kong: 379–389.

Hydrobiologia **512**: 97–108, 2004.
P.O. Ang, Jr. (ed.), Asian Pacific Phycology in the 21st Century: Prospects and Challenges.

Early stages of biofilm succession in a lentic freshwater environment

R. Sekar[1,2,*], V. P. Venugopalan, K. Nandakumar, K. V. K. Nair & V. N. R. Rao[1]
Water and Steam Chemistry Laboratory, Bhabha Atomic Research Centre Facilities, Kalpakkam 603 102, Tamil Nadu, India
[1] *Centre of Advanced Study in Botany, University of Madras, Guindy Campus, Chennai 600 025, India*
[2] *Present address: Dept of Molecular Ecology, Max-Planck Institute for Marine Microbiology, Celsiusstrasse 1, 28359 Bremen, Germany*
Author for correspondence; Tel: 0049-421-2028-545. Fax: 0049-421-2028-580. E-mail: sraju@mpi-bremen.de

Key words: colonization, succession, microalgae, biofilms, green algae, diatoms, cyanobacteria

Abstract

Initial events of biofilms development and succession were studied in a freshwater environment at Kalpakkam, East Coast of India. Biofilms were developed by suspending Perspex (Plexiglass) panels for 15 days at bimonthly intervals from January 1996 to January 1997. Changes in biofilm thickness, biomass, algal density, chlorophyll *a* concentration and species composition were monitored. The biofilm thickness, biomass, algal density and chlorophyll *a* concentration increased with biofilms age and colonization was greater during summer (March, May and July) than other months. The initial colonization was mainly composed of *Chlorella vulgaris, Chlorococcum humicolo* (green algae), *Achnanthes minutissima, Cocconeis scutellum, C. placentula* (diatoms) and *Chroococcus minutus* (cyanobacteria) followed by colonial green algae such as *Pediastrum tetras, P. boryanum* and *Coleochaete scutata*, cyanobacteria (*Gloeocapsa nigrescens*), low profile diatoms (*Amphora coffeaeformis, Nitzschia amphibia,* and *Gomphonema parvulum*) and long stalked diatoms (*Gomphoneis olivaceum* and *Gomphonema lanceolatum*). After the 10th day, the community consisted of filamentous green algae (*Klebshormidium subtile, Oedogonium* sp., *Stigeoclonium tenue* and *Ulothrix zonata*) and cyanobacteria (*Calothrix elenkinii, Oscillatoria tenuis* and *Phormidium tenue*). Based on the percentage composition of different groups in the biofilm, three phases of succession could be identified: the first phase was dominated by green algae, the second by diatoms and the third phase by cyanobacteria. Seasonal variation in species composition was observed but the sequence of colonization was similar throughout the study period.

Introduction

Microbial colonization on hard surfaces is a common phenomenon in natural aquatic environments which has both ecological and industrial significance (Ford et al., 1989). Submerged surfaces, including surfaces coated with toxic paints, are readily colonized by bacteria and microalgae (Callow, 1986; Cooksey & Cooksey, 1995), which cause problems to ship surfaces, cooling systems and other marine-based industries (Pederson, 1990; Hudson & Burke, 1994; Udayakumar et al., 1998). The formation of a primary biofilm over surfaces favours subsequent colonization by other organisms and facilitates corrosion. Microal-

gae are among the major components in the freshwater biofilms (Callow, 1993).

Microalgal colonization has been studied in different aquatic environments using various natural and artificial substrata (Brown, 1976; Hoagland et al., 1982; Oemke & Burton, 1986; Acs & Kiss, 1993; Lowe et al., 1996). A three dimensional microalgal succession was observed in biofilms by earlier workers and it was reported that microalgal succession is analagous to higher plant succession in terrestial environments (Hudon & Bourget, 1981; Hoagland et al., 1982; Korte & Blinn, 1983; Roemer et al.,1984). Among the diatoms, succession process has been found to be influenced by water velocity, size, immigration and reproduction rate of the organisms (Oemke & Burton, 1986; Steinman & McIntire, 1986; Stevenson & Peterson, 1989; Acs & Kiss, 1993; Johnson et al., 1997).

Most of the earlier studies on microalgal colonization on artificial substrata in freshwater environments have been focused on diatoms. Studies representing the complete microalgal assemblages are limited. Moreover, the previous succession studies have been carried out in lotic systems where water movement is likely to influence colonization. In the present work, we have studied the early events of colonization and succession of the biofilm in an undisturbed lentic freshwater system, taking into account all the biofilm components such as green algae, diatoms and cyanobacteria, to find out critical changes in microalgal colonization and succession, which take place during the early development of biofilm.

Materials and methods

Site description

This study was conducted in an open freshwater reservoir located at Kalpakkam (22° 33′ N, 80° 11′ E), 60 km south of Madras, on the East Coast of India. The reservoir serves as the source of cooling water for a power reactor located in the same campus, and is 1.7 hectares in area, with a maximum depth of 2.3 m at the overflow level. The reservoir receives water from a sub soil river bed. The physico-chemical and characteristics of the reservoir water during the study period was analyzed as per standard methods (APHA, 1989).

Panel preparation and immersion

The colonization and succession of the biofilm were studied at bimonthly intervals for the period of one year from January 1996 to January 1997 by suspending clean Perspex (Plexiglass) test panels of two sizes, namely small (7×3×0.3 cm) for microscopic observation and large (15×10×0.3 cm) for biomass and chlorophyll measurements. The panels were fastened on to a stainless steel frame and suspended vertically 0.5 m from the water surface.

Panel analysis

Duplicate panels of both sizes were retrieved periodically (i.e. after 1, 2, 3, 4, 5, 7, 10 and 15 days of exposure). The panels were rinsed with filtered reservoir water to remove loosely attached planktonic forms. The small panels were used for the measurement of biofilm thickness and for studying qualitative and quantitative distribution of biofilm species. After wiping one side (randomly chosen) of the small panels, the biofilm on the other side was observed directly under a Nikon Ophitiphot microscope. Biofilm thickness was measured by the method of Bakke & Olsson (1986). In this method, the distance travelled by the microscope stage (read off the stage micrometer) is measured while changing the focus from the base (biofilm substratum interface) to top of the biofilm. Thickness measurements were taken from 10 random fields and averaged to get mean wet film thickness. Algal density analyses were made at 200× or 400× magnification; a higher magnification was used for species identification, when required. The algal density and species composition were analyzed following the method of Brown (1976). As per this method, single cells, colonies and filamentous forms (as the case may be) were scored as individuals (Sekar et al., 1998). The algal species were identified using standard manuals (Hustedt, 1930; Desikachary, 1959; Philipose, 1967; Prescott, 1978). The total algal density was expressed as organisms cm^{-2}.

The biofilm on the larger panels was scraped with a soft, sterile nylon brush and made up to a known volume using distilled water. The samples were filtered through preweighed 0.45 μm Whatman GF/C filters and the filters were kept in Furnace at 450 °C for a minimum 6 h and again they were weighed and the biomass was expressed as ash free dry weight (APHA, 1989). Aliquot samples were also filtered through 0.45 μm Millipore filters and the chlorophyll was cold extracted with 90% acetone for 4–6 h in dark. After the complete extraction, the absorbance of clear supernatant was read at 750, 665, 664 and 630 nm using Spectrophotometer (Jeffrey & Humphrey, 1975).

Data analysis

The variation in biofilm thickness, algal density, biomass and chlorophyll a with respect to exposure time in days and between months were compared using one way and two way ANOVA (Sokal & Rohlf, 1987) after the data were log transformed. The correlation between algal density and biofilm thickness, biomass and chlorophyll a was calculated using Pearsons correlation test. Species diversity was calculated using the Shannon–Wiener index (H′) (Odum, 1971) and dominance index (D) (Margalef, 1958). Evenness of

Table 1. Water quality parameters (range) as measured during the study period (January 1996–January 1997)

Parameters	Range
Water temperature °C	31.0–33.1
pH	8.0–8.9
Conductivity (μS cm^{-1})	249–440
Total suspended solids (mg l^{-1})	15–20
Secchi disc transparency (m)	2.2–2.3
Water movement*	Negligible
Dissolved Oxygen (mg l^{-1})	7.2–9.2
Total alkalinity (mg CaCO$_3$ l^{-1})	70–128
Total hardness (mg CaCO$_3$ l^{-1})	58–104
Chloride (mg l^{-1})	26–58
Nitrate-N (μg l^{-1})	15–325
Nitrite-N (μg l^{-1})	1.5–8.8
Phosphate-P (μg l^{-1})	4–44
Silicate-Si (mg l^{-1})	1.5–4.2
Chlorophyll *a* (μg l^{-1})	9.1–32.6

*Measured with drift buoy.

species distribution was calculated using the evenness index of Pielou (1966).

Results

Biofilm characteristics

The data on limnological parameters during January 1996–January 1997 are given in Table 1. The changes in biofilm thickness, density, biomass and chlorophyll *a* in biofilms during the same period are given in Figure 1(a–d). Biofilm thickness varied significantly with exposure time (one way ANOVA, $F_{25.63} = 14$; $p < 0.001$), however the variation between months or the interaction between exposure time and months (two way ANOVA, $F_{0.83} = 28$; $p > 0.05$) were not significant. Thickness of the biofilm achieved a maximum during May 1996 and was minimum during January 1996. Biofilm thickness was significantly correlated with algal density ($r^2 = 0.582$; $p < 0.05$).

Algal density showed significant variations with exposure time (one way ANOVA, $F_{24.71} = 14$; $p < 0.001$) and between months (one way ANOVA, $F_{75.22} = 2$; $p < 0.001$). The interaction between exposure time and months was also significant (two way ANOVA, $F_{2.84} = 28$; $p < 0.001$). Density was highest during May 1996, followed by March and July 1996 and was lowest during September, November

and January 1996 and 1997. The algal density was correlated significantly with biomass ($r^2 = 0.546$; $p < 0.05$) and chlorophyll *a* ($r^2 = 0.768$; $p < 0.001$).

The biofilm biomass varied significantly with biofilm age (one way ANOVA, $F_{44.84} = 14$; $p < 0.001$) and between months (one way ANOVA, $F_{50.39} = 2$; $p < 0.001$), but the interaction between exposure time and months was not significant (two way ANOVA, $F_{1.54} = 28$; $p > 0.05$). The biomass was maximal during May 1996 followed by March and July 1996 and was minimal during September and November 1996. The biomass was significantly correlated with algal density ($r^2 = 0.55$; $p < 0.05$) and chlorophyll *a* ($r^2 = 0.934$; $p < 0.001$).

Chlorophyll *a* also showed similar results as biomass and algal density. The maximum concentration of chlorophyll was observed during May 1996 and concentration varied significantly with exposure time (one way ANOVA, $F_{55.9} = 14$; $p < 0.001$) and months (one way ANOVA, $F_{64.6} = 2$; $p < 0.001$). The interaction between exposure time and months was also significant (two way ANOVA, $F_{2.68} = 28$; $p < 0.05$).

Species composition and sequence of colonization/succession

The species observed in the biofilms are given in Table 2. A total of 108 species could be identified in the biofilms, comprising 20 genera and 38 species of Chlorophyceae, 22 genera and 41 species of Bacillariophyceae, 17 genera and 26 species of Cyanobacteria, 2 genera of Chrysophyceae and 1 genus of Dinophyceae. Chlorophyceae, Bacillariophyceae and Cyanobacteria were the dominant groups of organisms found on the substratum.

Ankistrodesmus convolutus, *A. falcatus*, *Chlorococcum humicolo*, *Chlorella vulgaris*, *Closterium* sp., *Cosmarium* spp., *Scenedesmus* spp. and *Staurastrum enorme* were the major green algal species observed during the initial phase, among which *Chlorella vulgaris*, *Chlorococcum humicolo* were relatively more abundant. *Coleochaete* spp., *Klebshormidium subtile*, *Oedogonium* sp., *Pediastrum boryanum*, *Stigeoclonium tenue* and *Ulothrix zonata* were observed during later phase.

Diatoms such as *Achnanthes minutissima*, *Cocconeis* spp., *Coscinodiscus* sp., *Gomphoneis olivaceum*, *Navicula* spp., *Nitzschia* spp. were observed during the early stages in the biofilms. Among them, *Achnanthes minutissima* and *Cocconeis scutellum* were more abundant. During the later phase,

Table 2. Algal species recorded in the biofilm during the study period

Algal species	Days of panel immersion							
	1	2	3	4	5	7	10	15
Chlorophyceae								
Ankistrodesmus convolutus Corda	+	+	+	−	−	−	−	−
A. falcatus (Corda) Ralfs.	−	−	+	+	−	−	−	−
Asterococcus limneticus G.M. Smith	−	−	−	+	−	−	−	−
Bulbochaete insignis Pringsh	−	−	−	−	+	−	−	−
Chlorella vulgaris Beijerinck	++	+++	+++	+	+	−	−	−
Chlorococcum humicolo (Näg.) Rebenh.	+	++	+++	+++	++	+	−	−
Chlorosarcina consociata (Klebs) G.M. Smith	−	−	−	−	+	−	−	−
Chlamydomonas sp.	−	+	−	−	−	−	−	−
Closterium sp.	−	+	+	−	−	+	−	−
Coelastrum microporum Näg	+	+	−	+	+	−	−	−
Coleochaete orbicularis Pringh	−	−	−	−	−	−	+	+
C. pulvinata A. Br.	−	−	−	−	−	−	+	+
C. scutata de Bréb.	−	−	−	−	−	+	+	+
Cosmarium granatum Bréb.	−	−	+	+	−	−	−	−
C. impressulum Elfv.	+	+	−	+	+	−	−	−
C. pyramidatum Bréb. in Ralfs.	−	+	+	−	−	−	−	−
C. reginelli Wille	−	−	+	+	−	−	−	−
C. subquadratum Nordst.	+	+	+	+	−	−	−	−
C. subtumidium Nordst.	−	+	+	+	−	−	−	−
Crucigena quadrata Morren	−	−	−	+	+	+	−	−
C. tetrapedia (Kirchn.) W. et. G.S. West	−	−	−	+	−	−	−	−
Klebshormidium subtile (Kirchner) Chodat	−	−	−	−	−	−	+	+
Oedogonium sp.	−	−	−	−	−	−	+	+
Pediastrum boryanum (Turp.) Meneghini	−	−	−	−	+	+	+	+
P. duplex Meyen	−	−	+	+	+	+	−	−
P. simplex Meyen	−	−	+	+	+	−	−	−
P. tetras (Ehr.) Ralfs.	−	−	+	+	+	+	−	−
Scenedesmus abundans (Kirchner) Chodat	+	−	+	+	−	−	−	−
S. acuminatus (Lagerh.) Chodat	−	+	−	−	−	−	−	−
S. bijugatus (Turpin) Kütz.	−	−	+	+	+	−	−	−
S. perforatus Lemm.	−	−	+	−	−	−	−	−
S. quadricauda (Turp.) Bréb.	−	−	+	+	+	−	−	−
S. quadricauda v. quadrispina Chodat	+	+	+	−	−	−	−	−
Selenastrum gracile Reinsch	−	−	+	+	+	−	−	−
Spirogyra sp.	−	−	−	−	−	+	+	+
Staurastrum enorme Ralfs.	+	+	−	−	−	−	−	−
Stigeoclonium tenue Rabenh.	−	−	−	−	−	+	+	+
Ulothrix zonata (Weber et Mohr) Kütz.	−	−	−	−	−	+	+	+
Bacillariophyceae								
Achnanthes microcephala Kütz.	+	+	+	−	−	−	−	−
A. minutissima Kütz.	++	++	++	+	+	−	−	−
Amphora coffeaformis (Agardh) Kütz.	+	+	+	+	++	+	+	+
A. ovalis Kütz.	−	−	+	+	+	+	+	+
Amphipleura pellucida Kütz.	−	−	+	−	−	−	−	−

Continued on p. 101

Table 2. Continued

Algal species	Days of panel immersion							
	1	2	3	4	5	7	10	15
Cocconeis disculus Schum	+	+	+	++	+	+	−	−
C. placentula (Ehr) Cleve	+	+	+	+	++	+	−	−
C. scutellum Ehr.v. *parva* Grun.	+	+	++	++	+	−	−	−
Coscinodiscus sp.	+	+	−	−	−	−	−	−
Cyclotella menginiana Kütz.	−	−	+	−	−	−	−	−
Cymbella affinis Kütz.	−	+	+	+	+	+	++	+
C. minuta Hilse	−	−	−	−	+	+	+	+
C. tumida (Breb) v. Heurck	−	+	+	+	+	++	++	−
Diatoma vulgare Bory	−	−	−	−	−	+	−	−
Diploneis ovalis (Hilse) Cleve	−	−	−	+	+	−	−	−
Eunotia sp.	−	−	+	+	+	−	−	−
Fraglaria construens (Ehr.) Grun.	−	+	+	+	+	+	+	+
F. vaucheriae (Kütz) Peter	−	−	−	+	+	+	−	−
Gomphoneis olivaceum (Lyngb.) Kütz.	−	+	+	+	++	++	+++	++
Gomphonema lanceolatum Ehr.	−	−	+	+	+	++	++	+
G. parvulum Kütz.	−	−	−	+	+	+	+	+
Gyrosigma acuminatum (Kutz.) Rabh.	−	+	+	−	−	−	−	−
Lichmophora sp.	−	−	+	+	+	+	−	−
Melosira amphigua (Grun.) O.Muller	−	−	+	+	+	−	−	−
M. granulata (Ehr.) Ralfs.	−	−	−	+	+	−	−	−
Meridion circulare (Grev.) Ag.	−	−	−	−	−	−	+	−
Navicula cryptocephala Kutz.	+	+	+	+	+	−	−	−
N. elegans Wm. Sm.	−	−	+	+	−	−	−	−
N. krasskei Hust.	+	+	+	+	+	+	+	−
N. pelliculosa (Breb) Hilse	−	−	−	−	+	−	−	−
Nitzschia amphibia Grun.	+	+	+	+	+	++	++	++
N. dissipata (Kütz.) Grun.	−	−	−	−	+	−	−	−
N. frustulum Kütz.	+	+	−	+	−	−	−	−
N. microcephala Grun.	−	−	−	−	+	−	−	−
N. ovalis Arnott	+	−	−	−	−	−	−	−
N. palea (Kütz.) Wm. Sm.	+	+	+	+	+	+	+	−
Opephora martyi Heribaud	−	−	−	−	−	−	+	−
Pinnularia subcapitata Gregory	−	−	−	−	+	+	−	−
P. subleniaris Grun.	−	−	−	−	+	−	−	−
Synedra radians Kütz.	−	−	−	−	−	+	+	+
S. ulna (Nitzsch) Ehr.	−	−	−	−	−	−	+	+
Cyanobacteria								
Anabaena circinalis Forti	−	−	−	−	−	−	−	+
Aphanocapsa elachista W. et. G.S. West	−	−	+	+	+	−	−	−
Aphanothece Näg.	−	−	−	−	−	−	+	−
Arthrospira Stizenb.	−	−	−	−	−	−	+	+
Aulosira Kirchner	−	−	−	−	−	+	−	−
Calothrix brevissima West, G.S.	+	+	−	+	+	+	+	+
C. elenkinii Koss.	−	−	−	+	++	+	+	++
Chroococcus minutus (Kütz.) Näg.	++	+	+	+	+	−	−	−
C. turgidus (Kütz.) Näg.	+	+	+	−	−	−	−	−
Cylinderospermum Kütz.	−	−	−	−	−	+	−	−
Gloeocapsa nigrescens Näg.	−	−	−	+	+	++	+	+
Lyngbya birgei Smith	−	−	−	−	−	−	+	+
Merismophedia glauca (Ehrenb.) Näg.	+	+	+	+	+	−	−	−

Continued on p. 102

Table 2. Continued

Algal species	Days of panel immersion							
	1	2	3	4	5	7	10	15
M. punctata Meyen	−	+	+	−	−	−	−	−
Nodularia spumigena Mertens	−	−	+	+	+	+	+	−
Nostoc commune Vaucher	+	−	−	−	−	−	−	−
N. muscorum Ag.	−	−	−	+	−	−	−	−
Oscillatoria annae Van Goor	−	−	−	−	−	−	+	+
O. princeps Vaucher *ex* Gomont	−	−	−	−	−	+	−	−
O. sancta (Kütz.) Gom.	−	−	−	−	−	−	+	−
O. tenuis Ag.	−	−	−	+	+	+	++	+++
Phormidium ambiguum Gom.	−	−	−	−	−	−	−	+
P. foveolarum Gom.	−	−	−	−	−	−	+	+
P. tenue (Menegh.) Gom.	−	−	−	−	−	+	++	++
Rivularia dura (Roth) *ex.* Born. *et.* Flah.	−	−	−	−	−	+	+	+
Scytomena hofmanii Ag.	−	−	−	−	−	−	−	+
Chrysophyceae (Chrysophyta)								
Dinobryon sp.	−	−	+	−	+	−	−	−
Rhyzochrysis sp.	−	+	−	−	−	−	−	−
Dinophyceae (Phyrrophyta)								
Peridinium sp.	−	−	+	−	−	−	−	−

+ = Present; ++ = Frequent; +++ = Abundant. − = Absent.

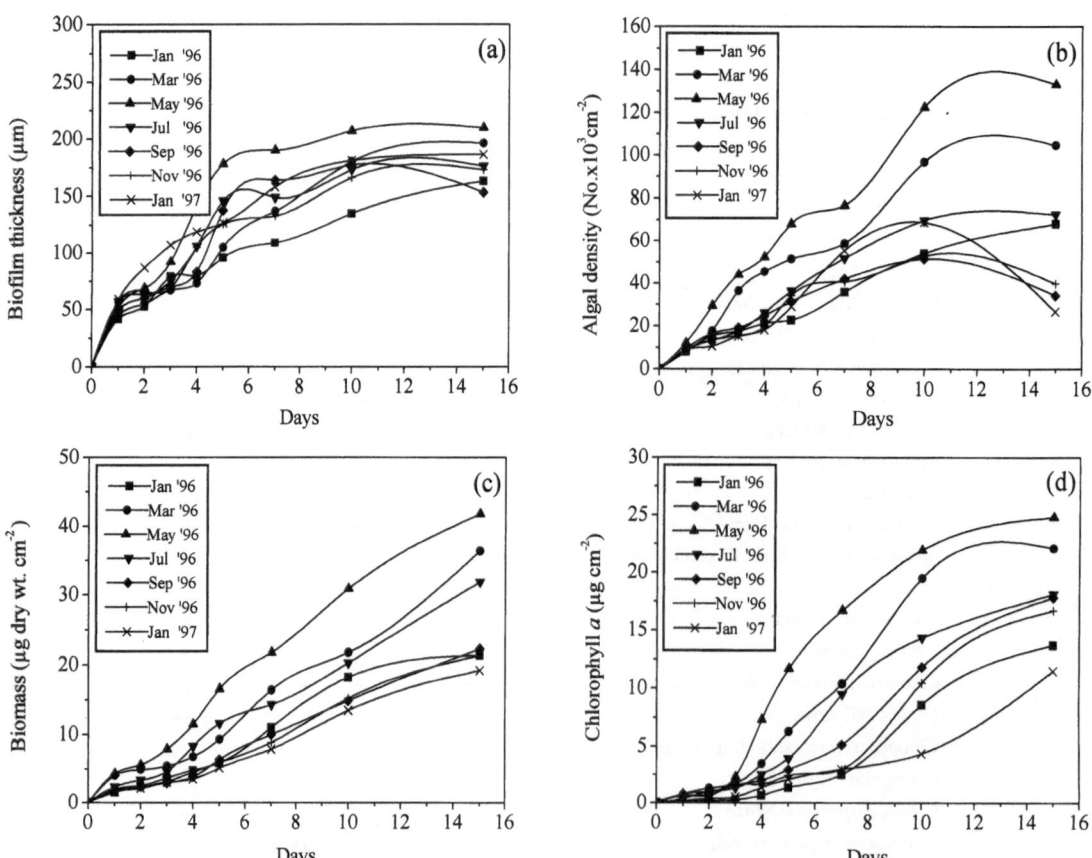

Figure 1. Changes in biofilm thickness (a), algal density (b), biomass (c) and Chlorophyll *a* (d) during the period of study.

diatoms such as *Gomphoneis olivaceum*, *Gomphonema* sp. and *Nitzschia amphibia* were observed more in the biofilm.

Cyanobacteria such as *Aphanocapsa elachista*, *Chroococcus minutus*, *C. turgidus*, *Merismophedia glauca*, *M. punctata* were observed during the early phase, among which *C. minutus* was dominant. Species such as *Calothix*, *Lyngbya*, *Oscillatoria*, *Phormidium*, *Rivularia* and *Scytonema* were observed during the later stage and among these, *Oscillatoria tenuis* and *Phormidium tenue* were relatively more abundant.

In general, the biofilm was initially dominated by *Chlorella vulgaris* with co-dominance of *Cocconeis scutellum* and *Chroococcus minutus*. This was followed by the emergence of *Gomphoneis olivaceum* along with *Nitzschia amphibia* (diatom), colonial forms such as *Pediastrum boryanum*, *Coleochaete scutata* (green algae) and *Gloeocapsa nigrescens* (Cyanobacteria). The later stages were dominated by *Oscillatoria tenuis* and *Phormidium tenue*, with *Nitzschia amphibia* and *Gomphoneis olivaceum* forming important constituents (Fig. 2).

Based on these results, the biofilms species succession could be divided into three distinct phases. The first phase (1–4 days) was dominated by green algae, the second phase (5–7 days) by diatoms and the third phase (10–15 days) by cyanobacteria. The percentage composition of each groups varied slightly with respect to months but the general trend in the sequence of colonization and succession pattern was comparable throughout the year (Fig. 3).

Species diversity

The Shannon–Wiener index showed much variation with exposure time and between months (Fig. 4). Relatively higher diversity of organisms in the biofilm during the early phases of colonization was evident from the higher Shannon–Wiener values that prevailed during 1–7 days. Evenness index values did not show much variation with respect to exposure time throughout the study period, indicating relatively even distribution of species (Fig. 4).

Discussion

Biofilm thickness, biomass build-up and microbial density on surfaces are described as functions of biofilm age (Christensen & Characklis, 1990). Biofilm thickness is an important parameter since it deter-mines the fluid frictional resistance and heat transfer efficiency when fouling occurs in pipes and heat exchangers (Christensen & Characklis, 1990). In the present study, parameters such as biofilm thickness, algal density, biomass and chlorophyll *a* concentrations increased with biofilm age (=exposure time). This is in agreement with reports by Rao et al. (1997) and Sekar et al. (1998). Thickness seems to be influenced by the species composition and season as reported by Christensen & Characklis (1990). Biofilms thickness, biomass, algal density and chlorophyll *a* showed variation with respect to different months and the algal density showed significant correlation with all other parameters.

The numerical density in the biofilm varied with age and between months. In general, the number increased up to 5–7 days after which it decreased. The biofilm was initially dominated by *Chlorella vulgaris* with the co-dominance of *Chlorococcum humicolo*, *Achnanthes minutissima*, *Cocconeis scutellum* and *Chroococcus minutus*. Patrick (1976) reported that prostrate diatoms, such as *Cocconeis* and *Achnanthes*, colonized surfaces during the initial period of biofilm development, followed by the attachment of genera possessing mucilagenous pads or stalks (e.g. *Fragilaria* and *Synedra*). Korte & Blinn (1983) also observed attachment of *Achnanthes minutissima* and *Cocconeis placentula* during the early phases of biofilms development, followed by the attachment of horizontally positioned species such as *Gomphonema*, *Nitzschia* and *Cymbella* in stream riffle zones. Miller et al. (1987), while studying diatom succession on sand grains, found increased attachment of low profile diatoms when compared to stalked diatoms. In the present study also, attachment of green algae and low profile diatoms was observed more during the early stages, whereas long-stalked diatoms, filamentous green algae and cyanobacteria colonized more during later stages. It has been reported that closely adhering nature and ability to produce mucilage are attributes that facilitate easy attachment of these organisms during the early stages of colonization (Siver, 1977; Hudon & Bourget, 1981; Korte & Blinn, 1983). It has been reported that certain species of *Cocconeis* are highly competitive as epiphytes in the presence of organic exudates (Tuchman & Blinn, 1979; Siver, 1980).

In the present study, stalked diatoms such as *Gomphoneis olivaceum*, *Cymbella tumida*, *Gomphonema* sp. and the rosette forming diatoms such as *Nitzschia amphibia* and *N. palea* were found in increased

104

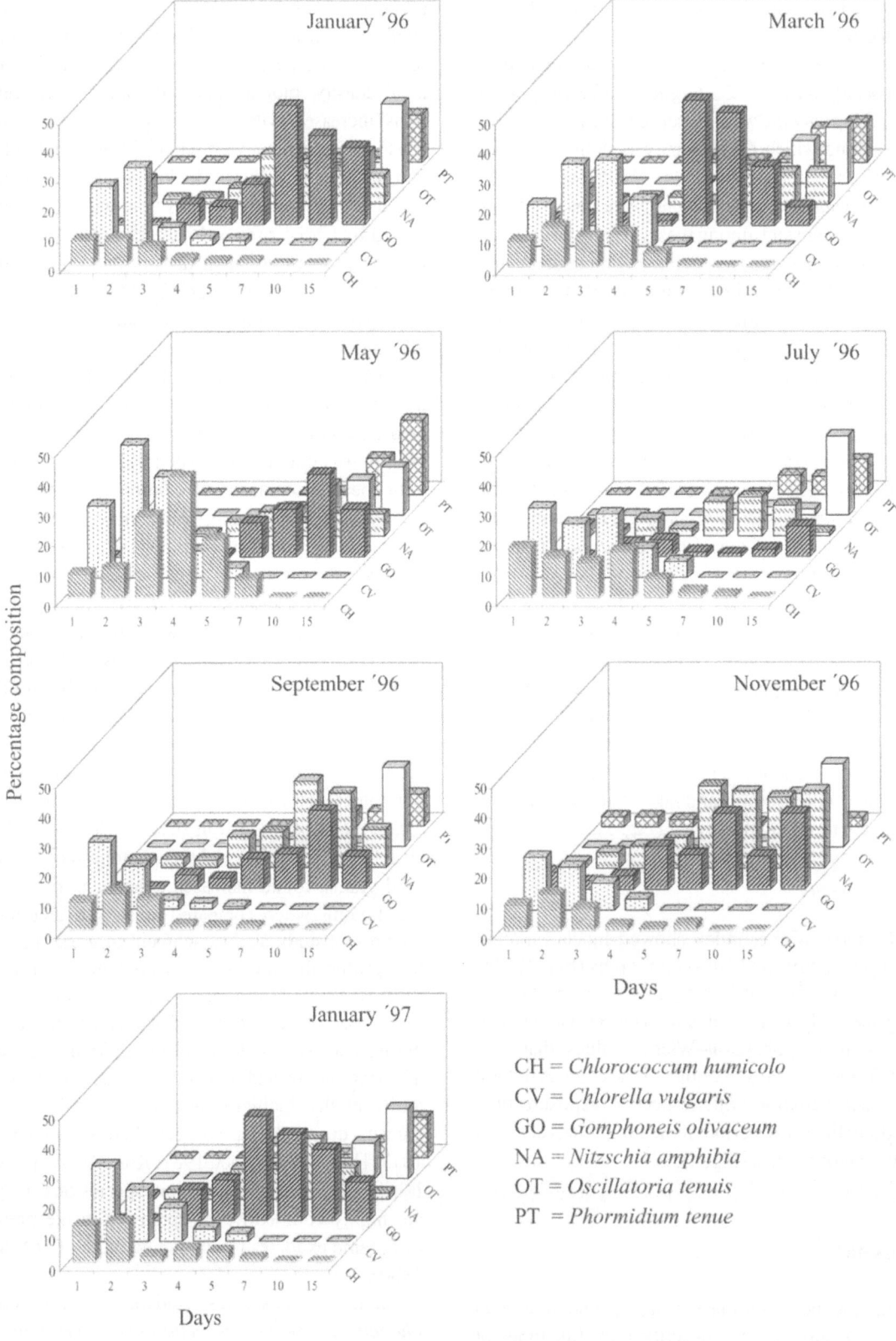

Figure 2. Percentage composition of major biofilms species during the study period.

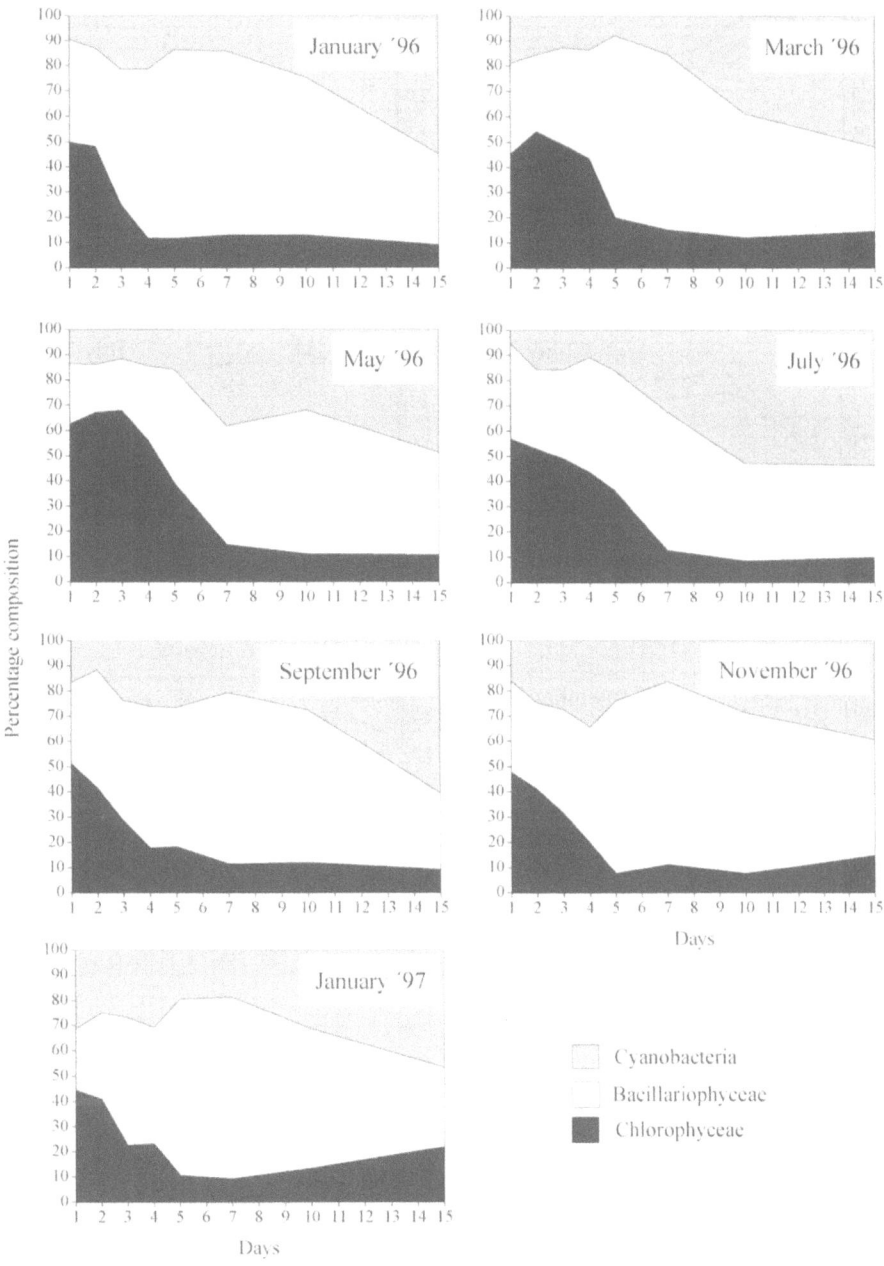

Figure 3. Succession pattern in the biofilms during different months.

numbers after the initial stages (5+ days) of biofilm development. Hoagland et al. (1982) reported that rosette forming diatoms such as *Nitzschia, Fragilaria* and *Synedra* colonized the substratum during the later successional stages, along with long-stalked diatoms such as *Gomphonema*. Korte & Blinn (1983) also observed greater attachment of *Gomphonema, Nitzschia* and *Cymbella* during later stages of biofilm development.

In contrast, Oemke & Burton (1986) and Stevenson & Peterson (1989) found that initial (up to 4 days) colonizers were *Cymbella minuta* and *Amphora* sp. in both riffle and pool zones, followed by the high abundance of *Fragilaria vaucheriae* and *Synedra ulna* in riffles. *Cocconeis* sp. was dominant in both the habitats due to their rapid doubling rate whereas *F. vaucheriae, Synedra ulna* and *Cymbella minuta*

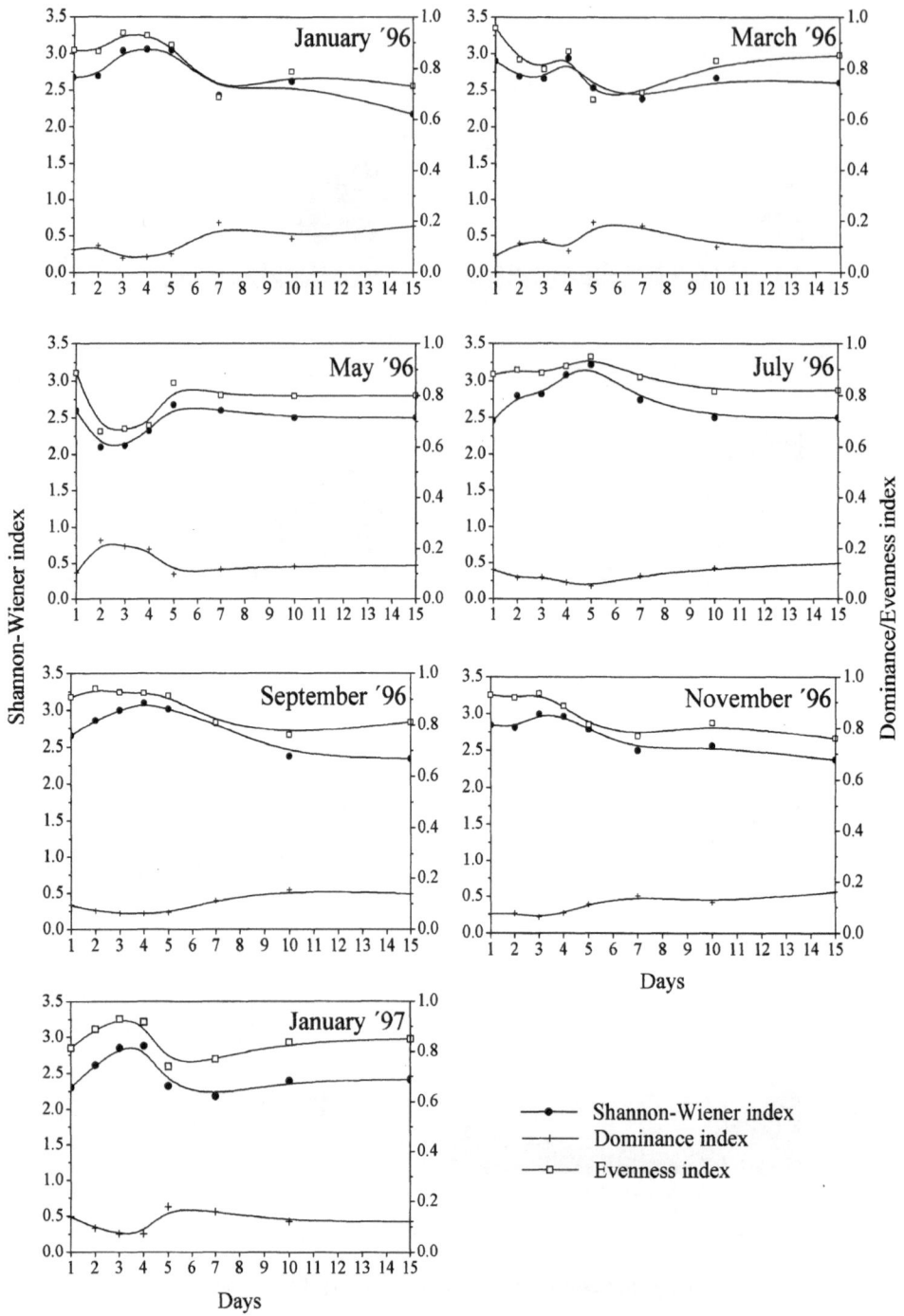

Figure 4. Bimonthly variations in Shannon–Wiener, Dominance and Evenness indices during the study period.

dramatically declined with increasing exposure time. Stevenson & Peterson (1989) stated that species which dispersed effectively were the initial colonizers while species which reproduced rapidly were the dominant forms in the later succession stages. Acs & Kiss (1993) reported that the pioneer colonists were mostly ar-

aphid diatoms of relatively large size. The large bodied species have an advantage during immigration, because they can settle more rapidly on the substratum; and as a result araphid diatoms are usually more active immigrants than mono and biraphid ones (Stevenson & Peterson, 1989). The intermediate stage colonizers are usually biraphid and monoraphid diatoms with relatively small sizes. Small species are fast reproducers and are better competitors in nutrient rich environments than larger species and the later colonizers are slow immigrants, but good competitors for available nutrients (Sommer, 1981).

The initial colonizers observed in the present study may be fast immigrants and also fast reproducers, as compared to the later colonizers. The colonial green algae such as *Pediastrum boryanum*, *Coleochaete scutata* and cyanobacteria such as *Gloeocapsa nigrescens* were found colonizing after 5 days due to their slow immigirant nature and growth in biofilm. The filamentous green algae such as *Klebshormidium subtile*, *Stigeoclonium tenue* and *Ulothrix zonata* and cyanobacteria such as *Calothrix elenkinii*, *Oscillatoria tenuis* and *Phormidium tenue* were found to attach after 5–7 days of immersion. Most of the filamentous green algae attach themselves using a holdfast whilst the cyanobacteria attach to the substrata by production of mucilage (Scott et al., 1996) and their growth rate is very slow as compared to the green algae and diatoms. The abundance of cyanobacterial filaments in the later stages could be related to their slow growth in the biofilm. Korte & Blinn (1983) also noted the dominance of cyanobacteria in the later stages of biofilm development. The reduction of early colonizers during the later stages is possibly due to the increased abundance of stalked diatoms and filamentous species, which restrict light penetration (to underlying species) or hinder nutrient transport by interrupting passage of water current to the underlying cells (Oemke & Burton, 1986).

Stevenson (1986) grouped periphyton community based on their immigration and growth rate: the pioneers have a high initial abundance but decrease with time, the relative abundance of the late colonizers' increases with time, and the intermediates have a relatively more stable abundance than the other groups. In the present study, these three basic types of organisms could be seen during the three different phases of biofilm development. *Chlorococcum humicolo*, *Chlorella vulgaris*, *Achnanthes minutissima*, *Cocconeis scutellum* and *Chroococcus minutus* were the pioneer colonizers and the duration of their abundance varied for each species. Filamentous green algae such as *Klebshormidium subtile*, *Stigeoclonium tenue*, *Ulothrix zonata* and the cyanobacterial filamentous forms such as *Calothrix elenkinii*, *Oscillatoria tenuis* and *Phormidium tenue* were the later colonizers. Colonial forms such as *Pediastrum boryanum*, *Coleocheate*

scutata and *Gloeocapsa nigrescens* and diatoms such as *Amphora coffeaeformis*, *Gomphoneis olivaceum*, *Nitzschia amphibia*, *Nitzschia palea* were intermediate colonizers, showing a more stable abundance.

Based on the percentage occurrence, a distinct three phase succession could be observed. The first phase was dominated by Chlorophyceae, the second phase by Bacillariophyceae and the third by cyanobacteria. This study showed that, in the low energy lentic freshwater environment in which it was carried out, biofilms development progressed in such a way that it was initially dominated by unicellular and small colonial green algae, cyanobacteria and horizontally positioned diatoms. It was followed by large colonial green algae and cyanobacteria and vertically positioned rosette and long stalked diatoms. The late phase community was mainly composed of filamentous green algae and cyanobacteria. Earlier work (Rao et al., 1997) in the same fresh water environment has shown that chemical conditions within the biofilm matrix, especially those relating to the nutrient dynamics, undergo changes depending on the age of the biofilms. Using light and dark experiments Rao et al. (1997) conclusively showed that photosynthesising components in the biofilms (microalgae and cyanobacteria) profoundly influence nutrient chemistry within the biofilms. It would be quite interesting to see how changes in species composition, brought about by successional changes, would influence the nutrient conditions within the biofilms matrix and how this, in turn, would further influence the biofilms progression.

Acknowledgements

We wish to express our sincere thanks to Prof. D. Lalithakumari, Director, Centre of Advanced Study in Botany, University of Madras, Chennai and Dr S.V. Narasimhan, Head, Water and Steam Chemistry Laboratory, BARC Facilities, Kalpakkam for providing laboratory facilities. This work was done under a project sanctioned to one of us (VNRR) by the Board of Research in Nuclear Sciences, Department of Atomic Energy, Government of India. One of the authors (RS) thank the Council of Scientific and Industrial Research (CSIR), Government of India for providing him with travel assistance to present this work at the Second Asian Pacific Phycological Forum held at the Chinese University of Hong Kong, Shatin, NT, Hong Kong.

References

Acs, E. & K. T. Kiss, 1993. Colonization process of diatoms on artificial substrates in the river Dunube near Budapest (Hungary). Hydrobiologia 269/270: 307–315.

APHA, 1989. Standard Methods for Estimation of Water and Waste Water. American Public Health Association, 17th edn., Washington D.C.

Bakke, R. & P. Q. Olsson, 1986. Biofilm thickness measurements by light microscopy. J. Microbiol. Methods. 5: 1–6.

Brown, H. D., 1976. A comparison of the attached algal communities of a natural and artificial substrates. J. Phycol. 12: 301–306.

Callow, M. E., 1986. A world wide survey of slime formation on antifouling paints. In Evans, L. V. & K. D. Hoagland (eds), Algal Biofouling. Elsevier Science Publishers B.V., Amsterdam: 1–20.

Callow, M. E., 1993. A review of fouling in freshwaters. Biofouling 7: 313–327.

Christensen, B. E. & W. G. Characklis, 1990. Physical and chemical properties of biofilms. In Characklis, W. G. & K. C. Marshall (eds), Biofilms. John Wiley & Sons, New York: 93–130.

Cooksey, K. E. & W. B. Cooksey, 1995. Adhesion of bacteria and diatoms to surfaces in the sea: A review. Aquat. Microbial Ecol. 9: 87–96.

Desikachary, T. V., 1959. Cyanophyta. New Delhi: Indian Council of Agricultural Research, New Delhi.

Ford, T. E., M. Walch, R. Mitchell, M. J. Kaufman, J. R.Vestal, S. A.Dither & M. A. Lock, 1989. Microbial film formation on metals in an enriched arctic river. Biofouling 1: 301–310.

Hoagland, K. D., S. C. Roemer & J. R. Rosowski, 1982. Colonisation and community structure of two periphyton assemblages, with emphasis on the Diatoms (Bacillariophyceae). Am. J. Bot. 69: 188–213.

Hudon, C. & E. Bourget, 1981. Initial colonization of artificial substrate: commuity development and structure studied by scanning electron microscopy. Can. J. Fish. aquat. Sci. 59: 1371–1384.

Hudson, S. T. & C. Burke. 1994. Microfouling of salmon cage netting: a preliminary investigation. Biofouling 8: 93–105.

Hustedt, F., 1930. Bacillariophyta (Diatomeae). In Pascher, A. (ed.), Die Susswasser-Flora Mittleuropas. Gustav, Fischer, Jena.

Jeffrey, S. W. & G. F. Humphrey, 1975. New spectrophotometric equations for determining chlorophyll a, b and c in higher plants, algae and natural phytoplankton. Biochem. Physiol. Pflanzen.167: 191–194.

Johnson, R. E., N. C. Tuchman & C. G. Peterson, 1997. Changes in the vertical microdistribution of diatoms within a developing periphyton mat. J. N. Am. Benthol. Soc. 16: 503–519.

Korte, V. L. & D. W. Blinn, 1983. Diatom colonization on artificial substrata in pool and riffle zones studied by light and scanning electron microscopy. J. Phycol. 19: 332–341.

Lowe, R. L., J. B. Guckert, S. E. Belanger, D. H. Davidson & D. W. Johnson, 1996. An evaluation of periphyton community structure and function on tile and cobble substrata in experimental stream mesocosms. Hydrobiologia 328: 135–146.

Margalef, R., 1958. Information theory in Ecology. Gen. Syst. 3: 36–71.

Miller, A. R., R. L. Lowe & J. T. Rotenberry. 1987. Succession of diatom communities on sand grains. J. Ecol. 75: 693–709.

Odum, E. P., 1971. Fundamentals of Ecology. W.B. Saunders Company, London.

Oemke, M. P. & T. M. Burton, 1986. Diatom colonization dynamics in a lotic system. Hydrobiologia 139: 153–166.

Patrick, R., 1976. The formation and maintenance of benthic diatom communities. Proc. am. phil. Soc. 120: 474–484.

Pederson, K., 1990. Biofilm development on stainless steel and PVC surfaces in drinking water. Wat. Res. 24: 239–246.

Philipose, M. T., 1967. Chlorococcales, ICAR, New Delhi.

Pielou, E. C., 1966. The measurement of diversity in different types of biological collections. J. theor. Biol. 13: 131–144.

Prescott, G. W., 1978. How to Know the Freshwater Algae. Wm. C. Brown Company Publishers, Dubuque, Iowa, U.S.A.

Rao, T. S., P. G. Rani, V. P. Venugopalan & K.V. K. Nair, 1997. Biofilm formation in photic and aphotic environments in fresh water system. Biofouling 11: 265–282.

Roemer, S. C., K. D. Hoagland & J. R. Rosowski, 1984. Development of a freshwater periphyton community as influenced by diatom mucilages. Can. J. Bot. 62: 1799–1813.

Scott, C., R. L. Fletcher & G. B. Bremer, 1996. Observations on the mechanisms of attachment of some marine fouling cyanobacteria. Biofouling 10: 161–173.

Sekar, R., K. Nandakumar, V. P. Venugopalan, K. V. K. Nair & V. N. R. Rao, 1998. Spatial variation in microalgal colonization on hard surfaces in a lentic freshwater environment. Biofouling 13: 177–195.

Siver, P. A., 1977. Comparison of attached diatom communities on natural and artificial substrates. J. Phycol. 13: 402–406.

Siver, P. A., 1980. Microattachment patterns of diatoms on leaves of Potamogeton robbinsii Oake. Trans. am. microsc. Soc. 99: 217–220.

Sokal, R. R. & J. Rohlf, 1987. Introduction to Biostatistics, 2nd edition. W.H. Freeman & Company, New York.

Sommer, U., 1981. The role of r- and K-selection in the succession of phytoplankton in Lake Constance. Acta. Oecol. Gener. 2: 327–342.

Steinman, A. D. & C. D. McIntire, 1986. Effects of current velocity and light energy on the structure of periphyton assemblages in laboratory streams. J. Phycol. 22: 352–361.

Stevenson, R. J., 1986. Importance of variation in algal immigration and growth rates estimated by modelling benthic algal colonisation. In Evans, L. V. & K. D. Hoagland (eds), Algal Biofouling. Elsevier Science Publisher, Amsterdam: 193–210.

Stevenson, R. J. & C. G. Peterson, 1989. Variation in benthic diatom (Bacillariophyceae) immigration with habitat characteristics and cell morphology. J. Phycol. 25: 120–129.

Tuchman, M. L. & D. W. Blinn, 1979. Comparison of attached algal communities on natural and artificial substrata along a thermal gradient. Br. phycol. J. 14: 243–254.

Udayakumar, M., S. Chongdar & R. B. Srivastava, 1998. Microfouling on austenitic stainless steel weldments immersed in Bombay harbour waters. Indian J. mar. Sci. 27: 230–232.

Hydrobiologia **512:** 109–116, 2004.
P.O. Ang, Jr. (ed.), Asian Pacific Phycology in the 21ˢᵗ Century: Prospects and Challenges.
© 2004 *Kluwer Academic Publishers.*

Laboratory studies on adhesion of microalgae to hard substrates

R. Sekar[1,2,*], V. P. Venugopalan, K. K. Satpathy, K. V. K. Nair & V. N. R. Rao[1]
Water and Steam Chemistry Laboratory, Bhabha Atomic Research Centre Facilities, IGCAR Campus,
Kalpakkam 603 102, India
[1]*Centre of Advanced Study in Botany, University of Madras, Guindy Campus, Chennai 600 025, India*
[2]*Present address: Dept of Molecular Ecology, Max-Planck Institute for Marine Microbiology,*
Celsiusstrasse 1, 28359 Bremen, Germany
*Author for correspondence; Tel: 0049-421-2028-545. Fax: 0049-421-2028-580. E-mail: sraju@mpi-bremen.de

Key words: adhesion, microalgae, biofilms, chlorophyceae, bacillariophyceae, cyanobacteria

Abstract

Adhesion of *Chlorella vulgaris* (chlorophyceae), *Nitzschia amphibia* (bacillariophceae) and *Chroococcus minutus* (cyanobacteria) to hydrophobic (perspex, titanium and stainless steel 316-L), hydrophilic (glass) and toxic (copper, aluminium brass and admiralty brass) substrata were studied in the laboratory. The influence of surface wettability, surface roughness, pH of the medium, culture age, culture density, cell viability and presence of organic and bacterial films on the adhesion of *Nitzschia amphibia* was also studied using titanium, stainless steel and glass surfaces. All three organisms attached more on titanium and stainless steel and less on copper and its alloys. The attachment varied significantly with respect to exposure time and different materials. The attachment was higher on rough surfaces when compared to smooth surfaces. Attachment was higher on pH 7 and above. The presence of organic film increased the attachment significantly when compared to control. The number of attached cells was found to be directly proportional to the culture density. Attachment by log phase cells was significantly higher when compared to stationary phase cells. Live cells attached more when compared to heat killed and formalin killed cells. Bacterial films of *Pseudomonas putida* increased the algal attachment significantly.

Introduction

Microorganisms attach on all submerged surfaces in the aquatic environment which leads to formation of 'biofilm' (Characklis & Cooksey, 1983). Biofilms cause damage to all water distribution systems including the cooling system of chemical, fertilizer and power plants. In cooling systems, biofilms reduce the thermal efficiency and increase the pressure drop in heat exchangers (Bott, 1990). Microalgae are one of the major components in the biofilms and problems due to algae in many industrial cooling systems have been reported (Ludyansky, 1991; Callow, 1993).

In cooling water systems, materials such as titanium, stainless steel 316-L, aluminum brass, admiralty brass are used as condenser tube materials. These materials are highly affected by the biofilm formation which leads to their corrosion. The cooling systems of Fast Breeder Test Reactor (FBTR) at Kalpakkam,

India faced many operational problems due to fouling and corrosion in the system. Preliminary studies showed that fouling and corrosion of construction material of the FBTR were mainly due to microoganisms present in the cooling water (Rao et al., 1993). Since microalgae are one of the major components in the biofilm it was felt necessary to study their attachment to surfaces and also influencing factors on their attachment.

Chlorella vulgaris (chlorophyceae), *Nitzschia amphibia* (bacillariophyceae) and *Chroococcus minutus* (cyanobacteria) were dominant forms during the early development of biofilms. In the present study, the attachment of these three organisms on perspex, glass, titanium, stainless steel, copper, aluminum brass and admiralty brass were studied. In addition, the influence of surface roughness, pH of the medium, culture density, age, cell viability and the presence of organic and bacterial films on the attachment of *Nitzschia am-*

phibia was studied to find out factors which influence the attachment of this diatom on the surfaces.

Materials and methods

Preparation of coupons

All the materials of size $2 \times 2 \times 0.5$ cm (coupons) were used in the experiment. The metal coupons were polished up to 400 grit, degreased with acetone and air dried.

Isolation, purification and maintenance of algae

Chlorella vulgaris, *Nitzschia amphibia* and *Chroococcus minutus* were isolated from the biofilms developed on perspex panels in the freshwater cooling system at Kalpakkam, India. The organisms were axenised by antibiotic treatment using Benzyl Penicillin, Streptomycin and Chloromphenicol following the method of Droop (1967). The cultures were maintained in modified Chu 10 medium (Gerloff et al., 1950) at $24 \pm 1 \,°C$ in a thermostatically controlled room illuminated with white fluorescent lamps (Philips 40 W) at an irradiance of 60 $\mu E \, m^{-2} \, s^{-1}$ in 12:12 hours L:D regime.

Adhesion assay

The organisms were grown in 500 ml conical flasks using Chu 10 medium (Gerloff et al., 1950). During log phase, the algae suspended in flasks were discarded and the remaining cells firmly attached on the flask walls were scraped with a soft brush (Sharma et al., 1990). The cells were suspended in half strength Chu 10 medium and the culture density was adjusted to 1–1.5 $\times 10^5$ cells ml^{-1} using the same medium. The coupons were placed in replicate large Petri dishes (15 cm diameter) and the medium (75 ml) was poured into the dishes. The dishes were then placed on an orbital shaker at 40 rpm and maintained at 12:12 L:D regime. Replicate coupons were retrieved after 2, 6, 12, 24 and 48 h and the attached cells in ten random fields were counted under an epiflourescence microscope after staining with 0.1% acridine orange (Holmes, 1986; Fukami et al., 1989).

Effect of surface wettability and roughness on adhesion

The wettability of the perspex, glass, titanium, stainless steel, aluminum brass, admiralty brass and copper coupons was studied by the drop spread method (Burchard et al., 1990). Replicate coupons were used for the experiment. The effect of surface roughness on adhesion was studied by using titanium and stainless steel coupons. Surfaces of different roughness were obtained by polishing the coupons using 120, 220, 320, 400, 600 and 800 grit of polishing papers. Log phase culture of *N. amphibia* was used and the experiment was conducted as mentioned earlier. The attached cells were counted after 24 h.

Effect of pH and organic film on adhesion

The half strength Chu 10 medium was adjusted to different pH's such as 6, 7, 8 and 9 using IN HCl or 1N NaOH. Log phase culture of *N. amphibia* was harvested and suspended in the media with different pH's. Replicate titanium (hydrophobic) and glass (hydrophilic) coupons were used for this experiment and the attachment was compared after 24 h. A suspension of gum arabic was used to test whether an organic film could influence the adhesion of algae to surfaces. Titanium and glass coupons were placed in the gum arabic solution (30 g 100 ml^{-1} distilled water) for 30 min and then dried at 30 °C according to the method of Kirchman et al. (1982). The adhesion of *N. amphibia* to organic film coated and control coupons (without film) was studied for 24 h.

Effect of culture density, age and viability on adhesion

Different culture densities of *N. amphibia* such as 2 $\times 10^2$, 2 $\times 10^3$, 2 $\times 10^4$, 2.3 $\times 10^5$ ml^{-1} were prepared in half strength Chu 10 medium. The effect of log and stationary phase cultures of *N. amphibia* on the adhesion was studied using titanium and glass coupons. The attachment of live and dead cells (heat killed and formalin killed) of *N. amphibia* on titanium and glass coupons was also studied. The cells were killed by high temperature (100 °C) or by treatment with formalin (5% solution) for 1 h, and their adhesion was compared with that of control (live cells).

Effect of bacterial film

The influence of the bacterial film (two strains of *Pseudomonas putida*) on the adhesion of *N. amphibia* was studied by using titanium and glass coupons. The bacteria were cultured in Nutrient broth (Peptone 10 g, Beef extract 10 g, Sodium chloride 5 g and Distilled water 1000 ml) and log phase cells were used for the experiment. Titanium and glass coupons were placed in Petri dishes containing *P. putida* suspension of 10^7 cells ml^{-1} for 12 h. Then the coupons were rinsed with sterile distilled water to remove the loosely adhered cells and placed in *N. amphibia* suspension. The density of the *P. putida* biofilm on coupons was 3×10^6 cells cm^{-2}. Coupons without the bacterial film served as the control. The attached cells on both the control and the bacterial film coated coupons were counted after 24 h.

Statistical analysis

The differential attachment of organisms on various materials and different incubation times was analyzed by one way and two way ANOVA (Sokal & Rohlf, 1987). The differential attachment of organisms on materials of different roughness, pH and cell density, cell viability and bacterial film coating was analyzed by one way ANOVA. Student's *t*-test was used to test differences in attachment in treatments such as organic film coating and culture age with their respective controls. The correlation between the attachment and the wettability of different materials was ascertained using Pearson's correlation test.

Results

Influence of wettability and material composition on adhesion

The wettability coefficients (W_c) of different materials used in the present study are given in Table 1. The results showed that glass is relatively hydrophilic whereas titanium, stainless steel 316-L and perspex are hydrophobic in nature.

The adhesion of *C. vulgaris* showed significant variation among the materials (one way ANOVA, F_{145} = 6; $p < 0.0001$) and with time (one way ANOVA, $F_{254.7}$ 4; $p < 0.0001$). The interaction between both the materials and time was also significant (two way ANOVA, $F_{17.2}$ = 24; $p < 0.0001$). At the end of the

Table 1. Wettability coefficient (W_c) of different materials used in the present study

Material	Wettability coefficient (W_c)
Glass	66.6
Copper	34.1
Stainless steel	29.9
Admiralty brass	29.2
Aluminum brass	28.5
Perspex	27.6
Titanium	22.5

48 h the attachment of *C. vulgaris* was maximum on stainless steel, followed by titanium, perspex and glass (Fig. 1a). The colonization was poor on aluminum brass, admiralty brass and copper. The attachment was not correlated with the surface wettability of the materials ($r = -0.1997$; $P = 0.3986$).

The adhesion of *N. amphibia* also varied significantly with materials (one way ANOVA, $F_{192.4}$ = 6; $p < 0.0001$) and with time (one way ANOVA, $F_{142.6}$ = 4; $p < 0.0001$). The interaction between the materials and time was also significant (two way ANOVA, $F_{6.5}$= 24; $p < 0.001$). Maximum colonization of *N. amphibia* occurred on titanium, followed by stainless steel, perspex and glass (Fig. 1b). The colonization was poor on copper and its alloys. Colonization increased rapidly up to 24 h and then stabilized/decreased. The rate of attachment was higher when compared to *C. vulgaris*. The attachment was significantly higher ($p < 0.01$) on titanium and stainless steel coupons when compared to glass and copper alloys. A low but significant negative correlation between the attachment and the wettability of the materials was observed ($r = -0.5581$; $p < 0.002$).

The adhesion of *C. minutus* varied significantly with materials (one way ANOVA, F_{114} = 6; $p < 0.0001$) and with time (one way ANOVA, $F_{101.5}$; $p < 0.0001$). The interaction between the materials and exposure time was also significant (two way ANOVA, $F_{2.2}$ = 24; $p < 0.0001$). *C. minutus* also colonized titanium panels better than stainless steel, perspex and glass (Fig. 1c). Its colonization was poor on copper and its alloys. The attachment was significantly higher ($p < 0.01$) on titanium panels when compared to glass and copper alloys. As in the earlier case, a significant negative correlation between the attachment and the wettability of the materials was observed ($r = -0.4940$; $p < 0.0075$).

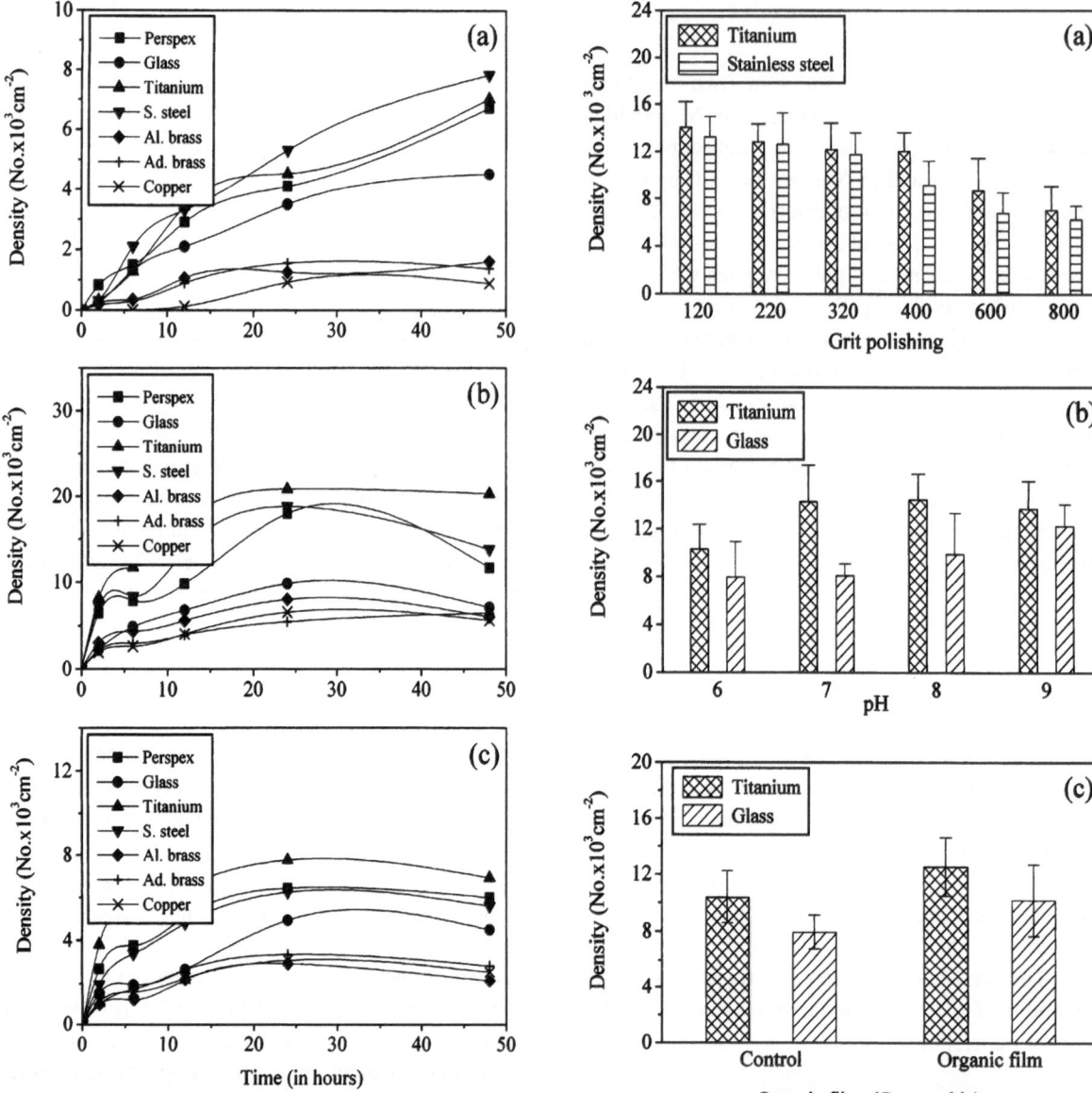

Figure 1. Adhesion of *Chlorella vulgaris* (a), *Nitzschia amphibia* (b) and *Chroococcus minutus* (c) to different hard substrata over time.

Figure 2. Influence of substratum roughness (a), pH (b) and organic film (c) on the mean (± S.D.) adhesion density of *Nitzschia amphibia*.

Influence of surface roughness on adhesion

The attachment was higher on 120 grit polished coupons, and the attachment decreased progressively with increasing smoothness (Fig. 2a). The results showed higher attachment on rough surfaces when compared to smooth surfaces in the case of both titanium and stainless steel coupons. The attachment varied significantly different between different roughnesses on both titanium (one way ANOVA, $F_{15.5} = 5$; $p < 0.0001$) and stainless steel (one way ANOVA, $F_{23.3} = 5$; $p < 0.0001$) coupons.

Influence of pH and organic film on adhesion

Adhesion of *Nitzschia amphibia* was studied on titanium and glass coupons immersed in media of different pH's (6, 7, 8 and 9). The results of a one way ANOVA showed that the attachment on both titanium (one way ANOVA, $F_{5.7} = 3$; $p < 0.05$) and glass (one way ANOVA, $F_{5.9} = 3$; $p < 0.0002$) varied significantly between different pH's. The attachment was significantly higher ($p < 0.01$) at pH 7, 8 and 9 when compared to pH 6 on titanium coupons (Fig. 2b). On

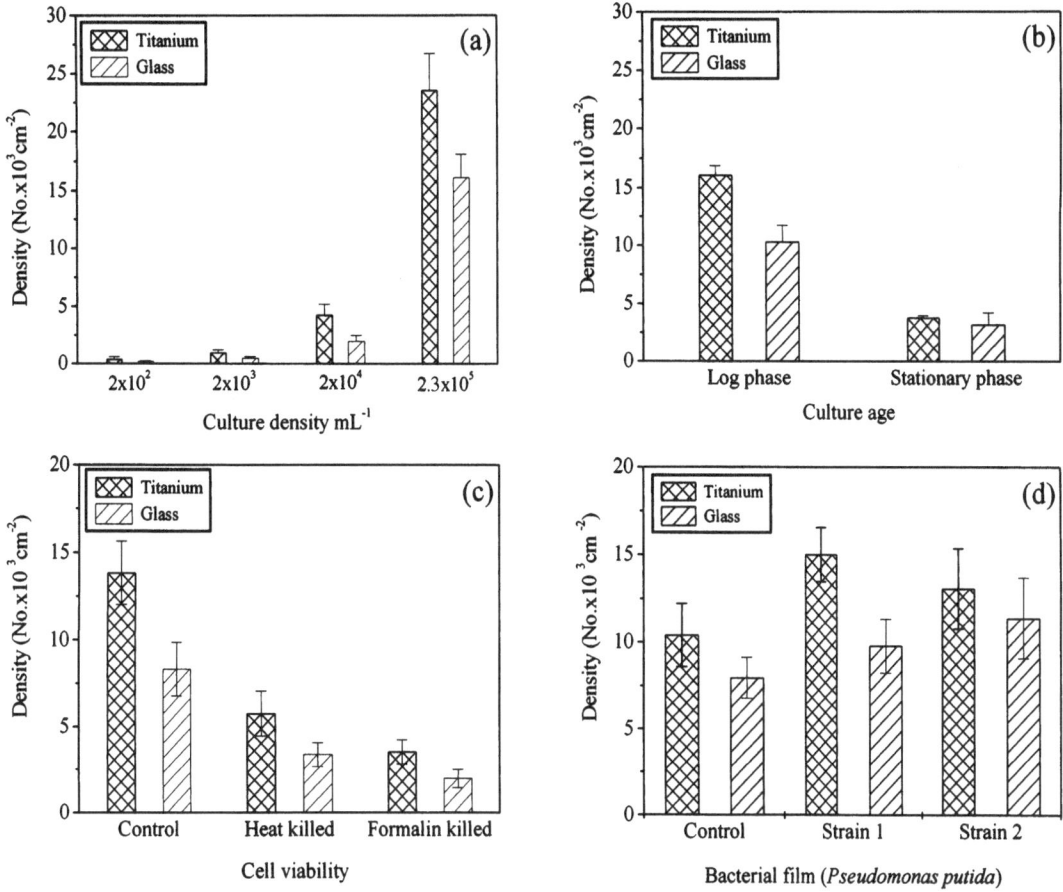

Figure 3. Influence of culture density (a), age (b), viability (c) and bacterial film (d) on the mean (± S.D.) adhesion density of *Nitzschia amphibia.*

glass, the attachment was significantly higher at pH 9 when compared to pH 7 ($p < 0.01$) and 6 ($p < 0.01$).

The adhesion of *N. amphibia* on titanium and glass coupons coated with organic films showed that the presence of organic films increased the algal attachment significantly over control on both titanium (Student's *t*-test, $t = 2.356$; $p < 0.05$) and glass ($t = 2.451$; $p < 0.05$) coupons (Fig. 2c).

Influences of culture density, culture age and cell viability on adhesion

Different culture densities of *Nitzschia amphibia* (2×10^2, 2×10^3, 2×10^4, 2.3×10^5 cells ml^{-1}) were used to study adhesion of the diatom on titanium and glass coupons. The adhesion increased with increase in the culture density. A maximum attachment was at 2.3×10^5 cells ml^{-1} followed by 2×10^4, 2×10^3 and minimum at 2×10^2 cells ml^{-1}, on both titanium and glass coupons. The results of one way ANOVA showed that the attachment was significantly different

among various densities on both titanium (one way ANOVA, $F_{293} = 3$; $p < 0.0001$) and glass ($F_{113} = 3$; $p < 0.0001$) coupons (Fig. 3a).

The attachment of log phase cultures on titanium and glass coupons was higher than that of stationary phase cultures (Fig. 3b). The differences were significant for both titanium (Student's *t*-test, $t = 18.6$; $p < 0.0001$) and glass ($t = 12.3$; $p < 0.0001$) coupons.

The adhesion of live, heat killed and formalin killed cells of *N. amphibia* on titanium and glass coupons showed significantly low attachment of heat killed and formalin killed cells on both titanium ($F_{147.5} = 2$; $p < 0.0001$) and glass ($F_{95.2} = 2$; $p < 0.0001$) coupons when compared to control (Fig. 3c).

Influence of bacterial films on adhesion

Surfaces coated with the bacterial film induced greater attachment of *N. amphibia* on both titanium and glass coupons, when compared to control. The attachment was more on biofilm coated titanium coupons than on glass. The results of one way ANOVA showed significant differences in attachment on coated coupons of titanium ($F_{13.2} = 2$; $p < 0.0001$) and glass ($F_9 = 2$; $p < 0.0001$) when compared to their respective controls (Fig. 3d).

Discussion

The development of microbial films on surfaces has recently received considerable attention. Ford et al. (1989) emphasized the importance of microbial film formation on industrial metals under natural conditions. Fattom & Shilo (1984) reported that cell surface hydrophobicity was a property aiding adhesion of cyanobacteria, similar to bacterial hydrophobicity. Becker & Wahl (1991) and Becker (1996), while studying the colonization of fouling organisms on materials of varying surface tensions, found that the attachment was more on hydrophobic surfaces when compared to hydrophilic surfaces (glass). While studying the bacterial attachment to surfaces Wrangstadeth et al. (1996) found that the higher attachment on hydrophobic surfaces is mediated by the water exclusion mechanism, whereas in hydrophilic substrata water is poorly excluded resulting in less attachment (Burchard et al., 1990). In the present study, the increased attachment observed by all three algae on hydrophobic surfaces (titanium, perspex and stainless steel) may be due to water exclusion mechanism. While studying the biofilm formation on various metals under natural conditions (Ford et al. 1989) found that the attachment was more on titanium and stainless steel, whereas it was poor on the copper alloys. In the present study, the attachment by all the organisms were also poor on copper and its alloys. Though copper, aluminum brass and admiralty brass are relatively hydrophobic, the attachment on those substrata was poor, possibly due to their toxic nature. Hence, both substratum wettability and material composition may be playing important role in the attachment of microalgae.

The density of attached algal cells varied with time among the three organisms. The attachment was highest in *N. amphibia,* followed by *C. minutus* and least in *C. vulgaris*. While studying the attachment strength and adhesion of four marine fouling diatoms *Amphora, Navicula* (procumbent species), *Achnanthes* and *Lichmophora* (stalked species), Woods & Fletcher (1991) found that the attachment rate varied with different species. The attachment of *Amphora* sp.

was more whereas the attachment of *Lichmophora* sp. (stalked) was less and they concluded that the variation was due to the cells' ability to produce EPS in a short time. Tosteson & Corpe (1975) reported that the adhesion of algal cells to solid surfaces depended on the ability of the organisms to secrete adhesive materials. The variation of attachment observed in the present study may also be related to attachment mechanism and their differential ability to produce EPS. Microalgal attachment mechanism varies with different groups of organisms. Most of the diatoms attach to the substrata by the production of EPS in the form of stalks, apical pads, mucilage pads and cell coatings (Hoagland et al., 1993) whereas in the case of filamentous green algae the attachment is mainly by the holdfast. Most of the cyanobacteria attach to the substrata by the production of EPS similar to that of bacteria and diatoms (Scott et al., 1996).

Woods & Fletcher (1991), while studying the adhesion of *Achnanthes minutissima, Amphora coffeaeformis* and *Navicula corymbosa* on smooth (mean amplitude Ra < 0.025 μm); fine (Ra $= 5.75$ μm) and coarse (Ra $= 35$ μm) glass surfaces, found that *A. minutissima* attached more on smooth surfaces, when compared to rough surfaces, whereas *Amphora coffeaeformis* and *Navicula corymbosa* attached more on coarse (rough) surfaces. Hunt & Parry (1998) studied the effect of surface roughness on biofilm development using P280 (rough) and P800 (smooth) grit sand papers for a period of 14 days and found that the bacterial attachment showed significant differences between rough and smooth surfaces, whereas the attachment of algae did not show much variation because of the formation of a uniform bacterial layer over both surfaces. In the present study, axenic cultures of *N. amphibia* were used for experiments on adhesion and hence bacterial film formation may not have preceded the algal attachment. Characklis et al. (1990) found increased bacterial attachment on rough surfaces when compared to smooth surfaces and explained that it was due to the increased convection associated with such surfaces. The rough surfaces provide more surface area for attachment when compared to smooth surfaces. In the present study, the increased attachment of *N. amphibia* observed on rough surfaces may also be due to this reason.

pH is an important factor in natural biofilms and may vary in different layers of the biofilm. It plays an important role in establishing the community (Liehr et al., 1988; Keithan & Barnese, 1989). Microalgae grow well in the pH range between 6 and 9. The effect of pH on the adhesion of *N. amphibia* showed that the attachment was favoured at pH 7 and above. There was a preferential attachment of diatoms in the alkaline range.

Nitzschia amphibia showed significant differences in attachment with variations in culture density. The attachment was directly proportional to the culture density, indicating that density of the organisms in the bulk water would be a factor influencing the rate of colonization by microalgae. The culture age of alga is also very important in its attachment. Log phase cells attached in more numbers when compared to stationary phase cells. Zaidi & Tosteson (1972) observed differential adhesion of *Chlorella* cells during different stages of the life cycle. They found that log phase cells attached vigorously when compared to cells in the other phases. Live cells of *N. amphibia* attached in greater numbers when compared to heat killed and formalin killed cells, showing the role of active process in its attachment.

Peterson & Stevenson (1989), while studying the effect of naturally 'conditioned' and 'unconditioned' ceramic tiles on adhesion of diatoms, found that the conditioning film did not have much influence on the attachment of both *Nitzschia* and *Synedra* sp. in both fast and slow current regimes. In contrast, Steinman & Parker (1990) observed a significant difference in biomass on conditioned ceramic tiles when compared to unconditioned ceramic tiles, for up to 9 days. They concluded that the influence of substratum conditioning on algal attachment was relatively a short-term effect. In the present study, organic film increased the attachment of *N. amphibia* significantly over control, indicating the role of organic polymers on algal attachment process.

Tosteson & Corpe (1975) found that the presence of bacterial films enhanced the attachment of *Chlorella vulgaris* to glass surfaces. In contrast, Kawamura et al. (1988) reported that a film of *Alcaligens* sp. had no effect on the attachment of *Synedra* sp. and concluded that a bacterial film was not always necessary for the attachment of diatoms to substrata. Fukami *et al.* (1989) found that film of *Alcaligens* sp. had a promoting effect on the attachment of *Nitzschia* sp. to surfaces and concluded that an ethanol-insoluble fraction of the bacterial culture (mainly polysaccharides) was most effective in promoting both the attachment and growth of diatoms on surfaces, especially when they were grown under unfavourable conditions. In the present study, both the strains of *Pseudomonas putida* promoted the at-

tachment of *N. amphibia* on both titanium and glass coupons. Interestingly, titanium collected more cells than glass when a bacterial film was present. The results would suggest that substratum properties exerted an influence on algal attachment even in the presence of bacterial films.

In conclusion, the surface property (wettability and roughness) and composition of the material play an important role in microalgal attachment to its surface. The attachment was also found to be influenced by pH, organic film, culture age, culture density, cell viability and bacterial films. Though all the above mentioned factors were found to influence the algal attachment, further experiments are required to find out the mechanisms involved in the attachment process.

Acknowledgements

We wish to express our sincere thanks to Prof. D. Lalithakumari, Director, Centre of Advanced Study in Botany, University of Madras and Dr S.V. Narasimhan, Head, Water and Steam Chemistry Laboratory, IGCAR for providing laboratory facilities. This work was done under a project sanctioned to one of us (VNRR) by the Board of Research in Nuclear Sciences, Department of Atomic Energy, Government of India. The senior author thanks the Council of Scientific and Industrial Research (CSIR) for providing him the travel assistance to present this paper at the 2nd Asian Pacific Phycological Forum held at the Chinese University of Hong Kong, Shatin, Hong Kong.

References

Becker, K., 1996. Exopolysaccharide production and attachment strength of bacteria and diatoms on substrates with different surface tensions. Microb. Ecol. 32: 23–33.

Becker, K. & M. Wahl, 1991. Influence of substratum surface tension on biofouling of artificial substrata in Kiel Bay (Western Baltic): in situ studies. Biofouling 4: 275–291.

Bott, T. R., 1990. Fouling Notebook. Institution of Chemical Engineers, England.

Burchard, R. P., D. Rittschof & J. Bonaventura, 1990. Adhesion and motility of gliding bacteria on substrata with different surface free energy. Appl. envir. Microbiol. 56: 2529–2534.

Callow, M. E., 1993. A review of fouling in freshwaters. Biofouling 7: 313–327.

Characklis, W. G. & K. E. Cooksey, 1983. Biofilms and microbial fouling. Adv. Appl. Microbiol. 29: 93–138.

Characklis, W. G., C. A. McFeters & K. C. Marshall, 1990. Physiological ecology in biofilm systems. In Characklis W. G.

116

&. K. C. Marshall (eds), Biofilms. John Wiley & Sons, Inc., New York: 341–394.

Droop, M.R., 1967. A procedure for routine purification of algal cultures with antibiotics. Br. Phycol. Bull. 3: 295–297.

Fattom, A. & M. Shilo, 1984. Hydrophobicity as an adhesion mechanism of benthic cyanobacteria. Appl. envir. Microbiol. 47: 135–143.

Ford, T. E., M. Walch, R. Mitchell, M. J. Kaufman, J. R. Vestal, S. A. Ditner & M. A. Lock, 1989. Microbial film formation on metals in an enriched arctic river. Biofouling 1: 301–311.

Fukami, K., T. Sakami, Y. Ishida & N. Tanaka, 1989. Effect of bacterial film on the growth of the attached diatom, *Nitzschia* sp. In Miyachi, S., I. Karube & Y. Isida (eds), Current Topics in Marine Biotechnology. The Japanese Society for Marine Biotechnology, Tokyo: 415–418.

Gerloff, G. C., G. P. Fitzerald & F. Skoog, 1950. The isolation, purification and culture of blue-green algae. Am. J. Bot. 37: 216–218.

Hoagland, K. D., J. R. Rosowski, M. R., Gretz & S. C. Roemer, 1993. Diatom extracellular polymeric substances: function, fine structure, chemistry and physiology. J. Phycol. 29: 537–566.

Holmes, P. E., 1986. Bacterial enhancement of vinyl fouling by algae. Appl. envir. Microbiol. 52: 1391–1393.

Hunt, A. & J. D. Parry, 1998. The effect of substratum roughness and river flow rate on the development of freshwater biofilm community. Biofouling 12: 287–303.

Kawamura, T., Y. Nimura & R. Hirano, 1988. Effects of bacterial films on diatom attachment in the initial phase of marine fouling. J. Oceanogr. Soc. Jap. 44: 1–5.

Keithan, E. & L. Barnese, 1989. Effects of pH and nutrients on periphyton colonization. J. Phycol. 25 suppl: 8.

Kirchman, D., S. Graham, D. Reish & R.Mitchell, 1982. Bacterial induce settlement and metamorphosis of *Janua* (*Dexiospira*) *brasiliensis* Grube (Polychaeta: Spiroribidae). J. exp. mar. Biol. Ecol., 56: 153–163.

Liehr, S. K., J. W. Eheart & M. T. Suidan, 1988. A modelling study of the effect of pH on carbon limited algal biofilms. Wat. Res. 22: 1033–1041.

Ludyansky, M. L., 1991. Algal fouling in the cooling system. Biofouling 3: 13–21.

Peterson, C. G. & R. J. Stevenson, 1989. Substratum conditioning and diatom colonization in different current regimes. J. Phycol. 25: 790–793.

Rao, T. S., M. S. Eswaran, V. P. Venugopalan, K. V. K. Nair & P. K. Mathur, 1993. Fouling and corrosion in an open recirculating cooling system. Biofouling 6: 245–259.

Scott, C., R. L. Fletcher & G. B. Bremer, 1996. Observations on the mechanisms of attachment of some marine fouling blue-green algae. Biofouling 10: 161–173.

Sharma, M. O., N. B. Bhosle & A. B. Wagh, 1990. Method of removal and estimation of microfouling biomass. Indian J. mar. Sci. 19: 174–176.

Sokal, R. R. & J. Rohlf, 1987. Introduction to Biostatistics. 2nd edn. W.H. Freeman & Company, New York.

Steinman, A. D. & A. F. Parker, 1990. Influence of substrate conditioning on periphytic growth in a heterotrophic woodland stream. J. N. Am. Benthol. Soc. 9: 170–179.

Tosteson, T. R. & W. A. Corpe, 1975. Enhancement of adhesion of the marine *Chlorella vulgaris* to glass. Can. J. Microbiol. 21: 1025–1031.

Woods, D. C. & R. L. Fletcher, 1991. Studies on the strength of adhesion of some common marine fouling diatoms. Biofouling 3: 287–303.

Wrangstadeth, M., P. L. Conway & S. Kjellberg, 1996. The release and production of extracellular polysaccharides during starvation of marine *Pseudomonas* sp. and the effect thereof on the adhesion. Arch. Microbiol. 145: 220–227.

Zaidi, B. R. & T. R. Tosteson, 1972. The differential adhesion of *Chlorella* cells during the life cycle. Proc. Int. Seaweed Symp. 7: 323–328.

Hydrobiologia **512**: 117–126, 2004.
P.O. Ang, Jr. (ed.), Asian Pacific Phycology in the 21st Century: Prospects and Challenges.
© 2004 *Kluwer Academic Publishers.*

Growth and production of Thai agarophyte cultured in natural pond using the effluent seawater from shrimp culture

Anong Chirapart & Khanjanapaj Lewmanomont
Department of Fishery Biology, Faculty of Fisheries, Kasetsart University, Chatuchak, Bangkok 10900, Thailand
E-mail: ffisanc@ku.ac.th

Key words: growth, natural culture pond, production, Thai agarophyte

Abstract

Growth rate determinations of the Thai agarophytes, *Gracilaria fisheri* (Xia *et* Abbott) Abbott, Zhang *et* Xia and *G. tenuistipitata* Chang *et* Xia var. *liui* Chang *et* Xia, were conducted by monoline method in natural earthen ponds (800 m^2 in area) using shrimp pond effluents (P1) and ambient seawater (P2), from January 1998 to July 1999. Generally, plants of both species cultured in P1 showed a better growth rate and total production than those cultured in P2. Growth rates and total production of the *Gracilaria* cultured in P1 increased in the rainy months and reached a maximum value of 3.08 ± 1.14% d^{-1} for *G. fisheri* and 2.68 ± 1.76% d^{-1} for *G. tenuistipitata* in January 1999. In contrast, growth of both species cultured in P2, projected a slight change in their growth rates, with a maximum value of 1.85 ± 1.00% d^{-1} for *G. fisheri* and 1.70 ± 0.49% d^{-1} for *G. tenuistipitata* attained in the rainy period (August 1998). All plants of *G. tenuistipitata* declined drastically in the following dry season. Total production of *G. fisheri* and *G. tenuistipitata* cultured in P1 showed the highest value of 1000 g wet wt and 961 g wet wt in January 1999, respectively. Plants of both species showed fluctuation in growth and total production, depending on specific cultured conditions of each pond, algal strain used, and on the season. The results suggest that *G. fisheri* can be grown all year round and is more suitable than *G. tenuistipitata* for earthen pond cultivation using shrimp pond effluents.

Introduction

Thailand is an entirely tropical monsoon country, pronounced wet and dry seasons characterize the climate. The rainy season of the south-west monsoon is usually established from May to September, whereas the dry season of the north-east monsoon begins in November and ends in February (Lewmanomont, 1998). Characteristics of the coast of Thailand are mainly mangrove swamps and muddy to sandy beaches.

In the past few years, intensive cultivation of shrimp has expanded considerably in many coastal areas of Thailand. The activities of shrimp cultivation continue almost throughout the year, except for the cold months (December and January) when there is a decrease in shrimp growth. The increase of the intensive farm activities has affected the coastal environment of the country. This is because waste waters from the sea farms usually contain large amounts of ni-

trogen excreted by the shrimps. Several investigations have considered the nutrient-rich effluents as a nitrogen source in mariculture (Vandermeulen & Gordin, 1990; Haglund & Pedersén, 1993; Jiménez del Rio et al., 1996; Yamasaki et al., 1997). They suggested that available nutrient-rich effluents could enhance the growth of marine algae.

Integrated cultivation of the Thai agarophyte *Gracilaria fisheri* (Xia *et* Abbott) Abbott, Zhang *et* Xia with fishes has been attempted in outdoor tanks and ponds (Singthaweesak, 1995, 1996). However, production of the agarophyte has not been successful. There have been very few reports on seaweed growth in Thailand, particularly on its seasonal changes in relation to environmental factors.

The present study was carried out to ascertain variations in growth of the Thai agarophytes, *Gracilaria fisheri* and *G. tenuistipitata* Chang *et* Xia var. *liui* Chang *et* Xia cultured under natural earthen pond

Table 1. Chemical and physical characteristics of the seawater used for cultivation of *Gracilaria fisheri* and *G. tenuistipitata* in natural earthen ponds from January 1998 to July 1999. P1= shrimp pond effluents, P2 = ambient seawater

Date		Temperature ($°C$)		Salinity (‰)		Turbidity ($mg\,l^{-1}$)		Alkalinity ($mg\,l^{-1}$)		PO_4-P ($mg\,l^{-1}$)		DIN ($mg\,l^{-1}$)	
		P1	P2	P1	P2	P1	P2	P1	P2	P1	P2	P1	P2
1998	Jan.	30.7	31.0	31.0	35.0	–	–	81	110	0.01	0.01	0.24	0.22
	Jan.	31.0	31.0	31.0	34.0	–	–	77	127	0.01	0.01	0.21	0.21
	Feb.	30.0	31.0	30.0	33.0	–	–	116	170	0.01	0.02	0.24	0.16
	Feb.	31.0	29.0	31.0	32.0	–	–	131	118	0.00	0.01	0.22	0.20
	Feb.	32.0	32.0	35.0	40.0	–	–	160	173	0.05	0.03	0.50	0.20
	Feb.	29.0	29.0	38.0	40.0	–	–	167	178	0.05	0.02	0.33	0.14
	Mar.	31.0	31.0	28.0	32.0	–	–	296	243	0.03	0.01	0.18	0.19
	Mar.	31.0	29.0	32.0	32.0	–	–	241	232	0.04	0.04	0.18	0.16
	Mar.	33.0	33.0	34.0	38.0	–	–	198	217	0.03	0.02	0.18	0.13
	Apr.	35.0	34.0	26.0	27.0	–	–	244	255	0.02	0.02	0.29	0.20
	May	34.5	34.0	34.0	30.0	–	–	147	234	0.04	0.02	0.33	0.22
	May	34.0	35.0	28.0	26.0	–	–	146	227	0.03	0.02	0.36	0.34
	Jun.	33.0	32.5	30.0	27.0	–	–	148	204	0.04	0.04	0.41	0.26
	Jun.	33.0	32.5	26.0	30.0	–	–	155	153	0.06	0.05	0.44	0.19
	Jun.	33.0	32.5	20.0	34.0	–	–	182	119	0.03	0.05	1.45	0.29
	Jul.	32.0	33.0	22.0	30.0	–	–	146	110	0.06	0.04	0.37	0.26
	Jul.	31.0	30.0	35.0	31.0	–	–	158	144	0.06	0.05	0.78	0.23
	Jul.	34.0	35.0	25.0	30.0	50.57	14.54	150	109	0.06	0.05	0.50	0.20
	Jul.	32.0	31.0	29.0	35.0	44.91	14.77	117	103	0.04	0.07	2.20	2.53
	Aug.	30.0	30.0	25.0	30.0	49.87	13.31	125	132	0.05	0.05	0.59	0.50
	Aug.	30.0	31.0	25.0	29.0	50.39	24.53	160	135	0.05	0.04	0.18	0.15
	Aug.	21.0	21.0	20.0	21.0	109.31	31.12	89	100	0.02	0.02	0.28	1.25
	Aug.	25.0	25.0	23.0	25.0	59.56	47.53	120	125	0.02	0.02	0.29	0.20
	Sep.	19.0	19.0	28.0	22.0	54.54	43.39	131	125	0.06	0.04	1.30	0.26
	Sep.	19.0	18.0	25.0	27.0	38.89	23.53	97	91	0.02	0.02	0.22	0.21
	Sep.	23.0	22.0	25.0	25.0	38.07	14.77	143	126	0.02	0.03	0.27	0.22
	Sep.	31.0	31.0	25.0	28.0	40.47	28.32	143	150	0.02	0.03	0.16	0.17
	Sep.	31.0	32.0	25.0	27.0	37.66	23.36	109	155	0.02	0.03	0.29	0.11
	Sep.	32.0	32.0	24.0	23.0	52.61	44.91	124	143	0.02	0.02	0.32	0.18
	Oct.	32.0	32.0	26.0	20.0	29.26	39.83	113	110	0.04	0.05	0.20	0.17
	Oct.	29.0	31.0	20.0	23.0	27.15	16.29	94	150	0.01	0.01	0.14	0.26
	Oct.	29.0	30.0	20.0	26.0	43.50	36.26	63	80	0.01	0.01	0.15	0.16
	Nov.	29.0	29.0	20.0	30.0	25.81	37.20	99	141	0.01	0.01	0.23	0.11
	Nov.	26.0	26.0	25.0	30.0	14.60	36.15	86	93	0.01	0.01	0.15	0.12
	Dec.	32.0	32.0	25.0	30.0	21.55	40.29	53	124	0.01	0.00	0.15	0.14
	Dec.	30.0	30.0	25.0	30.0	33.29	39.01	85	71	0.00	0.01	0.12	0.15
	Dec.	27.0	27.0	27.0	30.0	32.33	39.94	117	110	0.00	0.00	0.11	0.50
	Dec.	24.0	24.0	26.0	33.0	40.12	41.64	82	85	0.00	0.00	0.11	0.14
1999	Jan.	34.0	34.0	30.0	35.0	40.41	56.00	153	166	0.01	0.01	0.14	0.12
	Jan.	33.0	33.0	32.0	35.0	41.11	69.84	88	128	0.00	0.00	2.55	0.31
	Feb.	30.0	30.0	34.0	37.0	73.99	81.23	92	105	0.00	0.00	0.42	0.46
	Feb.	31.0	31.0	33.0	37.0	80.94	72.82	85	60	0.00	0.40	0.39	0.40
	Mar.	30.0	30.0	35.0	40.0	110.36	23.18	71	52	0.00	0.00	0.41	0.81
	Mar.	28.0	27.0	40.0	44.0	129.17	43.74	121	100	0.00	0.00	0.40	0.37
	Apr.	32.0	32.0	35.0	40.0	162.22	68.79	165	108	0.00	0.00	0.07	0.09
	Apr.	30.0	30.0	40.0	45.0	29.08	47.18	160	121	0.01	0.00	0.05	0.04

Continued on page 119

Table 1. Continued

Date	Temperature	(°C)	Salinity	‰)	Turbidity	(mg l^{-1})	Alkalinity	(mg l^{-1})	PO$_4$-P	(mg l^{-1})	DIN	(mg l^{-1})
	P1	P2	P1	P2	P1	P2	P1	P2	P1	P2	P1	P2
May	29.0	29.0	35.0	35.0	34.16	26.48	160	126	0.00	0.00	0.04	0.27
Jun.	30.0	30.0	38.0	40.0	46.48	61.14	77	105	0.00	0.00	0.02	0.02
Jun.	29.0	29.0	40.0	37.0	45.02	60.96	100	83	0.00	0.00	0.0 2	0.02
Jun.	35.0	35.0	40.0	38.0	25.58	69.66	139	112	0.01	0.00	0.11	0.12
Jun.	31.0	32.0	37.0	42.0	38.89	72.53	146	116	0.01	0.01	0.12	0.14
Jul.	33.5	34.0	35.0	38.0	34.69	70.60	133	106	0.01	0.00	0.11	0.09
Jul.	30.0	30.0	35.0	38.0	29.02	47.30	135	120	0.01	0.01	0.23	0.08
Jul.	32.0	31.0	32.0	33.0	32.29	38.42	135	118	0.01	0.01	0.14	0.06
Jul.	32.0	32.0	36.0	40.0	35.68	34.28	130	123	0.01	0.01	0.08	0.06

conditions, using shrimp pond effluents and ambient seawater. Annual growth and production of the algae with changes in environmental conditions, and their chemical compositions (chlorophyll, r-phycoerythrin, C:N ratio) were determined.

Materials and methods

Natural earthen pond conditions and plant material

Seaweed cultivation was conducted in natural earthen ponds at the Phetchaburi Coastal Aquaculture Station, Phetchaburi province (100° 05′ 15″ E, 13° 02′ 30″ N), 250 km south of Bangkok, Thailand. Two ponds (P1 and P2), originally constructed for cultivation of fish, were used in the experiments. The ponds were 800 m^2 in area and 150 cm deep. The seawater supply was from an extensive shrimp culture pond and from ambient seawater. In Experiment 1 the shrimp pond effluents of 350 m^3 were filtered with a net (2 mm mesh size) and mixed with 650 m^3 of ambient seawater in P1. In Experiment 2 the ambient seawater of 1000 m^3 was pumped directly through the filter net into P2. Aeration was provided with high-flow compressors supplying air to the different ponds, each having six PVC pipes (10 m length) with outlet, and placed in an interval of 1 m on the bottom. The water exchanges of 70% volume were conducted every 7–14 days.

Gracilaria fisheri and *G. tenuistipitata* var. *liui* were collected from Pattani Bay, in the south-east coast of the Gulf of Thailand. Plants were precultivated by monoline method (Trono, 1987) for 6 months under both pond conditions mentioned above,

prior to the start of the experiments. Fifty algal samples (5–10 g sample^{-1}) were inserted between the braids of the nylon rope (10 m length) at an interval of 15 cm sample^{-1}. Pre-culture and culture experiments were performed in the ponds with initial stocking weight of 500 and 250 g wet wt rope^{-1} or 5.0 and 2.5 kg wet wt pond^{-1}, respectively, at 100 cm water depth. Algae were suspended in the water column at 15 cm above the muddy bottom.

Harvesting, growth rate and production

Cultivars were harvested every 7–14 days and restocked to the same initial weight. Relative daily growth rates were calculated according to the equation (Penniman et al., 1986):

$$RGR = \{(W_t/W_0)^{1/t} - 1\} \times 100,$$

where RGR = relative daily growth rate; W_0 = initial fresh weight; W_t = final fresh weight; and t = time (days). Total production of the algae in g fresh wt was monitored monthly.

Laboratory analyses

During the experiments, determination of surface seawater temperature, salinity, turbidity, alkalinity, PO$_4$-P and total dissolved inorganic nitrogen (ammonia + nitrite + nitrate) were made as one point measurements every working day. Water temperature and salinity were measured using an alcohol thermometer, and refracto-salinometer, respectively. Turbidity of the seawater was recorded at 420 nm with a spectrophotometer (Shimadzu UV-1601). Alkalinity and ammonia-N were determined according to APHA, AWWA &

120

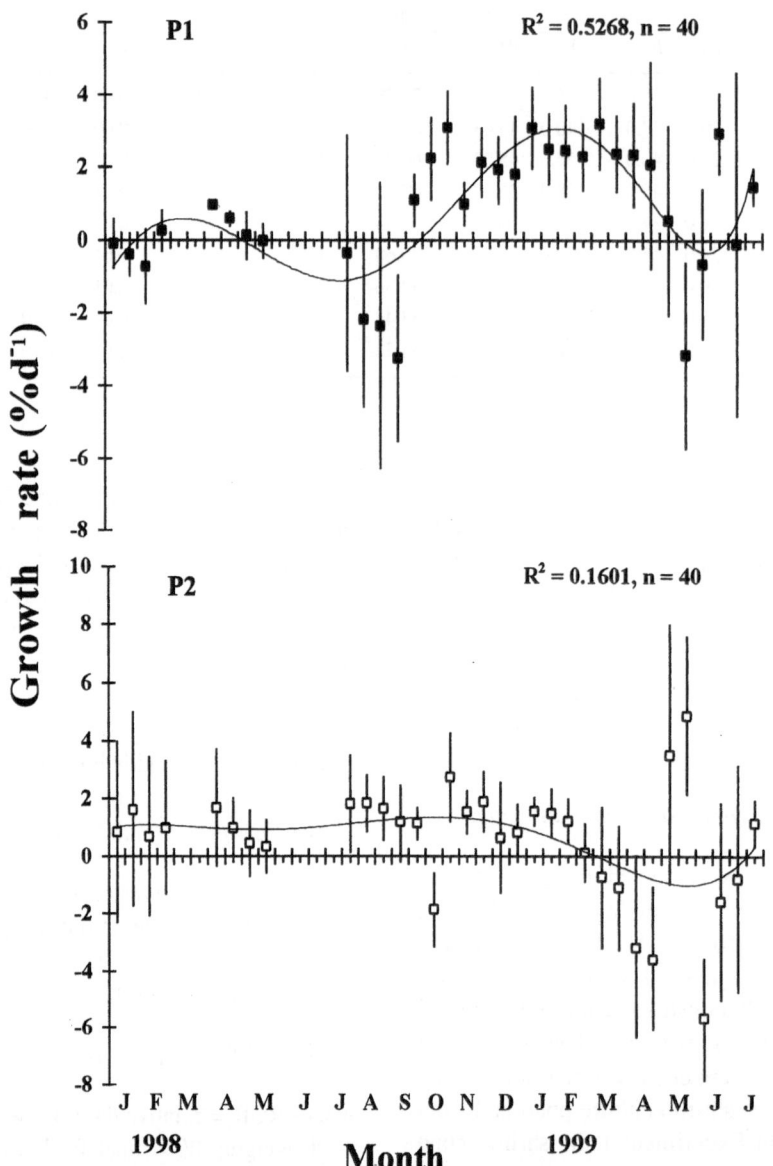

Figure 1. Mean growth rate (± SD) of *Gracilaria fisheri* cultured in natural earthen pond conditions using shrimp pond effluents (P1) and ambient seawater (P2) from January 1998 to July 1999. The growth curve (solid line) is best fitted using the polynomial regression ($n = 40$).

WPCF (1980) and PO_4-P, NO_3-N and NO_2-N according to Strickland & Parsons (1972). Determinations of nitrogen and organic carbon concentration, and C:N ratio in the algal tissue were performed with method described by Walsh & Beaton (1973). Samples of fresh material were analyzed for total chlorophyll (Mackinney, 1941) and for r-phycoerythrin (MacColl & Guard-Friar, 1987).

Results

The chemical and physical characteristics of the surface seawater in both experiments are shown in Table 1. Annual temperature of surface seawater in both P1 and P2 ranged from 18 °C in the late rainy season (September) to 35 °C in the dry season (April). Salinity generally varied from 20‰ to 35‰ in both P1 and P2, but increased drastically (36–45‰) during dry periods. Turbidity varied from 14.60 to 162.22 mg l^{-1}

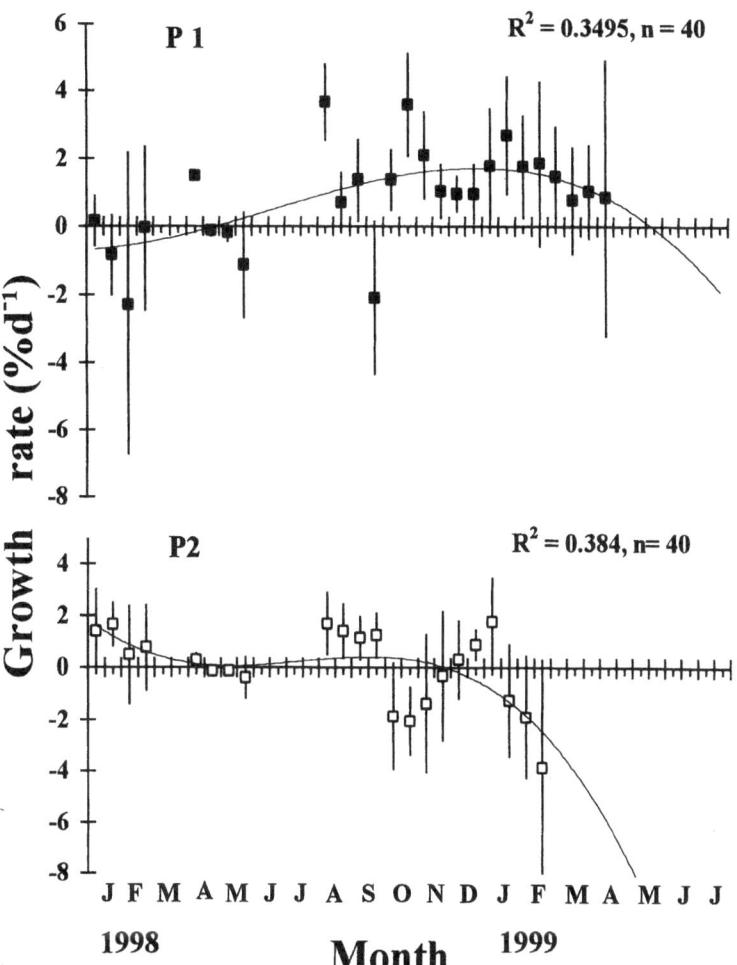

Figure 2. Mean growth rate (± SD) of *Gracilaria tenuistipitata* cultured in natural earthen pond conditions using shrimp pond effluents (P1) and ambient seawater (P2) from January 1998 to July 1999. The growth curve (solid line) is best fitted using the polynomial regression ($n = 40$).

in P1 and 13.31 to 81.23 mg l^{-1} in P2. In the natural earthen pond, total dissolved inorganic nitrogen (DIN) concentration was generally higher in P1 than in P2 with values ranging from 0.02 to 2.55 mg l^{-1} in P1 and 0.02 to 2.53 mg l^{-1} in P2. PO_4-P concentration was from <0.01 to 0.06 mg l^{-1} in P1 and <0.01 to 0.40 mg l^{-1} in P2. Alkalinity ranged from 53 to 296 mg l^{-1} in P1 and 52 to 255 mg l^{-1} in P2.

Cultivars of *Gracilaria fisheri* and *G. tenuistipitata* showed fluctuation in their growth rates throughout the year (Figs 1 and 2). Both agarophytes exhibited variations in growth depending on season, algal strain and specific culture conditions of each pond. Growth of the agarophytes was rather low at the beginning of culture period (January–April 1998), and continued to worsen during unusual dry period due to El Nino, increas-

ing water salinity (>45‰) and temperature (>30 °C). Cultivars grew again when water salinity and temperature decreased during rainy (August 1998) and cold months (November 1998–January 1999).

Higher growth rate of *Gracilaria* cultured in P1, as indicated by the fitted polynomial regression line, was obtained in the rainy period (August–October 1998) when salinity was between 20–28‰ in P1 and 20–30‰ in P2. A maximum value for *G. fisheri* (3.08 ± 1.14% d^{-1}; $r^2 = 0.53$; $n = 40$) and for *G. tenuistipitata* (2.68 ± 1.76% d^{-1}; $r^2 = 0.35$; $n = 40$) was reached in January 1999. Conversely, the regression growth curve of the *Gracilaria* cultured in P2 indicated a peak of maximum growth rate for *G. fisheri* (1.85 ± 1.00% d^{-1}; $r^2 = 0.16$; $n = 40$) and for *G. tenuistipitata* (1.70 ± 0.49% d^{-1}; $r^2 = 0.38$; $n = 40$)

122

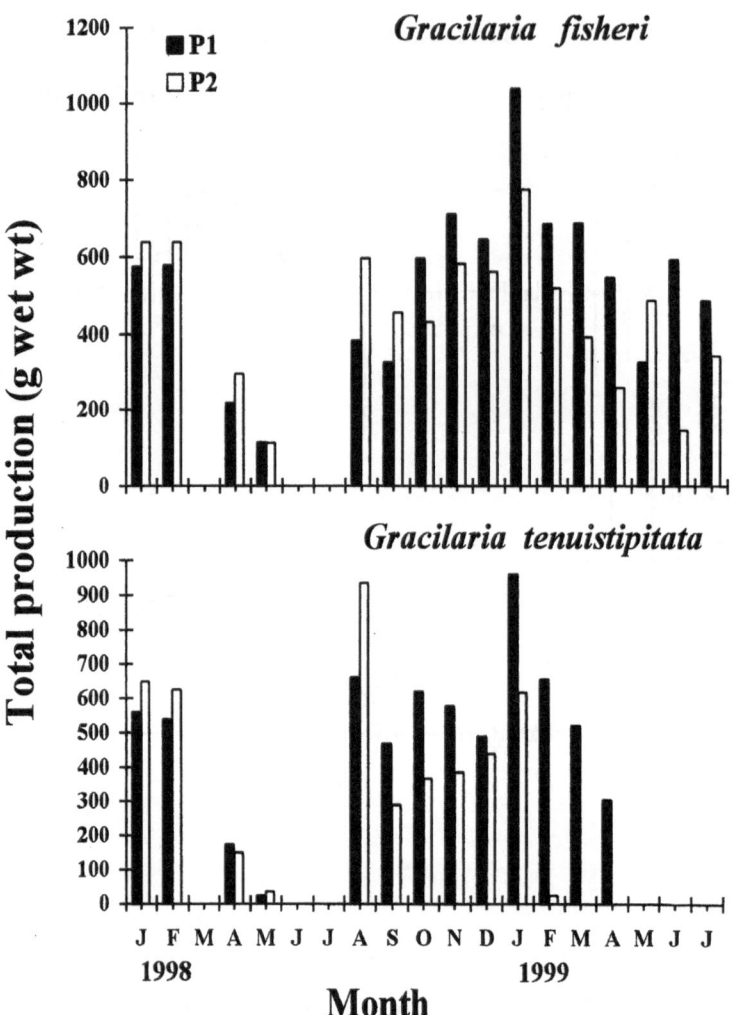

Figure 3. Changes in the total production of *Gracilaria fisheri* and *G. tenuistipitata* cultured in natural earthen pond conditions using shrimp pond effluents (P1) and ambient seawater (P2) from January 1998 to July 1999.

in the rainy season (August 1998). Growth of both species decreased gradually, then drastically when water salinity exceeded 35‰, temperature over 30 °C and alkalinity >160 mg l^{-1}. *G. fisheri* cultivars could grow and acclimatize to the new culture conditions, but not for *G. tenuistipitata*.

Total production (Fig. 3) of cultivars in P1 obtained higher value than those in P2. *Gracilaria fisheri* showed the highest production of 1000 g wet wt (P1) and 777 g wet wt (P2) in January 1999 while those of *G. tenuistipitata* were 936 g wet wt in August 1998 (P2) and 961 g wet wt in January 1999 (P1). However, production of the algal wet weight was higher in P2 than in P1 during the early period of the study (January–September 1998).

Total chlorophyll content (Fig. 4) was higher for the cultivars in P1 than those in P2. The value for

Gracilaria fisheri ranged from 0.25 to 1.19 mg g^{-1} tissue in P1, 0.27 to 0.97 mg g^{-1} tissue in P2, and for *G. tenuistipitata*, 0.35 to 1.37 mg g^{-1} tissue in P1 and 0.28 to 1.29 mg g^{-1} tissue in P2. The maximum values were obtained in the summer month of March 1999. The content of r-phycoerythrin (Fig. 5) in *G. tenuistipitata* increased during rainy and early cold months, and gradually decreased further towards summer. The polynomial regression curve of *G. fisheri* and *G. tenuistipitata* indicated the content of r-phycoerythrin to correspond with the growing period of the algae.

Changes in the tissue C:N ratios of *Gracilaria fisheri* and *G. tenuistipitata* are illustrated in Figure 6. The C:N ratio in *G. fisheri* ranged from 7.51 to 40.26 in P1 and 9.54 to 51.74 in P2, and in *G. tenuistipitata* from 5.51 to 14.90 in P1 and 6.35 to 19.03 in P2. These

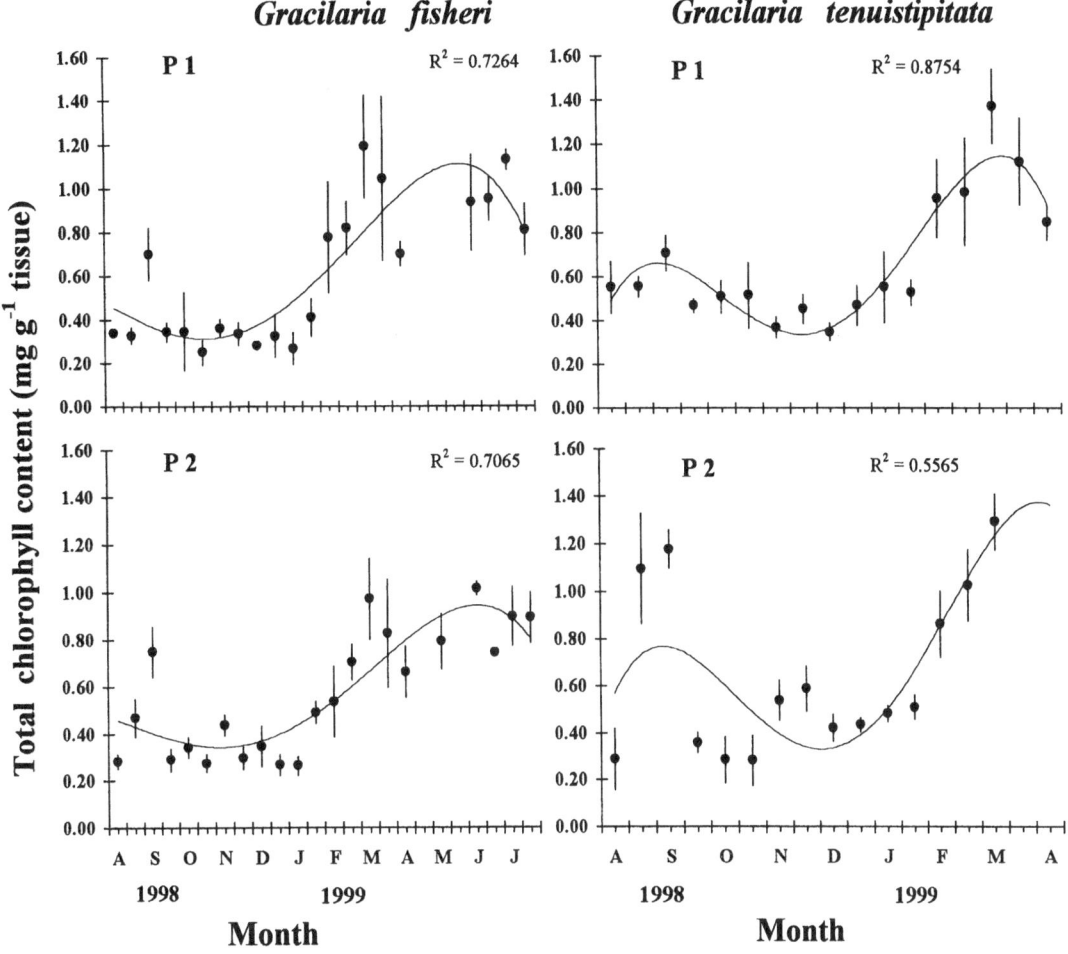

Figure 4. Mean total chlorophyll content (± SD) of *Gracilaria fisheri* and *G. tenuistipitata* cultured in natural earthen pond conditions using shrimp pond effluents (P1) and ambient seawater (P2), from August 1998 to July 1999. The overall pattern of changes in total chlorophyll content (solid line) is best fitted using the polynomial regression (*n* = 5).

ratios in both species were generally lower in P1 than in P2, which corresponded to the higher growth of the plant cultured in P1. However, relationship between the tissue C:N ratios, nutrient and growth rate was not clear among the *Gracilaria* species.

The main interfering epiphytes in our experiments were the blue green alga, *Lyngbya majuscula*, the green algae, *Enteromorpha clathrata*, *Chaetomorpha crassa*, and *Cladophora prolifera*, the seagrass, *Ruppia maritima* and other animals (e.g. tube worm, copepods). They proliferated on the *Gracilaria* and on the bottom of the culture ponds. The epiphytes entangled, entrapped and caused the *Gracilaria* to float, making the plants vulnerable to the high surface summer temperatures. The other animals also entrapped and damaged to tissue of the *Gracilaria*.

Discussion

Growth rate and chemical compositions of *Gracilaria* spp. in outdoor cultures have been reported as a function of salinity, temperature, irradiance, nitrogen and phosphate concentration, plant density, rate of water exchange, aeration, pH and inorganic carbon supply (Lapointe & Ryther, 1979; Lapointe, 1981, 1987; Friedlander et al., 1991; Friedlander et al., 1993; Gonen et al., 1993). Positive correlations between growth rates and nitrogen content have been shown in outdoor cultivation tanks of other *Gracilaria* species (Haglund & Pedersén, 1993; Jiménez del Rio et al., 1996; Yamasaki et al., 1997). With respect to the present study, growth rates of *G. fisheri* and *G. tenuistipitata* increased in response to the effluents from extensive shrimp culture pond. The C:N ratio indicated available nutrient in the shrimp pond effluents

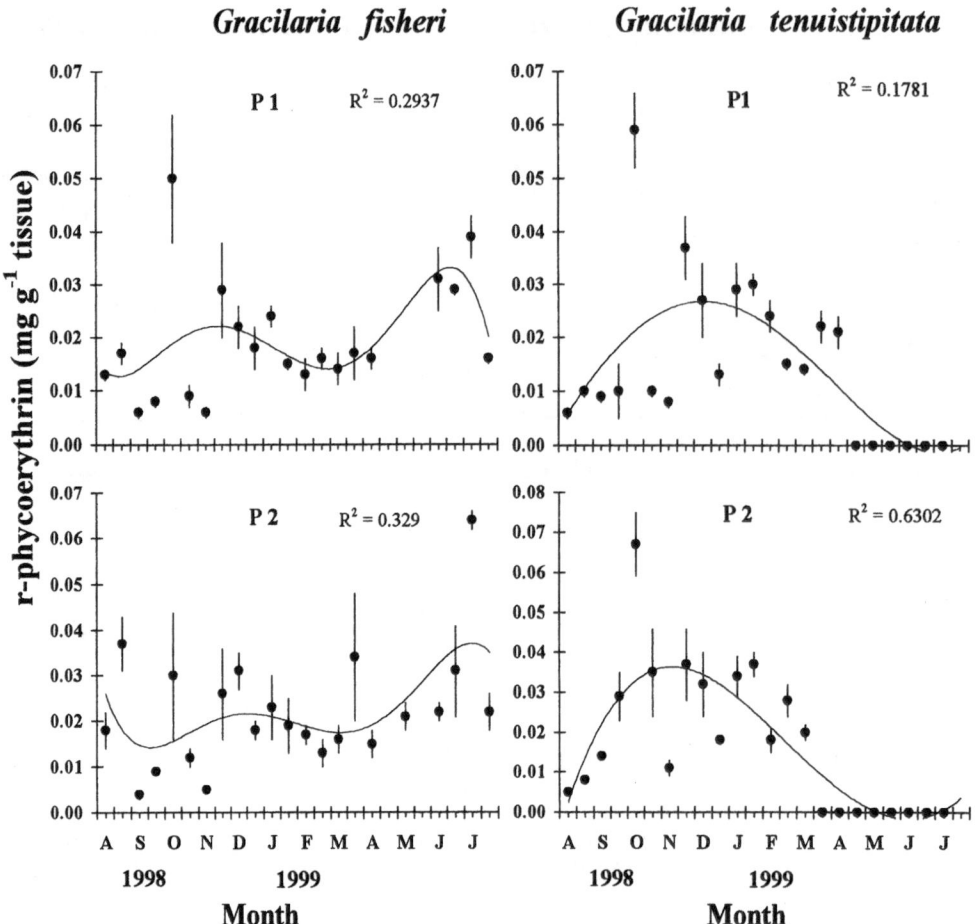

Figure 5. Mean concentration of r-phycoerythrin (± SD) in *Gracilaria fisheri* and *G. tenuistipitata* cultured in natural earthen pond conditions using shrimp pond effluents (P1) and ambient seawater (P2), from August 1998 to July 1999. The overall pattern of changes in concentration of r-phycoerythrin (solid line) is best fitted using the polynomial regression ($n = 5$).

sufficient for growing the Thai agarophytic *Gracilaria*. This result seems to have no N-limitation effect on the algal growth under the natural earthen pond conditions. Lower C:N ratios obtained in the tissues of both *Gracilaria* species in P1 than in P2, could be caused by the accumulation of nitrogen compounds in the tissues of plants which is related to higher nutrient-N in P1 (Friedlander & Levy, 1995). Tissue C:N ratio of about 6 indicated good growth in *Gracilaria secundata* grown at 25 °C under laboratory conditions in Uppsala, Sweden (Lignell & Pedersén, 1987). However, a C:N ratio of about 29 had been reported as an indicator of normal growth of *Kappaphycus alvarezii* in the open sea cultivation (Rui et al., 1990). A tissue C:N ratio in *G. tikvahiae* higher than 13.5 has been suggested as a sign of nitrogen deficiency which reduces growth (Hanisak, 1987). Friedlander et al. (1991) reported that the C:N ratio had an inverse relationship with the ammonium concentration of the

culture medium and with growth rate. Under the natural earthen pond conditions, however, the tissue C:N ratio of each *Gracilaria* species was not clearly related with the growth rate. An increment in growth rate and total production of the *Gracilaria* species in P1, observed in January 1999, could be caused by increments in DIN concentration during that period. In addition, higher production of the algae in P2 than in P1 obtained during the early period of culture (January–September 1998), may be caused by lower turbidity of the seawater used (higher light intensity) in P2 than in P1 associated with interrupted aeration in P1.

Furthermore, variations in the growth and population of both *Gracilaria* species were also dependent on season and strain of the algae used in the culture. Similar results had been reported in long-term experiments in outdoor culture of *Gracilaria verrucosa* strain G-16 in Florida (Bird & Ryther, 1990), *G. conferta* in Israel (Levy et al., 1990), and *G. tenuistipitata* in Sweden

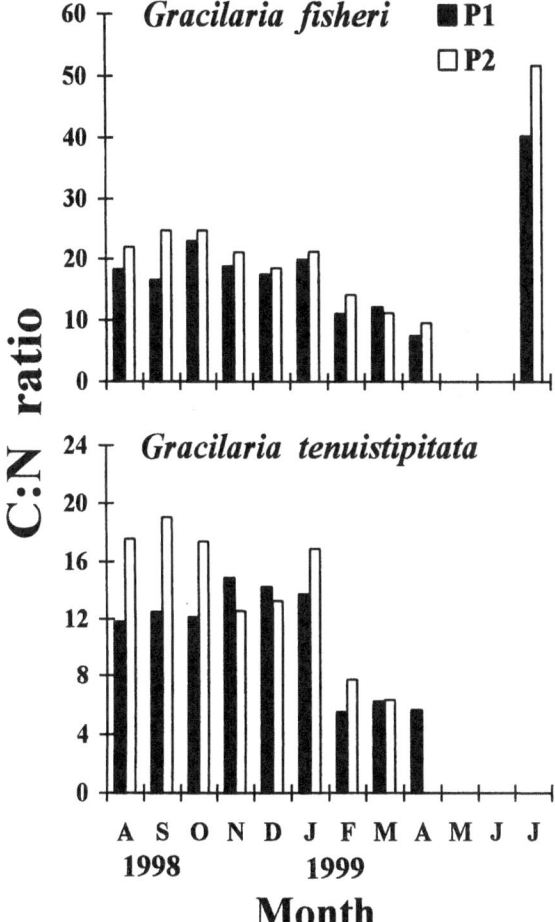

Figure 6. Tissue C:N ratio in *Gracilaria fisheri* and *G. tenuistipitata* cultured in natural earthen pond conditions using shrimp pond effluents (P1) and ambient seawater (P2), from August 1998 to July 1999.

in dry periods. A weekly addition of seawater and low rate of water exchange in some months, during growing period, may also be insufficient to maintain growth in the summer. However, a previous report by Vandermeulen & Gordin (1990) showed that continuous waste water flow through had not provided better growth of *Ulva* in culture tank.

These present results provide further evidence that the Thai agarophytes can be grown under natural pond conditions using shrimp pond effluents all year round. *Gracilaria fisheri* showed a better growth, suggesting that this species could be more suitable than *G. tenuistipitata* for outdoor cultivation in Thailand. However *G. tenuistipitata* can be cultivated under natural pond conditions for 6–7 months of the year. Growth of the algae may potentially be improved by a higher rate of water exchange as mentioned by Freidlander & Levy (1995).

Acknowledgements

This research was supported by a grant from the Thailand Research Fund (no. PDF/69/2540). The authors wish to thank the Phetchaburi Coastal Aquaculture Station for the permission to work in the station, the staff of the Fishery Genetic Section at the Phetchaburi Coastal Aquaculture Station for all their help with field work and laboratory support. We also thank the graduate students, Miss Jittima Mankit and Miss Nidsaraporn Pugdeepun for assistance in the experiment work.

(Haglund & Pedersén, 1993). Difference in pigment composition of the *Gracilaria* cultured in P1 and P2 likely arose in response to changes in environmental conditions (Fredriksen & Rueness, 1989). Although there may be general trends in growth due to seasonal changes in climatic factors, there also appears to be exceptions. Markedly higher growth rates were found in some months (Figs 1 and 2), probably due to the better growth of the freshly re-stocked samples after the older algae had declined in growth during the dry periods.

Weekly fluctuations in growth of *Gracilaria fisheri* and *G. tenuistipitata* may also be explained by the competitive growth of epiphytes (Hanisak, 1987; Friedlander et al., 1991). In addition, reduction in the growth rate and production of the algae may be caused by extremes in salinity, temperature and alkalinity,

References

American Public Health Association, American Water Works Association & Water Pollution Control Federation (APHA, AWWA & WPCF), 1980. Standard Method for the Examination of Water and Waste Water, 15th edn. APHA, Washington D.C.

Bird, K. T. & J. H. Ryther, 1990. Cultivation of *Gracilaria verrucosa* (Gracilariales, Rhodophyta) strain G-16 for agar. Proc. Int. Seaweed Symp. 13: 345–351.

Fredriksen, S. & J. Rueness, 1989. Culture studies of *Gelidium latifolium* (Grev.) Born et Thur. (Rhodophyta) from Norway. Growth and nitrogen storage in response to varying photon flux density, temperature and nitrogen availability. Bot. mar. 32: 539–546.

Friedlander, M. & I. Levy, 1995. Cultivation of *Gracilaria* in outdoor tanks and ponds. J. appl. Phycol. 7: 315–324.

Friedlander, M., C. Dawes & I. Levy, 1993. Exposure of *Gracilaria* to various environmental conditions, I. The effect on growth. Bot. mar. 36: 283–288.

Friedlander, M., M. D. Krom & A. Ben-Amotz, 1991. The effect of light and ammonium on growth, epiphytes and chemical consti-

126

tutents of *Gracilaria conferta* in outdoor cultures. Bot. mar. 34: 161–166.

Gonen, Y., E. Kimmel & M. Friedlander, 1993. Effect of relative water motion on photosynthetic rate of the red alga *Gracilaria conferta*. Proc. Int. Seaweed Symp. 14: 493–498.

Haglund, K. & M. Pedersén, 1993. Outdoor pond cultivation of the subtropical marine red alga *Gracilaria tenuistipitata* in brackish water in Sweden. Growth, nutrient uptake, co-cultivation with rainbow trout and epiphyte control. J. appl. Phycol. 5: 271–284.

Hanisak, M. D., 1987. Cultivation of *Gracilaria* and other macroalgae in Florida for energy production. In Bird, K. T. & P. H. Benson (eds), Seaweed Cultivation for Renewable Resources. Elsevier, Amsterdam: 191–218.

Jiménez del Rio, M., Z. Ramazanov & G. Garcia-Reina, 1996. *Ulva rigida* (Ulvales, Chlorophyta) tank culture as biofilters for dissolved inorganic nitrogen from fishpond effluents. Proc. Int. Seaweed Symp. 15: 67–73.

Lapointe, B. E., 1981. The effects of nitrogen and on growth, pigment content, and biochemical composition of *Gracilaria foliifera* var. *angustissima*. J. Phycol. 17: 90–95.

Lapointe, B. E., 1987. Phosphorus and nitrogen limited photosynthesis and growth of *Gracilaria tikvahiae* in the Florida Keys: an experimental field study. Mar. Biol. 93: 561–568.

Lapointe, B. E. & J. H. Ryther, 1979. The effect of nitrogen and seawater flow rate on the growth and biochemical composition of *Gracilaria foliifera* var. *angustissima* in mass outdoor culture. Bot. mar. 22: 529–537.

Levy, I., S. Beer & M. Friedlander, 1990. Growth, photosynthesis and agar in wild-type strains of *Gracilaria verrucosa* and *G. conferta* (Gracilariales, Rhodophyta), as a strain selection experiment. Proc. Int. Seaweed Symp. 13: 381–387.

Lewmanomont, K., 1998. The seaweed resources of Thailand. In Critchley, A. T. & M. Ohno (eds), Seaweed Resources of the World. Japan International Cooperation Agency, Yokosuka, Japan: 70–78.

Lignell, A. & M. Pedersén, 1987. Nitrogen metabolism in *Gracilaria secundata* Harv. Proc. Int. Seaweed Symp. 12: 431–441.

Mackinney, G., 1941. Absorption of light by chlorophyll solutions. J. Biol. Chem. 140: 315–322.

MacColl, R. & D. Guard-Friar, 1987. Phycobiliprotein. CRC Press, Florida: 218 pp.

Penniman, C. A., A. C Marthieson & C. E. Penniman, 1986. Reproductive phenology and growth of *Gracilaria tikvahiae* McLachlan (Gigartinales, Rhodophyta) in the Great Bay Estuary, New Hampshire. Bot. mar. 24: 147–154.

Rui, L., L. Jiajun & C. Y. Wu, 1990. Effect of ammonium on growth and carrageenan content in *Kappaphycus alvareziii* (Gigartinales, Rhodophyta). Proc. Int. Seaweed Symp. 13: 499–503.

Singthaweesak, W., 1995. Polyculture of agarophyte, *Gracilaria fisheri* (Xia & Abbott) Abbott, Zhang & Xia, with red tilapia, *Oreochromis niloticus* (Linn.) in cement tanks. Technical Paper no. 57, Chanthaburi Coastal Aquaculture Development Center, Department of Fisheries: 30 pp. (in Thai).

Singthaweesak, W., 1996. Effects of harvesting period and density on production of agarophyte, *Gracilaria fisheri* (Xia & Abbott) Abbott, Zhang & Xia, cultured in seabass tanks. Technical Paper no. 2, Chanthaburi Coastal Aquaculture Development Center, Department of Fisheries: 19 pp. (in Thai).

Strickland, J. D. H. & T. R. Parsons, 1972. A Practical Handbook of Sea Water Analysis, 2nd edn. Fisheries Research Board of Canada Bulletin No. 167, Ottawa: 310 pp.

Trono, G. C. Jr., 1986. Seaweed culture in the Asia Pacific region. RAPA Publication 1987/8. Regional Office for Asia and the Pacific (RAPA), Food and Agriculture Organization of the United Nations, Bangkok, Thailand: 41 pp.

Vandermeulen, H. & H. Gordin, 1990. Ammonium uptake using *Ulva* (Chlorophyta) in intensive fishpond systems: mass culture and treatment of effluent. J. appl. Phycol. 2: 363–374.

Walsh, L. M. & J. S. Beaton, 1973. Soil testing and plant analysis. Soil Science Society of America, Inc. Madison, Wisconsin, U.S.A.

Yamasaki, S., F. Ali & H. Hirata, 1997. Low water pollution rearing by means of polyculture of larvae of kuruma prawn *Penaeus japonicus* with a sea lettuce *Ulva pertusa*. Fish. Sci. 63: 1046–1047.

Hydrobiologia **512**: 127–131, 2004.
P.O. Ang, Jr. (ed.), Asian Pacific Phycology in the 21ˢᵗ Century: Prospects and Challenges.
© 2004 *Kluwer Academic Publishers.*

Seedling production using enzymatically isolated thallus cells and its application in *Porphyra* cultivation

Jixun Dai, Zhen Yang, Wanshun Liu, Zhenmin Bao, Baoqin Han, Songdong Shen & Liran Zhou
College of Marine Life Sciences, Ocean University of China, 5 Yushan Road, Qingdao 266003, China
E-mail: daijixun@public.qd.sd.cn

Key words: Porphyra yezoensis, seedling production, enzymatic isolation of cells, outdoor cultivation, China

Abstract

On the basis of previous achievements in *Porphyra* seedling production using enzymatically isolated thallus cells, we have investigated the influences of water content of thallus before freezing, storing temperature and enzymes on the viability of isolated cells. We have also tried outdoor cultivation of seedlings produced enzymatically in different seasons. It has been found that survival rate of isolated cells of thallus stored for two months can reach 90% if water content is 30% before freezing and storing temperature is constant at $-20\,°C$. The 80% cell survival rate can be reached after two years of storage under this condition. The fluctuation of storing temperature causes drastic decrease of cell survival rate. 84% of the cells have survived digestion of sea snail enzyme I, which is the highest among five enzymes used. Outdoor cultivation of seedlings produced on the first ten days of January can reach lengths of more than 20 cm upon harvesting in the middle of May. In addition, seedlings produced in spring can be stored frozen and cultivated in autumn. The harvesting date will advance 15 days in comparison with the conventional seedling producing approach. This promises multiple rounds of cultivation of *Porphyra* within a year.

Introduction

Traditional approach of *Porphyra* seedling production consists of the following steps: carpospore collection in spring, indoor cultivation of conch filament and outdoor cultivation of conchospore in autumn. The seedling production using this approach takes about 4–5 months (Fei, 1999). Since carpospores come from natural populations, traditional seedling production is restricted by seasons and plagued with diseases. In addition, both identity and quality of thalli harvested at different times from once seeded nets are low, reducing drastically the product quality and the economic benefit of laver cultivation.

In order to overcome such disadvantages of conventional seedling production, enzyme purification, single cell and protoplast isolation from thallus and *Porphyra* regeneration from isolated cells and protoplasts were conducted in the early 1980s in our laboratory (Tang, 1982; Liu et al., 1984). Following these preliminary trials, enzymatic isolation of cells

and protoplasts from thallus and *Porphyra* regeneration from isolated cells were conducted in the United States, Japan, Korea and other countries and regions (Polne-Fuller & Gibor, 1984; Chen, 1987; Araki et al., 1987; Song & Chung, 1988; Gall et al., 1993). We reported for the first time seedling production using isolated cells and successful outdoor cultivation of the cell seedlings in 1986 (Fang et al., 1986). This is followed by studies on cell isolation, cultivation, seedling regeneration and outdoor cultivation of *Porphyra* by other researchers (Wang et al., 1986, 1987). From that time on, we have carried out a series of basic studies on enzymatic seedling production of *Porphyra* (Dai & Bao, 1988; Dai et al., 1988, 1993). In order to utilize enzymatic seedling producing technique in *Porphyra* cultivation, we have optimized important factors that influence the viability of enzymatically isolated cells, and tried multiple round outdoor cultivation of cell seedlings in recent years. Our main findings in cell seedling production and its application in laver cultivation are presented in this paper.

Materials and methods

Collection and storage of Porphyra foliose thallus

The thalli of *Porphyra yezoensis* Ueda at nutritional growth stage were collected, air dried for various periods of time, sealed in plastic bags and stored in freezers preset at different temperatures. The water contents of thalli were assayed at 1 h interval.

Cultivation condition

Boiled seawater supplemented with 10 mg KNO_3-N and 1 mg KH_2PO_4-P per liter was used as medium. The irradiation was 2000 $\mu Em^{-2} s^{-1}$ for indoor cultivation. For cell seedling production, filtered seawater was used as the medium, and natural light was used as the light source.

Enzyme and cell isolation

Sea snail enzymes I and II were obtained from Ocean University of Qingdao. Three abalone (red, pink and black) enzyme acetone powders were purchased from Sigma Company.

The viability of freeze stored *Porphyra* thalli was recovered through incubation in sterilized seawater. When normal physiological condition was reached, the thalli were cleaned with pre-cooled boiled seawater, cut into pieces, and digested in solution containing 2% enzyme and 2 mol l^{-1} glucose for 2 h at 20 °C. The digestion solution was filtered through nylon mesh. The single cells and protoplasts in solution were collected by centrifugation, cultivated in laboratory and used for survival rate assaying. Collected cells and protoplasts were also diluted to proper concentration, sprayed onto nylon nets, cultivated in seedling producing ponds for a short period and then used for outdoor cultivation.

Results and discussion

Storage of cell-generating thallus

The influence of water content of thallus before freezing on cell viability

The storage of cell-generating thallus is one of the key steps for enzymatic seedling production of *Porphyra*. Seedlings with high viability can only be produced using properly stored thallus. Improper storage of thallus may cause the failure of the whole seedling producing process.

The water content of thallus drastically influences the viability of the isolated cells. Thalli with different water contents before freezing generated cells with variable viabilities after 2 months storage at −20 °C. All the cells from fresh thalli without drying died. The ideal water content of thallus was 20–40%, and cell survival rate can reach more than 90% when the pre-freezing water content of the thallus was about 30%. The majority of the cells died when water content was lower than 10%. The water contents out of the range of 20–40% were found to be harmful. High water content in cells may crystallize at low temperature, damaging cell structure and causing cell death. If water content is too low, cells may not be able to maintain normal metabolisms, triggering cell death as well.

Water content of thallus is adjusted by air-drying. Air-drying time influences cell viability as well. Since it was inconvenient to assay water content, we optimized air-drying time for practical use. At 20 °C and in normal air with moisture content ca. 65%, no significant changes of cell viability were observed within 12 h of drying. However, nearly half of the nutritional cells died if drying time was more than 24 h. An even larger proportion of these cells died when drying time was 48 h. The proper air drying time was found to be 4–8 h.

The influence of storing temperature on cell viability

Even when water content was properly maintained, changes of storing temperature could have drastic influences on cell viability. Almost all the cells of fresh thallus died within 3–5 days when storing temperature was about 20 °C, so did the cells of the thallus stored at 4 °C in about 15 days and at above −10 °C in 2 months. In contrast, the cell survival rate was very high when the thallus was stored at −20 °C. The storing temperature can vary between −20 °C and −30 °C without obvious cell survival rate changes. The proper storing temperature was found to be constant −20 °C. For example, the majority of the cells from thallus stored at −20 °C on Jan 10, 1996 were viable and could regenerate into blades when they were isolated 2 years later in October of 1998. However, the fluctuation of temperature during storage can cause drastic cell death.

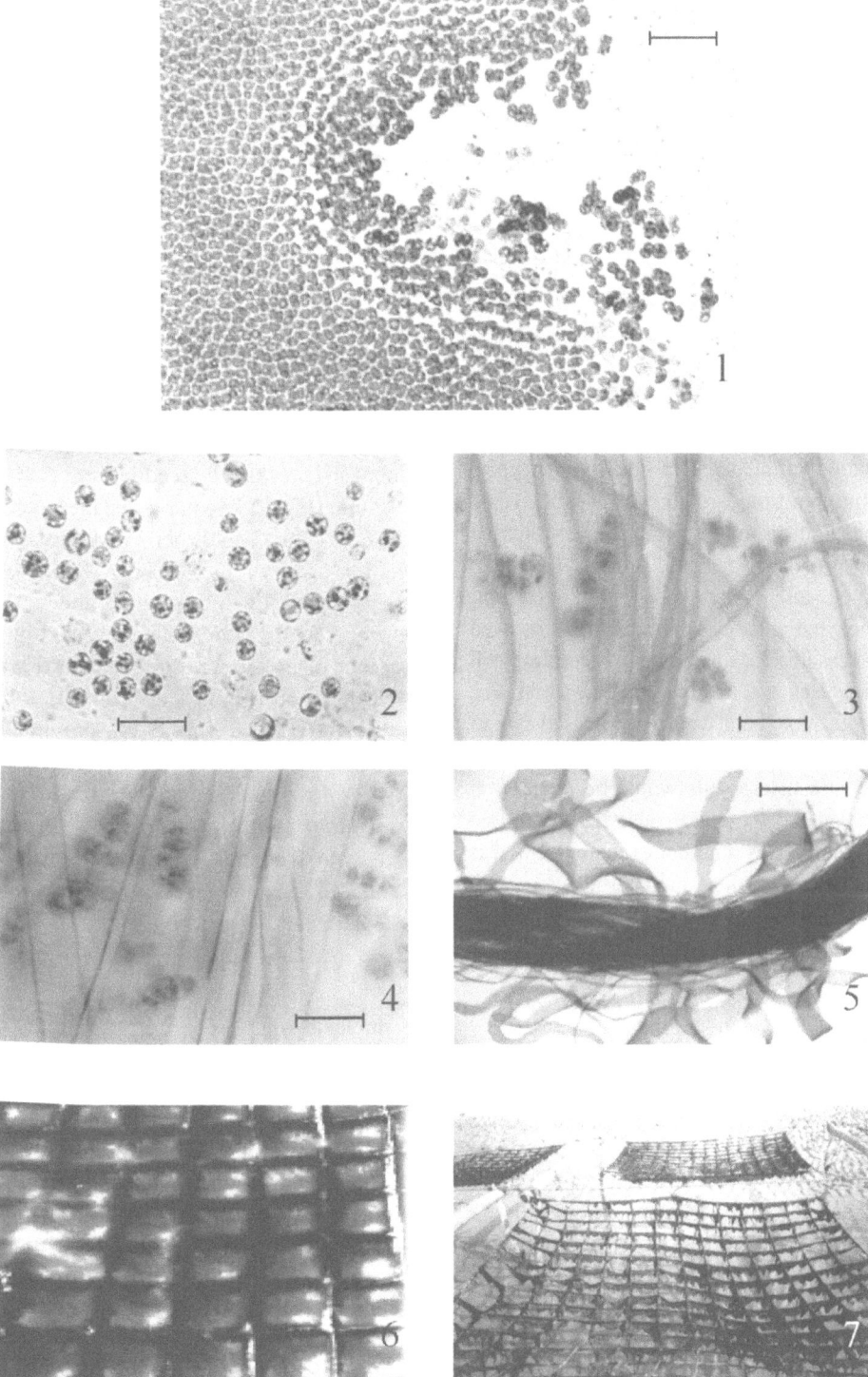

Figures 1–7. Seedling production and outdoor cultivation of *Porphyra*. 1. *Porphyra* cells released in one hour digestion using sea snail enzyme I, scale = 100 μm. 2. Isolated single cells and protoplasts, scale = 50 μm. 3. Cells divided into 2 cells in 2–3 days after adhering onto nylon nets, scale = 50 μm. 4. Young thalli with 3–4 cells after 3–4 days cultivation, scale = 50 μm. 5. Young seedlings with 1–2 mm lengths after 2 weeks cultivation, scale = 2 mm. 6. Thalli with 7–8 cm lengths after 40 days cultivation. 7. Lengths of *Porphyra* reaching 20–30 cm after 50 days cultivation.

Table 1. Comparison of average number of cell lines released and percentage survival rate in average of released cells among different enzymes

Enzymes	Number of cell lines released		Survival rate of cells after 2 days cultivation	
	5% concentration 1 h digestion	1% concentration 1 h digestion	5% concentration 3 h digestion	1% concentration 16 h digestion
Sea snail enzyme I	5–6	2–3	84	72
Sea snail enzyme II	3–5	1–2	55	45
Red abalone enzyme	6–7	2–3	81	68
Pink abalone enzyme	3	1	41	36
Black abalone enzyme	1	0	25*	48

*16 h digestion.

Effects of different enzymes on cell isolation

The cell isolation efficiencies of the five enzymes at different concentrations are shown in Table 1. At the same concentration, these five enzymes released different number of cell lines and produced variable number of viable cells. Among these five enzymes, red abalone enzyme and sea snail enzyme I showed best performances in releasing cells and retaining cell viability, while black abalone enzyme did worst. For the same enzyme, more cells would be released at high concentrations and these cells would retain high cell viability. In contrast, less cells would be released at low concentration and these cells also had decreased cell viability. The greatest number of viable cells was isolated with 5% red abalone enzyme and sea snail enzyme I in 3 h digestion.

Seedling production and outdoor cultivation

Cells in the outermost line around tissue pieces were released first, followed by those in the adjacent inner lines (Fig. 1). Within 2 h, somatic cells of *Porphyra* thallus pieces were digested completely into single cells and protoplasts (Fig. 2). Once being sprayed, the isolated cells started to adhere onto nylon nets in one day and to divide in 3–4 days (Fig. 3), forming young seedlings with 3–4 cells (Fig. 4). At the beginning of year 1994, we produced enzymatically seedlings from fresh *Porphyra* thallus and cultivated these seedlings in seedling producing ponds for about one month. On the last 10 days of January, the newly produced seedlings were transplanted into the sea and cultivated until the end of February. We checked the developmental performance of seedlings on Feb 28, and found that the seedlings had grown into sizes visible through naked eyes. At the beginning of May, all tender thalli reached

harvesting sizes. In April 1994, we produced 1600 m^2 (net area) of seedlings, and reared these seedlings outdoor for 2 weeks. The seedling size reached lengths of 1–2 mm (Fig. 5). We harvested these seedlings, stored them in $-20\,°C$ cold room, and transferred them into the sea again on Oct 8. Four weeks later, the thalli size reached 7.5–8 cm length and 2.5 cm width on average, with 12 cm as the longest (Fig. 6). The thalli reached sizes of harvesting standard within 5 weeks of cultivation (Fig. 7). As control, the conchospore seedlings reached only 3.5–4 cm length within the same period. These trials have shown that high quality *Porphyra* could be produced in advance of normal growing season of *Porphyra* by using freeze stored nets with seedlings. It is also feasible that seedlings produced in winter can be cultivated in spring, which is not the normal growing season of *Porphyra*.

Conclusions

Seedling production using enzymatically isolated thallus cells of *Porphyra* is a new and advanced method in comparison with the conventional approach of conch filament cultivation. It has omitted filament cultivation, thus shortening the seedling producing time from 4–5 months to about 5 days. It is an asexual and clonal process, which can avoid genetic recombination and facilitate the retaining of desirable characters of parental thalli and the identity of the adult thalli. It is noticeable that enzymatic method can get rid of seasonal restriction, producing seedlings throughout the year and meeting the demand for seedlings for cultivation anytime. It makes multiple round cultivation of *Porphyra* possible, and utilizes the natural light and temperature more efficiently. The shift from traditional multiple harvesting from once seeded nets to newly developed multiple harvesting from repeatedly seeded nets promises high quality, high yield and high economic benefit of *Porphyra* cultivation. Seedling production using enzymatically isolated somatic cells is a revolutionary advancement for *Porphyra* cultivation. With its consummation and popularization, this approach will certainly bring us the most economic benefits in the future.

References

Araki, T., T. Aoki & M. Kitamikado, 1987. Preparation and regeneration of protoplasts from wild-type of *Porphyra yezoensis* and green variant of *P. tenera*. Nippon Suisan Gakkaishi 53: 1623–1627.

Chen, L. C. M., 1987. Protoplast morphyogenesis of *Porphyra leucosticta* in culture. Bot. mar. 30: 399–403.

Dai, J. & Z. Bao, 1988. Developmental studies on protoplasts of *Porphyra haitanensis*. Chinese J. Genet. 15: 253–258.

Dai, J., Z. Bao, Y. Tang & W. Liu, 1988. Studies on isolation of *Porphyra* thallus cells with enzymes and cell cultivation. Chinese J. Biotechnol. 4: 127–133.

Dai, J., Q. Zhang & Z. Bao, 1993. Genetic breeding and seedling raising experiments with *Porphyra* protoplasts. Aquaculture 111: 139–145.

Fang, Z., J. Dai, Y. Tang, W. Liu & Z. Bao, 1986. Isolation of the vegetative cells of *Porphyra yezoensis* Ueda with enzymes and its application in aquaculture. Mar. Sci. (China) 10: 46–47 (in Chinese).

Fei, Xiugeng, 1999. Biology of *Porphyra* seedlings. In Zen Chengkui (ed.), Biology of Germplasms amd Seedlings of Economic Algae. Shandong Science and Technology Press, Jinan: 50–90 (in Chinese).

Gall, E. Ar, Y. M. Chiang & B. Kloareg, 1993. Isolation and regeneration of protoplasts from *Porphyra dentata* and *Porphyra crispata*. Eur. J. Phycol. 28: 277–283.

Liu, W., Y. Tang, X. Liu & T. C. Fang, 1984. Studies on the preparation and on the properties of sea snail enzymes. Hydrobiologia 116/117: 319–320.

Polne-Fuller, M. & A. Gibor, 1984. Developmental studies in *Porphyra*. I. Blade differentiation in *Porphyra perforata* as expressed by morphology, enzymatic digestion and protoplast regeneration. J. Phycol. 20: 609–616.

Song, S. H. & G. H. Chung, 1988. Isolation and purification of protoplasts from *Porphyra tenera* Thalli. J. Aquacult. (Korea) 1: 103–108.

Tang, Yanlin, 1982. Isolation and cultivation of the vegetative cells and protoplasts of *Porphyra suborbiculata* Kjellm. J. Shandong Coll. Oceanology 12: 37–50 (in Chinese).

Wang, S., X. Zhang, Z. Xu & Y. Sun, 1986. A study on the cultivation of the vegetative cells and protoplasts of *P. haitanensis* I. Oceanol. Limnol. Sin. 17: 217–221 (in Chinese).

Wang, S., Y. Sun, A. Lu & G. Wang, 1987. A study on the cultivation of the vegetative cells and protoplasts of *Porphyra haitanensis* II. Mar. Sci. (China) 11: 1–7 (in Chinese).

Hydrobiologia **512**: 133–140, 2004.
P.O. Ang, Jr. (ed.), Asian Pacific Phycology in the 21st Century: Prospects and Challenges.
© 2004 *Kluwer Academic Publishers.*

High monospore-producing mutants obtained by treatment with MNNG in *Porphyra yezoensis* Ueda (Bangiales, Rhodophyta)

Xing-Hong Yan[1,3]*, Yuji Fujita[1] & Yusho Aruga[2]
[1]*Faculty of Fisheries, Nagasaki University, 1-14 Bunkyo-Machi, Nagasaki 852-8521, Japan*
[2]*Tokyo University of Agriculture, Sakuragaoka 1-1-1, Setagaya-ku, Tokyo 156-8502, Japan*
[3]*Present Address: Fisheries College, Shanghai Fisheries University, 334 Jungong Road,*
Shanghai 200090, China
Author for correspondence; E-mail: xhyan@shfu.edu.cn

Key words: gametophytic blade, MNNG, monospore, mutant, mutation, *Porphyra yezoensis*

Abstract

Two high monospore-producing pigmentation mutants were obtained by treatment with MNNG in *Porphyra yezoensis* Ueda. The mutants produced many monospores in young gametophytic blades (1 month old) and old large blades (3 months old). Monospore production of the mutants was affected by the culture conditions. The higher the temperature and/or the light intensity, the more the number of monospores released. When the conchospores and monospores of the mutants were in a monoculture at 15 and 20 °C, they did not develop into large blades because their germlings repeatedly released many monospores. However, they developed into large blades when they were in a co-culture with the large gametophytic blades of the wild-type or other pigmentation mutants. One high monospore-producing red mutant (*rm-1*) was genetically characterized by crossing with the wild-type which does not easily release monospores. In the F$_1$ gametophytic blades from heterozygous conchocelis produced in the crosses, there were unsectored blades (2 types) and sectored blades (6 types) consisting of 2, 3 or 4 sectors having both parental color phenotypes. Sectors of both parental colors appeared in the F$_1$ sectored blades in the proportion 1W:1.03R, indicating that the mutant *rm-1* has a mutation for its color phenotype. In the 4-week-old F$_1$ blades, high monospore production occurred only in the unsectored red blades and the red sectors of the sectored blades. These results indicate that the tendency of high monospore production is associated with the color phenotype of the mutant, suggesting that high monospore production of the mutant *rm-1* is controlled by gene(s), which is closely linked with the gene for the mutant color.

Introduction

Porphyra yezoensis Ueda is a commercially important red alga, which has been extensively cultivated in Japan, China and South Korea. *P. yezoensis* has a biphasic life history that alternates between macroscopic, foliose, gametophytic blades and microscopic, shell-boring, sporophytic filaments referred to as the 'conchocelis' phase. Reproduction in *P. yezoensis* involves both sexual and vegetative reproduction cycles. Vegetative reproduction occurs by production of monospores (equal to archeospores) from young gametophytic blades. Monospores play an important role as secondary 'seeds" in large-scale cultivation of

P. yezoensis. Monospore germlings grow faster than conchospore germlings and the monospores can probably be used as a primary seed source (Li & Cui, 1980; Li, 1984; Chen et al., 1985). The primary source of spores for seeding nets in *P. yezoensis* has been conchospores. This requires culturing of conchocelis filaments on shells, a process that is relatively expensive and labor-intensive. In addition, unwanted genetic diversity is introduced at the time of meiosis in the first four cell divisions of the germinating conchospore. Vegetative propagation would solve these problems. By eliminating the conchocelis phase, production costs would be lowered and genetic diversity eliminated. However, the major difficulty in

134

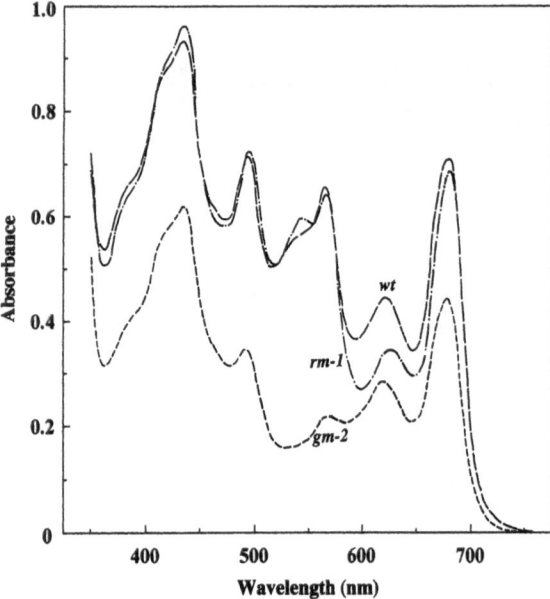

Figure 1. In vivo absorption spectra of gametophytic blades of mutants (*rm-1* and *gm-2*) and the wild-type (*wt*) in *Porphyra yezoensis* after 3 months in culture. Absorption maxima (λ_{max}) are numbered from left to right.

using *P. yezoensis* monospores as the primary seed source is obtaining sufficient quantities of them from the cultivated strains. Selection of high monospore-producing mutant strains will probably make a great progress in this problem.

Recently, we obtained a large number of *P. yezoensis* pigmentation mutants by treatment with a chemical mutagen (Yan & Aruga, 1996, 1997a,b, 1998). Simultaneously, several high monospore-producing mutants which released many monospores from the gametophytic blades as compared with the wild-type and other mutants, were selected. This study will primarily report monospore production of two high monospore-producing mutants in *P. yezoensis*, and the genetic transmission characteristics of one of the mutants.

Materials and methods

In the present study, a wild-type strain of *Porphyra yezoensis* Ueda (U-511, Ohme et al., 1986), whose free-living conchocelis has been maintained in the laboratory, was used. Stock culture of its free-living conchocelis was performed as described by Kato & Aruga (1984).

A chemical mutagen, N-methyl-N′-nitro-N-nitrosoguanidine (MNNG) (Wako Pure Chemical Industries, Ltd., Japan), was used to induce mutations by

dissolving it in the culture medium in a concentration of 30 ppm. Culture medium and methods for gametophytic blade mutagenesis with MNNG and mutant isolation were the same as those described in Yan & Aruga (1997a).

Two high monospore-producing mutants, red (*rm-1*) and green (*gm-2*) mutants obtained by treatment with MNNG, were used. The large gametophytic blades of the mutants were obtained by co-culturing their small conchospore germlings with the large gametophytic blades of the wild-type strain (5–15 cm long) at 15 °C under 90 μmol photons m^{-2} s^{-1} (10L:14D). After ca. 3 months in culture, when the blades of the mutant *rm-1* and the wild-type had grown to 10–15 cm in length, blade discs (9 mm in diameter) were cut out with a borer and cultured separately at different temperatures (10, 15, 20 and 23 °C) under 40 μmol photons m^{-2} s^{-1} (10L:14D), or at 15 °C under different light intensities (6, 12, 36, 72 and 98 μmol photons m^{-2} s^{-1}) with a 10L:14D photocycle, or co-cultured in different combinations of these two types at 15 °C under 80 μmol photons m^{-2} s^{-1} for examining monospore production. Ten vinylon monofilments (3 cm long) were added in each of the culture flasks. After 3 weeks in aerated culture, the blade discs were removed from the flasks but the cultures were kept up. After 3 d, the culture medium was refreshed and monospores attached on the flask wall and the vinylon monofilments were cultured for another 10 d. The monospore germlings were then detached and counted microscopically.

Crossing experiments between the mutant (*rm-1*) and the wild-type (*wt*) were done in a similar method as described by Yan & Aruga (2000). However, conchospores released from heterozygous conchocelis were cultured to obtain F$_1$ gametophytic blades at 20 °C under 40 μmol photons m^{-2} s^{-1} (10L:14D), because 20 °C is more favorable to the early development of *P. yezoensis* conchospores than 15 °C (Yan et al., 1999). After 4 weeks in culture, the color phenotypes and monospore production of the F$_1$ gametophytic blades from heterozygous conchocelis were examined microscopically and classified for genetic analysis.

Examinations of *in vivo* absorption spectra of gametophytic blades of the mutants and the wild-type were done as in Yan & Aruga (1997a).

Results

In vivo *absorption spectra of mutant gametophytic blades*

After 3 months in culture, conchospores of the mutants (*rm-1* and *gm-2*) and the wild-type (*wt*) of *P. yezoensis* developed into large gametophytic blades. The *in vivo* absorption spectra of the gametophytic blades, which were grown under the same culture conditions and at the same age, are illustrated in Figure 1. Both mutants showed 5 absorption peaks similar to those of the wild-type. The red mutant *rm-1* showed distinct differences in $_3\lambda_{max}$ and $_4\lambda_{max}$, the former being double-peaked and the latter being lower and shifting towards longer wavelengths by 3–4 nm as compared with the wild-type. The green mutant *gm-2* showed significantly lower $_2\lambda_{max}$ and $_3\lambda_{max}$ as compared with the wild-type.

Monospore production of mutants

Two-week-old gametophytic blades of both mutants (*rm-1* and *gm-2*) and the wild-type (*wt*) in *P. yezoensis* did not release any monospores (Fig. 2A–C). Four-week-old wild-type blades did not release monospores (Fig. 2D). However, the blades of the mutants *rm-1* and *gm-2* usually produced many monospores after 3 weeks in aerated culture (Fig. 2E,F). When the conchospore germlings of the mutants were cultured at 15 °C under 80 μmol photons m^{-2} s^{-1} (10L:14D), only the basal portion of the blades remained after monospore release (Fig. 3G,I). In monoculture, the mutant germlings did not develop into large blades because they repeatedly released many monospores (Fig. 3D,E). However, if the 2-week-old mutant germlings were co-cultured with the large wild-type blade (5–15 cm long), they developed into large blades (Fig. 3A–C). When the 4-week-old conchospore germlings, which were releasing monospores in monoculture, were transferred into a flask containing several large wild-type blades to co-culture, they stopped producing monospores in a week and finally developed into large blades. The mutant germlings also developed into large blades by co-culturing with large blades of other pigmentation mutants. The monospores released from the mutant blades developed into blades (Fig. 3F,H), which again produced many monospores in 3–4 weeks in monoculture.

Not only did the young mutant blades produce many monospores, but the large mutant blades (10–15 cm long, 3 months old) grown by co-culturing with the wild-type blades, also produced many monospores when they were in a monoculture. When the blade discs taken from the large blades of the mutant *rm-1* were cultured at different temperatures from 10 to 20 °C, the higher the temperature, the more the number of monospores released. However, almost no monospores were obtained when they were cultured at 23 °C (Fig. 4).

Monospore production of the mutant blade discs was also affected by light intensity. From 6 to 72 μmol photons m^{-2} s^{-1} (10L:14D), the higher the light intensity, the more the number of monospores released. When the blade discs were cultured under 6 μmol photons m^{-2} s^{-1}, very few monospores formed. At 98 μmol photons m^{-2} s^{-1}, the monospores production decreased (Fig. 5). Meanwhile, the wild-type blade discs in the same size were also examined under the same conditions, but almost no monospores were obtained. When the mutant blade discs were co-cultured with the wild-type blade discs of different numbers, the number of monospores released from the mutant discs decreased with increase in the numbers of the wild-type discs (Fig. 6). When the number of the wild-type discs was 2–3 times the mutant discs, the number of monospores released from the mutant significantly decreased. The mutant discs produced fewer monospores (approx. 1/100 time that produced in monoculture) when they were co-cultured with more than 4 times of the wild-type discs. However, almost no monospores were produced from the wild-type blade discs in either monoculture or mixed cultures (Fig. 6).

Coincidence of color and high monospore production

Reciprocal crosses between the mutant (*rm-1*) and the wild-type (*wt*) were performed to ascertain whether or not mutations of the mutant *rm-1* are nuclear mutations and whether the high monospore production coincides with the mutant color phenotype. Because the *P. yezoensis* gametophytic blade is monoecious, self- and cross-fertilization will possibly occur simultaneously in each blade in a co-culture. In crosses of *rm-1*×*wt*, carpospores produced by the mutant blade developed into two types of conchocelis, the red and the wild-type. Conchospores from the red conchocelis developed only into the red unsectored F_1 gametophytic blades, indicating that the red conchocelis was homozygous due to self-fertilization. On the other hand, conchospores from the wild-type conchocelis developed into both unsectored and sectored

136

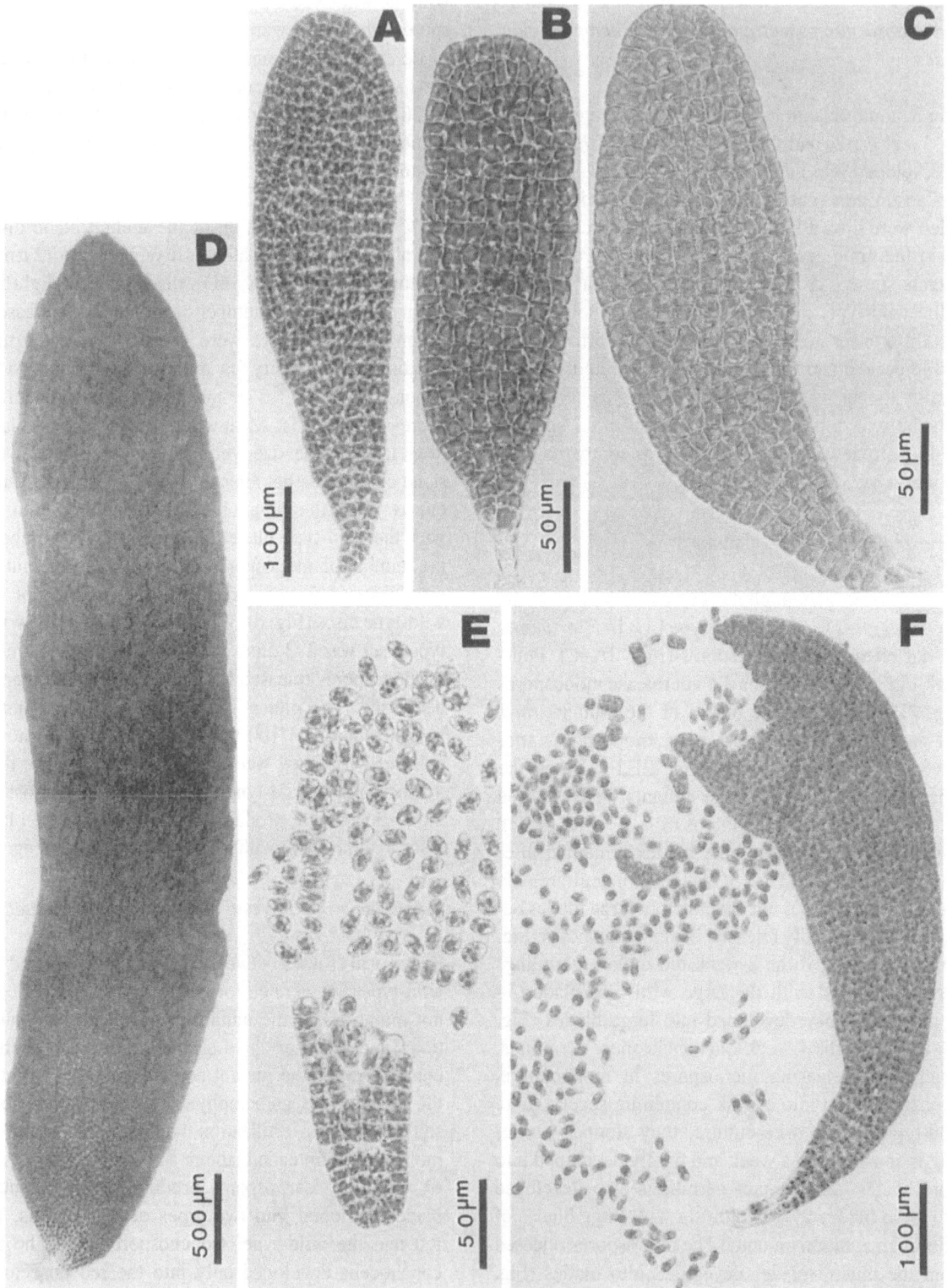

Figure 2. Monospore production of young blades in mutants (*rm-1* and *gm-2*) and the wild-type (*wt*). (A) *wt*; (B) *rm-1*; (C) *gm-2*; (D) *wt*; (E) *rm-1*; (F) *gm-2*. Age: 16 d old in A–C; 25 d old in D and 22 d old in E–F.

Figure 3. Growth and monospore production of mutant blades (*rm-1* and *gm-2*) in both monoculture and co-culture with the wild-type (*wt*). (A–C). Blades of *rm-1*, *gm-2* and *wt* grown in co-culture; (D) and (E) Blades of *rm-1* and *gm-2* grown in monoculture; (F) Monospore germlings of *rm-1* attached on the surface of blades shown in D; (G) Blade shown in D after monospore release; (H) Monospore germlings of *gm-2* attached on the surface of blades shown in E; (I) Blade shown in E after monospore release; A–E, at the same magnification. Age: 42 d old in A–E, G and I, and 3–5 d old in F and H.

F₁ gametophytic blades having both parental color phenotypes, indicating that the wild-type conchocelis was heterozygous due to cross-fertilization. However, in these crosses, carpospores from the wild-type blade only yielded wild-type conchocelis. Conchospores from some conchocelis colonies developed only into the wild-type unsectored F₁ gametophytic blades, indicating that these conchocelis were homozygous. On the other hand, conchospores from other conchocelis colonies developed into both unsectored and sectored F₁ gametophytic blades having both parental color

phenotypes, indicating that these conchocelis were heterozygous.

Color phenotypes and blade types of F₁ gametophytic blades from the heterozygous conchocelis produced in the cross *rm-1*♀×*wt*♂ are shown in Table 1. Only both the parental color phenotypes appeared in F₁ gametophytic blades, indicating that the mutant *rm-1* has a single mutation with respect to color. There were two types of unsectored blades and six types of sectored blades consisting of 2, 3 or 4 sectors. The sectored F₁ blades, reflecting the color segregation at meiosis, represented 98.9% of the total, giving a

Table 1. Color phenotypes, blade types and color segregation of F_1 gametophytic blades from heterozygous conchocelis in the cross $rm\text{-}1♀ \times wt♂$ in *Porphyra yezoensis*. W. wild-type; R. red

Blade types	Number of blades $rm\text{-}1♀ \times wt♂$
Unsectored	
W	9
R	3
Sectored[a]	
W+ R	472
R +W	392
W+ R +W	78
R +W+ R	100
W+ R +W+ R	7
R +W+ R +W	4
Unsectored	1.1%
Sectored	98.9%
Segregation ratio of colors	1152W : 1184 R
in sectored blades	(1W : 1.03R)

[a]Color sectors are shown in the order from the base to the apex.

nearly perfect 1:1 segregation ratio (1152W:1184R) in the sectored blades.

After 4 weeks in aerated culture, monospore production of the F_1 blades was examined microscopically. Many monospores released only from the unsectored red blades and the red sectors of the sectored blades. High monospore production was observed not only in the upper but also in middle and lower red sectors of the sectored blades when they were cultured with aeration. In the absence of large wild-type blades, the young red unsectored F_1 blades and the red sectors of the sectored F_1 blades often vanished after releasing many monospores, leaving only unsectored wild-type blades or wild-type sectors. However, when the F_1 blades were cultured without aeration, fewer monospores were usually released from the red unsectored blades and the upper red sectors of the sectored blades. Most (98.8%) of the monospore germlings developed from the monospores of the F_1 blades were the red one, which released monospores again in 3-4 weeks.

Discussion

Several high monospore-producing mutants were obtained by treatment with MNNG (Yan, 1997). Most of them were different in color compared to the wild-type. Out of seven such mutants, six were red, red

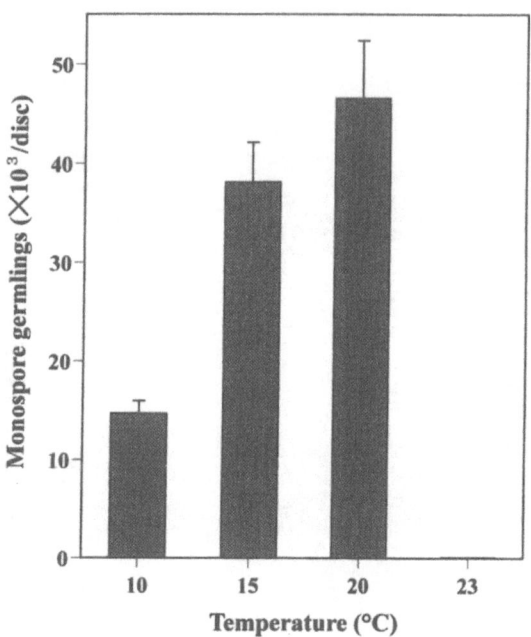

Figure 4. Effects of temperature on monospore production of the mutant (*rm-1*) blade discs. Monospore germlings developed from the monospores released from the mutant blade discs. Results represent the means of three independent experiments ± SE.

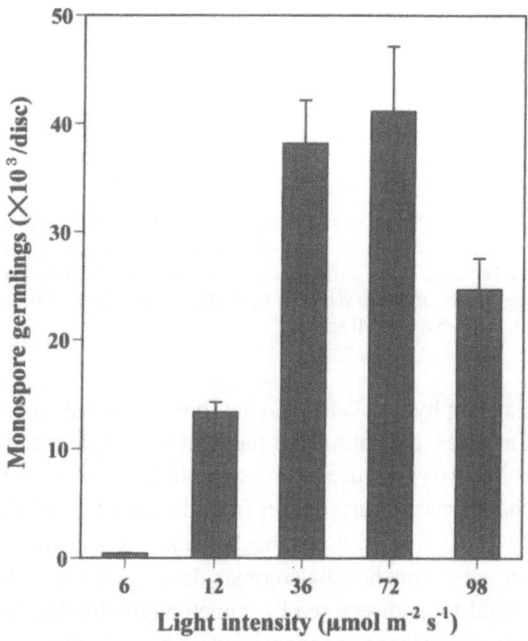

Figure 5. Effects of light intensity on monospore production of the mutant (*rm-1*) blade discs. Monospore germlings developed from the monospores released from the mutant blade discs. Results represent the means of three independent experiments ± SE.

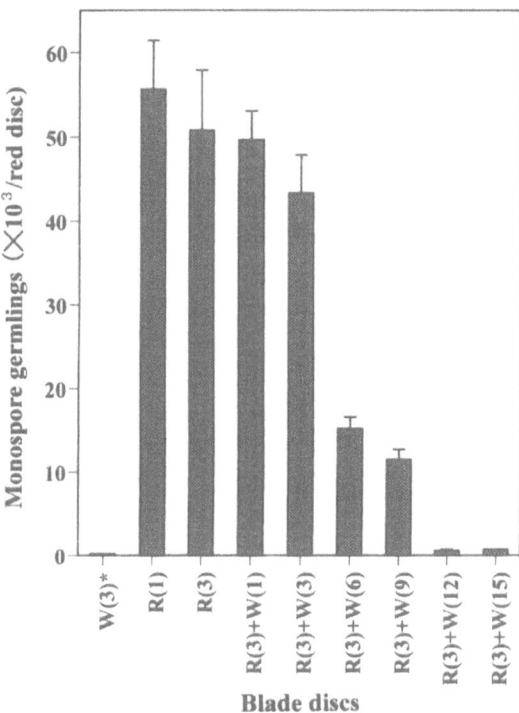

Figure 6. Monospore production of the mutant (*rm-1*) blade discs (R) in monoculture and co-culture with the wild-type blade discs (W). Monospore germlings developed from the monospores released from the mutant blade discs. *Monospore germlings of the wild-type blade discs. The number of the blade discs is shown in parentheses. Results represent the means of three independent experiments ± SE.

and other pigmentation mutants. In the sectored F_1 gametophytic blades from heterozygous conchocelis produced in the cross *rm-1*♀×*wt*♂, blade thickness of the red sector was only 72% of the wild-type sector. These results suggest that components and construction of both cell wall and cell wall matrix in these mutants might have been changed. When cultured under the same culture conditions, the gametophytic blades of the spontaneous red mutant (*C-22*) (Aruga & Miura, 1984), which showed very similar color and *in vivo* absorption spectra to the mutant *rm-1*, did not release many monospores, indicating that these two mutants are different.

Culture conditions, e.g., temperature, light intensity and aeration, significantly affected monospore production of the mutants. The young mutant blades, which developed from either conchospores or monospores, often began to release many monospores in the fourth week in aerated culture but released fewer monospores even in the sixth week of culture without aeration. When culture conditions were quickly changed, e.g., by changing all the culture media, applying aeration and increasing temperature and/or light intensity, the mutant blades usually released so many monospores that the upper portions of the blades quickly vanished. When the mutant blade discs were cultured at too high a temperature (23 °C) or under too low a light intensity (6 μmol photons m^{-2} s^{-1}), no substantive growth was observed and very few monospores were obtained. The mutant discs (*rm-1*) taken from the large blades (10–15 cm long, 3 months old) usually began to release monospores after the discs had grown to 1.5–2.5 times the original size in diameter after about 2 weeks of aerated culture. However, the mutant discs taken from the young blades (3–5 cm long, 5 weeks old) often released many monospores within a week under the same culture conditions. These results suggest that monospore production of the mutants varied according to variations in their physiological environment and the age of their blades.

The young blades and the large blade discs of the mutant *rm-1* often began to release monospores simultaneously in the periphery and the central part of the blade or blade disc. However, the wild-type blade usually released only very few monospores in the periphery of the blade. During the formation of monospores, most parts of the mutant blades faded 2–3 days before monospore release. After releasing many monospores, most parts of the blade vanished and the mucous substances remained.

orange, yellow orange and green, and one was like the wild-type in color. In general, the greenish mutants obtained with MNNG treatment tend to produce fewer monospores compared to the wild-type and other pigmentation mutants. Therefore, the high monospore-producing green mutant *gm-2* is particularly interesting. The high monospore-producing mutants *rm-1* and *gm-2* matured much later than the wild-type. Blade discs taken from the 3-month-old blades of both the mutant *rm-1* and the wild-type, which were cultured under the same conditions and at the same age, were cultured alone under the same conditions for growth test. After 20 d in culture, the mutant blade discs and the wild-type discs grew respectively to 11.6 and 2.2 times the original size in diameter, suggesting that the old mutant blade discs continued to exhibit fast growth rate.

Gametophytic blades of the high monospore-producing mutants *rm-1* and *gm-2* were thinner (their blade thickness was 69–74% of the wild-type) and lacked elasticity as compared with the wild-type

140

When the blades or blade discs of the mutant *rm-1* were co-cultured with large blades or blade discs of the wild-type, they stopped producing monospores and developed into large blades. The results suggest that the mutant probably needs some substance(s) released by the large wild-type blades to prevent monospore production. However, there is another possibility that the nutrients in the culture medium will be more quickly exhausted by adding the large wild-type blades or its blade discs into cultures of the mutant blades or blade discs, and let monospore production of the mutant blades be prevented.

In F_1 gametophytic blades from heterozygous conchocelis in the crosses of *rm-1*×*wt*, the high monospore production of the blades was associated with the mutant color. This result suggests that the high monospore production of the mutant is controlled by gene(s) that is closely linked with the gene for the mutant color. However, the number of genes involved in this high monospore production of the mutant was unclear.

The high monospore-producing mutants reported in the present study will be useful for studying the mechanism of formation and release of monospores in *P. yezoensis*. The mutant *rm-1* will also be directly used in obtaining large number of monospores as a primary source of spores for seeding nets in *P. yezoensis*. It will also be used in further breeding because it has several other desirable characters, including fast growth and late maturation relative to the wild-type.

Acknowledgements

Grateful acknowledgments are due to Japan Society for the Promotion of Science (JSPS) for granting X. H. Yan as a foreign research fellow in Japan. This study was supported by a grant-in-aid for JSPS Fellow from the Ministry of Education, Science, Sports and Culture, Japan.

References

Aruga, Y. & A. Miura, 1984. *In vivo* absorption spectra and pigment contents of the two types of color mutants of *Porphyra*. Jap. J. Phycol. 32: 243–250.

Chen, M. Q., B. F. Zheng & G. Z. Ren, 1985. The influence of temperature on the adherence of the conchospores and the monospores of *Porphyra yezoensis* Ueda. Trans. Oceanol. Limnol. 3: 66–69 [in Chinese with English abstract].

Kato, M. & Y. Aruga, 1984. Comparative studies on the growth and photosynthesis of the pigmentation mutants of *Porphyra yezoensis* in laboratory culture. Jap. J. Phycol. 32: 333–347.

Li, S. Y., 1984. The ecological characteristics of monospores of *Porphyra yezoensis* Ueda and their use in cultivation. Hydrobiologia 116/117: 255–258.

Li, S. Y. & G. F. Cui, 1980. An observation on the growth and development of sporelings from conchospores and monospores of *Porphyra yezoensis*. Chin. J. Oceanol. Limnol. 11: 370–374 [in Chinese with English abstract].

Ohme, M., Y. Kunifuji & A. Miura, 1986. Cross experiments of the color mutants in *Porphyra yezoensis* Ueda. Jap. J. Phycol. 34: 101–106.

Yan, X. H., 1997. Induction, isolation and characterization of pigmentation mutants in *Porphyra yezoensis* Ueda (Bangiales, Rhodophyta). Ph. D thesis, Tokyo University of Fisheries, Tokyo, Japan. 101 pp+ 42 Tables + 217 Figures.

Yan, X. H. & Y. Aruga, 1996. Induction and isolation of pigmentation mutants by treatment of *Porphyra yezoensis* Ueda (Rhodophyta) with NNG. First Asian Pacific Phycol. Forum, 22–25 July 1996, Sydney, Australia. p 34.

Yan, X. H. & Y. Aruga, 1997a. Induction of pigmentation mutants by treatment of monospore germlings with NNG in *Porphyra yezoensis* Ueda (Bangiales, Rhodophyta). Algae 12: 39–52.

Yan, X. H. & Y. Aruga, 1997b. Induction and isolation of pigmentation mutants by treatment of *Porphyra yezoensis* Ueda (Rhodophyta) with NNG. Phycologia 36: 128.

Yan, X. H. & Y. Aruga, 1998. Unstable pigmentation mutants obtained by NNG treatment in *Porphyra yezoensis* Ueda (Bangiales, Rhodophyta). Jpn. J. Phycol. 46: 89.

Yan, X. H. & Y. Aruga, 2000. Genetic analysis of artificial pigmentation mutants in *Porphyra yezoensis* Ueda (Bangiales, Rhodophyta). Phycol. Res. 48: 177–187.

Yan, X. H., Y. Fujita & Y. Aruga, 1999. Effect of culture conditions on color segregation in F_1 foliose thalli from heterozygous conchocelis in cross-experiments of *Porphyra yezoensis* Ueda. Jpn. J. Phycol. 47: 101.

Hydrobiologia **512**: 141–144, 2004.
P.O. Ang, Jr. (ed.), Asian Pacific Phycology in the 21st Century: Prospects and Challenges.
© 2004 *Kluwer Academic Publishers.*

An improved chromosome preparation from male gametophyte of *Laminaria japonica* (Heterokontophyta)

Li Ran Zhou[1,*], Ji Xun Dai & Song Dong Shen
Department of Marine Biotechnology, College of Marine Life Sciences, Ocean University of Qingdao, Qingdao, China
[1] *Present address: Department of Biology, The Chinese University of Hong Kong, Shatin, N.T. Hong Kong*
Author for correspondence; E-mail: liranzhou@hotmail.com

Key words: China, chromosome, male gametophyte, *Laminaria japonica*

Abstract

An improved method for the preparation of chromosomes from the male gametophyte of the alga *Laminaria japonica* Aresch. was described. The male gametophyte was pretreated with pDB (p-dichlorobenzene) and 8-hydroxyquinoline in order to clear cell wall and soften cytoplasm. The samples were treated by mordant iron alum [FeNH$_4$ (SO$_4$)$_2$·6H$_2$O] followed by staining with haemotoxylin. Well-spread and highly stained chromosomes were observed without precipitation. The chromosome number of male gametophyte of *L. japonica* was estimated to be 31.

Introduction

Parthenogenesis was observed in the gametophyte of *Laminaria japonica* Aresch. when gametophytes were cultured individually and grown vegetatively into sporophytes (Fang et al., 1978; Fang & Dai, 1980; Dai et al., 1993). Cytological observation on natural parthenogenesis of this species indicated that the doubling of chromosomes only occurred in female parthenogenetic sporophyte, while this phenomenon was absent in male parthenosporophyte (Fang & Dai, 1980; Dai et al., 1993). Therefore, unlike female parthenosporophytes, the males were sterile, for no active sperms were produced (Fang & Dai, 1980; Dai et al., 1993). It has been reported by Jiang & Tang (1979) that reproductive haploid sporophytes could be obtained from male gametophytes, however, these plants were abnormal in shape and repeated attempts to produce haploid sporophytes have failed. All thalli derived from male gametophytes were found to be haploid, the shapes of which were mostly abnormal and no spores could be released from these thalli (Dai et al., 1997). It was postulated by Dai et al. (1997) that a single set of chromosomes cannot lead to successful meiosis and no spores could thus be produced from haploid male parthenosporophytes. Theoretically, the

chromosome-doubling chemical, colchicine, could be used as a strategy to improve the propagation of the male pathenosporophyte. Furthermore, hybridization of diploid male and female parthenosporophytes, both of which possess good growing properties, would greatly promote the field output of this kelp (Dai et al., 1997). A chromosome number of $n = 22$ for *L. japonica* has been widely accepted for decades (Abe, 1939; Yabu, 1973; Tai & Fang, 1977; Lewis et al., 1993), but the latest observation revealed a new number of $n = 31$ (Yabu & Yasui, 1991). It must be emphasized that the chromosome number should be confirmed before any chromosome doubling experiment can be carried out.

The classic paraffin method was popular in cytological and chromosomal studies of algae before the appearance of the squashing method (Abe, 1939; Nishibayashi & Inoh, 1956; Yabu, 1964). After the squashing technique of aceto-iron-haematoxylin-chloral hydrate has been published (Wittmann, 1962, 1965), it was rapidly applied to algal chromosome preparation (Yabu & Tokida, 1966; Lewis, 1993). The chromosome numbers of *Laminaria japonica* (Yabu, 1973); *L. yendoana* Miyabe (Yasui, 1992); *Kjellmaniella crassifolia* Miyabe (Yabu et al., 1985) and *Ecklonia kurome* Okamura (Yabu & Taniguchi, 1990) were preliminarily identified with Wittman's method where chloral hydrate had a clearing function. However, chloral hydrate is highly toxic, and is particularly harmful when tests were performed on a large scale. Therefore, an improved chromosome preparation with equal efficiency but a higher safety factor is urgently needed.

The aims of this study were to improve the method of chromosome preparation from male gametophyte of *L. japonica* and to determine its chromosome number.

Material and methods

Male gametophyte of *L. japonica* used in this study was the strain lj 4-2.1982, isolated in 1982 from a sporophyte and preserved at 4 °C in the Laboratory of Genetics, Ocean University of Qingdao, China.

The male gametophyte was digested with gentle stirring in a multi-enzyme solution (5% cellulase: 4% pectinase = 1:1) for 3 h, and centrifuged at $1000 \times g$ for 5 min. The supernatant was discarded and the pellet was re-suspended in sterile seawater and filtered three times through a 40 μm mesh. Fragments that passed through the mesh were allowed to attach to the glass slides. The gametophyte fragments were cultured in enriched sterile seawater ([N] 3×10^{-4} mol l^{-1}, [P] 1×10^{-5} mol l^{-1}) in an illuminating incubator at 18 °C under cool-white fluorescent light at 2000 lx with a 12h:12h LD photoperiod. Two weeks later, when the fragments grew into numerous filaments, they were ready for pretreatment.

Three or four hours into the dark period, filaments were successively treated with 0.1% colchicine, saturated pDB (p-dichlorobenzene), and saturated 8-hydroxyquinoline for 3 h, 1 h and 15 min, respectively. After each treatment, samples were washed to remove chemical residues.

The filaments were fixed in Carnoy's fixative (100% alcohol: 100% acetic acid = 3:1) for at least 24 h. Prior to squashing, samples were mordant in 2% iron alum for 15 min to allow binding of iron on the chromosomes. The preparation was washed in tap water, and then stained in 5% haematoxylin for 5 min. At this time, the filaments were stained dark blue and it was impossible to differentiate the chromosomes from the cytoplasm. Thus, it is necessary to remove excess stain from the sample with a drop of 45% acetic acid. The color of the filaments then turned into pale blue, a cover slip was placed in position and a flat instrument was used for squashing. Squashing required

more pressure than with other materials because of solid nature of the filament cell wall.

Results and discussion

Chromosome preparations from male gametophyte were clear and countable (Figs 1 and 2). Chromosomes dispersed well without dye precipitation, and cytoplasm was stained light (Fig. 1). The chromosome numbers counted in male gametophyte cells ranged from 29 to 32, most of which were 31 (Fig. 2).

This study showed that the chemical pDB has a cytoplasm-decomposing effect on gametophyte cells. Previously, pDB was seldom applied to chromosome preparation of higher plants as pretreatment reagent

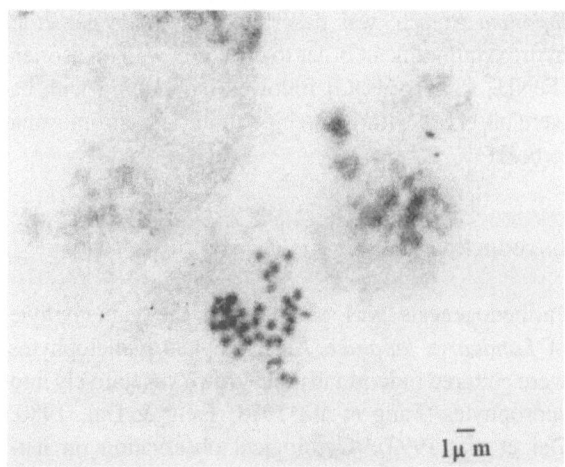

Figure 1. Chromosomes of the male gametophyte of *L. japonica* (partial view). Scale bar = 1 μm.

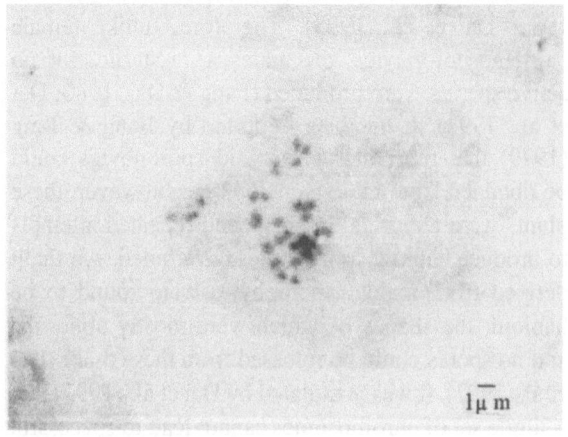

Figure 2. Chromosomes in a cell of the male gametophyte of *L. japonica* ($n = 31$). Scale bar = 1 μm.

because its strong scavenging effect may lead to chromosome fracture (Li & Zhang, 1991). In this study, however, it was demonstrated that the clearing property of pDB was effective in male gametophyte of lower plant *L. japonica* and cytoplasm staining has been reduced, leading to a light-colored background. Furthermore, using pDB not in working stain but as pretreatment has protected operators from contacting the harmful dye during squashing. 8-hydroxyquinoline is generally applied as a condensing reagent for chromosome preparation in higher plants (Li & Zhang, 1991). This study has shown that after 8-hydroxyquinoline treatment, the cell wall of male gametophyte became crisp and could be easily crushed but the mechanism of action of 8-hydroxyquinoline is still unclear.

The stain haematoxylin, combined with iron-mordant, yielded intense and selective chromosome staining in plant materials. Previous work involved mixing of iron alum and haematoxylin just before staining. Stain precipitation would rapidly occur from the binding of mordant and dye, the resulting light scatter would cause difficulty in chromosome counting. In this study, however, the mordant and stain were stored separately and used sequentially in order to avoid stain precipitation. Otherwise, complete washing of the mordant from samples could also help reduce possible precipitation. The present method was easy to perform and highly effective in producing chromosome preparation from the male gametophyte of *L. japonica*. The technique could equally be applied to chromosome preparation of other algal species.

A chromosome number of $n = 22$ was first observed in the sporangium of *L. japonica* (Abe, 1939), and this conclusion was supported by numerous reports (Yabu, 1973; Tai & Fang, 1977; Lewis et al., 1993). The evidence that nuclei in gametophytes of *Laminaria japonica* are haploid and those in sporophytes are diploid was put foward by Yabu (1973). Yabu (1973) suggested that it was hard to count the chromosome number in gametophyte and the number ranged from 16 to 22 occasionally in female gametophytes, rarely in males. Recently, Yabu & Yasui (1991) reported that the chromosome number of *L. japonica* was normally $n = 32$, which is similar to that of *L. angustata* Kjellm., *L. ochotensis* Miyabe and *L. religiosa* Miyabe. The chromosome number of male gametophytes examined in most samples in the present study was 31, which was different from those reported by Yabu (1973) ($n = 22$), Lewis et al. (1993) ($n = 22$), and by Nakahara (1984) (n = 28–35), but similar to

the latest report of $n = 32$ by Yabu & Yasui (1991). Results presented here mostly supported Yabu's conclusion with minor differences. Most of the former results were obtained from sporophytes of *L. japonica*, while this study used male gametophytes. It is reasonable to believe that the data obtained from the haploid is more reliable, because a smaller chromosome number can reduce the counting error. The reason for the number discrepancy is unknown, although it seems that differences in methodology may be responsible. Future effort will be directed at improving the squash method.

References

Abe, K., 1939. Mitosen in sporangium von *Laminaria japonica* Aresch. Sci. Rep. Tohoku Imp. Univ. 4th ser. (Bot.) 14: 327–329.

Dai, J. X., J. J. Cui, Y. L. Ou & Z. X. Fang, 1993. Genetical study on the parthenogenesis in *Laminaria japonica*. Acta Oceanol. Sinica 12: 295–298.

Dai, J. X., Y. L. Ou, J. J. Cui, Q. L. Gong, H. Wang, M. L Wang & J. E. Xu, 1997. Study on the development of male gametophytes of *Laminaria japonica* Aresch. J. Ocean Univ. of Qingdao 27: 41–44. [in Chinese with English abstract].

Evans, L.V., 1965. Cytological studies in the Laminariales. Ann. Bot. 29: 541–562.

Fang, Z. X. & J. X. Dai, 1980. The use of haploid phases in the genetic study of *Laminaria japonica*. Acta Gene Sinica 7: 19–25.

Fang, T. C., J. H. Tai, Y. L. Ou, C. C. Tsui & T. C. Chen, 1978. Some genetic observations on the monoploid breeding of *Laminaria japonica*. Sci. Sinica 21: 401–408.

Jiang, B. Y. & Z. J. Tang, 1979. Mature sporophytes cultured from the male gametophytes of *Laminaria japonica* Aresch. Kexue Tongbao. 24: 713–714 [in Chinese with English abstract].

Lewis, R. J., B. Y. Jiang, M. Neushul & X. G. Fei, 1993. Haploid parthenogenetic sporophytes of *Laminaria japonica* (Phaeophyceae). J. Phycol. 29: 363–369.

Li, M. X. & X. F. Zhang, 1991. Plant Chromosome Research Technique: Forestry University of Northeast China Press, Haerbin, China: 17–18 [in Chinese].

Nakahara, H., 1984. Alternation of generations of some brown algae in unialgal and axenic cultures. Sci. Pap. Inst. Algol. Res., Fac. Sci., Hokkaido Univ. 7: 277–292.

Nishibayashi, T. & S. Inoh, 1956. Morphogenetical studies in Laminariales. I. The development of zoosporangia and the formation of zoospores in *Laminaria angustata* Kjellm. Biol. J. Okayama Univ. 2: 147–158.

Tai, S. H. & T. C. Fang, 1977. The chromosomes of *Laminaria japonica* Aresch. Acta Gene Sinica 4: 325–328.

Wittmann, W., 1962. Aceto-iron-haematoxylin-chloral hydrate for staining chromosome in squashes of plant material. Stain Technol. 37: 27–30.

Wittmann, W., 1965. Aceto-iron-haematoxylin-chloral hydrate for chromosome staining. Stain Technol. 40: 161–164.

Yabu, H., 1964. Mitosis in the sporangium of *Agarum cribrosum* Bory and *Alaria praelonga* Kjellman. Bull. Fac. Hokkaido. Univ. 5: 1–4.

144

Yabu, H., 1973. Alternation of chromosome in the life history of *Laminaria japonica* Aresch. Bull. Fac. Fish. Hokkaido Univ. 23: 171–176.

Yabu, H. & K. Taniguchi, 1990. Mitosis in the young sporophytes of *Ecklonia kurome* Okamura. Bull. Fac. Fish., Hokkaido Univ. 41: 57–60 [in Japanese with English abstract].

Yabu, H. & J. Tokida, 1966. Application of aceto-iron-haematoxylin-chloral hydrate method to chromosome staining in marine algae. Bot. Mag. Tokyo. 79: 381–381.

Yabu, H. & H. Yasui, 1991. Chromosome number in four species of *Laminaria* (Phaeophyta). Jap. J. Phycol. 39: 185–187.

Yabu, H., H. Yamamoto & H. Yasui, 1985. Cytological observation on *Kjellmaniella crassifolia* and *Eisenia bicyclis* (Laminariales, Phaeophyceae). Bull. Fac. Fish., Hokkaido Univ. 36: 64–68 [in Japanese with English abstract].

Yasui, H., 1992. Chromosome number and a sex chromosome of*Laminaria yendoana* Miyabe (Phaeophyta). Nippon Suisan Gakkaishi 58: 1385–1385.

Hydrobiologia **512**: 145–151, 2004.
P.O. Ang, Jr. (ed.), Asian Pacific Phycology in the 21ˢᵗ Century: Prospects and Challenges.
© 2004 *Kluwer Academic Publishers.*

Solving the coastal eutrophication problem by large scale seaweed cultivation

Xiugeng Fei

Institute of Oceanology, Chinese Academy of Sciences, Qingdao 266071, China
E-mails: fei@ms.qdio.ac.cn

Key words: environment, eutrophication, *Laminaria*, *Gracilaria*, *Porphyra*, red tide, seaweed cultivation, China

Abstract

Eutrophication is becoming a serious problem in coastal waters in many parts of the world. It induces the phytoplankton blooms including 'Red Tides', followed by heavy economic losses to extensive aquaculture area. Some cultivated seaweeds have very high productivity and could absorb large quantities of N, P, CO_2 , produce large amount of O_2 and have excellent effect on decreasing eutrophication. The author believes that seaweed cultivation in large scale should be a good solution to the eutrophication problem in coastal waters. To put this idea into practice, four conditions should be fulfilled: (a) Large-scale cultivation could be conducted within the region experiencing eutrophication. (b) Fundamental scientific and technological problems for cultivation should have been solved. (c) Cultivation should not impose any harmful ecological effects. (d) Cultivation must be economically feasible and profitable. In northern China, large-scale cultivation of *Laminaria japonica* Aresch. has been encouraged for years to balance the negative effects from scallop cultivation. Preliminary research in recent years has shown that *Gracilaria lemaneiformis* (Bory) Daws. and *Porphyra haitanensis* Chang *et* Zheng are the two best candidates for this purpose along the Chinese southeast to southern coast from Fujian to Guangdong, Guangxi and Hong Kong. *Gracilaria tenuistipitata* var. *liui* Chang *et* Xia is promising for use in pond culture condition with shrimps and fish.

Introduction

Coastal environment is usually associated closely with intensive human economic activities. It is a very important part of the human society. Coastal marine environment receives all the water run off from land through river or ground water systems. With it are all the organic and inorganic wastes. In the later part of the last century, the coastal environment of China and many other parts of the world have experienced a tremendous increase in mariculture activities. These coastal zones have thus become an important center for the exploitation of marine fisheries resources. Because coastal environment is usually quite narrow, with shallow water depth and restricted water circulation, flushing rate is usually limited. One important consequence of this is that the heavy use of the coastal environment often resulted in the accumulation of wastes from different human activities that eventually lead to serious pollution problem. This in turn leads to the deterioration of the coastal environment that triggers a chain of negative reactions and consequences, including coastal eutrophication, frequent occurrence of red tides, and breakdown of sustainable economic development. Eutrophication, with excessive amount of N, P, CO_2 and insufficient amount of dissolved O_2, is becoming a serious problem in coastal seawater environment all over the world and has attracted much attention in recent years. Many researchers believe that it it possible but very difficult to address this problem.

For years, large-scale cultivation of *Laminaria japonica* Aresch. has been encouraged in northern China to balance the negative environmental effect due to scallop cultivation. This strategy works well at a preliminary stage. But to use this strategy as an ultimate solution to eutrophication due to mariculture, more basic and applied researches are needed. Per unit area production of cultivated seaweeds is high. Productivity of cultivated seaweeds is much higher than that in their natural habitats. Cultivated seaweed grows much better at those areas where N and P are abundant (Lobban & Wynne, 1981; Miura, 1992; Ohno & Critchley, 1993; Critchley & Ohno, 1998; Fei, 1996). For cultivation of *Porphyra yezoensis* Ueda in China, good harvest and good quality could only be achieved in those regions where N concentration (NO_3-N + NH_4-N) in the seawater is over 100 mg m^{-3} (>7.2 mM). When N-nutrient level is below 50 mg m^{-3} (3.6 mM), the quality and yield will go down and application of N fertilizers becomes necessary (S.E.P.E. & S.A.T.M., 1978). On the other hand, seaweed requirement for P is not as sensitive as that for N. From many years of practice and observation this author has found that this is true not only for cultivated *Laminaria* and *Porphyra* but also for cultivated *Gracilaria*. This author has thus proposed that large scale seaweed cultivation could be a possible solution to the problem of eutrophication by decreasing nutrients in seawater (Fei, 1978, 1983, 1996; Fei *et al.*, 1998, 1999). Cultivation experiments have been conducted since 1987 along the coastal region in South China. Most of the results were encouraging. This idea of using seaweed to reduce excessive nutrient, hence solving the problem of eutrophication, is thus very promising and further research efforts are now being continued to evaluate its potentials.

The sources of organic eutrophication in the sea and the coastal environment in mainland China

Fast growing large coastal cities and increasing population in these cities produce large amount of CO_2 and refuses containing large quantity of N and P. Large quantities of N and P fertilizers are applied in agriculture on land and only a part of these are absorbed by the plants, with the rest being eventually released to the environment. Culture of livestock releases huge amount of excrement to the environment as well. All these eutrophication products are created by human activities on land. They will eventually reach the coastal seawater via sewage and other pathways and become the main source of eutrophication in the sea.

The secondary pollution elements in coastal water are created by extensive mariculture of animals itself. The animal cultivation develops so fast in a vast area of coastal waters, such that in many places almost every space is occupied and exceptionally dense population of these cultivated animals has been formed. Large amount of N, P and CO_2 released from these dense populations of cultivated animals are being recycled back to the cultivation area, creating the secondary source of eutrophication (Zhang et al., 1994). Other than the above two sources, over feeding of shrimp in ponds and fish in cages also produces great amounts of nutrient rich sediment to their surrounding environment. This becomes another major input source of N, P and CO_2 to the mariculture environment (Zhu & Li, 1996).

Since the 80's, coastal mariculture development in China has increased tremendously. More than 40,000 km^2 of the coastal area are now being used for mariculture and more than three million people are directly engaged in mariculture related activities. The production output from mariculture in 1982 was 86 million tons. Ten years later in 1997, this reached 792 million tons, and in 1999, 974 million tons. For several years now, mariculture production in China, now stands at 39% of its total fisheries production, is the highest in the world. Over this same period, however, the occurrence of red tide event has increased from an average of 50 times annually in the 80's to 250 times annually in recent years. Because many aquaculture facilities are placed in the water for prolonged period of time, fouling by all sort of benthic organisms is serious. Heavy sedimentation of organic materials, e.g. from excessive feeds used in mariculture and from animal excrement, led not only to increase in the amount of organic nutrients in the surface water, but also to poor oxygen condition in the near bottom water. The quality of water in mariculature zones deteriorated, inducing diseases and red tide occurrence. This vicious cycle of events led to heavy economic losses. In 1998 alone, the occurrence of diseases in mariculture animals reached a high rate of 30%, with a direct loss of more than 10 billions Chinese Yuan.

In 1998, the coastal areas in China with water quality that was below the first class water quality standard reached 228,000 km^2. This was a 100% increase over that recorded in 1992. Of this, about 40,000 km^2 of the coastal areas were below the fourth class water quality standard. These areas could no longer satisfy the requirement for mariculture, water sports and coastal tourism activities. In some areas, the concentration of N is 10 times higher than that allowed for mariculture zone.

Pollution created by human activities induces eutrophication in the sea. The sources of pollution mainly come from the land and also from coastal waters by mariculture itself. This should be taken into consideration if this important problem is to be solved.

Effect of marine plants on eutrophication in the sea

Autotrophic and heterotrophic organisms are the two main inhabitants in the sea. All animals like fish and shrimp are heterotrophic; they feed on other organisms to obtain energy and raw material for their growth and release N, P and CO_2 via excrement and respiration. All marine plants like seaweeds or diatom are autotrophic; they feed themselves by absorbing N and P from the seawater for their nutrition and CO_2 also from the seawater for their photosynthesis. Living plants release O_2 via photosynthesis, release few N and P compounds via excretion but the amount is very small in comparison to those of the animals. Therefore, heterotrophic animals are the producers of N, P and CO_2 in the sea and autotrophic organisms are the removers. The activities of the autotrophic organisms balance those of the heterotrophic organisms (Lobban & Wynne, 1981; Tseng, 1990).

There is a large diversity of marine plants in the sea including the small sized micro-algae (such as diatoms) and the large sized macro-algae (such as *Laminaria* and *Porphyra*) (Radmer, 1996). Micro-algae grow well in nutrient rich seawater and some species grow very fast and could make a bloom in the sea within a few days during warm season. They do remove large quantity of nutrients from their surrounding seawater and perform as an eutrophication remover. However, they would not last very long because of their short life span of only a few days. They eventually die and huge amount of dead bodies would accumulate in the water. Consequently the bacteria will decompose their dead bodies, thus creating a low dissolved Oxygen environment. The nutrients from these dead bodies will eventually be released back into the seawater. The micro-algae started as a nutrient remover but turned to become the source of eutrophication. Macro-algae (seaweeds) on the other hand could accumulate considerable biomass within months and years. This is especially so for cultivated species during cultivation season. Based on this author's data so far, biomass (wet weight) of cultivated *Laminaria* just before harvest is 80–120 tons ha^{-1}, that of cultivated *Porphyra* is 30–60 tons ha^{-1} and that of cultivated *Gracilaria* 40–80 tons ha^{-1}. When they grow in large-scale, macro-algae accumulate large amount of C, N and P in their body.

Table 1 shows the C, N and P contents of some of the common cultivated species like *Laminaria*, *Porphyra* and *Gracilaria*. This shows that these cultivated species have a special ability to remove the main eutrophication elements of C, N and P from their surrounding seawater. Because their life span is at least several months longer than that of micro-algae, this makes them very effective removers of these elements. When cultivated macro-algae are harvested from their culture ground, 25–79 kg (mean 52 kg) of C and 2.5–6.2 kg (mean 4.4 kg) of N will be removed from the water with every ton wet weight of these algae harvested. As mentioned above the magnitude of productivity of cultivated seaweed is 30–120 wet ton ha^{-1}. That means that within 2 m depth of the surface seawater that is usually occupied by the cultivated algae, the cultivated population will remove 1560–6240 kg C or 132–528 kg N from 20 000 m^3 of seawater within one ha of cultivation area. Take N as an example, the removing rate will be 6600 mg N m^{-3}, 16.5 times higher than 400 mg N m^{-3} which is the indication level of N eutrophication. These figures make cultivated macro-algae ideal candidate as nutrient removers and point to their potential application for this purpose.

Possible ways to solve the problem of eutrophication

Given that the main elements causing eutrophication in the sea originally come from land through human activities, a solution to this is to control these activities at their source. Decreasing the export of terrestrial nutrients should be taken into first consideration. This is, however, not an easy task and needs huge amount of effort and investment on infrastructure development, research, careful planning and coordination before the goals could be achieved. This will take time and hence could only be considered as a long term target. Under the present situation, there may be palliative but workable short term solutions to these problems.

Seaweeds are nutrients remover. There is, therefore, a great potential to remove large amount of C, N, and P nutrients with extensive seaweed cultivation. Seaweeds produced from these cultivations have many commercial and industrial applications. They could be

Table 1. Nutrient contents (%) of cultivated seaweed[a]

Contents	Laminaria		Porphyra		Gracilaria	
	Dry	Wet	Dry	Wet	Dry	Wet
Protein	8.20	1.40	39.00	3.90	16.30	1.30
Carbohydrates[b]	62.70	10.50	43.00	4.30	78.80	6.30
C	23.60	7.90	27.30	2.70	31.30	2.50
N	1.30	0.22	6.20	0.62	2.60	0.25
P	0.20	0.03	0.58	0.06	0.03	0.03

[a] Resources Council, Science and Technology Agency, Japan (1982).
[b] Includes Carbohydrate, lipid and fiber.

widely used as food, as major raw material for the extraction of chemicals and pharmaceuticals, as fodder for livestock or as organic fertilizer. For instance, agar, carrageenans and alginates are the three major phycocolloids extracted from the seaweeds (Tseng & Fei, 1987; Yarish et al., 1988; Miura, 1992; Jensen, 1996). They are in great demand and have extensive commercial and industrial applications. It is thus possible to promote large-scale cultivation of commercially important species. This will not only be a powerful method for removing large amount of nutrients, it could also benefit the local economy by providing jobs and sources of marketable products.

Seaweed cultivation would complement the cultivation of animal ecologically. At present, in most Chinese cultivation regions, animal cultivation prevails over seaweed cultivation and in extreme cases animal is the only organism under cultivation. There is an economic explanation for this phenomenon as animal products tend to command a higher price than seaweed products. However, the author believes that adding seaweed cultivation to these regions as much as possible will help to form a better ecologically balanced culture environment. This is far more critical than purely economic consideration alone.

From idea to real application and prerequisite conditions

Theoretically, to apply seaweed cultivation for eutrophication improvement is possible. To put this idea into practice, some prerequisite conditions should be fulfilled. Firstly, oceanographic conditions should be suitable for large-scale seaweed cultivation. Usually, near-shore or offshore area with 20 m water depth is ideal. Social acceptance is also an important condition for conducting cultivation activities in these areas.

Second, in order to prevent failure, scientific and technological problems for cultivation of the species should have been fundamentally solved before taking on large scale cultivation. Third, transplantation and introduction of 'good' species or strains from outside is often necessary. Harmful ecological effects caused by invading species should be carefully considered and avoided. Finally, There should be a market for the product of seaweed cultivation in order to make the whole exercise economically viable.

Promising cultivated seaweed species as candidates for eutrophication cleaner

Based on extensive research experiences on cultivation biology of economically important seaweeds, this author believes that all cultivated seaweeds have actually played and are continue to play the role of cleaning their surrounding environment by removing large quantity of nutrients. In north China, *Laminaria* and *Undaria* have been the most important seaweeds traditionally cultivated. *Porphyra* is the most important seaweed cultivated traditionally in central China and *Gracilaria* is a newly developed seaweed species cultivated along the southern coast of China in a much smaller scale. China is producing about 3 million tons of *Laminaria* each year. This means about 237 thousand tons of C and 660 tons of N are fixed within a year in cultivated *Laminaria*. Scientific and technological problems for large-scale cultivation of these species have already been solved. It is not difficult to use them as a tool for solving the coastal eutrophication problem. For this new application, the cultivation scale should be much larger than the original. The main problem is how to deal with the huge amount of product from these large-scale cultivations. In northern China, *Laminaria* has been under cultivation for

Table 2. Results of transplantation experiment on *Gracilaria lemaneiformis*. The algae were transplanted from Qingdao and grown in Lianjiang, Fujian, SE China

Culture rope length (m)	Start 11/12/98	Finish 10/05/99	Increase (times)	Growth (% d^{-1})
92	3.22 kg	356.5 kg	110.7	3.14
273	8.19 kg	814.4 kg	99.4	3.06
Total 365	11.4 kg	1170.9 kg	102.6	3.09

many years and is still the best candidate for eutrophication treatment. In southern China, more seaweed cultivation is needed because the problems of eutrophication are more serious. But seaweed cultivation in southern China is largely under-developed at present. Our recent research and experiment has shown that *Gracilaria lemaneiformis* and *Porphyra haitanensis* are the two promising candidates for southern China. The main characters of the four most promising species are described briefly below.

Gracilaria lemaneiformis

This species belongs to Rhodophyta (red algae), growing well along the coast of Shandong. Standing crop in natural habitat is very low. Research and experimental cultivation has been carried out since 1981. In Qingdao it grows in spring and autumn with a growth rate of 6.6% d^{-1} and a 20 times biomass increase in 45 days has been recorded. Its favorable growing temperature is 12–23 °C. Because the cold season is very long in Qingdao, the fast growing period for this species is very short. Twenty times increase in biomass is not good enough for initiating a good cultivation industry in the Qingdao region. But if one could transplant it to southern provinces where favorable culture period in the winter and spring is much longer, we expect much higher harvests. On December 11, 1998 an experiment on transplanting *Gracilaria lemaneiformis* from Qingdao to Lianjiang, Fujian was started. The experiment was completed 150 days later on May 10, 1999 and the result is summarized in Table 2. The result shows a 100 times increase in the biomass of *G. lemaneiformis* within 150 days. In 1997 similar experiment on transplanting this species from Qingdao to Zhanjiang, Guangdong was carried out and we obtained 63 times increase of biomass within 45 days. In early 1999, to research on eutrophication improvement we transplanted *G. lemaneiformis* from Qingdao

to Santou, Guangdong and obtained 207 times biomass increase within 64 days. All these transplanting results have convinced us that this species introduced from further North to the South will grow much better in its new environment than in its native old region (Brawley & Fei, 1988; Zhang et al., 1993; Fei, 1996; Fei et al., 1998). Because of high temperature during most time of the year, *G. lemaneiformis* could not pass the summer in the southern provinces in China. Harmful ecological effects by invading species could not happen. *Gracilaria lemaneiformis* is a high quality raw material for agar industry and a good fodder for industrialized abalone cultivation. It is in big demand in southern China. This makes it a very promising species at present for both purposes, i.e., for removing nutrient from coastal water and for providing the product to feed the mariculture abalone.

Gracilaria tenuistipitata *var.* liui

This is another red alga and was introduced recently from Taiwan. This can grow well under high temperature and in relatively stable water conditions like in a fish or shrimp pond. Its ability to absorb large amount of nutrient is also evident. In January 2002, introduction of this species in ponds experiencing red tide in Dongshan, Fujian Province in SE China quickly suppressed red tide bloom within one or two days. Similar to *Gracilaria lemaneiformis*, it is easy to harvest this species. Because this species is also restricted in its seasonal growth due to temperature, it is unlikely to cause negative ecological effect as an introduced species. In fact, it can alternately grow with *G. lemaneiformis*, which grows in colder months of the year, thus ensuring a year round availability of seaweed resources.

Porphyra haitanensis

This species, a traditional edible red alga, grows along the coast of Fujian in southeastern China. Artificial cultivation of this species was initiated in the sixties and large-scale cultivation has been developed mainly in Fujian and Zhejiang, SE China. Favorable temperature for its growth is 12–25 °C. Transplanting experiment was conducted in 1997 from Jiangsu to Zhanjiang, Guangdong where *P. haitanensis* has never been grown. Our experimental result showed that one harvest from a 30 m^2 net was 50 kg wet weight. That is equivalent to about 16.7 tons ha^{-1}, a level acceptable to promote cultivation in this new area (Li et al., 1992). Because of high temperature during most time of the year, *P. haitanensis* could not pass the summer there and the negative ecological effects could not happen. Therefore, *P. haitanensis* becomes the second promising species for removing eutrophication nutrients especially in southern China, including Hong Kong.

Laminaria japonica

Laminaria japonica is not a native of Chinese shores though it has been a well-known food commodity in the Chinese market for more than one thousand years. It is a native species of the Japan Sea. The species was accidentally introduced to Dalian in 1927. Modern cultivation of this species started in 1952 and has now developed greatly, growing well by the Chinese summer sporeling method. It is now under cultivation from Liaoning and Shandong to Zejiang and Fujian Provinces with annual production of more than 300 million tons wet weight. From 47 years successful history of *Laminaria* cultivation in China, effects on eutrophication have all been positive and harmful or damage examples caused by the cultivation activity have not been found. This is a good candidate for removing eutrophication nutrients especially in northern China.

Conclusions

The fast-growing coastal cities and fast-developing animal mariculture industries have resulted in eutrophication along the coastal regions of China. Actually, it is the human activity on land and in the sea, which brings about this problem. China is the world largest mariculture country and is also one that is most seriously suffering from disasters caused by eutrophication. The first and most serious victim is mariculture itself. To find a solution to this problem is a very urgent task. It is a difficult, complicated and long-term task. To use seaweed cultivation for the solution of eutrophication is just one of the ways, but it is the promising and workable one at present time. Cultivated seaweed does not only grow fast but can also be easily harvested from the water. Because the cultivation technique is already mature, it is easy to manipulate. Because the investment cost for seaweed cultivation is comparatively low it is feasible (Tseng &

Fei, 1987; Crichley & Ohno, 1988; Miura, 1992; Ohno & Critchley, 1993; Fei, 1996; Yarish et al., 1998). Cultivation of seaweed can remove nutrients efficiently and could be profitable at the same time. *Gracilaria lemaneiformis* is good for sub-tidal shallow water area. *Porphyra haitanensis* is good for both inter-tidal and sub-tidal shallow water. *Gracelaria tenuistipitata* var. *Liui* is promising for shrimp pounds or fish cages. We strongly believe that large-scale seaweed cultivation in the sea will help to solve the eutrophication problem at the present stage.

Acknowledgements

This work was sponsored by the National Climbing Project B PD-B 6-4-2 of the State Science and Technology Commission of China and by the National Science Foundation of China H C0205 39770593. The author extends sincere thanks to Prof. C. K. Tseng for his encouragement, valuable comments and for reviewing this manuscript. This is Contribution No. 3360 from the Institute of Oceanology, Chinese Academy of Sciences and Contribution No. 183 from the Experimental Marine Biology Laboratory, Chinese Academy of Sciences (EMBLC).

References

Brawley, S. H. & X. G. Fei, 1988. Ecological studies of *Gracilaria asiatica* and *Gracilaria lemaneiformis* in Zhanshan Bay, Qingdao. Chinese J. Oceanol. Limnol. 6: 20–34.

Critchley, A. & M. Ohno, 1998. Seaweed Resources of the World. Japan International Cooperation Agency (JICA), Tokyo.

Fei, X. G., 1978. Principle results on the experimental phyco-ecology research of Haidai (*Laminaria japonica*) and Zicai (*Porphyra yezoensis*), Haiyang Kexue (Supplement): 55–56 [in Chinese].

Fei, X. G., 1983. Macroalgal culture in California and China. In Laura Mckay (ed.), Seaweed Raft and Farm Design in the United States and China. New York Sea Grant Institute, New York: 301–308.

Fei, X. G., 1996. Domestication and cultivation of seaweed. In Li Deshang (ed.), Proceedings of the International Symposium on Aquaculture. Qingdao Ocean University Press, Qingdao: 31–37.

Fei, X. G., Y. Bao & S. Lu, 1999. Seaweed cultivation: traditional way and its reformation. Chinese J. Oceanol. Limnol. 7: 193–199.

Fei, X. G., S. Lu, Y. Bao, R. Wilkes & C. Yarish, 1998. Seaweed cultivation in China. World Aquacult. 29: 22–24.

Jensen, A., 1993. Present and future needs for algae and alga products. Hydrobiologia 260/261: 15–23.

Li, S. Y., B. F. Zheng & X. G. Fei, 1992. Northward transplantation of *Porphyra haitanensis* in China. Oceanol. Limnol. Sinica 23: 297–301 [in Chinese with English abstract].

Lobban, C. S. & M. J. Wynne, 1981. The Biology of Seaweeds. University of California Press, Berkeley, CA.

Miura Akio, 1992. Cultivation of edible algae in Japan. Serial in Fisheries (88) [in Japanese].

Ohno, M. & A. T. Critchley, 1993. Seaweed Cultivation and Marine ranching. Kanagawa International Fisheries Training Center, Japan International Cooperation Agency (JICA), Tokyo.

Radmer, R. J., 1996. Algal diversity and commercial algal products. BioScience 45: 263–270.

Resources Council, Science amd Technology Agency, Japan, 1982. Standard Tables of Food Composition in Japan, fourth revised edition. Resources Council, Science amd Technology Agency, Japan [in Japanese].

S.E.P.E & S.A.T.M. (Section of Experimental Phyco-ecology and Section of Algal Taxonomy and Morphology, Institute of Oceanology Academia Sinica), 1978. Manual of Tiaoban Zicei (*Porphyra yezoensis* Ueda) cultivation. Science Press, Beijing [in Chinese].

Tseng, C. K., 1990. The theory and practice of phycoculture in China. In Rajarao, V. N. (ed.), Perspectives in Phycology (Prof. M.O.P. Lyenger Centenary Celebration Volume). Today and Tomorrow's Printers and Publishers, New Delhi, India: 227–246.

Tseng, C. K. & X. G. Fei, 1987. Macroalgal commercialization in the Orient. Hydrobiologia 151/152: 167–172.

Yarish, C. R., Wilkes, T, Chopin, X. G. Fei, A. C. Mathieson, A. S. Klein, C .D. Neefus, G. G. Mitman & I. Levine, 1998. Domestication of indigenous *Porphyra* (Nori) species for commercial cultivation in Northeast America. World Aquacult. 29: 26–29.

Zhang, S. Q., Q. L. Yang, H. H. Qiu & H .H. Lin, 1994. Red tide and it's prevention and cure. China Ocean Press. Beijing [in Chinese].

Zhang, X. C., Y. X. Wang, X. N. Wu & X. G. Fei, 1993. A comparative study on photosynthetic pigments of *Gracilaria lemaneiformis* from different habitats. Trans. Oceanol. Limnol. 1993: 52–59 [in Chinese with English abstract].

Zhu, M. Y. & R. X. Li, 1996. The monitoring of water environment and red tide in prawn ponds. In Zhu, M. Y. (ed.), Research of Red Tide in China. (Proceedings of the Second Meeting of Chinese Committee of SCOR-IOC HAB Working Group). Qingdao Press, Qingdao: 60–68.

Hydrobiologia **512:** 153–156, 2004.
P.O. Ang, Jr. (ed.), Asian Pacific Phycology in the 21st Century: Prospects and Challenges.
© 2004 *Kluwer Academic Publishers.*

Mass culture of *Undaria* gametophyte clones and their use in sporeling culture

Chaoyuan Wu, Dapeng Li, Haihang Liu, Guang Peng & Jianxin Liu
Institute of Oceanology, Chinese Academy of Sciences, No. 7 Nanhai Rd., Qingdao 266071, P.R. China
E-mail: cywu@ms.qdio.ac.cn

Key words: Undaria pinnatifida, gametophyte clones, mass culture, sporeling raising, China

Abstract

Undaria pinnatifida (Harv.) Sur. is one of the three main seaweed species under commercial cultivation in China. In the mid-1990s the annual production was about 20 000 tons dry. The supply of healthy sporelings is key to the success of commercial cultivation of *Undaria*. Previous studies demonstrated that instead of the zoospore collection method, sporelings can be cultured through the use of gametophyte clones. This paper reports the experimental results on mass culture of clones and sporeling raising in commercial scale. Light had an obvious effect on growth of gametophyte clones. Under an irradiance of 80 μmol m^{-2} s^{-1} and favorable temperature of 22–25 °C, mean daily growth rate may reach as high as 37%. Several celled gametophyte fragments were sprayed onto the palm rope frame. Gametogenesis occurred after 4–6 days. Juvenile sporeling growth experiments showed that nitrate and phosphate concentrations of 2.9 10^{-4} mol l^{-1} and 1.7 10^{-5} mol l^{-1} were sufficient to enable the sporelings to maintain a high daily growth rate. Sporelings can reach a length of 1 cm in a month. Since 1997, extension of the clone technique has been carried out in Shandong Province. Large-scale production of sporelings for commercial cultivation of 14 and 31 hectares in 1997 and 1998 had been conducted successfully.

Introduction

Commercial cultivation of *Undaria* began in China only in the last decade. Today, *Undaria* production ranks third, after *Laminaria* and *Porphyra*, in the seaweed cultivation industry in China, where *Undaria pinnnatifida* (Harv.) Sur. is the main species under cultivation concentrated in two northern provinces, Liaoning and Shandong. The annual production in the mid-nineties was about 20 000 tons dry weight (Wu, 1998), five times the pre-1980 figures. The per hectare yield is generally around 11 tons dry weight.

All the sporelings used for cultivation were cultured in a greenhouse with ambient temperature. The key step in the zoospore collection sporeling culture technique developed in the mid-eighties is to collect zoospores in the summertime. The mature sporophylls are kept in a dark moist container for several hours to induce mass discharge of zoospores which attach themselves onto ropes and give rise to male and female gametophytes. As soon as the temperature decreases to

20–22 °C, discharge of female and male gametes takes place and consequently zygotes are formed. After some 3 months of being cultured in the greenhouse, the sporelings grow up to about 1 cm in length and can be moved to the open sea in late autumn. Long term cultivation in greenhouse is very expensive and laborious and the sporelings are inevitably attacked by diseases caused by certain kinds of saprophytic bacteria (Wu et al., 1979) and alginic acid decomposing bacteria (Chen et al., 1984). The rot disease due to alginic acid decomposing bacteria e.g. *Pseudomonas* causes decay of the holdfast by enzymatic action of alginase, finally resulting in the detachment and loss of the sporelings. The disease is serious, sometimes destroying virtually the entire crop. Therefore, a new method of culturing sporelings by the use of gametophyte clones was developed (Pang & Wu, 1996). This paper reports the extension work of culturing sporelings by the use of gametophyte clones in 2 sporeling culture stations.

In general, the gametophyte clone method includes four steps:

1. Isolation of single male and female gametophytes and induction of gametophyte clones.
2. Mass culture of male and female gametophyte clones.
3. Sporophyte induction and sporeling culture in greenhouse.
4. Sporeling culture in the open sea.

Materials and methods

Collection of zoospores, formation of gametophytes and induction of clones

Sporophyll from selected strain No. 10 with mature sporangial sori was thoroughly sterilized by immersion in a 2% (w/v) sodium hypochorite seawater solution for 10 min, washed, and then subjected to partial drying at 15 °C for 2–4 h to induce mass discharge of zoospores. The dense zoospore suspension was diluted until several zoospores could be seen per field under microscope (100×); slides were then put into the dilute suspension to allow the zoospores to attach. The attached zoospores were cultured under 20 °C in Provasoli enriched seawater (PES) (Provasoli, 1968) at irradiance of 16 μmol m^{-2} s^{-1}. Gametophytes formed after 3 d. Male and female gametophytes were picked up one by one by sterilized glass pipette. The isolated unialgal gametophytes grew into small, visible clones after a month's culture. These gametophyte clones were then propagated vegetatively in bigger containers.

Effect of light on the growth of gametophyte clones

Male and female gametophyte clones were cut by a blender into several celled small fragments, washed thoroughly several times with sterilized seawater, and then cultured separately in 20 l cylindrical glass bottles with inner diameter of 45 cm. The culture room was air-conditioned to maintain a temperature of 25±0.5 °C. Cool white fluorescent tubes (40w) were used as light source with a light period L:D of 12:12 and the distance between the bottle and the light source was adjusted so that the exterior surface of the bottle could receive different irradiances. Six treatments of 20, 40, 60, 80, 100, and 120 μmol m^{-2} s^{-1} were set up. Provasoli enriched seawater (PES) was used as culture medium and was renewed every seven days.

The medium was bubbled with filtered air to provide sufficient oxygen and carbon dioxide. The transmitted lights and the weight increments were measured periodically during the experiment.

Commercial scale cultivation of sporelings and disease control

After a period of growth, the gametophyte clones grew up to small balls of 0.5–2 mm diameter. Male and female clones were mixed by the ratio of 4:1, and then cut into several celled fragments by a blender. The suspension was then sprayed onto palm rope frames and kept motionless for several days to allow the gametophytes to attach on the substratum. The culture was kept in ambient temperature ranging from 20 to 15 °C in a greenhouse. Seawater was enriched with NO_3-N and PO_4-P to concentrations of 2×10^{-1} mol m^{-3} and 2×10^{-2} mol m^{-3}, respectively, and was renewed every 4–5 days. Light was carefully controlled at about 80–100 μmol m^{-2} s^{-1}.

In 1997, a commercial scale sporeling cultivation experiment was conducted at Old Stone Man Sporeling Culture Station in Qingdao. A total of 315 palm rope frames (50 cm × 100 cm) were used for the first crop cultivation experiment (Li et al., 1998). In 1998, 300 frames were used for the second crop cultivation experiment in Longxudao Sporeling Culture Station, Rongcheng City.

It has been proved that long term cultivation in greenhouse may cause the detachment of sporelings. The experiment was, therefore, designed to shorten the period of sporeling culture by the use of gametophyte clones instead of the routine method of zoospore collection. In order to control the multiplication of alginic acid decomposing bacteria, clones were cultured in the above mentioned seawater media supplemented with chloromycetin at concentration of 1 g l^{-1}.

Results and discussion

Mass culture of gametophyte clones

Experiments showed that clones grew fast vegetatively under 25±0.5 °C. Growth rate declined sharply above 25 °C. Male and female gametes could be formed and discharged only in a temperature range of 5–22 °C.

Table 1 shows that on day 6 the average daily growth rates reached as high as 37.1% and 38.6% under irradiances of 80 and 100 μmol m^{-2} s^{-1}. The

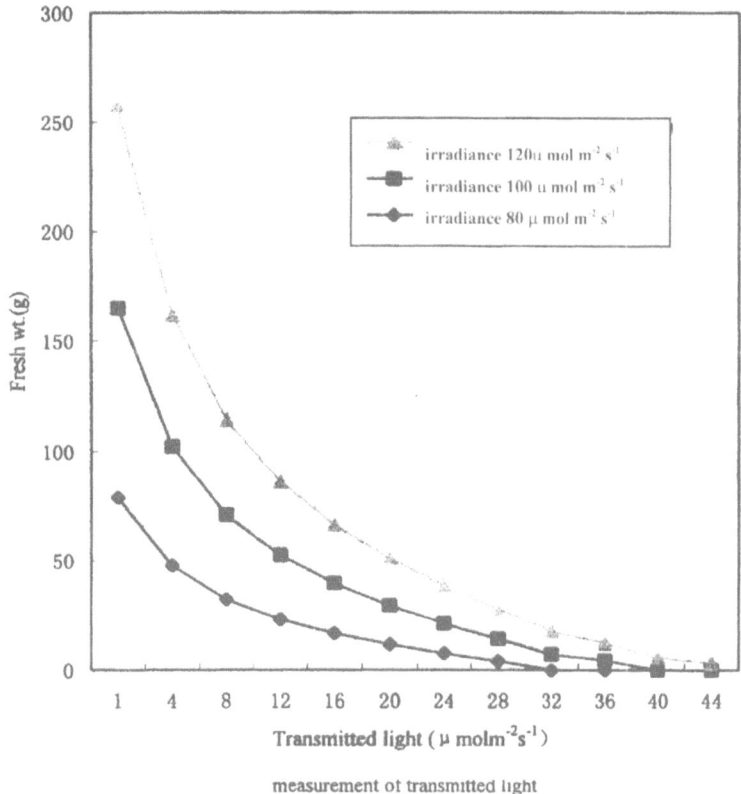

measurement of transmitted light

Figure 1. Detection of growth of *Undaria pinnatifida* gametophyte clones through the measurement of transmitted light.

growth rates declined sharply after day 13. On day 13 the growth rates approached 11.0 and 12.8, 3.5 and 6 times higher than that on day 27. This means that fresh weight of clones can be doubled in 3 d during the first 6 d. It is therefore suggested that clones should be harvested on day 6 to day 13. Experimental result of Ms. Hu Xiaoyan and L. Popoba in 1997 (pers. comm.) showed a photosynthetic rate of 33.9 μmol O_2 g^{-1} h^{-1} under irradiance of 80 μmol m^{-2} s^{-1} in *Undaria pinnatifida* gametophyte clones. Such a high rate coincides with the high increment in fresh weight shown in this experiment. This very high growth rate makes clone culture on commercial scale possible.

A better understanding of the relationship between the increment of fresh weight and the decline of transmitted light led us to believe that transmitted light can be used as an index to estimate the fresh weight of clones in the bottle. Figure 1 was plotted so that one may find the fresh weight through the measurement of transmitted light. For example under irradiance of 100 μmol m^{-2} s^{-1}, if the transmitted light is 4 μmol m^{-2} s^{-1}, the clones in the bottle should be 54 g or so.

This makes the estimation easier and diminishes the possibility of contamination.

Commercial scale cultivation of sporelings

Experimental results showed that male and female gametes will be formed 5–6 d after blender cutting, consequently sporophytes will appear the next day. Therefore, it is believed wounding promotes the formation of oogonium and antheridium. Similar to that of *Laminaria*, eggs discharge only after a period of darkness. Sperms will release immediately after the discharge of eggs and fertilization takes place afterwards (Wu et al., 1979). One day after fertilization, two celled sporophytes appeared. Mitosis took place in darkness too, once every 24 h. Three days after fertilization, seven celled sporophytes were formed. Juvenile sporeling growth experiments showed that nitrate and phosphate concentration of 2.9×10^{-4} mol l^{-1} and 1.7×10^{-5} mol l^{-1} were sufficient to enable the sporelings to maintain a high growth rate. Under such favorable conditions, sporelings grow fast in 1 month to 1 cm in length, which is the right size for

Table 1. Growth of gametophyte clones of *Undaria pinnatifida* under different light regimes

Irradiance (μmolm^{-2}s^{-1})	Day 6		Day 13		Day 20		Day 27	
	Fresh wt.(g)	Average daily growth rate (%) after last measurement	Fresh wt.(g)	Average daily growth rate (%) after last measurement	Fresh wt.(g)	Average daily growth rate (%) after last measurement	Fresh wt.(g)	Average daily growth rate (%) after last measurement
20	23.3	22.2	34.1	6.6	50.6	6.9	68.5	5.0
40	31.3	36.2	48.2	7.8	69.9	6.5	91.2	4.3
60	29.9	33.2	48.4	8.9	70.7	6.6	91.1	4.1
80	32.2	37.1	57.0	11.0	84.6	6.9	102.7	3.1
100	33.1	38.6	62.9	12.8	88.0	5.7	101.0	2.1
120	30.8	34.6	59.5	13.3	81.2	5.2	95.2	2.5

10 g of clones were inoculated to each bottle in the beginning of experiment.

their transplantation to the open sea. After transplantation, the daily growth rate in length of the young sporophytes was measured and calculated to be 15%. Since 1997, extension of the clone technique has been carried out in Shandong Province. Large scale production of sporelings for commercial cultivation of 14 and 31 hectares in 1997 and 1998 has been conducted successfully.

Contamination of clone culture by alginic acid decomposing bacteria and long term cultivation of sporelings in the greenhouse are believed to be the cause of the detachment disease of sporelings. The addition of chloromycetin to the clone culture medium showed obvious inhibitory effect on these bacteria. Moreover, the use of the clone method of raising sporelings shortened their culture period from 3 months to 1 month. This resulted in more effective control of the detachment disease. Since 1997, extension of the disease control technique has been carried out in the Old Stone Man Station in Qingdao and Longxudao, and large scale production of healthy sporelings for commercial use has been achieved successfully.

Acknowledgements

This is contribution No. 3751 from the Institute of Oceanology, Chinese Academy of Sciences. This study was supported by the Project of Bio-Engineering Center of China SSTC 96-C01-05-01, National Climbing Plan B PD-B 6-4-2, and is a part of the Project of Science & Technology Commission of Shandong Province. The authors would like to thank Mr Zheng Shaoxiong for reading the manuscript.

References

Chen Dou, Guangheng Lin & Shize Shen, 1984. Studies on alginic acid decomposing bacteria III. The cause of the rot disease and detaching of *Laminaria* sporophytes in sporeling culture stations and their preventive measures. Oceanol. Limnol. Sinica 15: 581–589.

Li Dapeng, Haihang Liu, Guang Peng & Chaoyuan Wu, 1998. Application of gametophyte clone of japanese strain *Undaria pinnatifida* to large-scale sporeling production. Mar. Sci. No. 5: 4–5.

Pang Shaojun & Chaoyuan Wu, 1996. Study on gametaphyte vegetative growth of *Undaria pinnatifida* and its application. Chin. J. Oceanol. Limnol. 14: 205–210.

Provasoli, L., 1968. Media and prospect for the cultivation of marine algae. In Watanabe, A. & A. Hattori (eds), Cultures and Collections of Algae. Proceedings of the U.S.–Japan Conference. Hakone, Japanese Society of Plant Physiology: 63–75.

Wu Chaoyuan, 1998. Seaweed resources of China. In Critchley, A. T. & M. Ohno (eds), Seaweed Resources of the World. Japan International Cooperation Agency, Tokyo, Japan: 34–45.

Wu Chaoyuan, Nansheng Gao, Decheng Chen, Baicheng Chou, Peixian Cai, Shuxi Dong, Zhongcuen Wen & Renyi Cong, 1979. On the malformation disease of *Laminaria japonica*. Oceanol. et Limnol. Sinica 10: 238–251.

Hydrobiologia **512**: 157–164, 2004.
P.O. Ang, Jr. (ed.), Asian Pacific Phycology in the 21st Century: Prospects and Challenges.
© 2004 *Kluwer Academic Publishers.*

Karyology and sex determination in *Aglaothamnion oosumiense* Itono (Ceramiaceae, Rhodophyta)

Ok-Kyong Chah, In Kyu Lee & Gwang Hoon Kim[1],*
Department of Biology, Seoul National University, Shillimdong, Seoul 151-742, Korea
[1]*Department of Biology, Kongju National University, Shingwandong, Kongjushi, Chungnam 314-701, Korea*
Author for correspondence; E-mail: ghkim@kongju.ac.kr

Key words: Aglaothamnion oosumiense, chromosome, karyology, sex determination, post-fertilization, Rhodophyta

Abstract

A cytogenetic investigation on male and female reproductive cells of *Aglaothamnion oosumiense* Itono indicates that the sexuality of this species might be determined by a sex chromosome. Chromosome counts in female and male gametophytes gave 37 and 36, respectively. Sex ratio of gametophytes was 1:1. Both male-derived and female-derived bisexual plants were observed. Bisexual plants were different in gross morphology and position of carpogonial branches from normal unisexual gametophytes. The chromosome number of female-derived bisexual plants was $N = 37$ and male-derived bisexual plants was $N = 36$. Some male plants developed parasporangia in addition. The paraspore germlings showed the same chromosome number as the male plants. The fertilized carpogonium and gonimoblast cells had $2N =$ ca. 70 chromosomes.

Introduction

Chromosome number and morphology can be used as critical taxonomic features in red algal taxonomy (e.g. Kim et al., 1999). Cytogenetic details, however, other than chromosome numbers are unavailable for most marine red algae because of the small size of nuclei, difficulty in obtaining adequate nuclear staining, and rarity of nuclear divisions in collected material (Kapraun, 1989). Recently, improved cytogenetic techniques and use of periodically fixed material through culture have made karyological studies more rewarding (Kapraun, 1989, 1993; Cole, 1990).

Some irregular sequences in the life history of red algae such as mixed phases or bisexuality have been reported for more than a century, and new reports appear almost every year (Chah & Kim, 1998). Although the cause and meaning of the occurrence of bisexuality in normally dioecious species are still under debate (Hawkes, 1990; Choi & Lee, 1996), these irregular sequences provide a good opportunity for studying sex determination mechanism.

So far, the only satisfactory explanation for the sex determination mechanism in red algae is that reported in *Gracilaria tikvahiae* McLaclan (Van der Meer & Todd, 1977; Van der Meer, 1981, 1986). From genetic experiments, they demonstrated that the sexuality of this species is controlled by a pair of alleles rather than a pair of sex chromosomes, and heterozygosity for mating type rather than the diploid state triggers development of the tetrasporophytic phase (Van der Meer & Todd, 1977). They further suggested that the bisexuality could arise by a recessive mutation of a gene other than the primary sex determining locus (e.g., Van der Meer, 1990).

Drew (1955) suggested that segregation of sex chromosomes during meiosis is responsible for the production of equal numbers of male and female gametophytes in *Antithamnion spirographidis* Schiffner. The presence of sex chromosomes in red algae was first reported by Rao (1970, 1971) in *Wrangelia argus* Mont. Observing heteropycnosis of four prophase I chromosomes during tetrasporogenesis, he suggested that they might be sex chromosomes. There appears to be no other reports of sex chromosomes in the red algae (Cole, 1990).

In this study, we examined the life history and karyology of *Aglaothamnion oosumiense* and tried to elucidate the sex determining mechanism from crossing experiments between bisexual plants.

Materials and methods

Tetrasporic plants of *Aglaothamnion oosumiense* were collected from 5 to 10 m depth at Auchungdo, the south-western coast of Korea and were maintained in modified f/2-enriched seawater (Kim & Fritz, 1993). The plants were kept at 15 °C under a 16:8 h LD cycle with 10 μmol·m^{-2}·s^{-1} cool-white fluorescent light. Tetraspores were released and developed into male and female plants four weeks after germination. The sexual plants were maintained separately.

To observe the effects of environmental factors on vegetative morphology and sex ratio of the species, some isolates were maintained under the following combinations of temperature, LD cycle and photofluence rate; 15 °C + 8:16 h LD, 15 °C + 12:12 h LD, 15 °C + 16:8 h LD at 5–10 μmol·m^{-2}·s^{-1}, 20 °C + 8:16 h LD, 20 °C + 16:8 h LD at 10–30 μmol·m^{-2}·s^{-1}, and 25 °C?8:16 h LD at 25 μmol·m^{-2}·s^{-1}. Procedures for crossing experiments were the same as those described by Kim et al. (1996).

For chromosome observation, plants were fixed in 3:1 absolute ethanol-glacial acetic acid and left overnight (Austin, 1959). Fixed material was stored in 70% ethanol, hydrolysed in 1N HCl for 10 min at room temperature to soften tissue, rinsed in distilled water, and stained in 2% aceto-carmine for 2–3 h prior to squash preparation. Slides were heated until the aceto-carmine solution formed small bubbles under the cover slip, and cells were than squashed with teaspoon. Representative karyotypes were made from various stages of prophase contraction. Chromosome numbers were based on ten or more well-spread mid- to late-prophase nuclei in both vegetative and reproductive cells.

All specimens were examined with an Olympus BX-50 microscope equipped with differential interference optics. Micrographs were taken with Kodak TMAX 100 film.

Results

Culture experiments were initiated to get the material for cytogenetical investigation from excised vegetative apices of field-collected tetrasporophyte of *Aglaothamnion oosumiense*. New sporangia developed after two weeks. Tetraspores grew into male and female gametophytes in a month. The sex ratio was 1:1 regardless of the environmental conditions. After fertilization, two carposporophytes developed at a fertile segment. Carposporophytes released carpospores 1 month after fertilization. Carpospores germinated to form mature tetrasporophytes in 2 months.

Chromosome numbers of haploid and diploid nuclear phases were obtained (Figs 1 and 2). Ten identical counts of $N = 37$ were made for female plants and 18 of $N = 36$ were made for male plants (Fig. 1). Female plants, therefore, had one more chromosome than male. Chromosomes in dividing carpogonial branch cells were generally elongate, sausage-shaped or spherical, and were relatively bigger, 2 μm (small) to 8 μm (large) in length, compared to other red algae (Fig. 1). Centromere position could be discerned in some big chromosomes but not in all. Chromosomes in dividing spermatangial mother cells were similar in shape and size to those of carpogonial branch cells except for having one less (Figs 1C, D, F). The missing chromosome in male cells was not discernible because it belonged to a group of small chromosomes (Figs 1E, F). No heteropycnosis of sex chromosome was observed during mitotic or meiotic divisions. First mitotic division of the fertilized carpogonium was observed (Fig. 2A), and an intercalary cell division was observed during the gonimoblast development (Fig. 2B). The chromosome number was $2N =$ ca. 70 during this post-fertilization process.

Some unusual reproduction was observed during culture (Fig. 3). About 2% of male plants developed parasporangia in addition to spermatangia. The paraspores released and developed into male plants or parasporangiate male plants again. Parasporagium was more irregularly shaped than carposporangium. The paraspore germlings had the same chromosome number as male plants. Bisexual plants were also observed (Fig. 4). Less than 1% of the tetraspore germlings developed into bisexual plants. They never became unisexual again, regardless of environmental conditions. However, parasporangiate plants often did not produce parasporangia any more, and became normal male.

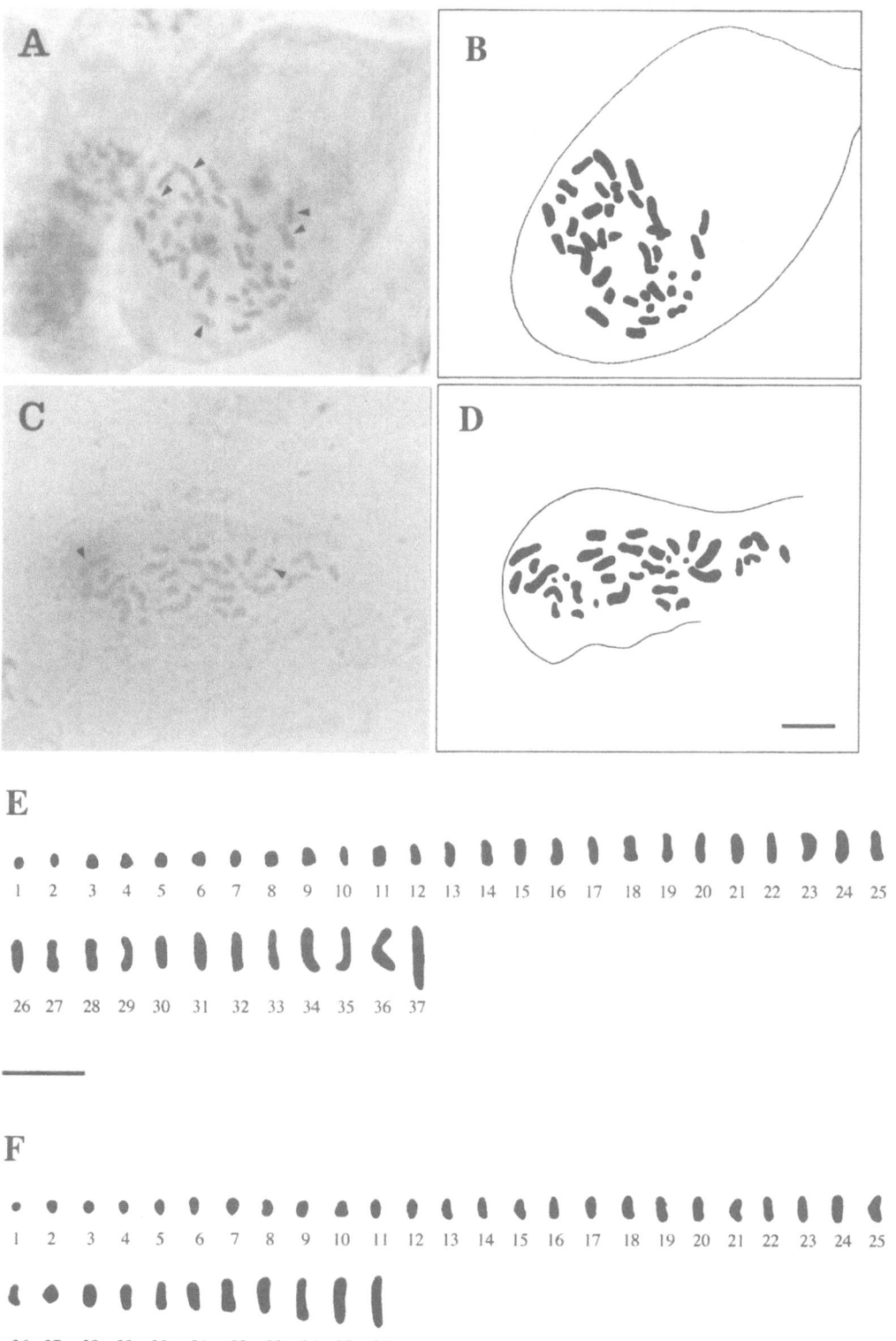

Figure 1. Karyology of *Aglaothamnion oosumiense* Itono. (A, B, E) Chromosomes in carpogonial branch cell of female plant. (C, D, F) Chromosomes in spermatangial mother cell of male plant. Arrow head indicates discernible centromere (scale bar = 10 μm).

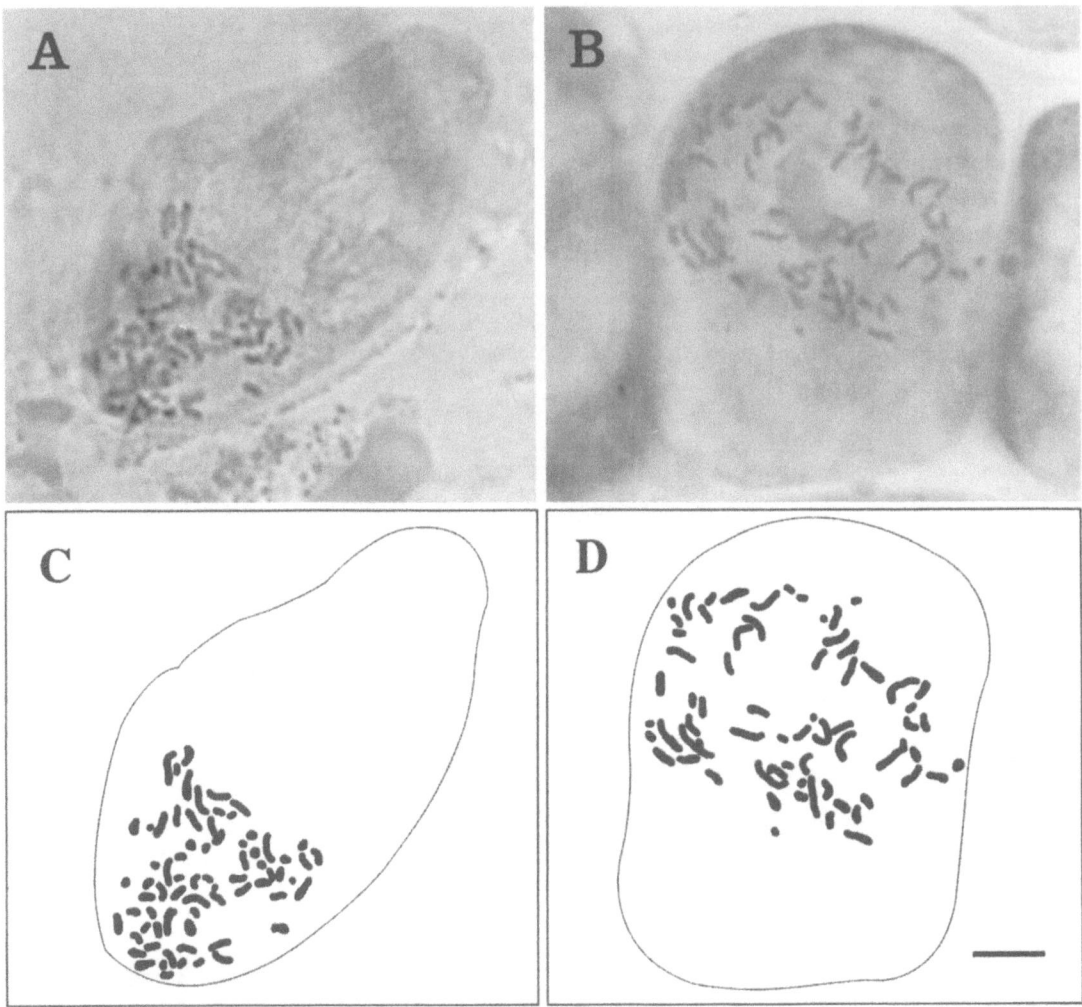

Figure 2. Chromosomes of *Aglaothamnion oosumiense* Itono. (A, C) First cell division in the fertilized carpogonium. (B, D) Intercalary division of the gonimoblast cell (Scale bar = 15 μm).

Table 1. The results of crossing experiment among the isolates of *Aglaothamnion oosumiense* Itono

Crossing experiment			Carposporophyte development (%)
Male plants (1010)	×	female plants (1011)	94
Female-derived bisexual plants (2010)	×	female-derived bisexual plants (2010)	70
Female-derived bisexual plants (2010)	×	female plants (1011)	80
Male-derived bisexual plants (3010)	×	male-derived bisexual plants (3010)	85
Male-derived bisexual plants (3010)	×	female plants (1011)	75

There were two types of bisexual plants; female-derived and male-derived (Fig. 3). They were easily distinguished from normal gametophytes in gross morphology, abundance of sexual reproductive struc-tures, the position of carpogonial branches, and sper-matangial clusters (Fig. 3). Eight identical counts of $N = 37$ chromosomes were made for female-derived bisexual plants and seven of N = 36 were

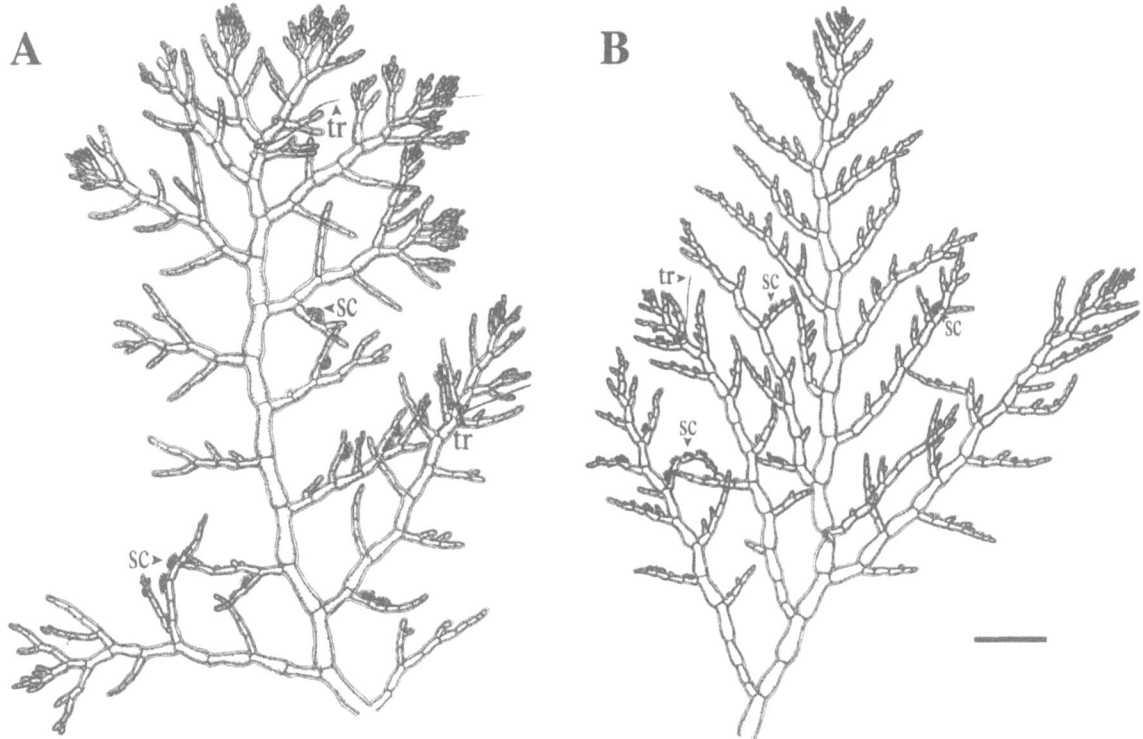

Figure 3. Two types of bisexual plant of *Aglaothamnion oosumiense* Itono. (A) Female-derived bisexual plant. (B) Male-derived bisexual plant (sc; spermatangial cluster, tr; trichogyne, scale bar = 390 μm).

made for male-derived bisexual plants (Figs 4A, B). Female-derived bisexual plants, therefore, had one more chromosome than male-derived ones. No significant difference in shape and size of chromosomes was observed in the both bisexual plants (Figs 4C, D).

The results of crossing experiment between bisexual and unisexual plants are summarized in Table 1. There was no significant difference in the success of fertilization. However, the fate of offspring differed a lot according to the combination of crossing experiments (Fig. 5). Self-crossing of the male-derived bisexual plants resulted in diploid male plants (Fig. 5A). Cross between normal female and male-derived bisexual plants resulted in tetrasporophyte, and the released tetraspores developed into male and female plants (Fig. 5B). The sex ratio was 1:1. Self-crossing of the female-derived bisexual plants resulted in tetrasporophyte, while the released tetraspores developed into female or bisexual plants (Fig. 5C). Cross between normal female and female-derived bisexual plants gave the same result as self-crossing of the female-derived bisexual plants (Fig. 5D).

Discussion

Our data suggest that the sexuality of *Aglaothamnion oosumiense* could be determined by a sex chromosome. Female plants of this species had one more chromosome than male. Male and female gametophytes showed a 1:1 Mendelian segregation regardless of the environmental conditions. The result of crossing experiment and cytogenetical investigation on bisexual plants indicates that there are two types of bisexual mutants in this species, one male-derived and the other female-derived. Both the bisexual plants are haploid. Tetrasprophyte is derived only when the crosses including female or female-derived bisexual plants, implying that genetic loci of tetrasporangia might be associated with the sex chromosome, the extra chromosome observed in the female plant.

So far, there appears to be only one report of sex chromosomes in red algae (Cole, 1990). Using standard Feulgen staining of early and late meiotic stages in tetrasporangia of *Wrangelia argus*, Rao (1970, 1971) observed heteropycnosis of four prophase I chromosomes that formed a ring-like configuration indicative of a translocation of heterozygote. Three of

162

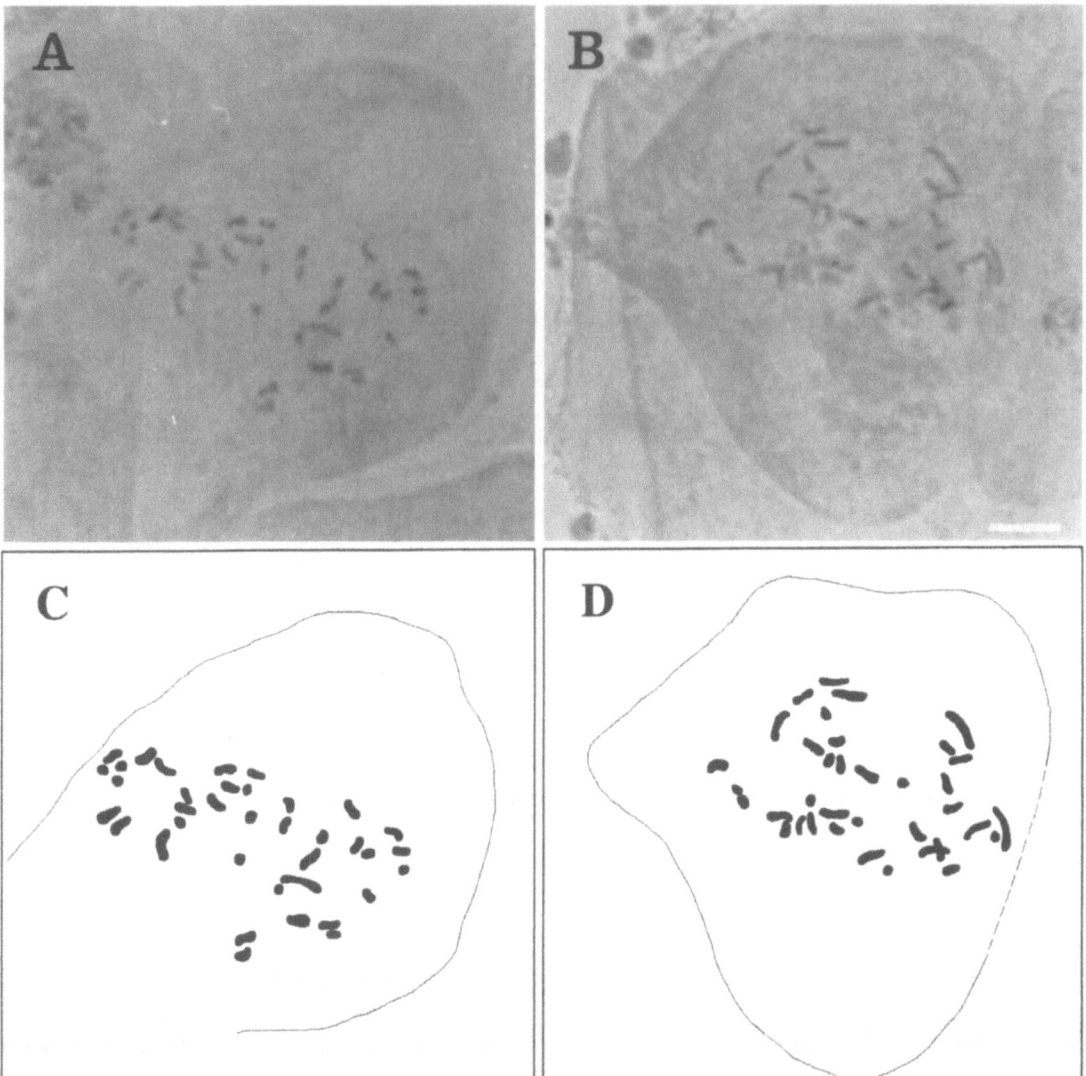

Figure 4. Chromosomes of bisexual plants of *Aglaothamnion oosumiense* Itono. (A) Female-derived bisexual plant. (B) Male-derived bisexual plant (scale bar = 15 μm).

the chromosomes were the largest in the diploid complement, one being particularly long. Each of the four was distinguishable on the basis of length, centromere position, distribution of light and dark staining blocks, and knobs in the chromosome arms (e.g., Cole, 1990). Rao (1971) proposed that these may be sex chromosomes X1, X2, Y1 and Y2, likening the longest to the large heteropycnotic X chromosome present in some laminarialean (Phaeophyta) gametophytes (e.g., Evans, 1965; Yabu & Sanbonsuga, 1981).

In *Aglaothamnion oosumiense* we could identify centromere position and banding pattern in some large chromosomes. However, it was difficult to point out

the sex chromosome, because it belonged to a group of the small and spherical chromosomes, and no heteropycnosis was observed. The results of crossing experiment also suggest that the structure and the role of the sex chromosome of *Aglaothamnion oosumiense* may be different from those of *Wrangelia argus*.

How the sexuality in red algae is determined will require an explanation through more genetic investigations on an allelic basis (Rueness & Rueness, 1985). At present, however, very little is known about the organization of the mating type locus in red algae. From genetical data on segregation of sex determining elements after mitotic recombination in *Gracilaria*

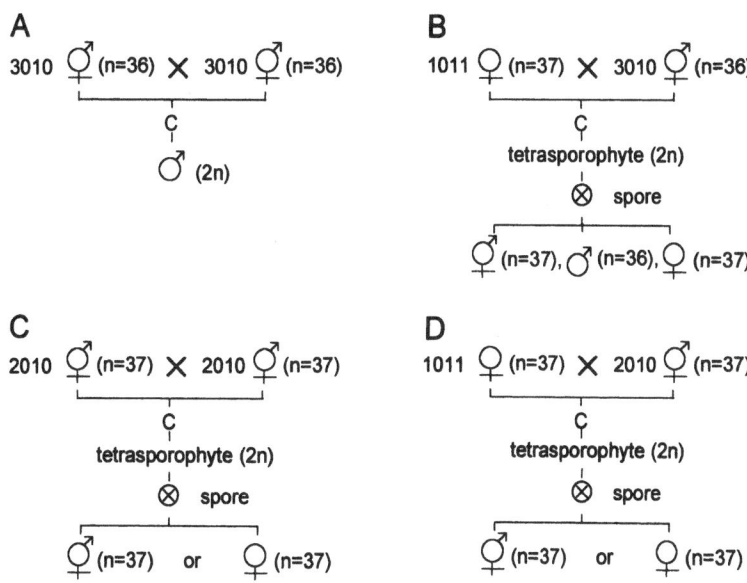

Figure 5. Summarized crossing experiments of *Aglaothamnion oosumiense* Itono.

tikvhiae, Van der Meer & Todd (1977) suggested that sexuality was controlled by a pair of alleles rather than a pair of sex chromosomes, and bisexuality could arise by a recessive mutation of a gene other than the primary sex determining locus. However, the mixed phases shown in many species of Antithamnieae (West & Norris, 1966; Rueness & Rueness, 1973, 1985; Notoya & Yabu, 1981; Kim & Lee, 1989; Kim, 1990) and also Dasyaceae (Choi & Lee, 1996) cannot be explained by mitotic recombination as described by Van der Meer (1990). In regard to sexuality of these species, especially of *Antithamnion tenuissimum* (Rueness & Rueness, 1973), Van der Meer & Todd (1977) suggested that it might be controlled by complex mating type loci like in yeast (Hawthorne, 1963) and *Chlamydomonas* (Gillham, 1969).

The development of equal numbers of male and female plants from tetraspores of *Aglaothamnion oosumiense* demonstrates that the primary control of sex determination is through a single pair of Mendelian factors (e.g., Van der Meer, 1990). The occurrence of bisexuality through a genetic recombination indicates that the sex-determining alleles may be present in a limited region of a chromosome (van der Meer & Todd, 1977; Van der Meer, 1981). The difference between male-derived and female-derived bisexual plants appears to be due to the presence of sex chromosome. The role of sex chromosome in *A. oosumiense* is of interest because both sexes could be expressed even though there was no sex chromosome in male-derived bisexual plants. The sex-determining allele, therefore, appears to be located in an autosome and there may be another regulatory gene in the sex chromosome. The occurrence of two types of bisexual plants may support Van der Meer's suggestion that the bisexuality could arise by a recessive mutation of a gene other than the primary sex determining locus. The difference of self-crossing results of male-derived and female-derived bisexual plants may be due to the presence of sex chromosome in the latter.

From self-crossing experiment of bisexual mutant, Van der Meer & Todd (1977) suggested that heterozygosity for mating type rather than the diploid state triggers the development of the tetrasporophytic phase. Our results support their idea in part; self-crossing of male-derived bisexual plant of *Aglaothamnion oosumiense* resulted in diploid male plants. Although carpospores from self-crossing of female-derived bisexual plants were also a homozygous mating type, they developed into tetrasporophytes, suggesting that the presence of sex chromosome may be more important for the development of tetrasporophyte phase.

More detailed crossing experiments using genetic markers are presently underway to elucidate further allelic basis for gametophytic sex determination in *Aglaothamnion oosumiense*.

Acknowledgements

The authors extend sincere thanks to J. West for his careful review and very useful comments. This work

164

has been partially supported by Ministry of Education, BSRI-98-4416 to G. H. Kim and National Research Laboratory grant from KISTEP.

References

Austin, A. P., 1959. Iron-alum aceto-carmine staining for chromosomes and other anatomical features of Rhodophyceae. Stain. Technol. 34: 69–75.

Chah, O.-K. & G. H. Kim, 1998. Life history and taxonomy of *Aglaothamnion oosumiense* Itono (Ceramiaceae, Rhodophyta). Algae 13: 199–206.

Choi, H.-G. & I. K. Lee, 1996. Mixed phase reproduction in *Dasysiphonia chejuensis* (Rhodophyta) from Korea. Phycologia 35: 9–18.

Cole, K. M., 1990. Chromosomes. In Cole, K. M. & R. G. Sheath (eds), Biology of The Red Algae. Cambridge University Press, Cambridge: 73–102.

Drew, K. M., 1955. Sequence of sexual and asexual phases in *Antithamnion spirographidis* Schiffner. Nature 175: 813–814.

Evans, L. V., 1965. Cytological studies in the Laminariales. Ann. Bot. N. S. 29: 541–62.

Gillham, N. W., 1969. Uniparental inheritance in *Chlamydomonas reinhardi*. Am. Nat. 103: 355–388.

Hawkes, M. W., 1990. Reproductive strategies. In Cole, K. M. & R. G. Sheath (eds), Biology of The Red Algae. Cambridge University Press, Cambridge: 455–476.

Hawthorne, D. C., 1963. A deletion in yeast and its bearing on the structure of the mating type locus. Genetics 48: 1727–1729.

Kapraun, D. F., 1989. Karyological investigation of chromosome variation pattern associated with speciation in some Rhodophyta. In George, E. Y. R. & A. W. Hulbert (eds), Carolina Coastral Oceanography Symposium. National Undersea Res. Prog. Res. Report. Washington. DC. 892: 65–76.

Kapraun, D. F., 1993. Karyology and cytogenetic estimation of nuclear DNA variation in several species of *Polysiphonia* (Rhodophyta, Ceramiales). Bot. mar. 36: 507–516.

Kim, G. H., 1990. A Biosystematic Study on Fourteen Species of Ceramiaceae (Rhodophyta) in Korea. Ph. D. Thesis, Seoul National University, Seoul, Korea: 359 pp. [in Korean].

Kim, G. H., O.-K. Chah & I. K. Lee, 1996. *Antithamnion aglandum* (Rhodophyta, Ceramiacea): a new species from Korea. Nova Hedwigia 63: 203–214.

Kim, G. H. & L. Fritz, 1993. Ultrastructure and cytochemistry of early spermatangial development in *Antithamnion nipponicum* (Ceramiaceae, Rhodophyta). J. Phycol. 29: 797–805.

Kim, G. H. & I. K. Lee, 1989. Mixed phases reproduction of *Platythamnion yezoense* Inagaki in culture. Kor. J. Phycol. 4: 34–42.

Kim, M. S., Y.-S. Keum & I. K. Lee, 1999. Chromosome counts in three species of *Polysiphonia* (Ceramiales, Rhodophyta). Phycologia 38: 66–69.

Notoya, M. & H. Yabu, 1981. *Platythamnion yezoense* Inagaki (Rhodophyta, Ceramiales) in culture. Jpn. J. Phycol. 29: 39–46.

Rao, C. S. P., 1970. Morphology of the chromosomes of red algae. Indian Biol. 2: 37–40.

Rao, C. S. P., 1971. Sex chromosomes of *Wrangelia argus* Mont. Bot. mar. 14: 113–115.

Rueness, J. & M. Rueness, 1973. Life history and nuclear phases of *Antithamnion tenuissimum* with special reference to plants bearing both tetrasporangia and spermatangia. Norw. J. Bot. 20: 205–210.

Rueness, J. & M. Rueness, 1985. Regular and irregular sequences in the life history of *Callithamnion tetragonum* (Rhodophyta, Ceramiales). Br. Phycol. J. 20: 329–334.

Van der Meer, J. P., 1981. Genetics of *Gracilaria tikvahiae* (Rhodophyceae). VII. Further observations of mitotic recombination and the construction of polyploidy. Can. J. Bot. 59: 787–792.

Van der Meer, J. P., 1986. Genetics of *Gracilaria tikvahiae* (Rhodophyceae). XI. Further characterization of a bisexual mutant. J. Phycol. 22: 151–158.

Van der Meer, J. P., 1990. Genetics. In Cole, K. M. & R. G. Sheath (eds), Biology of the Red Algae. Cambridge University Press, Cambridge: 103–121.

Van der Meer, J. P. & E. R. Todd, 1977. Genetics of *Gracilaria* sp. (Rhodophycea, Gigaertinales). II. The life history and genetic implications of cytokinetic failure during tetraspore formation. Phycologia 16: 159–161.

West, J. A. & R. E. Norris, 1966. Unusual phenomena in the life histories of Florideae in culture. J. Phycol. 2: 54–57.

Yabu, H. & Y. Sanbonsuga, 1981. A sex chromosome in *Cymathaere japonica* Miyabe et Nagai. Jpn. J. Phycol. 29: 79–80.

Hydrobiologia **512**: 165–170, 2004.
P.O. Ang, Jr. (ed.), Asian Pacific Phycology in the 21st Century: Prospects and Challenges.
© 2004 *Kluwer Academic Publishers.*

Cytological damage to the red alga *Griffithsia pacifica* from ultraviolet radiation

David J. Garbary[1,*], Kwang Young Kim[2] & Jennie Hoffman[3]
[1]*Department of Biology, St. Francis Xavier University, Antigonish, Nova Scotia, B2G 2W5, Canada*
[2]*Department of Oceanography, Chonnam National University, Kwangju 500-757, Korea*
[3]*Department of Zoology, University of Washington, Seattle, WA, U.S.A.*
Author for correspondence; E-mail: dgarbary@juliet.stfx.ca

Key words: cytology, development, *Griffithsia*, rhizoids, Rhodophyta, ultraviolet radiation, UV

Abstract

Continuous exposure for 7–10 days to 60% of ambient levels (sea level at mid-day in December) of UV-A and UV-B radiation caused cytological damage to regenerating fragments of *Griffithsia pacifica* under laboratory conditions. There was high mortality of individual cells and entire fragments in UV treated filaments. Rhizoid initiation was slower and rhizoids grew more slowly following UV treatment. After 7 days, UV radiated thalli showed chloroplast and nuclear degeneration. In addition, filaments tended to disarticulate so that single or groups of apparently healthy cells were common in the medium. These data suggest that the subtidal habitat of *G. pacifica* is based in part on lack of tolerance to UV radiation, and that UV protection mechanisms are not inducible or insufficient to prevent the accumulation of damage in this species.

Introduction

Increasing levels of ultraviolet B radiation (UVBR) reaching the earth's surface as a consequence of atmospheric ozone depletion is a demonstrated hazard for biological systems (Caldwell et al., 1995; Madronich et al., 1995). Many organisms, especially those in aerial habitats, have constitutive protection mechanisms, whereas other organisms have inducible mechanisms that provide protection. Subtidal organisms (e.g., attached seaweeds) may be especially prone to the negative impacts from UVR (Häder & Figueroa, 1997) because their habitats have been subjected to very low levels of UVR over evolutionary time scales. In addition, organisms in habitats where UVR has been low historically may not have evolved inducible protection mechanisms to escape from, or to repair damage caused by increases in UVR. Most research on effects of UVR on macroalgae describe biochemical or physiological impacts, especially those associated with photosynthesis (Franklin & Forster, 1997; Häder & Figueroa, 1997; Franklin et al., 1998) or damage to DNA (Pakker & Breeman, 1997). Here we examine developmental and cytological changes induced by

UVR on regenerating fragments of *Griffithsia pacifica* Kylin. Preliminary experiments identified UV dosages such that several days of exposure resulted in sublethal effects.

Material and methods

Algal culture

Stock culture of single clonal individual of *Griffithsia pacifica* from San Juan Island, Washington (ca. 49° N) was maintained for several months in an indoor running seawater table at Friday Harbor Laboratories. The plant was grown under natural light from south facing windows supplemented during working hours with room lighting provided by fluorescent lights. During experiments fragments from a single thallus were grown under continuous light provided by cool-white fluorescent tubes with PAR (photosynthetically active radiation) irradiance of 7–8 μmol m^{-2} s^{-1} as measured with a LI-COR 185 meter. PAR was supplemented with UV radiation as described below. Plants were grown at 13 °C using a modified von Stosch medium (Guiry & Cunningham, 1984) supplemented with 5 mg l^{-1} GeO$_2$ to inhibit diatoms.

Ultraviolet radiation and filters

A pair of UVA-340 fluorescent lamps (Q-Panel, Cleveland, Ohio) was used to provide UV radiation. The UV lamps have the same emission spectrum as sunlight between 295 and 365 nm, and there is no emission below 295 nm. Cultures were exposed to 130–168 μW cm^{-2} UV-A and 85–112 μW cm^{-2} UV-B as measured with a newly calibrated UVX Digital Radiometer (UVP, San Gabriel, CA). Total UV radiance (250–400 nm) was 1.7× greater in zenith sky on a sunny day (December 3, 1997) relative to the culture bench, and was equivalent to radiance at 50–60 cm water depth at the dock of Friday Harbor Laboratories. UVR treatment is considerably higher than in nature because of the continuous nature of the laboratory illumination. Radiance was measured with an Ocean Optics radiometer (Dunedin, Florida). Culture dishes were covered with a cellulose acetate filter or a lexan filter. Cellulose acetate is almost transparent to PAR and has high transmission of UV-A and UV-B radiation, (i.e., between 290 and 400 nm). Lexan has a sharp transmission cut off at about 390 nm, and has the same transmission spectrum of cellulose acetate with respect to PAR. Absorption spectra for filters were determined at the beginning and end of each experiment, and there was no change in filter transmission.

Experimental design

In each experiment, 3–7 celled apical fragments were excised from a large plant and placed singly in each well of 6-well, polystyrene, tissue culture plates (FALCON 3224) in 10 ml of medium (10 mm depth). Above each well the lid had been removed and replaced with the appropriate filter (either cellulose acetate or lexan). Two experiments were carried out. In experiment 1, a UV well was paired with a control well, and three fragments were placed in each well and grown for seven days (n=18). In experiment 2, wells were randomly assigned filter type and only a single fragment was placed in each dish (n=10). Rhizoid length and number were determined each day from day 3 to day 10. At day 10 each fragment was assessed for rhizoid health and erect cell health on a five point scale (0=dead, 5=no different from best regenerates in controls). If two rhizoids were present only the longest was measured. Cytological changes in nuclei, chloroplasts and cell walls resulting from

Table 1. *Griffithsia pacifica*. Quantitative changes in regenerating fragments after 10 days in UV-exposed (cellulose acetate filter) and UV-protected fragments (lexan filter) (experiment 2). Figures indicate $x\pm$S.E.

Character	UV-exposed	UV-protected	Significance (p)
Rhizoid length (μm)	407 ± 80	1053 ± 91	<0.01
Rhizoid number	0.8 ± 0.1	1.8 ± 0.2	<0.01
Erect filament health	2.2 ± 0.42	4.9 ± 0.08	<0.01
Rhizoid health	2.2 ± 0.37	4.5 ± 0.20	<0.01
Number of living cells	2.7 ± 0.66	5.6 ± 0.23	<0.01

exposure to UVR were determined using light and epifluorescence microscopy, and DAPI to stain nuclei (Garbary & McDonald, 1998).

Results

UV effects on erect axes

The UV exposure caused 10–25% mortality among fragments. Qualitative cytological changes were induced in cells of upright axes by UVR (Figs 1–11). In the UV-exposed fragments considerable cell death occurred (Figs 1 and 3). This occurred as complete mortality of a single fragment before or after rhizoid initiation. Cell death occurred as a more or less uniform bleaching through each fragment or as the bleaching of individual cells (Fig. 1). In some cells, deterioration began at one end of a cell and progressed to the other. At the end of the 10 days there were only half as many living (i.e., unbleached) cells in UV treated regenerates as controls (Table 1) and some cells were conspicuously plasmolysed (Fig. 7). Following bleaching neither chloroplasts (Fig. 1) nor nuclei (after staining with DAPI) was visible within cells. Where differential bleaching occurred within a fragment, it was typically the apical cells that retained pigmentation the longest.

A conspicuous effect of UV-radiation was the disarticulation of filaments (Figs 2 and 3). This preceded cell death and in some cases single, apparently living (i.e., pigmented) cells were present as free-floating cells. This breakup is partly reflected in the smaller number of living cells in the axes, although cell death and disarticulation may also be independent events. When disarticulation occurs there is an initial break in the cell walls (Fig. 2). This leaves the cells attached at the pit connection that also breaks. The detached

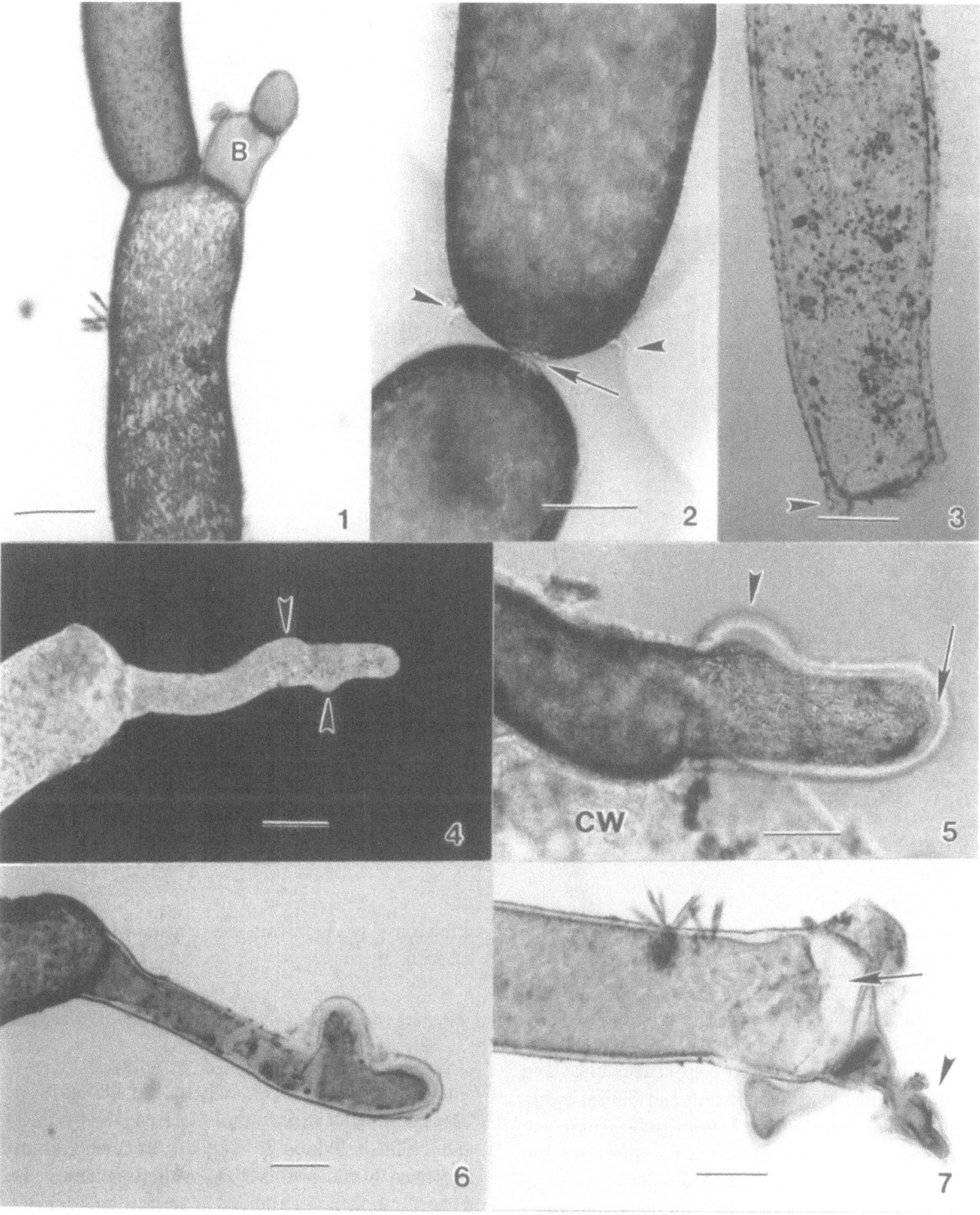

Figures 1–7. Griffithsia pacifica following UV radiation. **Figure 1**. Fragment with small bleached cell (B) and large cell with chloroplasts in initial stages of degradation. Scale bar=100 μm. **Figure 2**. Cells remaining attached at pit connection (arrow) and with remnants of adjoining cell walls (arrow heads). Scale bar=50 μm. **Figure 3**. Bleached cell with only cytoplasmic remnants showing and with deteriorated end walls (arrow head). Scale bar=50 μm. **Figure 4**. Rhizoid with two lateral swellings. Scale bar=100 μm. **Figure 5**. Tip of rhizoid without apical dome and with remnants of cell wall (CW) of excised cell. Scale bar=50 μm. **Figure 6**. Irregular rhizoid with lateral branch. Scale bar=100 μm. **Figure 7**. Regenerating rhizoid with highly irregular outline (arrow head) and with partially plasmolysed supporting cell (arrow). Scale bar=100 μm.

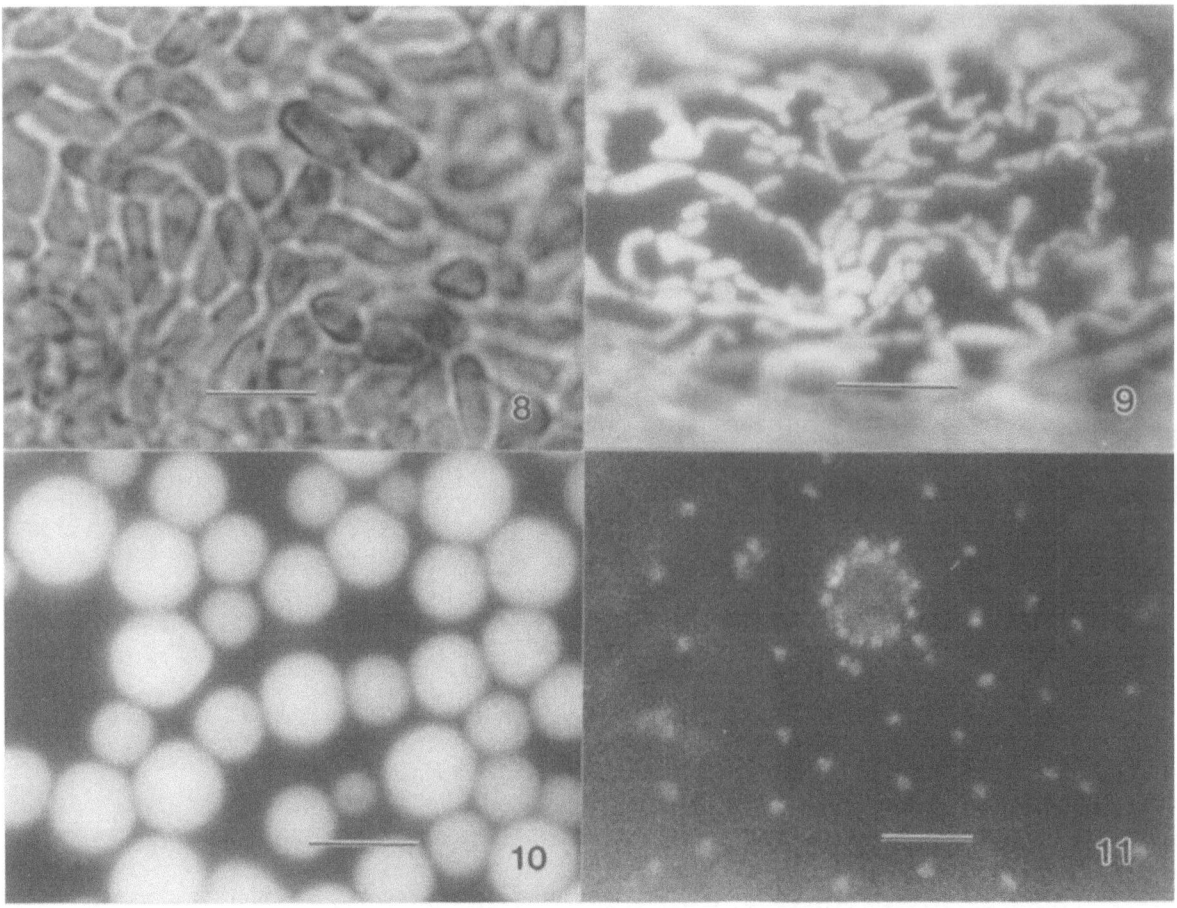

Figures 8–11. Griffithsia pacifica. **Figure 8**. Portion of control cell with densely packed discoid to cylindrical chloroplasts. Scale bar=5 μm. **Figure 9**. UV radiated cell with chloroplasts assuming irregular shapes and forming linear arrays. Scale bar=10 μm. **Figure 10**. UV treated cell with enlarged, globular chloroplasts prior to bleaching and cell death. Scale bar=5 μm. **Figure 11**. DAPI stained cell with irregular cluster of nuclei and more normal dispersed nuclei. Scale bar=20 μm.

cell can appear to be healthy, although regeneration of these cells was not observed.

Chloroplast degeneration was an evident effect of UV treatment. Chloroplasts in controls were densely packed and ovoid, with bright red pigmentation (Fig. 8). UV treated thalli showed conspicuous bleaching beginning at day 3, with many cells and thalli showing complete bleaching by day 7. Prior to bleaching and complete disintegration, chloroplast arrangement and morphology changed such that they aligned in more or less linear arrays and then became slightly enlarged and rounded (Figs 9 and 10), with continual fading occurring throughout this process.

DAPI staining for nuclei was unsuccessful in bleached cells and they were presumably degraded. Nuclei of unbleached cells in UV treatments were often clumped rather than having a regular spatial dis-tribution (Fig. 11). This was apparent in both erect axes and rhizoids.

Rhizoid initiation and growth

Among remaining regenerates rhizoid initiation was slowed. Rhizoid initiation was delayed 24–48 h in UV treated thalli. Within 24 h control fragments formed localized swelling at the base of the fragment with a concentration of chloroplasts in preparation for exten-sion of the rhizoid which formed within 48 h. After 2 d UV treated thalli showed only slight swelling and little concentration of cytoplasm. At day 3 only 50% of the UV-treated fragments had initiated rhizoids whereas 100% of control fragments did. At day 4, rhizoids in control plants were over twice as long, and by 7–10 d they were 2.5–3 times as long (Fig. 1, Table 1).

Growth rates peaked in both conditions in days 3–4 with rates of ca. 100 and 400 μm d^{-1} in the UV-treated and control thalli, respectively. In addition to slower rhizoid initiation and growth, UV treated fragments formed fewer rhizoids (0.8 vs. 2.4 and 0.8 vs. 1.8 rhizoids per regenerate in experiments 1 and 2, respectively), and the overall health of the rhizoids was diminished (Table 1).

Rhizoids produced in the UV treatment often had irregular lobes or swellings (Figs 4–7) and typically had no apical dome (i.e., an area at the tip devoid of chloroplasts). The chloroplast bleaching, irregular nuclear distribution and nuclear degeneration were also apparent in rhizoids.

Discussion

In these experiments G. pacifica was exposed to continuous UVR equivalent to those at 0.5 m depth based on mid-day on a clear day in early December. This resulted in considerable negative effects on thallus growth, cytology and survival. UV delayed rhizoid initiation, suggesting that damage occurred within the first 24 h. Even after rhizoid formation, elongation rates were slowed by factors of two to six. The slowing of the rhizoidal extension rate after day four is considered natural regulation (i.e., rhizoids may be approaching a maximum size for the size of the thallus), whereas it is interpreted as a sign of damage in the UV condition.

Algae have several mechanisms for protection from UVR. These include the production of mycosporine-like amino acids (MAAs) (Karentz et al., 1991; Karentz, 1994; Karsten et al., 1998), and deposition of sporopollenin (Xiong et al., 1997) and phlorotannins (Pavia et al., 1997). Some of these are inducible (e.g., MAAs, phlorotannins) whereas others are constituent (e.g., sporopollenin). Subtidal red algae, such as Delesseria sanguinea (Hudson) Lamouroux, Polyneura hilliae Greville (Kylin) and G. pacifica are easily damaged by UVR (Dring et al., 1996; Pakker & Breeman, 1997, this study) and may not have the protection mechanisms cited above. Avoidance from UVR may be the mechanism for survival.

The rhizoid apex of G. pacifica is comparable to many other tip-growing plant systems (Heath, 1990) in that there is a radiating pattern of microfilaments (MF) in a chloroplast-free zone at the rhizoid tip (Garbary et al., 1992). The fact that UV exposure disrupts the apical dome and is associated with reduced growth is consistent with an effect of UVR on rhizoid MF. The reduction in growth is consistent with an effect on microfilaments especially since chloroplast movement in G. pacifica is mediated through MF (Russell et al., 1996). In some fungi, ultraviolet microbeams effect microtubules and nuclear movement in tips of fungal hyphae (McKerracher & Heath, 1986). The nuclear clumping in erect axes is consistent with this phenomenon as Garbary et al. (1992) showed that the nuclei in G. pacifica are associated with a meshwork of microtubules.

Morphological changes in chloroplasts associated with UV damage are similar to those observed by Koslowsky & Waaland (1987) in G. pacifica following somatic cell fusion of incompatible strains. Although we did not observe chloroplast fusion during the degradation process, this would explain the presence of enlarged spherical chloroplasts prior to final bleaching and cell death. The wave of chloroplast destruction that we observed in some cells from one end of the cell to the other was also a feature of the incompatibility reaction (Koslowsky & Waaland, 1984) and may reflect equivalent damage to DNA, albeit from different sources.

Here we describe the effects of UVR on regenerating fragments of G. pacifica. The negative effects we demonstrate suggest that the ability to grow intertidally and in shallow subtidal areas may reflect not only desiccation and high light and temperature intolerance, but also a restricted tolerance for UVR. Indeed G. pacifica is typically found at depths of 3–5 m or more in the San Juan Islands. The sensitivity of G. pacifica is such that environmental increases in UVR may further limit distribution of the species. Although G. pacifica does not seem to produce UV absorbing substances (e.g., MAAs) that can protect it from UVR, other substances in the environment may play a similar role. Elsewhere we (Garbary et al., in preparation) show that phlorotannins derived from Laminariales can function as UV absorbing substances and protect G. pacifica from ambient levels of UVR. Thus it is difficult to generalize the particular effects of a single variable such as UVR without considering additional phenomena in the natural environment.

Acknowledgements

This work was carried out while D.J.G. and K.Y.K. were visiting scientists at Friday Harbor Laboratories, University of Washington. We thank Brian Clarke for

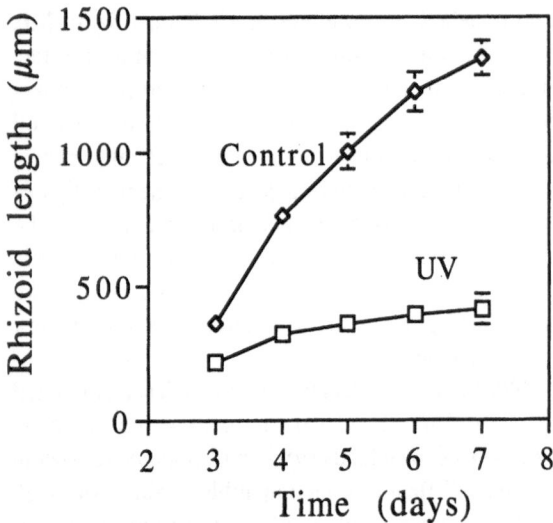

Figure 12. Griffithsia pacifica. Growth of rhizoids under UV-exposed (UV) and UV-protected (Control) conditions over 7 days (experiment 1). Note: data included only regenerates that initiated rhizoids (*n*=12–18).

technical assistance, Dr Claudia Mills for providing the plants of *Griffithsia pacifica* and Dr Ken McFarlane for carrying out the UV radiometry. This work was supported by grants from the Natural Sciences and Engineering Research Council of Canada to D.J.G. and the Korean Research Foundation (KRF-2001-041-H00021) to K.Y.K.

References

Caldwell, M. M., A. H. Teramura, M. Tevini, J. F. Bornman, L. O. Björn & G. Kulandaivelu, 1995. Effects of increased solar ultraviolet radiation on terrestrial plants. Ambio 24: 166–173.

Dring, M. J., A. Wagner, J. Boeskov & K. Lüning, 1996. Sensitivity of intertidal and subtidal red algae to UVA and UVB radiation, as monitored by chlorophyll fluorescence measurements: influence of collection depth and season, and length of irradiation. Eur. J. Phycol. 31: 293–302.

Franklin, L. A. & R. M. Forster, 1997. The changing irradiance environment: consequences for marine macrophyte physiology, productivity and ecology. Eur. J. Phycol. 32: 207–232.

Franklin, L. A., R. M. Forster & K. Lüning, 1998. UVB radiation and macroalgae: present effects and future directions. Eur. Soc. Photobiol., Proc. 1998: 1–12.

Garbary, D. J. & A. R. McDonald, 1998. Molecules, organelles and cells: fluorescence microscopy and red algal development. In Cooksey, K. E. (ed.), Molecular Approaches to the Study of the Ocean. Chapman & Hall, London: 409–422.

Garbary, D. J., A. R. McDonald & J. G. Duckett, 1992. Visualization of the cytoskeleton in red algae using fluorescent labelling. New Phytol. 120: 435–444.

Guiry, M. D. & E. Cunningham, 1984. Photoperiodic and temperature responses in the reduction of north-eastern Atlantic *Gigartina acicularis* (Rhodophyta: Gigartinales). Phycologia 23: 357–367.

Häder, D.-P. & F. L. Figueroa, 1997. Photoecophysiology of marine macroalgae. Photochem. Photobiol. 66: 1–14.

Heath, I. B., 1990. Tip Growth in Plant and Fungal Cells. Academic Press, San Diego.

Karentz, D., 1994. Ultraviolet tolerance mechanisms in Antarctic marine organisms. Ant. Res. Ser. 62: 93–110.

Karentz, D., F. S. McEuen, M. C. Land & W. C. Dunlap, 1991. Survey of mycosporine-like amino acid compounds in Antarctic marine organisms: potential protection from ultraviolet exposure. Mar. Biol. 108: 157–166.

Karsten, U., L. A. Franklin, K. Lüning & C. Wiencke, 1998. Natural ultraviolet-radiation and photosynthetically active radiation induce formation of mycosporine-like amino-acids in the marine macroalga *Chondrus-crispus* (Rhodophyta). Planta 205: 257–262.

Koslowsky, D. J. & S. D. Waaland, 1984. Cytoplasmic incompatibility following somatic cell fusion in *Griffithsia pacifica* Kylin, a red alga. Protoplasma 123: 8–17.

Koslowsky, D. J. & S. D. Waaland, 1987. Ultrastructure of selective chloroplast destruction after somatic cell fusion in *Griffithsia pacifica* Kylin (Rhodophyta). J. Phycol. 23: 638–648.

Madronich, S., R. L. McKenzie, M. Caldwell & L. O. Björn, 1995. Changes in ultraviolet radiation reaching the earth surface. Ambio 24: 143–152.

McKerracher, L. J. & I. B. Heath, 1986. Fungal nuclear behavior analysed by ultraviolet microbeam irradiation. Cell Motil. Cytoskel. 6: 35–47.

Pakker, H. & A. M. Breeman, 1997. Effects of ultraviolet-B radiation on macroalgae: DNA damage and repair. Phycologia 36 (4, supplement): 82–83.

Pavia, H., G. Cervin, A. Lindgren & P. Åberg, 1997. Effects of UV-B radiation and simulated herbivory on phlorotannins in the brown alga *Ascophyllum nodosum*. Mar. Ecol. Prog. Ser. 157: 139–146.

Russell, C. A., M. D. Guiry, A. R. McDonald & D. J. Garbary, 1996. Actin-mediated chloroplast movement in *Griffithsia pacifica* (Ceramiales, Rhodophyta). Phycol. Res. 44: 57–61.

Xiong, F., J. Komenda, J. Kopecky & L. Nedbal, 1997. Strategies of ultraviolet-B protection in microscopic algae. Physiol. Plant. 11: 378–388.

Hydrobiologia **512:** 171–176, 2004.
P.O. Ang, Jr. (ed.), Asian Pacific Phycology in the 21st Century: Prospects and Challenges.
© 2004 *Kluwer Academic Publishers.*

The first spindle formation in brown algal zygotes

Taizo Motomura* & Chikako Nagasato
*Muroran Marine Station, Field Science Center for Northern Biosphere, Hokkaido University,
Muroran 051-0003, Japan*
Author for correspondence; E-mail: motomura@bio.sci.hokudai.ac.jp

Key words: anisogamy, brown algae, centriole, centrosome, fertilization, isogamy, mitotic spindle, oogamy

Abstract

Regulation of the first spindle formation in brown algal zygotes was described. It is well known that there are three types of sexual reproduction in brown algae; isogamy, anisogamy and oogamy. Paternal inheritance of centrioles can be observed in all these cases, similar to animal fertilization. In isogamy and anisogamy, female centrioles (= flagellar basal bodies) selectively disappear and male centrioles remain after fertilization. In a typical oogamy (e.g. fucoid members), liberated egg does not have centrioles, and sperm centrioles are introduced in zygote. Participation of sperm centrioles to the spindle formation in zygotes was also described using *Fucus distichus* as a model system. Sperm centrioles function as a part of centrosome, namely microtubule organizing center, in zygote. Therefore, they have a crucial role in the spindle formation. Observations on the spindle formation in polygyny and karyogamy-blocked zygotes strongly suggest that egg nucleus can form a mitotic spindle by itself without centrosome, even though the resulting spindles are of abnormal shapes.

Introduction

Mitosis in brown algae has been observed in detail by using electron microscopy; for example, plurilocular sporangia of *Pylaiella littoralis* (L.) Kjellm. (Markey & Wilce, 1975), apical cells of *Sphacelaria tribuloides* Meneghini (Katsaros et al., 1983), *Halopteris filicina* (Grateloup) Kützing (Katsaros & Galatis, 1990), *Zonaria farlowii* S. and G., *Dictyopteris zonarioides* Farlow, *Padina pavonia* (L.) Gaill., and *Dictyota dichotoma* (Hudson) L. (Neushul & Dahl, 1972), vegetative cells of *Dictyota dichotoma* (Katsaros & Galatis, 1985) and *Dictyopteris membranacea* (Stackh.) Batt. (Katsaros & Galatis, 1988), male gametangia of *Cutleria hancokii* Dawson (La Claire & West, 1979), trichothallic meristem of *Cutleria cylindrica* Okamura (La Claire, 1982) and *Carpomitra cabrerae* Kützing (Motomura & Sakai, 1985), and antheridium of *Fucus serratus* L. (Berkaloff & Rousseau, 1979). It becomes apparent that a pair of centrioles exists in interphase cells of brown algae, and it duplicates and each of them migrates to both poles prior to mitosis. Moreover, recent immunofluorescence studies using anti-tubulin antibody (Motomura, 1991; Katsaros, 1992; Katsaros & Galatis, 1992; Rusig et

al., 1993; 1994) have clearly shown that microtubule (MT) cytoskeleton in brown algal cells is quite different from that of green algae and land plants (Lloyd, 1991). Characteristic MT arrays like cortical MTs during interphase and preprophase band just before mitosis in land plant cells are not detected. Almost all MTs elongate from the defined centrosome, which is composed of a pair of centrioles and pericentriolar materials, and located near to the nucleus (Figs 1–3). Therefore, in this respect, centrosome structure and MT cytoskeleton in brown algal cells are similar to those in animal cells (Wheatley, 1982; Kimble & Kuriyama, 1992; Balczon, 1996).

Different from the vegetative cells, a pair of centrioles changes into flagellar basal bodies in swarmers in the brown algae, i.e. gametes and zoospores. Characteristic, well-regulated MT flagellar rootlets (O'Kelly, 1989) are differentiated, instead of the random MT radiation from the centrosome in vegetative cells. Therefore, it can be speculated that several regulations including centrosome activation must have occurred before the mitosis in zygote after fertilization of male and female gametes. In this paper, we will discuss the regulation of the first spindle formation in brown algal zygotes with regards to centriole and centrosome.

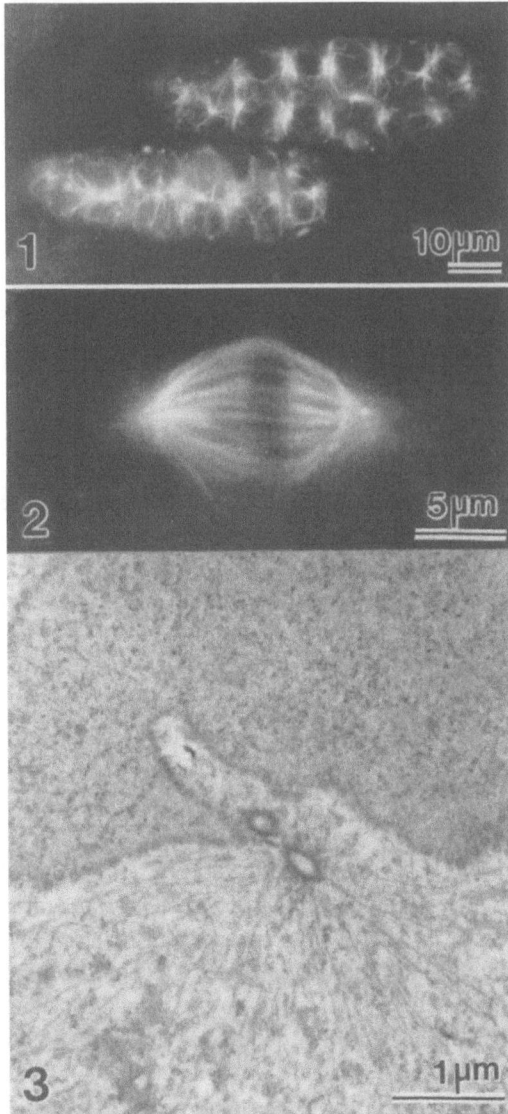

Figures 1–3. Microtubule (MT) cytoskeleton, mitotic spindle and a pair of centrioles in brown algal cells. **1.** MT cytoskeleton of two juvenile sporophytes of *Laminaria angustata*. Note that almost all MTs elongate from centrosome. **2.** Mitotic spindle in metaphase of *Dictyopteris divaricata* cell. Spindle MTs converge to centrosomes at both poles. **3.** A pair of centrioles just nearby a nucleus in *Fucus distichus* zygote. Note that numerous MTs radiate from the periphery of centrioles.

Paternal inheritance of centrioles in the fertilization of brown algae

It is well known that in most animal fertilization, centrioles in zygote are derived from sperm (Schatten, 1994). In oogamy in animals, including those in sea urchin, *Drosophila*, *Xenopus*, and mammals,

centrioles disappear in the process of oogenesis. For example, in starfish oocytes, each division pole in meiosis I has a pair of centrioles and each in meiosis II has only one centriole (Sluder et al., 1989; Kato et al., 1990). Contrary to animal fertilization, it is a characteristic feature that three types of sexual reproduction have been confirmed in the brown algae: isogamy, anisogamy and oogamy (Wynne & Loiseaux, 1976). We have examined fertilization in the brown algae and the zygote development using electron and immunofluorescence microscopy, and recently confirmed that paternal inheritance of centrioles is universal in brown algae (Fig. 4; Nagasato et al., 1998), similar to animal fertilization.

In isogamous brown algae, e.g. *Colpomenia bullosa* (Saunders) Yamada and *Scytosiphon lomentaria* (Lyngbye) Link (Scytosiphonales), one of two pairs of centrioles (= flagellar basal bodies) disappeared in zygote after plasmogamy. In isogamy, it is difficult to distinguish the cell organelle including nucleus, chloroplast, mitochondria and centrioles between male and female gametes. But observations on the polyspermic zygotes showed that only one pair of centrioles disappeared. Therefore, we concluded that centrioles from female gamete selectively disappeared (Motomura, 1992).

In anisogamous brown algae, e.g. *Cutleria cylindrica* Okamura (Cutleriales), one of two pairs of centrioles disappeared after plasmogamy as well as isogamy. In anisogamy, sizes between male and female gametes are quite different. Therefore, we can distinguish male and female nuclei; male nucleus is a little smaller and the chromatins are more condense than those of the female. From observations using electron and immunofluorescence microscopy on *C. cylindrica* zygotes, in which karyogamy had not yet occurred, we confirmed that the centrioles near the male nucleus remained and the ones near the female nucleus disappeared (Nagasato et al., 1998).

In the extremely anisogamous *Laminaria angustata* Kjellman (Laminariales) (because eggs have vestigial flagella, Motomura & Sakai, 1988), there are two pairs of centrioles derived from egg and sperm in zygote. We are able to distinguish egg centrioles and sperm centrioles in the zygote since the sperm flagellar basal bodies (= centrioles) are associated with a characteristic electron dense connecting band (Motomura, 1989). Observation using serial sections on electron microscopy clearly showed that egg centrioles selectively disappeared and sperm centrioles remained in the zygotes (Motomura, 1990).

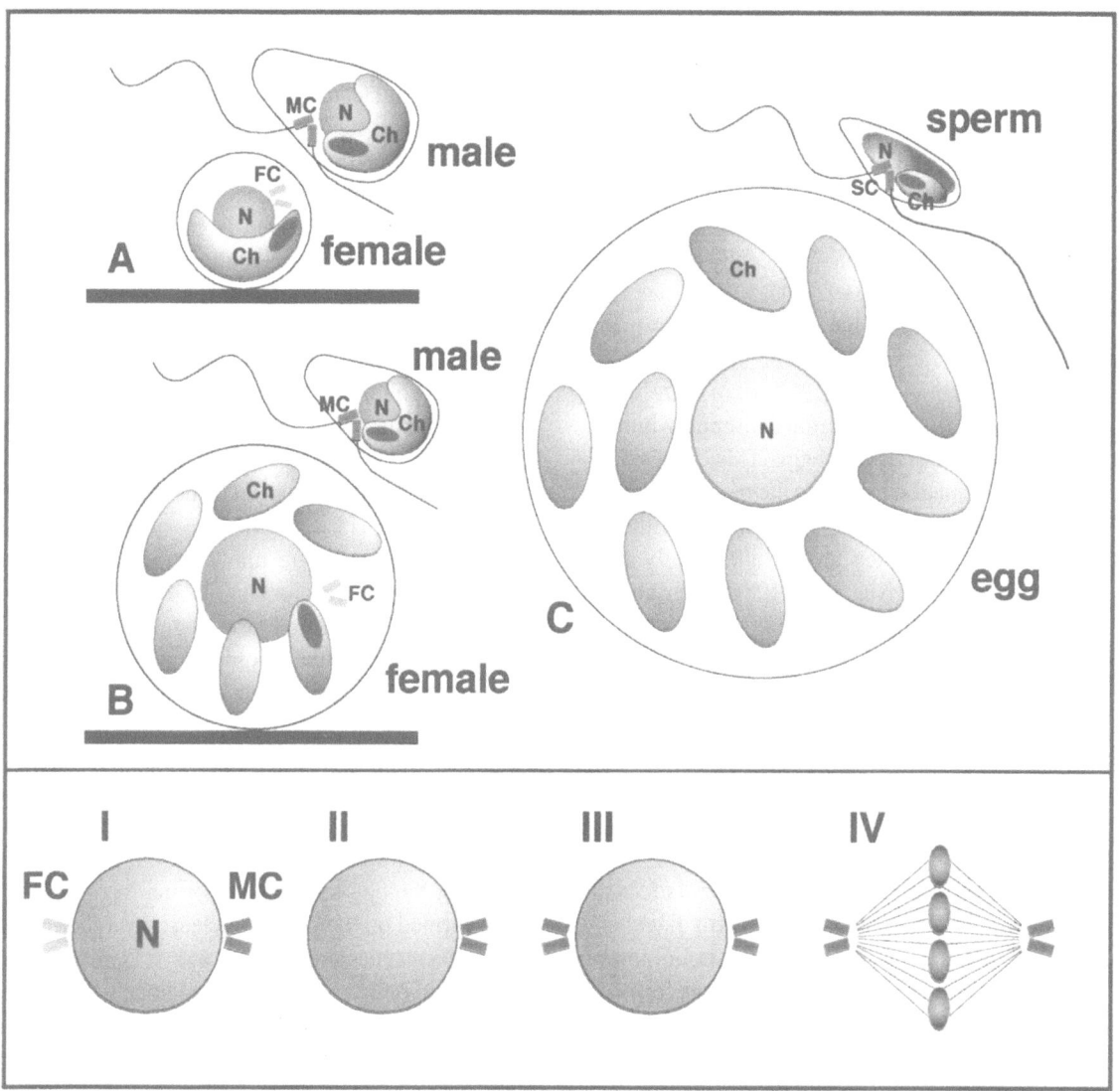

Figure 4. Schematic representation showing the paternal inheritance of centrioles in the fertilization of brown algae. (A) isogamy (*e.g. Colpomenia* and *Scytosiphon*), (B) anisogamy (e.g. *Cutleria*) and (C) oogamy (e.g. *Fucus* and *Pelvetia*). In isogamy and anisogamy, after female gamete settles on substratum and drew the flagella, male gamete approaches the female gamete as it is attracted by the sexual pheromone, and plasmogamy occurs. Therefore, in these cases, two pairs of centrioles (= flagellar basal bodies) from female and male gamete exist in the zygote at first (I). Afterwards, centrioles from female gamete disappear and ones from male gamete remain selectively (paternal inheritance of centrioles, II). In oogamy, egg does not contain centrioles originally, therefore, only a pair of centrioles from sperm exists in the zygote (II). A pair of centrioles from sperm duplicates and migrates to both mitotic poles (III). Finally, bipolar spindle with centrioles at both poles is formed (IV). N, nucleus; Ch, chloroplast; FC, female centrioles; MC, male centrioles; SC, sperm centrioles.

In the typical oogamy (Fucales), e.g. *Fucus distichus* Linnaeus and *Pelvetia fastigiat* (J. Ag.) De Toni, centrioles disappear during oogenesis. They do not exist originally in mature eggs (Motomura, 1994; Bisgrove et al., 1997; Nagasato et al., 1999). Therefore, centrioles in zygotes are introduced as the sperm flagellar basal bodies as that occurring during fertilization in most animals (Schatten, 1994). Of course,

judging from the recent molecular phylogenetic data, we must remind that the typical oogamy in Fucales is not directly derived from isogamy (Scytosiphonales) and anisogamy (Cutleriales and Laminariales).

At present, it is clear that karyogamy does not affect the selective disappearance of female centrioles in *Colpomenia bullosa* (Motomura, 1992) and *Cutleria cylindrica* (Nagasato et al., 1998), but the morphological, physiological and biochemical differences between female and male centrioles are still obscure. We are now proposing that, in the case of isogamy and anisogamy, plasmogamy would trigger a signal that induces the selective disappearance of female centrioles, because they always disappear 4–8 h after plasmogamy in *C. bullosa* and *C. cylindrica*. Moreover, unfertilized female gametes of *C. cylindrica* can develop parthenogenetically as well as zygotes (Kitayama et al., 1992), and centrioles do not disappear and a normal bipolar spindle is formed during parthenogenetic development of the female gametes (Nagasato et al., 1998). While, in extremely anisogamous *L. angustata*, centrioles disappear in the parthenogenesis of unfertilized eggs. This causes abnormal spindle formation including the monopolar and multipolar ones (Motomura, 1991). Most of these parthenogenetic sporophytes develop abnormally and abort at the early stage (Nakahara, 1984; Motomura, 1991; Lewis et al., 1993).

Spindle formation in (I) polyspermic zygotes, (II) zygotes from multinucleate eggs, and (III) karyogamy-blocked zygotes in *Fucus distichus*

When a normal fertilization occurs in *Fucus distichus*, each of the centrioles derived from sperm flagellar basal bodies separates, migrates to the mitotic poles and duplicates. A normal bipolar spindle (I) is formed and chromosomes are arranged at the equator (Motomura, 1994; Bisgrove et al., 1997). Contrary to the normal fertilized zygotes, the mitotic spindles in (II) polyspermic zygotes, (III) zygotes from multinucleate eggs (polygyny) and (IV) karyogamy-blocked zygotes possess unusual forms as shown in Figure 5 (Nagasato et al., 1999). Using these experimentally induced systems in *Fucus distichus*, we tried to clarify in details the participation of several factors, including sperm centrioles, in the spindle formation.

In polyspermic zygotes, all sperm nuclei that are incorporated into an egg fuse to the egg nucleus and a multipolar spindle is formed. Each sperm brings a pair of centrioles, and each of these centrioles duplicates, migrates to the mitotic poles and participates in spindle formation. Therefore, in the case of polyspermy in *Fucus*, as well as in animal polyspermy (Holy & Schatten, 1991; Navara et al., 1994), the number of mitotic poles is double that of the number of sperm that is incorporated into an egg.

Eight eggs are formed in an oogonium after meiosis and one mitosis in *Fucus*. However, final cytoplasmic cleavage occurs abnormally or not at all, producing multinucleate eggs. When sperms inseminated these multinucleate eggs (polygyny), a sperm nucleus fuse with one of the egg nuclei. The mitotic spindle possesses a multipolar, eccentric form in many cases. Although centrioles derived from the sperm exist at two mitotic poles in the spindle, several supplementary mitotic poles without centrioles exist. As a result, chromosomes cannot be arranged at the equator.

When karyogamy between sperm and egg nuclei is blocked by inhibiting the migration of sperm nucleus with colchicine (Brawley & Quatrano, 1979; Swope & Kropf ,1993; Motomura, 1995), we can see two types of the spindle formation in the egg nucleus. First, the sperm nucleus is located close to the egg nucleus, and sperm centrioles are drawn towards the egg nucleus from the sperm nucleus. In this case, a normal mitotic spindle is formed in the egg nucleus, and the associated MTs elongate from one pole of the spindle towards the scattered sperm chromosomes. Second, the egg and the sperm nuclei are sufficiently separated, and sperm centrioles cannot be drawn to the egg nucleus. In this case, monopolar or bipolar barrel-shaped (this means that spindle MTs do not converge at a spot on both mitotic poles) spindles without centrioles are formed at the egg nucleus. A normal bipolar spindle is not formed on the sperm nucleus, even though two pairs of centrioles are located and MTs radiate from there. Motomura (1995) reported that when karyogamy is blocked, the egg chromosomes condense after DNA synthesis but the sperm chromosomes condense prematurely without DNA synthesis. Therefore, the absence of a normal bipolar spindle on the sperm nucleus might be related to the premature condensation of the sperm chromosomes.

These observations strongly suggest that sperm centrioles certainly have a crucial role in the mitotic spindle formation in *Fucus distichus* zygotes. In contrast, monopolar or bipolar barrel-shaped spindles without centrioles in the egg nucleus of karyogamy-blocked zygotes and several supplementary mitotic poles without centrioles in polygyny are characteristic features. This indicates that the egg nucleus in *Fucus* can form mitotic spindle by itself, even though the shapes are abnormal, irrespective of the participation of centrioles from sperm. We have confirmed that γ-tubulin, which serves as the nucleation of MT growth at the centrosome in eukaryotic cells (Oakley et al., 1990; Joshi et al., 1992), is not localized at

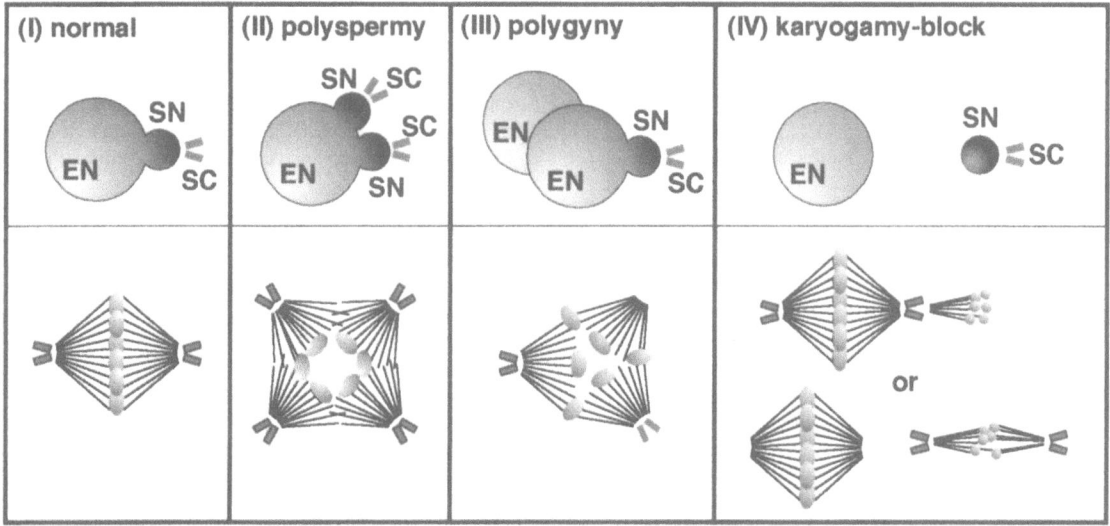

Figure 5. Schematic representation showing the spindle formation in (I) normal, (II) polyspermy, (III) polygyny and (IV) karyogamy-block with colchicine in *Fucus distichus*. In normal fertilization, after karyogamy between egg nucleus (EN) and sperm nucleus (SN) and duplication of a pair of centrioles from sperm, a bipolar spindle is formed. In polyspermy, incorporated sperm nuclei fuse to egg nucleus, and a multipolar spindle is formed. In this case, each of the mitotic poles contains a pair of centrioles. In polygyny, sperm nucleus fuses to one among several egg nuclei, and a multipolar spindle is formed. In this case, two mitotic poles contain a pair of centrioles, while the others do not. In the case of karyogamy-block using colchicine, there are basically two types of spindle formation. When centrioles migrate to egg nucleus from sperm nucleus, a normal bipolar spindle is formed at egg nucleus and MTs elongate to sperm nucleus from one mitotic pole. When centrioles do not migrate to egg nucleus, barrel-shaped bipolar or monopolar spindle without centrioles is formed at egg nucleus. Spindle-like MT array is formed at sperm nucleus. EN, egg nucleus; SN, sperm nucleus; SC, sperm centrioles.

these mitotic poles without centrioles (Nagasato et al., unpublished data). Recently, it is becoming clear that non-centrosomal proteins, NuMA (nuclear protein associated with a mitotic apparatus), dynactin and cytoplasmic dynein, are essential in organizing the minus ends of MTs for spindle formation (Gaglio et al., 1997; Heald et al., 1997; Compton, 1998). Therefore, we are now considering that the formation of the mitotic spindle poles without centrosomes in the egg nucleus of *Fucus distichus* would be induced by these noncentrosomal proteins which are hidden behind the luster of the centrosomes.

Acknowledgements

We express our thanks to C. S. Vairappan, Hokkaido University for his critical reading of the manuscript. This study was supported by Grant-in-Aid for Scientific Research from the Ministry of Education, Science, and Culture of Japan (11874121 and 12440232). Chikako Nagasato was supported by the Research Fellowships of the Japan Society for the Promotion of Science for Young Scientists.

References

Balczon. R., 1996. The centrosome in animal cells and its functional homologs in plant and yeast cells. Int. Rev. Cytol. 169: 25–82.

Berkaloff, C. & B. Rousseau, 1979. Ultrastructure of male gametogenesis in *Fucus serratus* (Phaeophyceae). J. Phycol. 15: 163–173.

Bisgrove, S. R., C. Nagasato, T. Motomura & D. L. Kropf, 1997. Immunolocalization of centrin during fertilization and the first cell cycle in *Fucus distichus* and *Pelvetia compressa* (Fucales, Phaeophyceae). J. Phycol. 33: 823–829.

Brawley, S. H. & R. S. Quatrano, 1979. Effects of microtubule inhibitors on pronuclear migration and embryogenesis in *Fucus distichus* (Phaeophyta). J. Phycol. 15: 266–272.

Compton, D. A., 1998. Focusing on spindle poles. J. Cell Sci. 111: 1477–1481.

Gaglio, T., M. A. Dionne & D. A. Compton, 1997. Mitotic spindle poles are organized by structural and motor proteins in addition to centrosomes. J. Cell Biol. 135: 399–414.

Heald, R., R. Tournebize, A. Habermann, E. Karsenti & A. Hyman, 1997. Spindle assembly in *Xenopus* egg extracts: Respective roles of centrosomes and microtubule self-organization. J. Cell Biol. 138: 615–628.

Holy, J. & G. Schatten, 1991. Spindle pole centrosomes of sea urchin embryos are partially composed of material recruited from maternal stores. Dev. Biol. 147: 343–353.

Joshi, H. C., M. J. Palacios, L. McNamara & D. W. Cleveland, 1992. γ-tubulin is a centrosomal protein required for cell cycle-dependent microtubule nucleation. Nature 356: 80–83.

176

Kato, K. H., S. Washitani-Nemoto, A. Hino & S. Nemoto, 1990. Ultrastructural studies on the behavior of centrioles during meiosis of starfish oocytes. Dev. Growth and Differ. 32: 41–49.

Katsaros, C., 1992. Immunofluorescence study of microtubule organization in some polarized cell types of selected brown algae. Bot. Acta 105: 400–406.

Katsaros, C. & B. Galatis, 1985. Ultrastructural studies on thallus development in *Dictyota dichotoma* (Phaeophyta, Dictyotales). Br. Phycol. J. 20: 263–276.

Katsaros, C. & B. Galatis, 1988. Thallus development in *Dictyopteris membranacea* (Phaeophyta, Dictyotales). Br. Phycol. J. 23: 71–88.

Katsaros, C. & B. Galatis, 1990. Thallus development in *Halopteris filicina* (Phaeophyceae, Sphacelariales). Br. Phycol. J. 25: 63–74.

Katsaros, C. & B. Galatis, 1992. Immunofluorescence and electron microscope studies of microtubule organization during the cell cycle of *Dictyota dichotoma* (Phaeophyta, Dityotales). Protoplasma 169: 75–84.

Katsaros, C., B. Garatis & K. Mitrakos, 1983. Fine structural studies on the interphase and dividing apical cells of *Sphacelaria tribuloides* (Phaeophyta). J. Phycol. 19: 16–30.

Kimble, M. & R. Kuriyama, 1992. Functional components of microtubule-organizing centers. Int. Rev. Cytol. 136: 1–50.

Kitayama, T., H. Kawai & T. Yoshida, 1992. Dominance of female gametophytes in field populations of *Cutleria cylindrica* (Cutleriales, Phaeophyceae) in the Tsugaru Strait, Japan. Phycologia 31: 449–461.

La Claire, J. W., II, 1982. Light and electron microscopic studies of growth and reproduction in *Cutleria* (Phaeophyta). III. Nuclear division in the trichothallic meristem of *C. cylindrica*. Phycologia 21: 273–287.

La Claire, J. W. II & J. A. West, 1979. Light- and electron-microscopic studies of growth and reproduction in *Cutleria* (Phaeophyta). II. Gametogenesis in the male plant of *C. hancokii*. Protoplasma 101: 247–267.

Lewis, R. J., B. Jiang, Y. M. Neushul & X. G. Fei, 1993. Haploid parthenogenetic sporophytes of *Laminaria japonica* (Phaeophyceae). J. Phycol. 29: 363–369.

Lloyd, C. W., 1991. The Cytoskeletal Basis of Plant Growth and Form. Academic Press, London: 330 pp.

Markey, D. R. & R. T. Wilce, 1975. The ultrastructure of reproduction in the brown alga *Pylaiella littoralis*. I. Mitosis and cytokinesis in the plurilocular gametangia. Protoplasma 85: 219–241.

Motomura, T., 1989. Ultrastructural study of sperm in *Laminaria angustata* (Laminariales, Phaeophyta), especially on the flagellar apparatus. Jpn. J. Phycol. 37: 105–116.

Motomura, T., 1990. Ultrastructure of fertilization in *Laminaria angustata* (Phaeophyta, Laminariales) with emphasis on the behavior of centrioles, mitochondria and chloroplasts of the sperm. J. Phycol. 26: 80–89.

Motomura, T., 1991. Immunofluorescence microscopy of fertilization and parthenogenesis in *Laminaria angustata* (Phaeophyta). J. Phycol. 27: 248–257.

Motomura, T., 1992. Disappearance of centrioles derived from female gametes in zygotes of *Colpomenia bullosa* (Phaeophyceae). Jpn. J. Phycol. 40: 207–214.

Motomura, T., 1994. Electron and immunofluorescence microscopy on the fertilization of *Fucus distichus* (Fucales, Phaeophyceae). Protoplasma 178: 97–110.

Motomura, T., 1995. Premature chromosome condensation of the karyogamy-blocked sperm pronucleus in the fertilization of *Fucus distichus* (Fucales, Phaeophyceae). J. Phycol. 31: 108–113.

Motomura, T. & Y. Sakai, 1985. Ultrastuctural studies on nuclear division in the sporophyte of *Carpomitra cabrerae* (Clemente) Kützing (Phaeophyta, Sporochnales). Jpn. J. Phycol. 33: 199–209.

Motomura, T. & Y. Sakai, 1988. The occurrence of flagellated eggs in *Laminaria angustata* (Phaeophyta, Laminariales) J. Phycol. 24: 282–285.

Nagasato, C., T. Motomura & T. Ichimura, 1998. Selective disappearance of maternal centrioles after fertilization in the anisogamous brown alga *Cutleria cylindrica* (Cutleriales, Phaeophyceae): paternal inheritance of centrioles is universal in the brown algae. Phycol. Res. 46: 191–198.

Nagasato, C., T. Motomura & T. Ichimura, 1999. Influence of centriole behavior on the first spindle formation in zygotes of the brown alga *Fucus distichus* (Fucales, Phaeophyceae). Dev. Biol. 208: 200–209.

Nakahara, H., 1984. Alternation of generations of some brown algae in unialgal and axenic culture. Sci. Pap. Inst. Algol. Res. Fac. Sci. Hokkaido Univ. 7: 77–194

Navara, C. S., N. L. First & G. Schatten, 1994. Microtubule organization in the cow during fertilization, polyspermy, parthenogenesis, and nuclear transfer: the role of the sperm aster. Dev. Biol. 162: 29–40.

Neushul, M. & A. L. Dahl, 1972. Ultrastructural studies of brown algal nuclei. Am. J. Bot. 59: 401–410.

Oakley, B. R., C. E. Oakley, Y. Yoon & M. K. Jung, 1990. γ-tubulin is a component of the spindle pole body that is essential for microtubule function in *Aspergillus nidulans*. Cell 61: 1289–1301.

O'Kelly, C. J., 1989. The evolutionary origin of the brown algae: information from studies of motile cell ultrastructure. In Green, J. C., B. S. C. Leadbeater & W. L. Diver (eds), The Chromophyte Algae. Problems and Perspectives. Oxford University Press, Oxford: 55–278.

Rusig, A. M., H. Le Guyader & G. Ducreux, 1993. Microtubule organization in the apical cell of *Sphacelaria* (Phaeophyceae) and its related protoplast. Hydrobiologia 260/261: 167–172.

Rusig, A. M., H. Le Guyader & G. Ducreux, 1994. Dedifferentiation and microtubule reorganization in the apical cell protoplast of *Sphacelaria* (Phaeophyceae). Protoplasma 179: 83–94.

Schatten, G., 1994. Centrosome and its mode of inheritance: the reduction of the centrosome during gametogenesis and its restoration during fertilization. Dev. Biol. 165: 299–335.

Sluder, G. F., F. J. Miller, K. Lewis, E. D. Davidson & C. L. Rieder, 1989. Centrosome inheritance in starfish zygotes: Selective loss of the maternal centrosome after fertilization. Dev. Biol. 131: 567–579.

Swope, R. E. & D. L. Kropf, 1993. Pronuclear positioning and migration during fertilization in *Pelvetia*. Dev. Biol. 157: 269–276.

Wheatley, D. N., 1982. The Centriole: A Central Enigma of Cell Biology. Elsevier Biomedical Press, Amsterdam.

Wynne, M. J. & S. Loiseaux, 1976. Recent advances in life history studies of the Phaeophyta. Phycologia 15: 435–452.

Hydrobiologia **512**: 177–183, 2004.
P.O. Ang, Jr. (ed.), Asian Pacific Phycology in the 21ˢᵗ Century: Prospects and Challenges.
© 2004 *Kluwer Academic Publishers.*

The generic delimitation of *Rhodella* (Porphyridiales, Rhodophyta) with emphasis on ultrastructure and molecular phylogeny

Akiko Yokoyama[1,3,*], Kazumichi Sato[2] & Yoshiaki Hara[2]
[1]*Regional Joint Research Project of Yamagata Prefecture, Yamagata Technopolis Foundation*, Matsuei 2-2-1, Yamagata 990-2473, Japan*
[2]*Department of Biology, Faculty of Science, Yamagata University. Kojirakawa 1-4-12, Yamagata 990-8560, Japan*
[3]*Present address: Department of Biology, Faculty of Science, Yamagata University, Japan*
*Author for correspondence; E-mail: akiko@sbiol.kj.yamagata-u.ac.jp

Key words: Rhodella, R. cyanea, 18SrDNA, molecular phylogeny, taxonomy, ultrastructural features

Abstract

We investigated the cellular features and molecular phylogeny of *Rhodella* species and related unicellular red algae including undescribed species that we isolated. Results provide a new taxonomic interpretation at both generic and specific levels. The genus *Rhodella* is defined by its pyrenoid that is free from any internal structures. Based on phylogenetic analysis using 18SrDNA, there are two possibilities for the generic delimitation of *Rhodella*: *Rhodella sensu stricto* and *Rhodella sensu lato*. The generic autonomy of *Dixoniella* and the taxonomic position of *R. cyanea* were also discussed.

Introduction

When Evans (1970) established the genus *Rhodella* with the type species *R. maculata*, he considered that the ultrastructural characteristics were sufficient to separate it from the other unicellular reds, such as *Porphyridium*, *Rhodosorus* and *Cyanidium*. He was the first to do such an ultrastructural taxonomic study on unicellular reds. According to him, the following morphological and ultrastructural characters appear in *R. maculata* (Evans 1970); (1) the chloroplast is axile and highly-lobed stellate with a naked pyrenoid; (2) the pyrenoid is free from any structures, except a tongue-shaped nuclear invagination and is covered with floridean starch shells; (3) the nucleus is not situated in the central portion of the cell. However, this initial diagnosis did not provide any clear delimitation between the generic and specific characters in *Rhodella*.

A second species was reported by Wehrmeyer (1971), who transferred *Porphyridium violaceum* Kornmann to *Rhodella* as *R. violacea*, since its ul-

trastructural features were basically the same as those of *R. maculata* with only one distinction; the apparent absence of a nuclear projection into the pyrenoid. However, Patrone et al. (1991) found a small nuclear projection into the pyrenoid of *R. violacea* from the type culture. It shows the necessity to re-examine the two species to determine whether they are conspecific or not.

A third species, *Rhodella reticulata* Deason, Bulter *et* Rhyne described by Deason et al. (1983), possesses a highly-lobed chloroplast with a naked pyrenoid containing convoluted thylakoids in the matrix. Fresnel et al. (1989) pointed out that this species is conspecific with *Porphyridium griseum* Geitler since the same chloroplast ultrastructure was found and made a new combination of *Rhodella grisea* (Geitler) Fresnel, Bellard, Hindák *et* Pekárková. According to Scott et al. (1992), however, the ultrastructural features of this species, such as pyrenoid structure, dictyosome arrangement, nuclear position in the cell, relative configuration of pyrenoid and nucleus, are different from those of the former two *Rhodella* species. Scott et al. (1992) consequently established the genus *Dixoniella* and placed *R. reticulata* under this genus in synonymy

---

*Present name is Yamagata Public Corporation for the Development of Industry R&D Department.

Table 1. List of *Rhodella* species and related taxa examined in this study

Species	Strain name	Collection site/Source	TEM	Accession no.
Rhodella cyanea	Tobishima	Tobishima Island, Yamagata, Japan	+	AB045605
R. maculata	Amami	Amami Island, Kagoshima, Japan	+	AB045608
R. violacea	BRW	Point Barrow, Elson Lagoon, Alaska, USA	+	AB045604
	B115.79	SAG[*1]		AB045580
Rhodella species	Nagura 1	Nagura River, Ishigaki Island Okinawa, Japan	+	AB045598
Rhodella sp.	Gamou T4	Gamou Tideland, Miyagi, Japan	+	AB045591
Rhodella sp.	Mexico BC73C	Mulege River, Baja California Sur, Mexico	+	AB045594
Dixoniella grisea	B39.94	SAG[*1]		AB045581
	Ogasawara	Chichi Island, the Ogasawara Islands, Tokyo, Japan	+	AB045583
Glaucosphaera vacuolata	1662	UTEX[*2]		AB045583
Porphyridium purpureum	R-1	IAM[*3]		AB045584

TEM=Transmission Electron Microscopy. Plus (+): studied. [*1] SAG: Sammlung von Algenkulturen at the University of Göttingen. [*2] UTEX: The Culture Collection of Algae at the University of Texas at Austin, Department of Botany, University of Texas. [*3] IAM: Institute of Applied Microbiology, The University of Tokyo (the present name: Institute of Molecular and Cellular Biosciences (IMCB), The University of Tokyo).

with *Dixoniella grisea* Scott, Broadwater, Saunders, Thomas *et* Gabrielson.

A fourth species, *Rhodella cyanea,* was described by Billard & Fresnel (1986). The configurations of the pyrenoid, nucleus and other organelles are quite unique. However, the naked pyrenoid contains convoluted thylakoids in the matrix, like those of *D. grisea* (Scott et al., 1992). The subcellular organization, with a central nucleus, perinuclear dictyosomes and multilobed chloroplast, is similar to those of *D. grisea* and *Glaucosphaera vacuolata* Korshikov (Broadwater et al., 1995). Because of the blue-green cell color, *R. cyanea* was thought to lack phycoerythrin (Billard & Fresnel, 1986), whereas *R. maculata* and *R. violacea* contain Bangiophycean type of phycoerythrin (Koller & Wehrmeyer, 1975; Billard & Fresnel, 1986). For the reason stated above it appears incorrect to assign *R. cyanea* to *Rhodella*.

From the historical background of unicellular red algal investigations focused on *Rhodella*, the following taxonomic questions remain; How is the genus *Rhodella* defined? What is the taxonomic position of *R. cyanea*? What is the relationship among *Rhodella*, *Dixoniella* and *Glaucosphaera*?

In order to elucidate these questions, we observed the cellular features of *Rhodella* including several undescribed species originally isolated by us, and determined their 18SrDNA sequences and constructed a phylogenetic tree.

Materials and methods

The Porphyridialean algae examined in this study are listed in Table 1, together with their culture names and collection sites or institutions maintaining them. All of them were grown in ESM medium (Okaichi et al., 1982) and maintained at 20 °C under 14:10 L:D cycle. Light was provided by cool white fluorescent lamps at ca. 20 μmol m^{-2} s^{-1}. Light microscope studies were made on cultured live cells, and transmission electron microscope studies were done on fixed materials by the methods published in a previous report (Hara & Chihara, 1985).

Total DNAs were extracted by modified 2xCTAB method developed by Hasebe & Iwatsuki (1990). PCR amplification was performed by the method of Hasebe et al. (1994). The genes of 18SrDNA were amplified with SR primers (Nakayama et al., 1996). The sequences were determined from the PCR products with Dye Terminator Cycle Sequencing FS Ready Reaction Kit (Applied Biosystems) on ABI 373A or ABI 310 sequencers. Sequences obtained in this study were aligned manually, including the published data of *Erythrotrichia carnea* L26188, *Dixoniella grisea* L26187 and *Rhodella maculata* U21217 (Ragan et al., 1994). The phylogenetic tree was constructed by fastDNAml (Olsen et al., 1994) and bootstrap values were calculated by PAUP ver. 3.1.1 (Swofford, 1993).

Phycobiliproteins were extracted and analyzed by the following procedure. Cultured cells were pelleted by centrifugation (1500 *g*, 5 min). The pellets were suspended in 10 mM potassium phosphate buffer

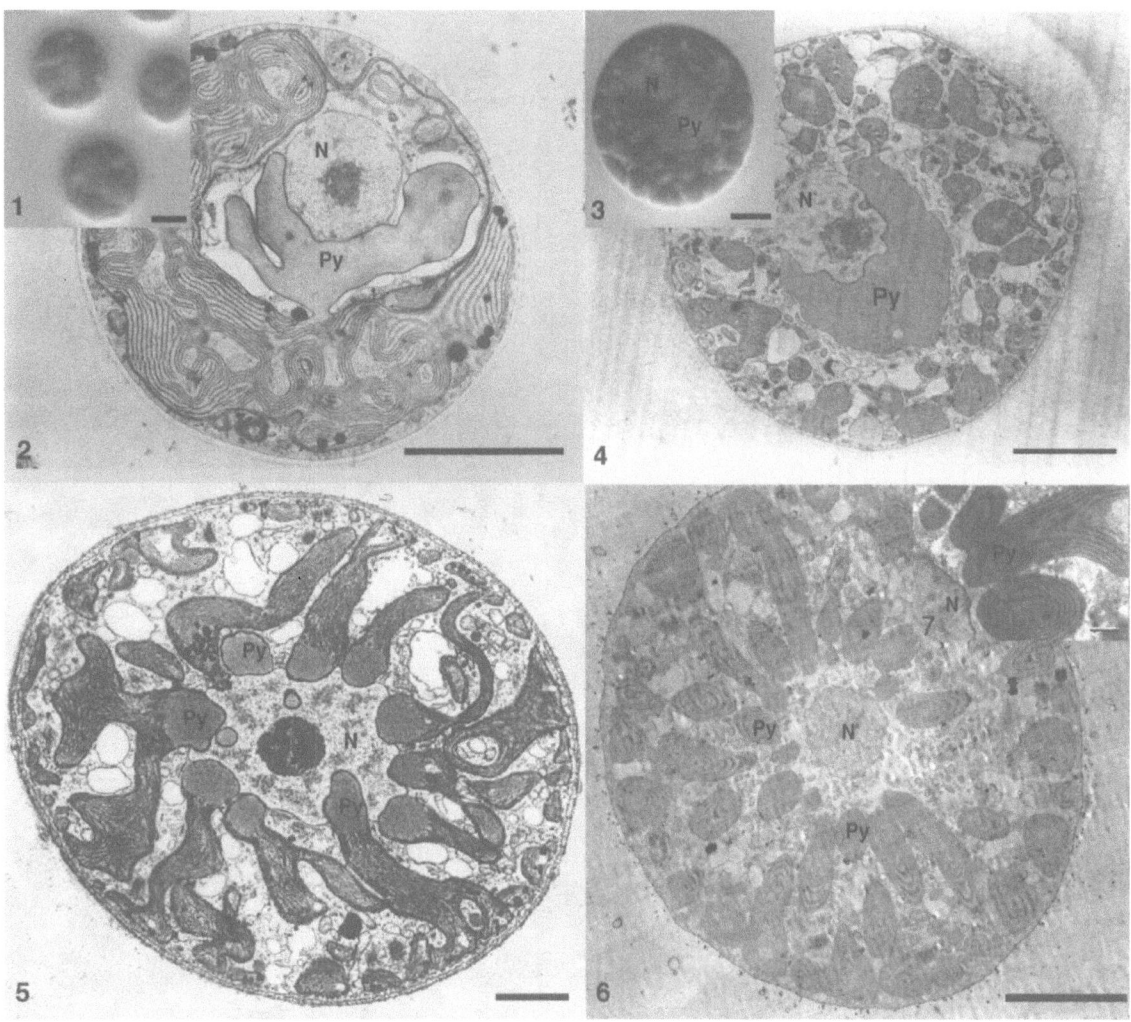

Figures 1–7. Light (Figs 1 and 3) and transmission electron (Figs 2 and 4–7) micrographs of *Rhodella* species. Bars: 5 µm (Figs 1–6), 0.5 µm (Fig. 7). **Figure 1**. Cells of *Rhodella* sp. (Nagura strain) showing a highly lobed stellate chloroplast. **Figure 2**. A cell of *Rhodella* sp. (Nagura st.) showing the subcellular features, in particular, configurational relation between nucleus (N) and pyrenoid (Py). **Figure 3**. Cells of *Rhodella* sp. (Gamou st.) showing the location and shapes of a nucleus and chloroplast with pyrenoid. **Figure 4**. A cell of *Rhodella* sp. (Gamou st.) showing the subcellular features similar to those of Nagura strain. **Figure 5**. A cell of *Rhodella* sp. (Mexico st.) showing the unique configuration between a nucleus and chloroplast-pyrenoid. **Figure 6**. A cell of *R. cyanea* (Tobishima st.) showing similar subcellular features as holotype of *R. cyanea* (Billard & Fresnel, 1986). **Figure 7**. A central part of the cell of *R. cyanea* (Tobishima st.) showing thylakoids in the chloroplast entering into the pyrenoid matrix.

(pH 6.85). The cells were broken by ultra sonicator UD-200 (TOMY SEIKO Co. Ltd., Tokyo) or were vortexed with glass beads. After centrifugation (20 000 g, 15 min, 4 °C), the supernatant was transferred to a new tube and saturated with ammonium sulfate. After stirring 2 h at 4 °C, the samples were dialyzed in 5 mM potassium phosphate buffer (pH 6.85). Chromatography was performed by DEAE-Cellulofine A-500 (Seikagaku Co., Tokyo) gradient from 5 mM to 150 mM potassium phosphate buf-

fer. Phycobiliproteins of all fractions were measured by a spectrophotometer UV-200 (SHIMADZU Co., Kyoto).

Results and discussion

Previously described species of *Rhodella* (*R. maculata, R. cyanea, R. violacea*) and three undescribed species (provisionally named as *Rhodella* sp. and followed by Nagura, Gamou and Mexico strains in the

Table 2. Morphological and ultrastructural comparison of *Rhodella* and related taxa

	Cell size (μm)	Chloroplast Location	Pyrenoid Type	Pyrenoid Internal structures	Nucleus Location	Nucleus Projection into pyrenoid	Dictyosome[1] Association	Pigment Phycoerythrin type	Reference
Rhodella maculata	7–24	Axile	Naked	Absent	Peripheral	Present	ER	B	Evans (1970)
R. violacea	8–30	Axile	Naked	Absent	Peripheral	Absent[2*]	ER	B	Wehmeyer (1971)
R. sp. (Nagura st.)	7–13	Peripheral	Naked	Absent	Peripheral	Absent	ER?	B	This study
R. sp. (Gamou st.)	19–40	Peripheral	Naked	Absent	Peripheral	Absent	ER?	B	This study
R. sp. (Mexico st.)	18–35	Axile	Naked	Absent	Central	Absent	ER?	B	This study
R. cyanea (Tobishima st.)	20–28	Axile	Naked	Thylakoid	Central	Absent	Nu?	B	This study
R. cyanea	22–40	Axile	Naked	Thylakoid	Central	Absent	Nu	ND?	Billard & Fresnel (1986)
Dixoniella grisea	8.5–17	Axile	Naked	Thylakoid	Peripheral	Absent	Nu	C[*3]	Scott et al. (1992)
Glaucosphaera vacuolata	14–22	Peripheral	Absent	–	Central	Absent	Nu	ND	Broadwater et al. (1995)
Porphyridium purpureum	6–12[*4]	Axile	Embedded	Thylakoid	Peripheral	Absent	ER/M	B,b	Schornstein & Scott (1982)

[*1]Dictyosome data except for ER? and Nu? were referred to Broadwater & Scott 1994. [*2]Nuclear projection into pyrenoid was observed in *R. violacea* by Patrone et al. (1991). [*3]Phycoerythrin type of *Dixoniella* was determined in this study. [*4]Cell size of *P. purpureum* was examined in this study from *P. purpureum* R-1 strain. ER: endoplasmic reticurum, ER?: Dictyosomes are not faced to nucleus but probably to ER, Nu: Nucleus, B: Bangiophycean type of phycoerythrin, b: modified type of B-phycoerythrin, C: Cyanophyoean type of phycerythrin, ND? or ND: Not detected.

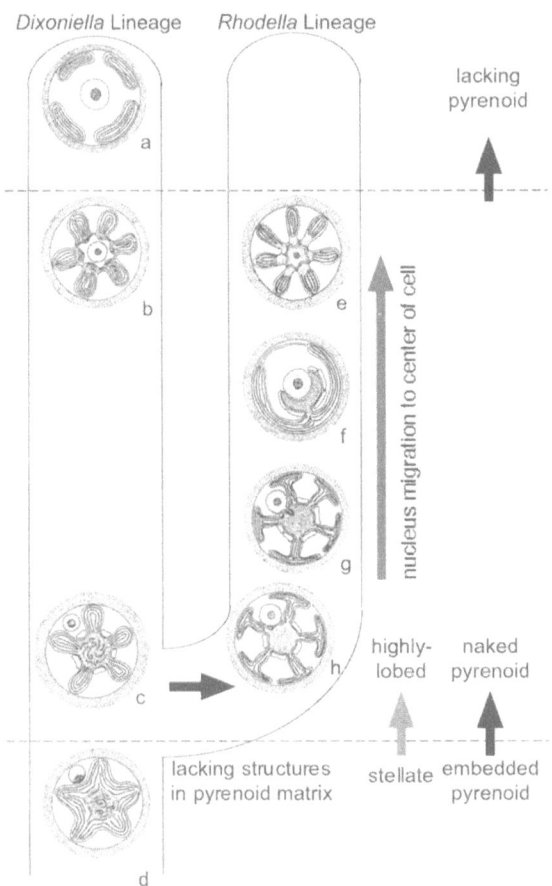

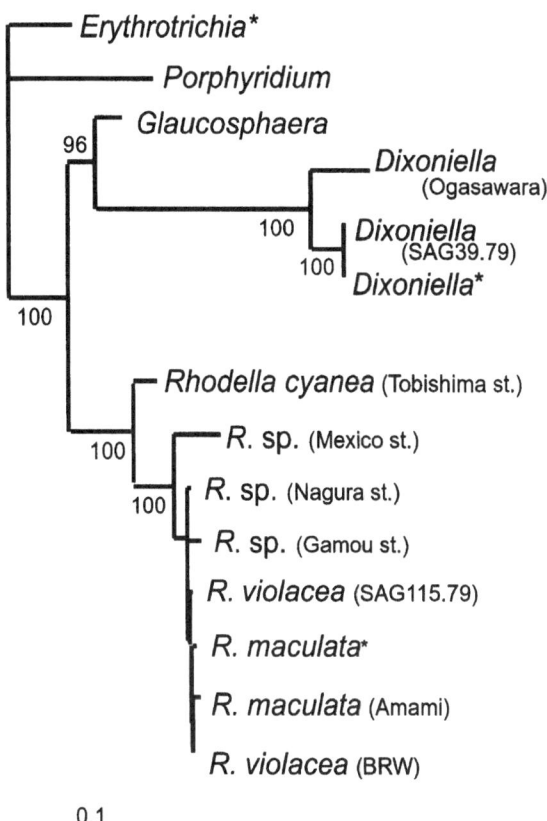

Figure 8. A hypothetical scheme of the relationship of *Rhodella* and its related species based on the structural characteristics. Arrows show the structural clines focused on chloroplast, pyrenoids and nucleus. (a) *Glaucosphaera vacuolata*, (b) *Rhodella cyanea*, (c) *Dixoniella grisea*, (d) *Porphyridium purpureum*, (e) *Rhodella* species (Mexico st.), (f) *Rhodella* sp. (Nagura st.) and *Rhodella* sp. (Gamou st.), (g) *R. maculata*, (h) *R. violacea*.

Figure 9. Maximum likelihood tree (fastDNAml) of 18SrDNA sequences, based on the conservative regions (1458 nucleotides). The numbers on branches indicate bootstrap percentages (1000 replicates, without less than 50%) which are calculated by PAUP 3.1.1 (Swofford, 1993). Three species indicated by the asterisks were sequenced by Ragan et al. (1994).

parentheses) were investigated by combining their cellular features with molecular phylogenetic analysis using 18SrDNA to provide a new interpretation of their generic and specific taxonomy.

In *Rhodella* sp. (Nagura strain), a unique configuration and structures of organelles were found (Fig. 2). The chloroplast with a naked pyrenoid is parietal and situated in the cell periphery. The amorphous pyrenoid is connected to the chloroplast by a narrow isthmus and surrounds half the nuclear surface. The pyrenoid matrix is free from any internal structures such as thylakoids and tubular structures, a ultrastructural characteristic that is basically shared with *Rhodella maculata* (Evans, 1970) and *Rhodella violacea* (Wehrmeyer, 1971).

An organellar configuration and pyrenoid ultrastructure similar to those of *Rhodella* sp. (Nagura st.) were recognized in *Rhodella* sp. (Gamou st.). The surface of the nucleus is partly enclosed by the pyrenoid and the pyrenoid matrix lacks thylakoids (Fig. 4). The cell size range (20–40 μm diam.) of *Rhodella* sp. (Gamou st.), however, is considerably larger than that (7–13 μm diam.) of *Rhodella* sp. (Nagura st.) (Figs 1 & 3).

Rhodella sp. (Mexico st.) also has a unique organellar configuration and pyrenoid structure (Fig. 5). A stellate chloroplast is highly lobed; the pyrenoid and nucleus configuration closely resembles that of *R. cyanea* (Billard & Fresnel, 1986) and the Tobishima strain examined in this study (Figs 6 and 7). However, some differences were found between *Rhodella* sp. (Mexico st.) and *R. cyanea*. The pyrenoid matrix of *Rhodella* sp. (Mexico st.) is free from any internal structures (Fig. 5), but that of *R. cyanea* is invaded by

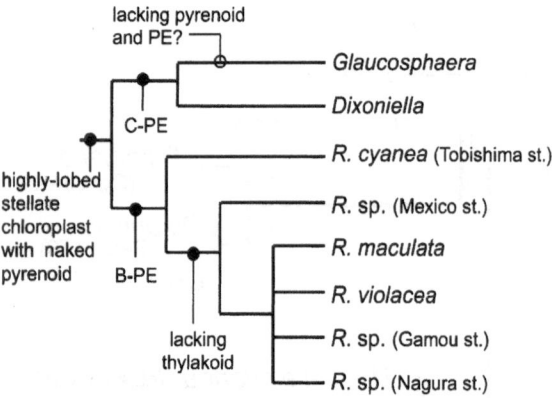

Figure 10. Structural and physiological characters are mapped onto the molecular phylogenetic tree inferred from 18SrDNA sequences. C-PE: Cyanophycean type of phycoerythrin; B-PE: Bangiophycean type of phycoerythrin; PE: phycoerythrin; lacking thylakoid: lacking thylakoid in the pyrenoid matrix.

thylakoids. No cytoplasmic space between the nucleus and the pyrenoid was found in *Rhodella* sp. (Mexico st.), while the nucleus of *R. cyanea* (Billard & Fresnel, 1986) and that of Tobishima st. (Fig. 6) were surrounded by cytoplasm and sometimes associated with dictyosomes.

The organellelar configuration and pyrenoid structure of Tobishima st. examined here are closely similar to those of *R. cyanea* (Billard & Fresnel, 1986). Initially we considered these to be conspecific. But a color difference exists between Tobishima st. (purple-red) and *R. cyanea* (blue-green). In our pigment analysis, Bangiophycean type of phycoerythrin was detected in the Tobishima st. *R. cyanea* was speculated to lack phycoerythrin from its blue-green cell color (Billard & Fresnel, 1986).

The cellular features of all species examined in this study and related species published so far are summarized in Table 2, for their taxonomic analysis. Species of *Rhodella* and *D. grisea* with a naked pyrenoid free from internal structures are primarily divided into two groups, depending mainly on the presence or absence of thylakoids in the pyrenoid matrix. If the presence of thylakoids in the pyrenoid matrix is adopted for the generic delimitation of *Rhodella*, *R. cyanea* has to be excluded from *Rhodella*. It is noteworthy that another structural characteristic of dictyosomes, which are located around the nucleus and associated with the nuclear envelope from their *cis* face, is common in *R. cyanea*, *D. grisea* and *Glaucosphaera vacuolata*.

The hypothetical scheme in Figure 8 based on Hara et al. (1989) and the results obtained here, including *Rhodella* species derived from *Porphyridium*,

show two lineages: *Dixoniella*-lineage and *Rhodella*-lineage. As previously reported (Hara et al., 1989), the structural clines shown below are commonly recognized among Porphyridialean algal groups: (1) chloroplast; axile to peripheral, stellate to highly-lobed, presence or absence of pyrenoids, (2) pyrenoid; embedded to naked type, with or without thylakoids or tubular structures in the matrix, (3) nucleus; peripheral to central. Some of these are cited in Figure 8. However, lacking internal structures in the pyrenoid matrix (indicated by a horizontal arrow in Fig. 8) may give rise to a group intermediate between the two lineages.

Based on cellular features mentioned above, it is understandable that the genus *Rhodella* is delimited by possessing a pyrenoid without any internal structures in the matrix and dictyosomes directed and associated with ER by their *cis* faces (Table 2). In this sense, we have a problem because *R. cyanea* possesses convoluted thylakoids in the pyrenoid matrix which is common with *D. grisea*, and dictyosomes associated with nucleus which is common with *D. grisea* and *G. vacuolata* (Table 2).

For better understanding of the taxonomy and phylogeny of *Rhodella* and other related algae, the phylogenetic tree inferred from the 18SrDNA was constructed with *Porphyridium* and *Erythrotrichia* selected as the outgroups (Fig. 9). All species of *Rhodella* formed a clade whereas *D. grisea* and *G. vacuolata* made up another clade (Fig. 9). Consequently, members in each clade show monophyletic divergence. Our three undescribed species were phylogenetically supported to belong to the genus *Rhodella*. The generic autonomy of *Dixoniella* reported by Scott et al. (1992) was also supported. Based on the phylogenetic analysis, *R. cyanea* may remain in *Rhodella*, although it may also be excluded from *Rhodella*.

The results of cellular and pigment analyses obtained here and published elsewhere are mapped onto a simplified tree (Fig. 10) derived from Figure 9. All species examined in this study, except for *G. vacuolata*, commonly possess a highly-lobed stellate chloroplast with a naked pyrenoid, which is considered a symplesiomorphic character when compared with a stellate chloroplast having an embedded pyrenoid. The lack of a pyrenoid in *Glaucosphaera* is considered as autapomorphy. It is not so curious that the Cyanophycean type of phycoerythrin (C-PE) might occur in *D. grisea* but lacking in *G. vacuolata* (Table 2), while acquisition of Bangiophycean type of phycoerythrin (B-PE) might also occur at the branch indicated by a solid circle (marked with B-

PE, Fig. 10). Acquisition of B-PE can be understood as a synapomorphic character of *Rhodella* and lacking thylakoids from the pyrenoid matrix is also a synapomorphic character of *Rhodella*, except for *R. cyanea*.

The following opinion can be proposed based on these cellular and molecular phylogenetic analyses. There are two possibilities to delimit the genus *Rhodella*, *Rhodella sensu lato* and *Rhodella sensu stricto*. In the former case, the genus is defined by possessing B-PE and includes all species of *Rhodella* published so far, except *R. grisea/R. reticulata* which was transferred to *Dixoniella* as *D. grisea*. In the latter case, *Rhodella* is defined by possessing a pyrenoid free from any internal structures in the matrix and dictyosomes associated with ER. It includes all species of *Rhodella* except *R. cyanea*. These two possibilities are both supported by the molecular phylogenetic analysis (Fig. 9).

Although it is necessary to discuss the taxonomic position of *R. cyanea* and the question of whether Tobishima strain is conspecific with *R. cyanea* or not, unfortunately, we had no chance to examine the type strain used by Billard & Fresnel (1986). It is only after investigations of the pigment and molecular phylogenetic analyses, using the type strain of *R. cyanea*, could the generic delimitation of *Rhodella* and the taxonomic position of *R. cyanea* be finally confirmed.

Acknowledgements

We sincerely thank Prof. J. A. West, University of Melbourne for kindly reading the manuscript and providing many living cultures. We also thank M. Iwataki, University of Tokyo, and Prof. I. Inouye, University of Tsukuba, for assisting with cultures. This work was supported in part by Grant-in-Aid for Scientific Researches (Nos. 03041017 and 10440245) from the Japanese Ministry of Education, Science and Culture, and Research Fund for Regional Joint Research Project of Yamagata Prefecture.

References

Billard, C. & J. Fresnel, 1986. Plant systematics. *Rhodella cyanea* nov. sp., a new unicellular Rhodophytan alga. C. R. Acad. Sc. Paris, t. 302, Série III: 271–276.

Broadwater, S. T. & J. L. Scott, 1994. Ultrastructure of unicellular red algae. In Seckbach, J. (ed.), Evolutionary Pathways and Enigmatic Algae: *Cyanidium caldarium* (Rhodophyta) and Related Cells. Kluwer Academic Publishers, Dordrecht: 215–230.

Broadwater, S. T., J. L. Scott, S. P. A. Gross & B. D. Saunders, 1995. Ultrastructure of vegetative organization and cell division in *Glaucosphaera vacuolata* Korshikov (Porphyridiales, Rhodophyta). Phycologia 34: 351–361.

Deason, T. D., G. L. Bulter & C. Rhyne, 1983. *Rhodella reticulata* sp. nov., a new coccoid rhodophytan alga (Porphyridiales). J. Phycol. 19: 104–111.

Evans, L. V., 1970. Electron microscopical observations on a new red algal unicell, *Rhodella maculata* gen. nov., sp. nov. Br. phycol. J. 5: 1–13.

Fresnel, J., C. Billard, F. Hindák & B. Pekárková. 1989. New observations on *Porphyridium griseum* Geitler = *Rhodella grisea* (Geitler) comb. nova (Porphyridiales, Rhodophyta). Pl. Syst. Evol. 164: 253–262.

Hara, Y. & M. Chihara, 1985. Ultrastructure and taxonomy of *Fibrocapsa japonica* (Class Raphidophyceae). Arch. Protistenk. 130: 133–141.

Hara, Y., T. Nagumo & S. Kato, 1989. Microalgal flora of mangrove forests in Japan. Proceedings of the 5th International Symposium on Microbial Ecology (ISME5): 292–296.

Hasebe, M. & K. Iwatsuki, 1990. *Adiantum capillus-veneris* Chloroplast DNA clone bank as useful heterologous probes in the systematics of the leptosporangiate ferns. Am. Fern J. 80: 20–25.

Hasebe, M., T. Omori, M. Nakazawa, T. Sano, M. Kato & K. Iwatsuki, 1994. *rbcL* gene sequences provide evidence for the evolutionary lineages of leptosporangiate ferns. Proc. Natl. Acad. Sci. U.S.A. 91: 5730–5734.

Koller, K. P. & W. Wehrmeyer, 1975. B-phycoerythrin from *Rhodella violacea*; characterization of two isoproteins. Arch. Microbiol. 104: 255–261.

Nakayama, T., S. Watanabe, K. Mitsui, H. Uchida & I. Inouye, 1996. The phylogenetic relationship between the Chlamydomonadales and Chlorococcales inferred from 18SrDNA sequence data. Phycol. Res. 44: 47–55.

Okaichi, T., S. Nishino & Y. Imatomi, 1982. Collection and mass culture. In Bull. Jap. Soc. Sci. Fish. (ed.), Toxic Phytoplankton Occurrence, Mode of Action and Toxins. Kôseisha-Kôseikaku, Tokyo: 23–34 [in Japanese].

Olsen, G. J., H. Matsuda, R. Hagstrom & R. Overbeek, 1994. fastDNAml: a tool for construction of phylogenetic trees of DNA sequences using maximum likelihood. CABIOS 10: 41–48.

Patrone, L. M., S. T. Broadwater & J. L. Scott, 1991. Ultrastructure of vegetative and dividing cells of the unicellular red algae *Rhodella violacea* and *Rhodella maculata*. J. Phycol. 27: 742–753.

Ragan, M. A., C. J. Bird, E. L. Rice, R. R. Gutell, C. A. Murphy & R. K. Singth, 1994. A molecular phylogeny of the marine red algae (Rhodophyta) based on the nuclear small-subunit rRNA gene. Proc. Natl. Acad. Sci. U.S.A. 91: 7276–7280.

Schornstein, K. L. & J. Scott, 1982. Ultrastructure of cell division in the unicellular red alga *Porphyridium purpureum*. Can. J. Bot. 60: 85–97.

Scott, J. L., S. T. Broadwater, B. D. Saunders, J. P. Thomas & P. W. Gabrielson, 1992. Ultrastructure of vegetative organization and cell division in the unicellular red alga *Dixoniella grisea* gen. nov. (Rhodophyta) and a consideration of the genus *Rhodella*. J. Phycol. 28: 649–660.

Swofford, D. L., 1993. PAUP; Phylogenetic Analysis Using Parsimony, version 3.1.1. Computer program and manual. Illinois Natural History Survey, Champaign, Illinois.

Wehrmeyer, W., 1971. Elektronmikroskopische Untersuchung zur Feinstructur von *Porphyridium violaceum* Kornmann mit Bemerkungen über seine taxonomische Stellung. Arch. Mikrobiol. 75: 121–139.

Hydrobiologia **512**: 185–192, 2004.
P.O. Ang, Jr. (ed.), Asian Pacific Phycology in the 21st Century: Prospects and Challenges.
© 2004 *Kluwer Academic Publishers.*

Four new species of the Genera *Eudesme* and *Sphaerotrichia* (Chordariaceae, Heterokontophyta) from the Chinese Coast

Lanping Ding[1,2,*] & Baoren Lu[1]
[1]*Institute of Oceanology, Chinese Academy of Sciences, Qingdao, 266071, P. R. China*
[2]*Graduate School of the Chinese Academy of Sciences, P.R. China*
Author for correspondence; E-mail: dinglp@ms.qdio.ac.cn

Key words: new species, *Eudesme*, *Sphaerotrichia*

Abstract

Four new species, *Eudesme huanghaiensis* Ding et Lu, *E. qingdaoensis* Ding et Lu, *E. shandongensis* Ding et Lu and *Sphaerotrichia huanghaiensis* Ding et Lu, from the western Yellow Sea coast of China are described. *Eudesme huanghaiensis* is mainly characterized by its spherical or sub-spherical sub-cortical cells, its rhizoidal filaments developing from the basal cells of sub-cortex and its broad sub-cortical and medullary layers. *E. qingdaoensis* is mainly characterized by its long medullary cells, generally hollow center of the medulla, short sub-cortex with only 3–4 cylindrical cells and long, slender and clavate terminal cells of the rhizoidal filaments. *E. shandongensis* is mainly characterized by its hollow frond, thick cell walls of both medulla and inner sub-cortical layers and the spherical terminal cells of the rhizoid filaments. *Sphaerotrichia huanghaiensis* is mainly characterized by its cylindrical, sparsely branched frond with acute angle, and its thick 5–6 layered sub-cortex with long assimilating filaments of 6–10 cells.

Introduction

The genera *Eudesme* Agardh (1880) and *Sphaerotrichia* Kylin (1940) belong to the family Chordariaceae. In China, Tseng & Bailing Zheng (1954) first reported *E. virescens* (Carm.) J. Agardh, collected from Qingdao City, Shandong Province, eastern China. Lu & Tseng (1983) reported *E. virescens* from north China coasts. In our research on the genus *Eudesme* from China, we identified four species including *E. virescens* and three other species which we believed have never been found elsewhere before.

Gepp (1904) reported a new species belonging to the genus *Chordaria*, collected from Weihai City, Shandong Province, and named it *Chordaria firma* Gepp. Collins (1919) reported two species from the coast of Beidaihe, Hebei Province, *C. flagelliformis* (Fl. Dan) Ag. and *C. cladosiphon sensu* Collin. Cowdry (1922) reported two species from the coast of Beidaihe again, *C. firma* and *C. cladosiphon sensu* Okam. Howe (1924) reported *C. firma* and *C. chordaria* (Harv.) Howe from Yantai City, Shandong Province and Beidaihe, Hebei Province. Howe

believed that the *C. cladosiphon* Kuetz. reported by Cowdry (1922) is *C. chordaria*. Howe (1934) reported *C. firma* again from Qingdao City and Penglai County, Shandong Province. In 1958, J.JI 3nhoba studied some specimens of the genus *Chordaria* from Qingdao and concluded that the report of *Chordaria* from China should be changed to *Sphaerotrichia* (3nhoba, 1958). According to her views, she divided these materials into two species, *Sphaerotrichia dissessa* (S. *et* G.) Zinova and *S. firma* (Gepp) Zinova. Tseng & Chang (1964) agreed with her and Lu & Tseng (1983) reported *S. firma* again from Huanghai Sea coast of China. Previous studies of the genus *Sphaerotrichia* in China were fragmentary. We carried out an exhaustive study and identified three species of *Sphaerotrichia*, including a new record and a new species. The latter is described in this paper.

Materials and methods

Dried and wet herbarium specimens from the Herbarium of the Institute of Oceanology, Chinese Academy

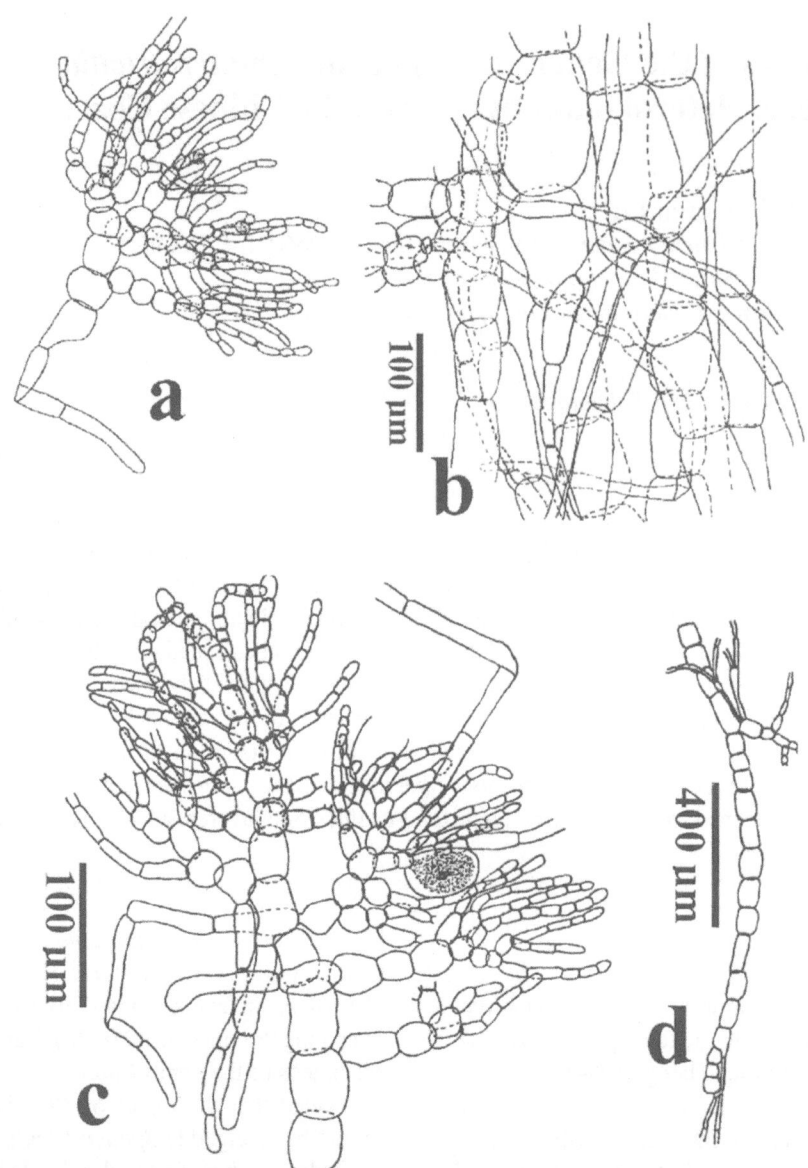

Figure 1. Eudesme huanghaiensis Ding *et* Lu sp. nov. Based on the holotype AST 56-3699. (**a**) whole branch of cortical filaments; (**b**) medullary filaments and part of the sub-cortical filaments; (**c**) longitudinal section of frond top; (**d**) medullary filament and sub-cortical filaments.

of Sciences (AST) were studied. Sections of the specimens were cut by hand using a razor blade or by a freezing microtome. Sections of the specimens were observed under an Olympus microscope.

Description of the new species

1. *Eudesme huanghaiensis* **Ding** *et* **Lu sp. nov.** (Fig. 1, Plate I: 1)

Frons mediocris, 5–20 cm alta, 1.5–3 mm diametro. Cellulae subcorticales sphaericae vel subsphaericae. Fila rhizoidea ex cellulis basalibus sub-corticalis oriunda. Strata sub-corticalia et medullosa lata.

Holotype: AST 56-3699, collected by Zheng Shudong and Xia Enzhan at Xiazeizi, Rongcheng, Shangdong Province, China, on May 23, 1956.

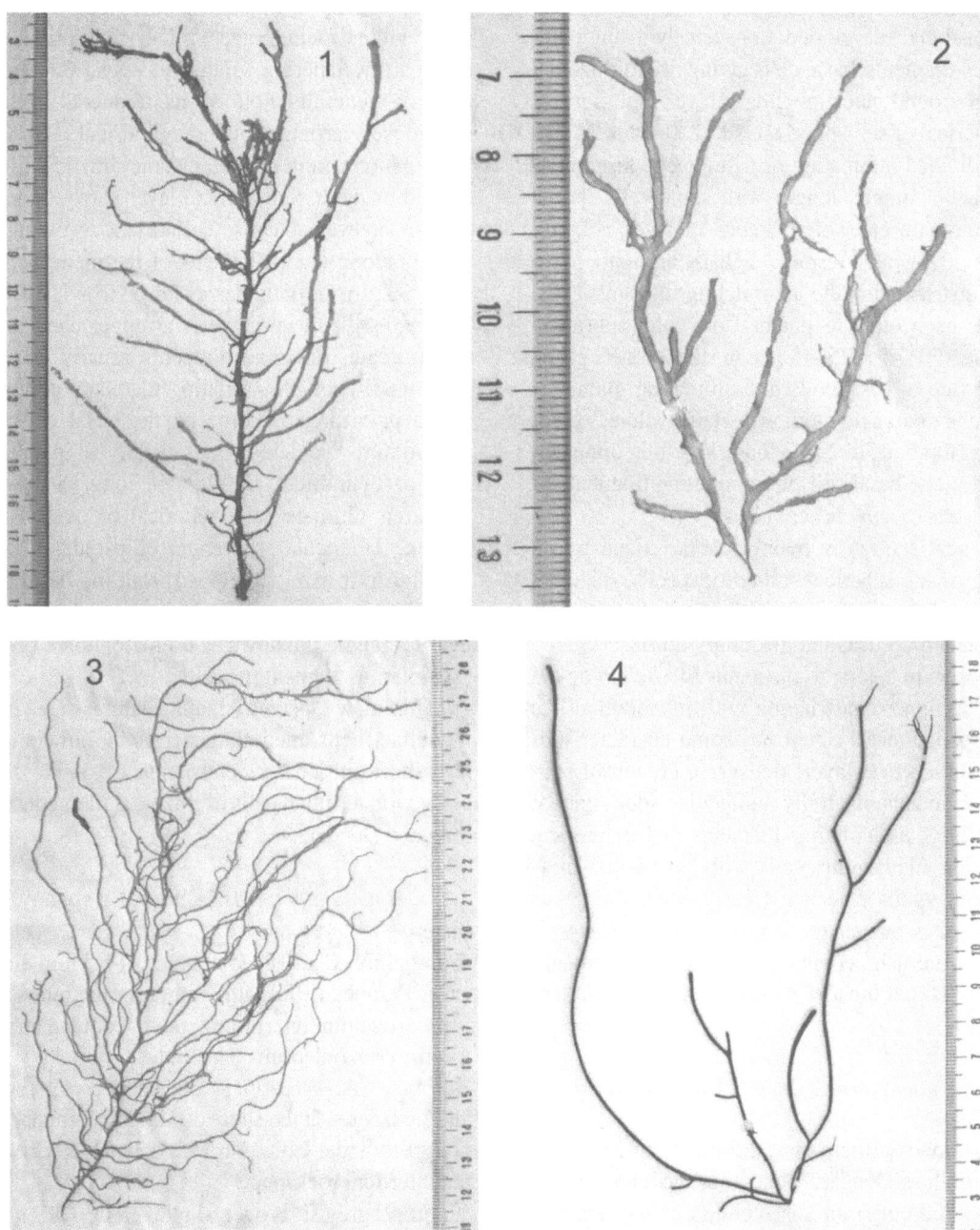

Plate I. 1. Habit of *Eudesme huanghaiensis* Ding *et* Lu sp. nov. (AST 56-3699). 2. Habit of *Eudesme qingdaoensis* Ding *et* Lu sp. nov. (AST 63-0586) 3. Habit of *Eudesme shandongensis* Ding *et* Lu sp. nov. (AST 63-0519). 4. Habit of *Sphaerotrichia huanghaiensis* Ding *et* Lu sp. nov. (AST 57-0534).

Frond yellowish brown, adhering completely to paper when dried, single or tufted, erect, 5–20 cm high, 1.5–3.0 mm thick, solid, embedded in gelatine, slimy, soft, single main axis, alternately branched 1–3 times. Holdfast discoid. Branches and branchlets patent, dense or sparse, 1.5–8.0 cm long, blunt at apex. Medullary layer composed of a bundle of cells arranged in lengthwise rows, polysiphonous, loosely, easily separated from each other. Medullary cells cylindrical, 80–110 μm long, few up to 130 μm, 39–78 μm in diameter. Sub-cortical filaments narrow, several cells long, developed transversely from outer medullary filaments, basal cells giving rise to rhizoidal filaments around outer medullary layer. Sub-cortical cells spherical or sub-spherical, 20×20–75×45 μm in size, divided bifurcately or trifurcately. Assimilating filaments simple, length with some cells, borne on the ultimate cells of sub-cortical layer, slightly curved, with swollen terminal cells. Chromatophores densely distributed in the assimilating filaments cells, gathering each other to patch. Unilocular sporangia ellipsoidal, 50–74×25–45 μm in size, sessile, growing on a side of basal cells of assimilating filaments. Plurilocular sporangia unknown. Hair hyaline, except its base colored, up to 2 mm long, 11 μm in diameter, growing on the basal cell of assimilating filaments or ultimate cells of sub-cortical layer.

This new species is mainly characterized by its spherical or sub-spherical sub-cortical cells, rhizoidal filaments developing from the sub-cortical basal cells and broad sub-cortical and medullary layers.

It appears to belong to the genus *Mesogloia* on the basis of its inner construction with monopodial-like medullary filaments. But it has some characteristics such as sub-cortical layer, transverse growth of sub-cortical filaments, laterally unilocular sporangia of basal cells of assimilating filaments and cylindrical assimilating filamentous cells with no swollen ultimate cells, which placed it clearly under the genus *Eudesme*. It is mainly related to *Eudesme virescens*, differing from it in its spherical or subspherical sub-cortical cells and broader sub-cortical and medullary layers.

2. *Eudesme qingdaoensis* **Ding** *et* **Lu sp. nov.** (Fig. 2, Plate I: 2)

Frons mediocris, 10 cm alta, 2.5 mm diametro. Cellulae medullosae longae. Generatim medulla cava ad centrum. Sub-cortex brevis, cellulis cylindricis non nisi 3–4. Cellulae terminales filiorum rhizoideorum langae, graciles et clavatae.

Holotype: AST 63-0586, collected by Xia Enzhan and Hua Maosen at Jianggezhuang, Qingdao, Shandong Province on May 24, 1963. Growing in intertidal rock-pool.

Frond greyish-brown, single, erect, embedded in gelatine, adhering to paper except basal stem when dried, 10 cm high, 2.5 mm in diameter, main axis evident, with branches. Holdfast discoid, small, stem 4–9 mm long and 0.4 mm in diameter. Branches

sparse, usually with large angle, sometimes nearly right angle, main branch thick, cylindrical, upper part narrow tapering, obtuse at apex. Center of the medulla generally hollow, its filaments longitudinally loosely arranged, cells cylindrical, 90–230 μm long, 38–65 μm in diameter, some rhizoidal filaments around medulla. Sub-cortical layer narrow, consisting of 2–3 or even 4 cells, cylindrical, growing transversely along outer filaments of the medulla, 75 μm long, 35 μm in diameter at lower frond, 30–40 μm long, 20–30 μm in diameter at upper part, bifurcate or trifurcate, rhizoidal filaments usually descending from basal cells. Assimilating filaments very simple, single or bifurcate, borne on the basal cells of the assimilating filaments, cells mostly ellipsoidal, oblong or cylindrical, 20–25 μm long, 8–12 μm in diameter, ultimate cells spherical or oval. Hair unknown. Unilocular sporangia ellipsoidal, 70–75×28–41 μm in size, developed from the basal cells of assimilating filaments, sessile or petiolate. Plurilocular sporangia unknown. Chromatophores present in assimilating filamentous cells.

This new species' main features are its long medullary cells, medulla generally hollow in the center; short sub-cortex, consisting of 3–4 cylindrical cells; with terminal cells of rhizoidal filaments clavate, long and slender.

3. *Eudesme shandongensis* **Ding** *et* **Lu sp. nov.** (Fig. 3, Plate I: 3)

Frons simplex, ad 18–25 cm alta, 1–1.5 mm diametro, cava. Parietes cellularum stratorum medullosorum et sub-corticalium interiorum crassi. Cellulae terminales filiorum rhizoideorum sphaericae.

Holotype: AST 63-0519, collected by Zhang Junfu and Lu Baoren at the shore of Maidao, Qingdao, Shandong Province, China on May 22, 1963. Growing in the intertidal rock-pool.

Frond greyish-brown, single, erect, with branches completely adhering to paper when dried, 18–25 cm high, 1–1.5 mm in diameter, generally hollow, embedded in gelatine, surface smooth, soft, alternately furcate 1-3 times. Holdfast discoid. Main axis evidently elongating to the apex. Branches long or short, densely arranged. Primary branches growing from main axis mostly at right angle, 15 cm long, 1 mm in diameter. Secondary branches short, slender, 3 cm long. Terminal branchlets very short, usually with 1–2 ramulis, 5 mm long, 0.4–0.5 mm in diameter. Branches or branchlets usually curved, blunt at apex. Medullary

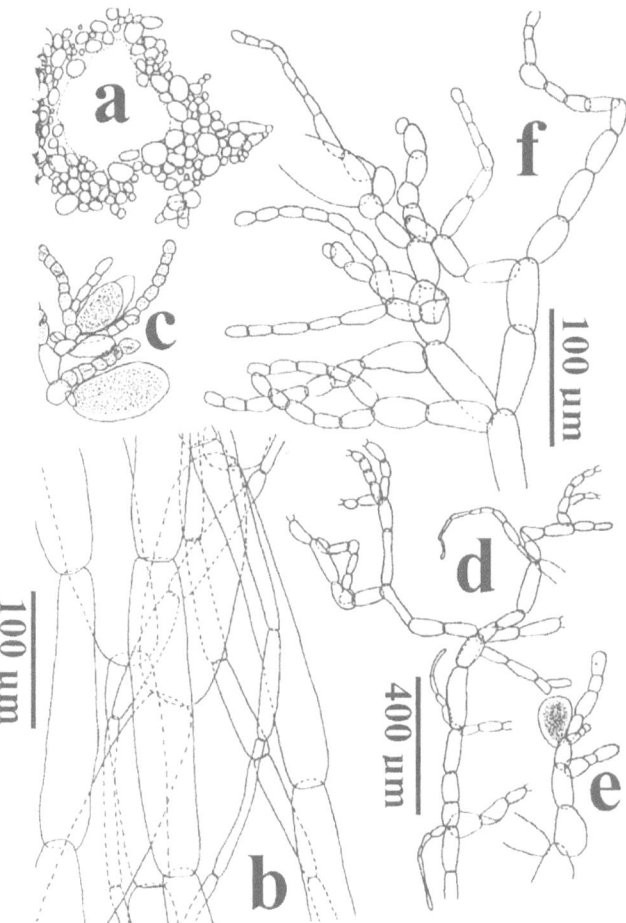

Figure 2. Eudesme qingdaoensis Ding *et* Lu sp. nov. Based on the holotype AST 63-0586. (**a**) transverse section of frond; (**b**) medullary filaments; (**c**) growth point and unilocular sporangia; (**d**) medullary filaments and its sub-cortical filaments and rhizoidal filaments; (**e**) unilocular sporangia; (**f**) assimilating filaments.

layer consisting of longitudinal bundle of cylindrical filaments. Medullary cells, 80–195 μm long, 32–85 μm in diameter, cell wall thick. Medullary filaments loosely arranged, easily separable from each other. Sub-cortical layer not evident. Sub-cortical filaments growing out transversely from outer medulla, bifurcate, trifurcate or multi-furcate, its inner cells with thick wall larger than outer ones, cylindrical or ellipsoidal, somewhat spherical, 20–60 μm long, 21–35 μm in diameter, the rhizoid filaments descending from basal cells generally transversely developed, ultimate cell short and near to spherical. Assimilating filaments simple, 2–8 cells, cylindrical or monoform or subspherical, ultimate cells mostly ordinary. Hair few, colorless except its base, 12 μm in diameter, borne on terminal cells of sub-cortical filaments.

Chromatophores mostly containing in assimilating filaments cells, sub-cortical filaments also contain few chormatophores. Unilocular sporangia ellipsoidal or ovate, borne on the basal cell of assimilating filaments, sessile or pedicellate with 1–2 cells, 42–60 μm long, 30-40 μm in diameter. Plurilocular sporangia unknown.

This new species is mainly characterized by its hollow frond, cell walls of both medullary and inner subcortical layers thick, terminal cells of rhizoid filaments spherical. It is closely related to *E. virescens*, differing from it in its slender frond, nearly at right angle between branches and main axis, hollow medulla, thick wall of medullary and sub-cortical cells, and spherical terminal cells of rhizoid filaments.

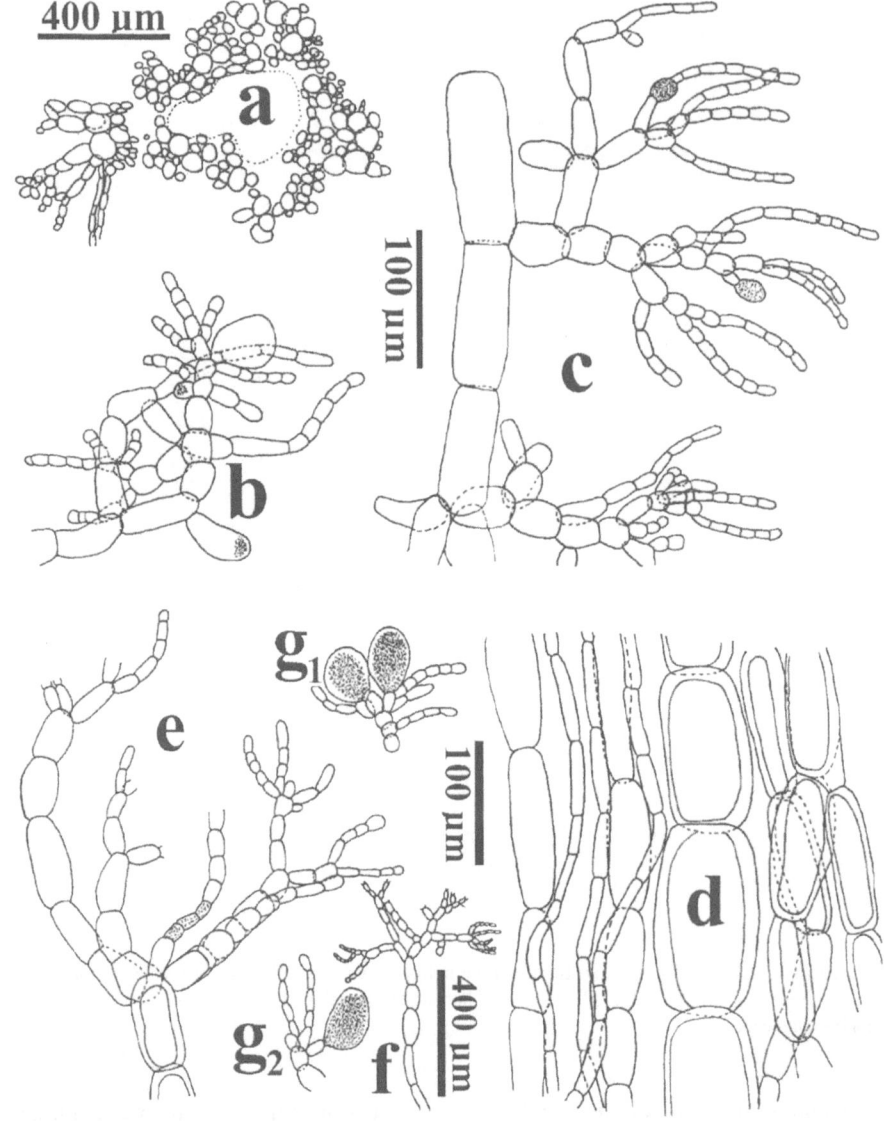

Figure 3. Eudesme shandongensis Ding *et* Lu sp. nov. Based on the holotype AST 63-0519. **(a)** transverse section of frond; **(b)** growth point; **(c)** medullary filaments with some of them developing into cortical filaments; **(d)** rhizoid filaments descending from medullary filaments; **(e)** hair; **(f)** whole cortical filaments; **(g)** (**g**₁, **g**₂) unilocular sporangia.

4. *Sphaerotrichia huanghaiensis* **Ding** *et* **Lu sp. nov.**
(Fig. 4, Plate I: 4)
Frons cylindrica, ramis sparsis acutatis. Sub-cortex crassus 5–6 stratis et filis assimilantibus longis ex 6–10 cellulis constatis.
Holotype: AST 57-0534. Collected by Zhang Junfu and Wang Liming at Maidao, Qingdao, Shandong Province, on 28 Aug., 1957. Growing on low-tidal rocks washed by strong wave. Other materials examined: AST 53-1147, 55-0362, 55-0402.

Frond brown or light brown, tufted, sparsely branched, filiform, cylindrical, adhering to paper when dried, 6–23 cm high, slightly soft, gelatinous, smooth. Holdfast small discoid. Main branch evident, usually flexible, 1–1.5 mm in diameter, alternately or laterally branched 1–2 times. Primary branch alternate, ordinarily trifurcate or multifurcate at upper portion, branching few, short and evidently curved, the tips of the branch more or less swollen or giving rise to slender divaricated branchlets. Medulla hollow, except terminal and base of frond, composed of longitudinal

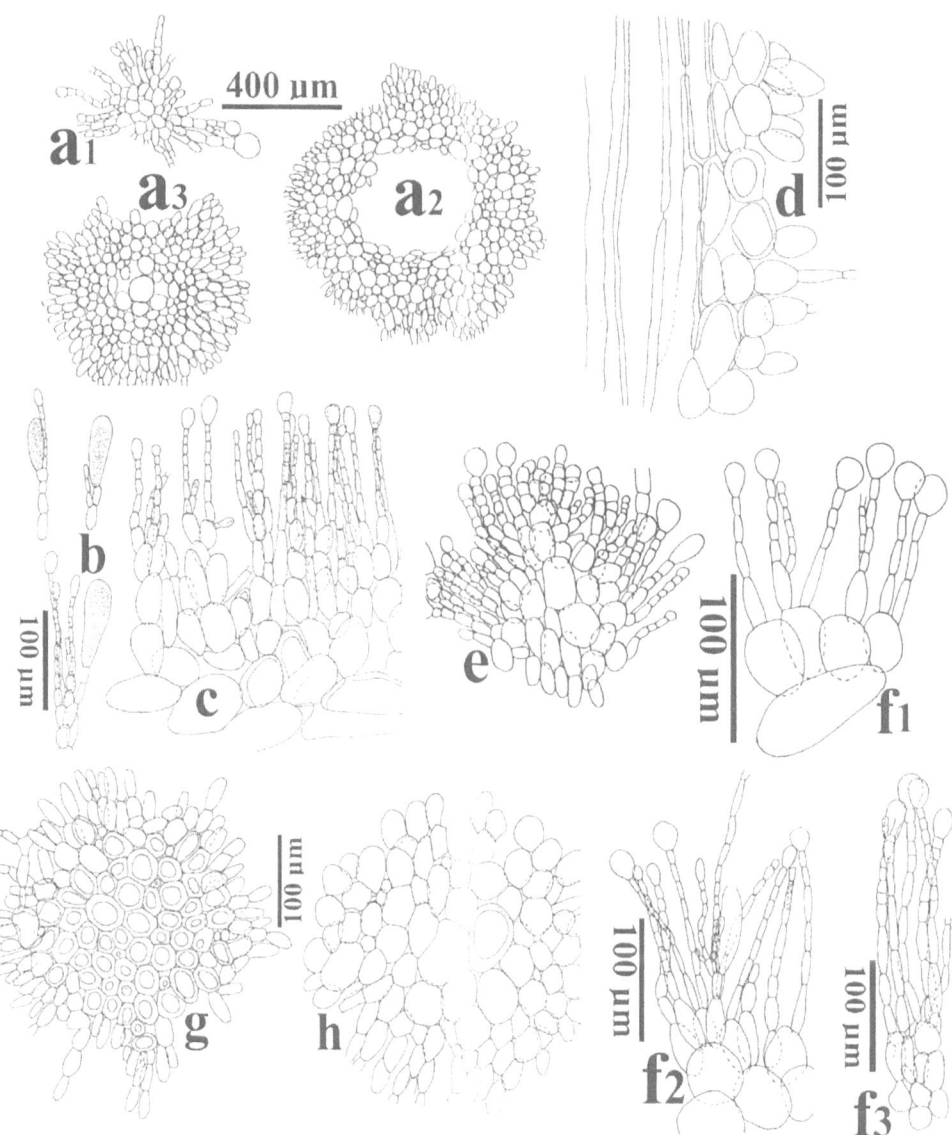

Figure 4. Sphaerotrichia huanghaiensis Ding *et* Lu sp. nov. Based on the holotype AST 57-0534. (**a**) transverse section of frond (**a₁**. terminal part, **a₂**. middle part, **a₃**. basal part); (**b**) assimilating filaments and unilocular sporangia; (**c**) cortical filaments; (**d**) middle longitudinal section of frond; (**e**) terminal longitudinal section of frond; (**f**) assimilating filaments (**f₁**. younger frond, **f₂**. middle part of frond, **f₃**. basal part of frond); (**g**) transverse section near; (**h**) transverse section near tip.

long fusiform cells, 160–270 μm long or more, 29–32 μm in diameter, cell walls thick and pits near the inner lateral of sub-cortical layer. Sub-cortical layer at adult frond thick, consisting of 5–6 layers. The cells of the sub-cortical layer polygonous, sub-ellipsoidal or sub-spherical, ellipsoidal near outer side, 30–60 μm long, 30–40 μm in diameter. Assimilating filaments very simple, densely arranged, consisted of 6–10 cells, cylindrical or monoform, about 10–30 μm long, a few up to 40 μm long, 5–8 μm in diameter, except

terminal cell varying in size; terminal cell swollen, cylindrical or pyriform, generally 24–22 × 30–26 μm in size. Hair hyaline, very long, up to 2 mm, 10 μm in diameter, basal cells slender and colored. Unilocular sporangia borne on the base of assimilating filaments, sessile or short stipe, pyriform or obovate, 50–90 μm long, 20–30 μm in diameter. Plurilocular sporangia unknown.

This new species is mainly characterized by its cylindrical frond, sparse branches with a sharp angle, thick subcortical layer, 5–6 layers with very long assimilating filaments. This species is closely related to *S. japonica* Kylin, differing from it in its very long assimilating filaments generally composed of 6–10 cells.

Acknowledgements

The paper is contribution No. 3783 from the Institute of Oceanology, Chinese Academy of Sciences, Qingdao, China. We thank Prof. C. K. Tseng for help with the preparation of the manuscript and Put O. Ang, Jr. (Chinese University of Hong Kong) for reading and carefully editing the manuscript. Thanks also to Prof. Karla McDermid (University of Hawaii) for help with the Latin translation.

References

Agardh, J. G., 1880. Till Algernas Systematik: IV Chordariaceae, V Dictyoteae. Lunds Univ. Arsskrift, Bd 17, Lund: 31.

Collins, F. S., 1919. Chinese marine algae. Rhodora (Journal New England Bot. Club) 21: 203–207.

Cowdry, N. H., 1922. Algae in plants of Peitaiho. J. N. China Branch, Roy. Asia. Soc. 53: 180–181.

Gepp, E. S., 1904. Chinese marine algae. J. Bot. London 42: 161–165, Tab. 460, 7, 8.

Howe, M. A., 1924. Chinese marine algae. Bull. Torrey Bot. Club. 51: 133–144, Pls. 1, 2.

Howe, M. A., 1934. Some marine algae of the Shantung Peninsula. Lingn. Sci. J. 13: 667–670, f. 1.

Kylin, H., 1940. Die Phaeophyceenordnung Chordariales. Lunds Univ. Arsskr., N. R. Avd. 2, Bd 36(9), Lund: 31–32, 38–41, Figures 16, 20–21.

Lu, Baoren & C. K. Tseng, 1983. Phaeophyta. In Tseng, C. K. (ed.), Common Algae of China. Science Press, Beijing, China: 176, Pl. 89, Figures 1, 4.

Tseng, C. K. & C. F. Chang, 1964. A critical review of the records of the benthic marine algae as reported from the western Yellow Sea coast. Stud. mar. Sin. 6: 1–26.

Tseng, C. K. & B. Zhen, 1954. Studied on the marine algae from Qingdao. J. Bot. 3: 105–120, Figures I, 6.

3nhoba, A. Jl., 1958. Knohmahnio Bnaobpoaa *Sphaerotrichia* Kyl. botahnyecknn Kyphaji. 43: 1462–1469, Pnc, 1–7.

Hydrobiologia **512**: 193–199, 2004.
P.O. Ang, Jr. (ed.), Asian Pacific Phycology in the 21st Century: Prospects and Challenges.
© 2004 *Kluwer Academic Publishers.*

Studies on four new species of the malacocarpic *Sargassum* (Sargassaceae, Heterokontophyta) in China

Baoren Lu* & C. K. Tseng

Institute of Oceanology, Chinese Academy of Sciences, Qingdao 266071, P.R. China
Author for correspondence; E-mail: brlu@ms.qdio.ac.cn

Key words: Sargassum, Section Malacocarpicae, China

Abstract

In this paper, four new species of malacocarpic *Sargassum* are described: *S. fuliginosoides* Tseng *et* Lu sp. nov. is characterized by its discoid holdfast, the presence of the bulbs on the basal parts of the primary branches and very thick, lanceolate leaves. *S. gemmiphorum* Tseng *et* Lu sp. nov. is characterized by its conical holdfast, usually branched axis and very long, narrow, thin, denticulate leaves. *S. shandongense* Tseng, Zhang *et* Lu sp. nov. is characterized by its discoid holdfast, flattened primary branches, its leaves mostly entire or wavy at the margins and its racemose to paniculate receptacles. *S. qingdaoense* Tseng *et* Lu sp. nov. is characterized by the presence of the glandular dots on the ultimate branches and elongated lanceolate, acute, mostly wavy leaves.

Introduction

Members of the Malacocarpiceae Section of the subgenus *Sargassum* are characterized by their subracemosely or racemosely arranged smooth receptacles. In China, 15 species of Malacocarpic *Sargassum* have so far been reported (Tseng & Lu, 1992). In this paper, four new species are described. All materials examined are deposited at the Herbarium of the Institute of Oceanology, Academy of Sciences (AST) in Qingdao, China.

Description of the new species

1. *Sargassum fuliginosoides* **Tseng** *et* **Lu sp. nov.** (Fig. 1; Plate I: 1)
Species nova, *S. fuliginoso* Kuetzing proxima. Differt bulbis basi ramorum primariorum; foliis magnis, crassis, coriaceis marginibus integris; vesiculis obverse-ovatis vel ovatis stipitibus teretibus; et receptaculis dioeciis.
Holotype: AST 70-0047, collected by Liu Qingchen on April 24, 1970, in Huilai, Guangdong Province. Growing on the subtidal rocks below 0.5 m.
   Frond dark brown, to 70 cm in height. Holdfast discoid, 1 cm in diameter. Main axes cylindrical, 2 cm

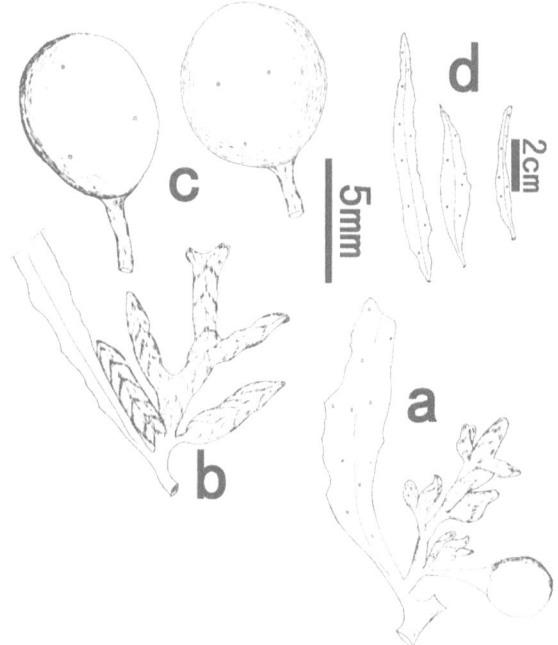

Figure 1. Sargassum fuliginosoides Tseng *et* Lu sp. nov. (Based on **AST** 70-0047): **a.** female receptacular branches with leaves, vesicles and receptacles; **b.** male receptacular branches with leaves, and receptacles; **c.** vesicles; **d.** leaves.

in height, 3 mm in diameter, verrucose on the surface, usually 2–4 primary branches arising from upper position of the axis. Primary branches compressed or ancipate below, subcylindrical and cylindrical in middle and upper portion, smooth on the surface, to 68 cm long, 3 mm wide, often with 2–3 bulbs at the base of the primary branches. Secondary branches arising from the axils of the primary branches, subcylindrical or cylindrical, to 25 cm long, about 1–1.2 mm in diameter, smooth, at intervals of 2–3 cm. Ultimate branchlets shorter and slender, cylindrical, about 1–1.5 cm long, less than 1 mm in diameter, beset with leaves, vesicles and receptacles. Leaves on the primary branches large and thick, lanceolate, about 8–10 cm long, 8–10 mm wide, wavy, sometimes incomplete at the margins, acute at the apex, obliquely cuneate at the base, with conspicuous percurrent midrib, obscure cryptostomata irregularly scattered on both sides of the midrib. Leaves on the secondary branches similar to those on the primary branches in shape, only shorter and narrower, to 5 cm long, 5–7 mm wide, entire at the margins: leaves on the ultimate branchlets very short and slender, narrow-lanceolate, or linear, about 3–4 cm long, 2–2.5 mm wide, entire at margins below, only few denticulate at margins above. Most vesicles ovate or obovate, rounded at the apex, only a few cryptostomata on the surface, to 6 mm long, 5 mm in diameter, some vesicles smaller, to 3 mm long, 2 mm in diameter, with cylindrical stipes, its upper parts larger than below, to 7 mm long.

Plant dioecious. Female receptacles conical, verrucose on the surface, simple or 2–3 forked, about 3–4 mm long, 1.5–2.0 mm in diameter, usually forked at the apex; male receptacles longer, cylindrical, smooth, usually 2–3 times forked with slender stipe, to 10 mm long, 1.2 mm in diameter. Several receptacles racemosely arranged on fertile branches.

This new species is mainly characterized by its compressed primary branches, ancipate below, subcylindrical or cylindrical above; some bulbs at the base of the primary branches; leaves on the primary branches lanceolate, very thick and large, wavy, sometimes incomplete at the margins; vesicles obovate or ovate with cylindrical stipes; plant dioecious, female and male receptacles cylindrical, smooth, only female receptacles verrucose on the surface, male receptacles longer than female receptacles.

This new species is nearest to *S. fuliginosum* Kuetzing (1849). It differs by having bulbs at the base of primary branches; large, thick, coriaceous leaves with

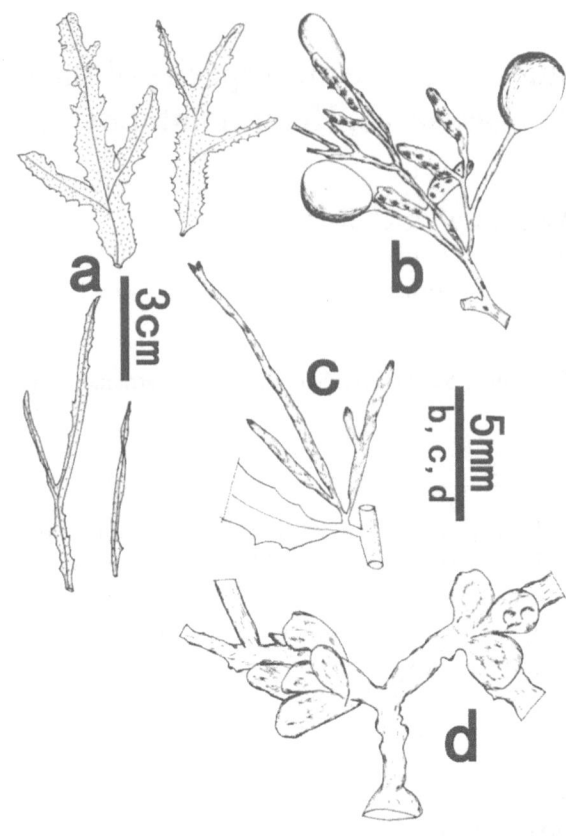

Figure 2. Sargassum gemmiphorum Tseng *et* Lu sp. nov. (Based on **AST** 55-2112): **a.** leaves; **b.** vesicles and female receptacles, **c.** male receptacles; **d.** holdfast and primary axis with bulbs.

entire margins; obovate or ovate vesicles with terete stipes; and dioecious receptacles.

2. *Sargassum gemmiphorum* **Tseng** *et* **Lu sp. nov.**
(Figure 2; Plate I: 2)
Species nova, *S. bulbifero* Yoshida affinis. Differt haptero conico, ramis primariis complanatis; et foliis tenuibus, angustis marginibus denticulatis.
Holotype: AST 55-2112, collected by Zheng Shudong on May 11, 1955, at Fangchen beach, Guangxi Province. Growing on the subtidal rocks.

Fronds yellow brown, more than 55 cm in height. Holdfast conical, 1 cm in diameter. Axes cylindrical, to 2.5 cm long, 2 mm in diameter, simple or sometimes 1–2 forked, verrucose on the surface, several primary branches arising from the upper parts of the axes, with leaves dropped. Primary branches flattened below, compressed above, smooth, to 52 cm long, 2 mm wide, with bulbs arising from the base of the primary branches, some subspherical, other oblong, 5 mm long, 4 mm wide, verrucose, several gathered, flesh. Secondary branches arising from axils of the primary branches, compressed, short, to 13 cm long, 1.2 mm wide, at intervals of 2–3 cm. Ultimate branches shorter, slender, with a few not raised glands on the surface, to 3 cm long, 1 mm in diameter, beset with leaves, vesicles and receptacles. Basal leaves long -lanceolate, thin and membranous, usually 2–3 pinnately forked, to 7 cm long, 5 mm wide, acute at the apex, more regularly cuneate at the base, with percurrent midrib slightly raised, conspicuous cryptostomata, mostly in two series, arranged on both sides of the midrib, denticulate at the margins. Upper leaves slender, linear, sometimes one or twice pinnately divided, to 5 cm long, 2–5 mm wide. Vesicles small, subspherical when adult or ovate when young, about 3–4 mm in diameter, rounded at apex, with 2–3 cryptostomata on the surface, and slender, cylindrical stipe, 6 mm long, 1 mm in diameter.

Plant dioecious. Female receptacles conical, smooth, verrucose on the surface, to 4 mm long, 1.3 mm in diameter, simple or forked; male receptacles very long, cylindrical smooth, 20 mm long, 1 mm in diameter, simple, sometimes forked, female and male receptacles racemosely arranged on fertile branchlets.

This new species is mainly characterized by its conical holdfast, cylindrical axes, often 1–2 divided, verrucose on the surface; subspherical or oblong bulbs in the base of the primary branches; flattened primary branches; ultimate branchlets with a few not raised glands; thin and narrow leaves, often 1–2 pinnately divided, denticulate at the margins; dioecious plant with smooth, receptacles, and very long male receptacles.

This new species is closely related to *S. bulbiferum* Yoshida (1994: 48). It differs by its conical holdfast; flattened primary branches; and thin, narrow leaves with denticulate margins.

3. *Sargassum shandongense* **Tseng, Z. F. Zhang** *et* **Lu sp. nov.** (Fig. 3, Plate I: 3)

Ad S. paniculato J. Agardh aemulans, differt ramis primariis complanatis et foliis basalibus saepe pinnatim divisis.

Holotype: AST 54-0339c, collected by Zhang Jingfu on September 17, 1954, on the beach of Daheilan Qingdao City, Shandong Province. Growing in the intertidal rock pool and on subtidal rocks.

Fronds dark brown when dried, to 50 cm in height. Holdfast discoid, verrucose on the surface, very ir-

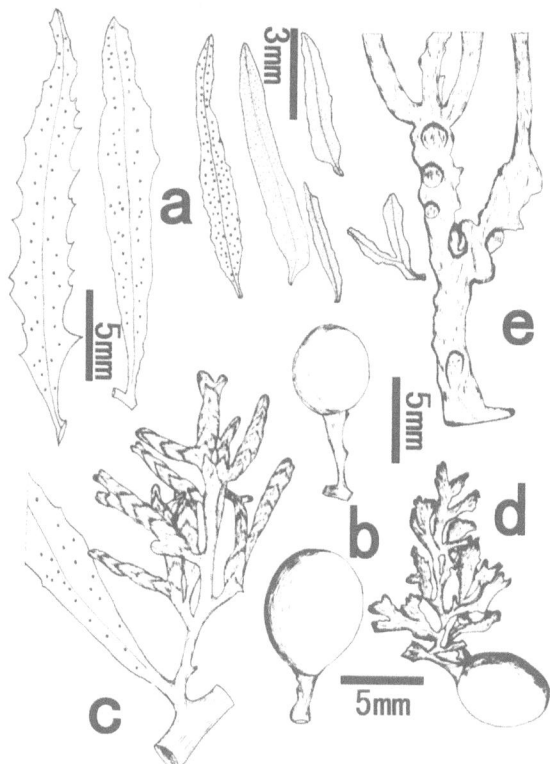

Figure 3. *Sargassum shandongense* Tseng, Z.F. Zhang *et* Lu sp. nov. (Based on **AST** 54-0339c): **a.** leaves; **b.** vesicles; **c.** male receptacles; **d.** female receptacles; **e.** primary axis.

regular at the margins, 2–2.5 cm in diameter, giving rise to several axes from the same holdfast. Axes cylindrical, sometimes divided, verrucose on the surface, to 2 cm long, 2.5 mm in diameter. Primary branches arising from upper part of the axes, flattened below, compressed or subcylindrical above, smooth on the surface, to 48 cm long, 2 mm wide: Secondary branches subcylindrical, smooth, arising from the axil of the primary branches, shorter, to 8 cm long, 1.3 mm diameter: Ultimate branches cylindrical, very short, about 2–3 cm long, 1 mm in diameter, beset with leaves, vesicles and receptacles. Basal leaves large, 1–2 pinnately divided, alternate, to 13 cm long, 4 mm wide, wavy or few dentate at the margins, with conspicuous percurrent midrib and slightly raised, obscure cryptostomata. Leaves on the primary branches lanceolate, some pinnately divided, to 5 cm long, 7 mm wide, acute at the apex, cuneate at the base, wavy or entire, sometimes slightly dentate at the margins; with conspicuous percurrent midrib and raised cryptostomata irregularly arranged on both sides of the midrib. Leaves on the secondary and ultimate branches

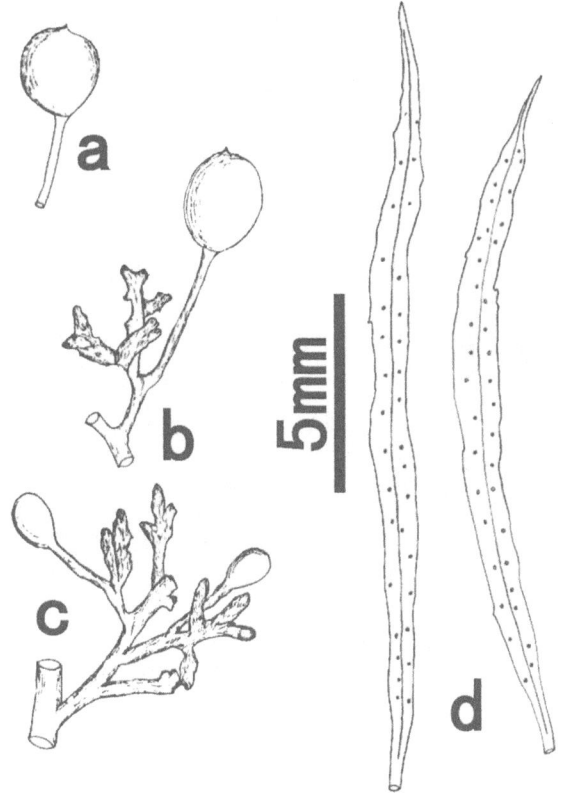

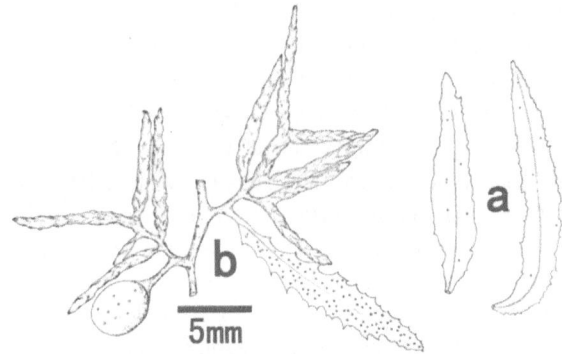

Figure 5. Sargassum qingdaoense Tseng *et* Lu sp. nov. (Based on **AST** 49-0002): **a.** leaves on the primary branches; **b.** fertile branchlet with male receptacles, leaves and vesicles.

Figure 4. Sargassum shandongense f. *linearium* Tseng, Z. F. Zhang *et* Lu f. nov. (Based on **AST** 54-0338a): **a.** vesicle; **b.** female receptacles; **c.** male receptacles; **d.** elongated leaves.

very similar to those of the primary branches in shape, only smaller in size, sometimes sparsely denticulate at the margins. Vesicles spherical or subspherical, to 5 mm in diameter, rounded at the apex, with a few cryptostomata on the surface and mostly cylindrical stipes, usually longer than the vesicles, to 6 mm long, 1 mm in diameter.

Plant dioecious. Female and male receptacles smooth, racemosely or paniculately arranged on fertile branches. Female receptacles conical, verrucose on the surface, forked, to 10 mm long, 1.2 mm in diameter; male receptacles cylindrical, smooth, forked, to 15 mm long, 1 mm in diameter,

This new species is mainly characterized by its discoid holdfast; primary branches flattened below, compressed above; basal leaves large, sometimes pinnately divided, upper leaves lanceolate, mostly wavy or entire.

It is closely related to *S. paniculatum* J. Agardh (1848: 315; 1889: 122). It differs by its flattened

primary branches and often pinnately divided basal leaves.

Forma *linearium* **Tseng, Z. F. Zhang** *et* **Lu forma nov.** (Fig. 4)

Foliis linearis, 9 cm longis, 4 mm latis, crassis, marginibus integris vel undulatis.

Holotype: AST 54-0338a, collected by Z. F. Zhang on August 14, 1954, at beach of Dahelan Qingdao, China. Growing on subtidal rocks.

This new form, *linearium*, is mainly characterized by its very narrow leaves, thick, linear, to 9 cm long, 4 mm wide, with obscure midrib and dark in color when dried.

Plant dioecious. Female and male receptacles and vesicles similar to *S. shandongense* Tseng, Z. F. Zhang *et* Lu.

4. *Sargassum qingdaoense* **Tseng** *et* **Lu sp. nov.** (Fig. 5; Plate II: 1, 2)

Species nova, *S. siliquoso* J. Agardh affinis. Differt longissimis, magnis, lanceolatis foliis ramis primariis, marginibus plerumque undulatis vel interdum denticulatis; foliis superis parvis, lanceolatis, marginibus denticulatis; omnibus foliis apicibus acutis et conspicue obliquis.

Holotype: AST 49-0002, collected by C.K.Tseng, August 11, 1949, Xiaoqingdao Qingdao City, Shandong Province.

Fronds yellow brown, to 70 cm in height. Axes cylindrical, verrucose on the surface because of the scar of the basal leaves. Primary branches arising from the upper part of the axes, cylindrical, smooth, to 68 cm long, 2 mm in diameter. Secondary branches very similar to the primary branches in shape, but shorter and slender, alternately arising from the axil of the

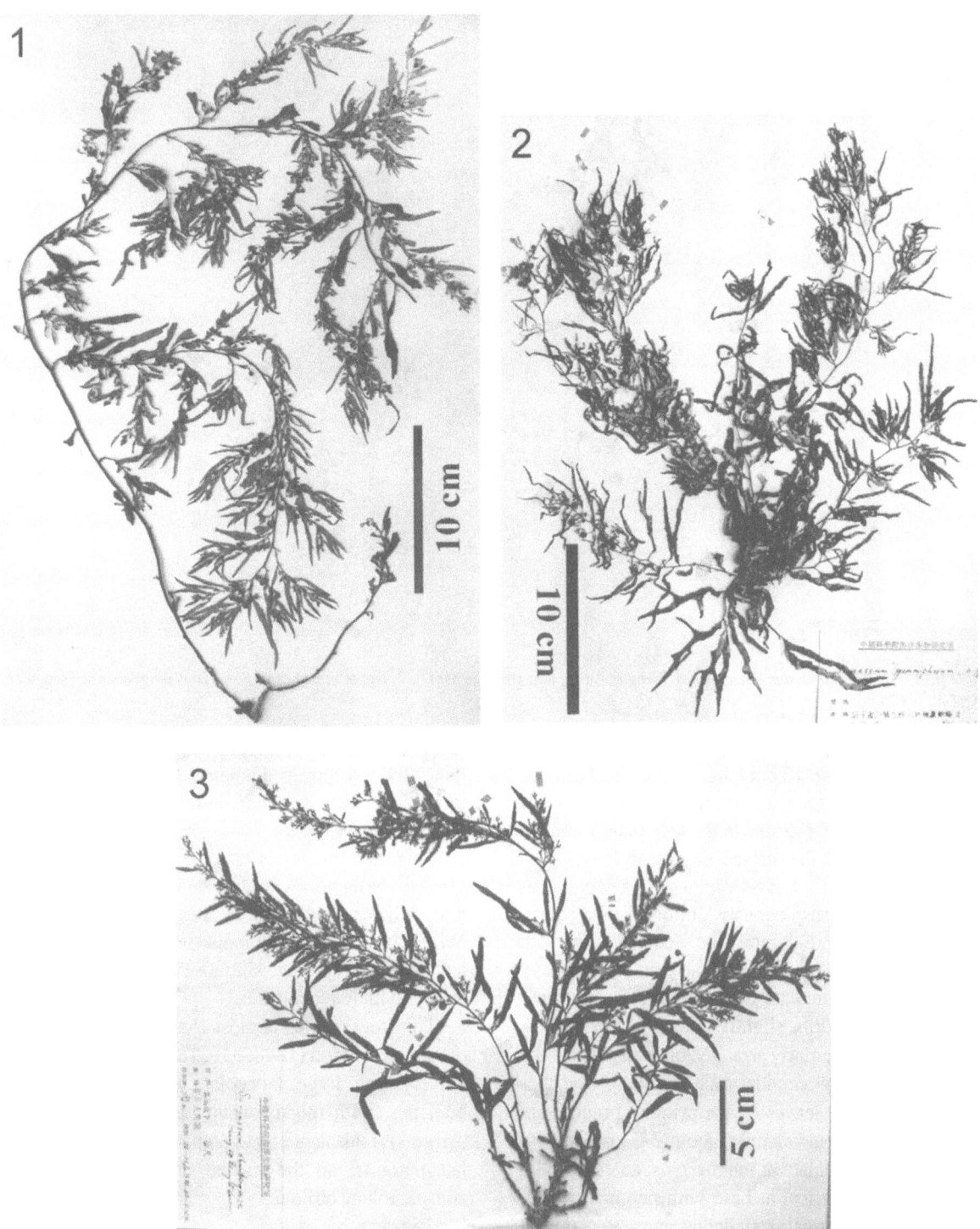

Plate I. 1. Habit of *Sargassum fuliginosoides* Tseng *et* Lu sp. nov. (AST 70-0047). 2. Habit of *Sargassum gemmiphorum* Tseng *et* Lu sp. nov. (AST 55-2112). 3. Habit of*Sargassum shandongense* Tseng, Z.F. Zhang *et* Lu sp. nov. (AST 54- 0339c).

198

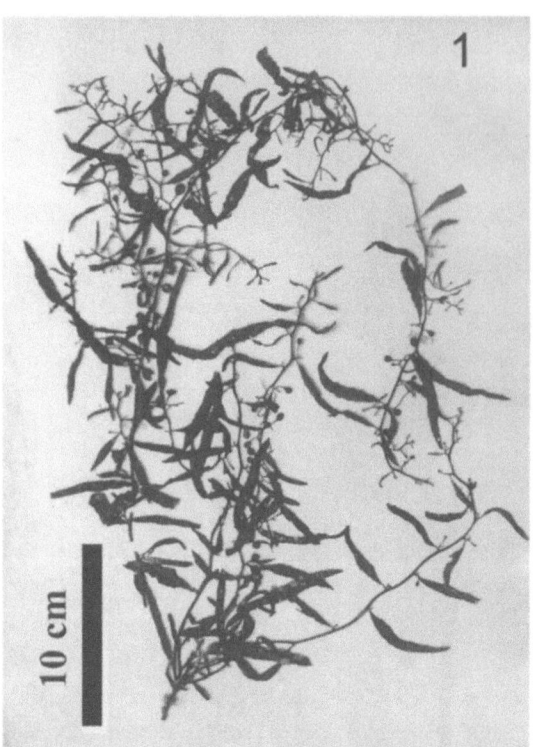

Plate II. 1. Habit of *Sargassum qingdaoense* Tseng *et* Lu sp. nov. (AST 49-0002). 2. Upper branches of *Sargassum qingdaoense* Tseng *et* Lu (AST 49-0002).

primary branches, to 10 cm long, 1.5 mm in diameter, at intervals of 2–2.5 cm. Ultimate branchlets short and slender, about 1–2 cm long, less than 1 mm in diameter, glands on the surface, beset with leaves, vesicles and receptacles, Leaves on the primary branches larger, thicker, conspicuously obliquely lanceolate, about 4–5 cm long, 6–8 mm wide, acute at apex, with conspicuously oblique base, very asymmetrical, wavy or irregularly denticulate at the margins; conspicuous percurrent midrib, slightly raised on the surface, cryptostomata irregularly arranged on both sides of the midrib. Leaves on secondary and ultimate branches very similar to the leaves on the primary branches in shape, but shorter and slender, about 2–3 cm long, 3–4 mm wide, denticulate at the margins, acute at apex, obliquely asymmetrical at base with percurrent midribs. Vesicles spherical, rounded at apex, to 5 mm in diameter, with only a few cryptostomata on the surface and mostly terete stipes, about 4–5 mm long.

Plant dioeceous. Male receptacles elongate, cylindrical, siliquose, 2–3 forked, about 10–15 mm long, 0.4–0.6 mm in diameter, verrucose on the surface, racemosely arranged on fertile branches; female receptacles not found.

This new species is characterized by its elongate primary branches with leaves large and thick, obliquely lanceolate, acute at apex, very conspicuously oblique base; its glandular dots on the ultimate branchlets and its very long, siliquose, smooth, racemosely arranged male receptacles.

This new species is most closely related to *S. siliquosum* J. Agardh (1848: 316; 1889: 121). It differs by its very long, large, lanceolate leaves on the primary branches, with mostly wavy, sometimes denticulate margins; small, lanceolate, upper leaves with denticulate margins; all the leaves with acute apices, and conspicuously oblique.

Acknowledgements

The paper is contribution No. 3786 from the Institute of Oceanology, Chinese Academy of Sciences,

Qingdao. We thank Prof. Put O. Ang, Jr. for reading and editing our manuscript, Prof. Karla McDermid of the University of Hawaii for help with the Latin descriptions and Mrs Li Shiling for inking our drawings.

References

Agardh, J. G., 1848. Species, Genera et Ordines Algarum. I. Species, Genera et Ordines Fucoidearum. Gleerup, Lund: viii + 363 pp.

Agardh, J. G., 1889. Species Sargassorum Australiae Descriptae et Dispositate. Kgl. Svenska Vet.-Akad. Handl. 23: 1–133, Pls. 1–31.

Kuetzing, F. T., 1849. Species Algarum. Brockhaus, Leipzig: 922 pp.

Tseng, C. K. & Baoren Lu, 1992. Studies on the Malacocarpic *Sargassum* of China. II. *Racemosae* J. Agardh. In Abbott, I. (ed.), Taxonomy of Economic Seaweed Vol 3: 11–33, Figures 1–28.

Yoshida, T., 1994. Three new species of *Sargassum* (Sargassaceae, Phaeophyta) from Japan. Jpn. J. Phycol. 42: 43–51.

Hydrobiologia **512**: 201–207, 2004.
P.O. Ang, Jr. (ed.), Asian Pacific Phycology in the 21st Century: Prospects and Challenges.
© 2004 *Kluwer Academic Publishers.*

201

Studies on Chinese species of *Gelidiella* and *Pterocladiella* (Gelidiales, Rhodophyta)

Bang-Mei Xia, C. K. Tseng & Yong-Qiang Wang
Institute of Oceanology, Chinese Academy of Sciences, Qingdao 266071, P.R. China
E-mail: bmxia@ms.qdio.ac.cn

Key words: seaweed, *Gelidiella*, *Pterocladiella*, taxonomy, China

Abstract

A detailed study of materials from our herbarium shows the presence of two species of *Gelidiella* and three species of *Pterocladiella*, including two new records and a new species. Previously recorded species include *G. acerosa* and *P. capillacea*. *G. bornetii* and *P. caerulescens* are new records for China and *P. yinggehaiensis* is believed to be a new species. This new species is characterized by its small thalli, 1.5–2 cm high, that are more densely branched on the middle to upper parts for one half their lengths. These characteristics are so distinctive that they can be used to separate this species readily from other species now reported in this genus.

Introduction

Gelidiella was proposed by Feldmann & Hamel (1934) and *Pterocladiella* by Santelices & Hommersand (1997). Recently, we studied the Gelidiales from the herbarium of the Institute of Oceanology, Academia Sinica (AST) and found two species of *Gelidiella* and three species of *Pterocladiella*. Previously recorded species included *Gelidiella acerosa* (Forsskål) Feldmann *et* Hamel and *Pterocladiella capillacea* (Gmelin) Santelices *et* Hommersand. *Gelidiella bornetii* (Weber–van Bosse) Feldmann et Hamel and *Pterocladiella caerulescens* (Kützing) Santelices are new records for China, and *P. yinggehaiensis* sp. nov. is described here for the first time. These materials are important economic seaweeds, and together with *Gelidium* are used as raw materials for industrial or domestic production of agar in China.

Key to the Chinese species of *Gelidiella* and *Pterocladiella*

1. Thallus with rhizoidal filaments within medullary tissue 3
1. Thallus lacking rhizoidal filaments within medullary tissue 2

2. Thallus small, erect parts 2–5 mm high, 139–158 μm broad, 83–86 μm thick . . . *G. bornetii*
2. Thallus large, erect parts 5–7 cm high, about 1 mm broad, 797–930 μm thick . . . *G. acerosa*
3. Thallus large, 5–15 cm high, consisting of one to several distichously, pinnately branched percurrent axes rising from a holdfast of entangled stolons, with pyramidal outlook . . . . *P. capillacea*
3. Thallus small, less than 4 cm high, consisting of creeping parts and erect axes 4
4. Thallus blackish, 2–3.6 cm high, erect axes subcylindrical below, flattened or ligulate above, branching from pinnate to alternate . . . . *P. caerulescens*
4. Thallus purple red, 1.5–2 cm high, compressed lower axes often naked, and dense palmate branches on the middle to upper parts of axes . . . . *P. yinggehaeiensis*

Description of the species

Gelidiella bornetii (**Weber -van Bosse) Feldmann** *et* **Hamel** (Figs 1–3)
 Feldmann & Hamel, Rev. Gen. Bot. 46: 535, 1934.
 Basionym: *Gelidium bornetii* Weber–van Bosse, Vidensk. Medd. Dan. Naturhish. Foren. Kobenhavn 81: 107, 1926.

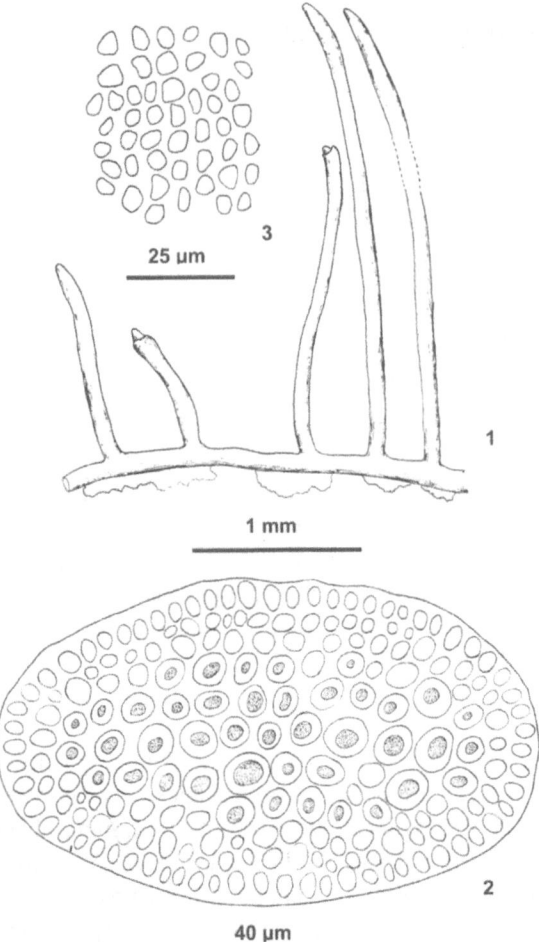

25 µm

3

1

1 mm

2

40 µm

Figures 1–3. Gelidiella bornetii (Weber–van Bosse) Feldmann *et* Hamel (**AST** I Ph-99). **1.** Habit sketch of frond; **2.** Transection of erect branch; **3.** Surface view of part of cortex cells of frond.

Thallus is very small, purple red, the decumbent creeping filaments with vigorous rhizoids fixed to the substratum, from the basal part the erect filaments arise reaching a height of about 2–5 mm. The filaments are about 139–158 µm broad, erect axes usually unbranched, apices obtuse. In transverse section of erect filaments, they are more or less oval, showing that the thallus is cartilaginous and somewhat compressed, consisting of a medulla of roundish cells, 7–10 µm diam., one to two layers of smaller cortical cells, filaments 83–86 µm thick.

Only a few sterile specimens were found which agree well in size, manner of growth, general habit and transverse section of frond with Børgesen's description (1938: 210).

Habitat: Creeping on a piece of dead coral. Xiaodonghai, Hainan province, China.

Distribution: India, Indonesia (Type locality).

Remarks: Transverse sections of the thallii of Hainan and South Indian materials are more or less oval or somewhat compressed. This differs from the original description of Weber –van Bosse (1926:107) and more recent listing by Kraft & Abbott (1998:53), which indicate that axis symmetry is flattened throughout.

A new record for China.

Gelidiella acerosa (**Forsskål**) **Feldmann** *et* **Hamel** (Figs 4–8)

Feldmann & Hamel, Rev. Gen. Bot. 46: 533, 1934; Zhang & Xia, Studia Marina Sinica, 15: 21, Figure 1, Pl. I: 9, 1979; Xia, Xia & Zhang, In Tseng C. K. (ed.), Comm. Seaweeds China: 64, Pl. 35, Figure 4, 1983.

Basionym: Fucus acerosus Forsskål, Flor a Aegyptiaco-arabica: Post mortem auctoris edidit Carsten Niebuhr: 190, 1775.

Thallus 5–7 cm high, with several tufted, entangled, cylindrical, erect axes rising from creeping axes that are decumbent and arcuate, attached to the substratum by stoloniferous rhizoids; erect axes cylindrical, 3–6 cm long, about 1 mm broad, frequently incurved abaxially, normally with opposite or subopposite pinnate branch sometimes secondly branched, up to 15 mm long, generally shorter apically. Thallus in transverse section consisting of medulla of irregular roundish cells, 22–26 µm in diam., surrounded by small cortex cells, 6–8 µm × 3–5 µm; dull purplish, cartilaginous.

Tetrasporangia borne around the apical swollen part of the branchlets, in surface view, roundish, 23–30 µm × 17–26 µm, in transverse section, obovate, 48–70 µm × 26–32 µm, surrounded by slightly modified cortical cells, cruciately divided. Cystocarp and spermatangia unknown.

Habitat: Growing on intertidal or subtidal dead coral. Hainan and Taiwan Provinces.

Distribution: Common in the tropics, Red Sea (Type locality).

Pterocladiella caerulescens (**Kützing**) **Santelices** *et* **Hommersand** (Figs 9–16)

Santelices & Hommersand, Phycologia 36: 118, 1997.

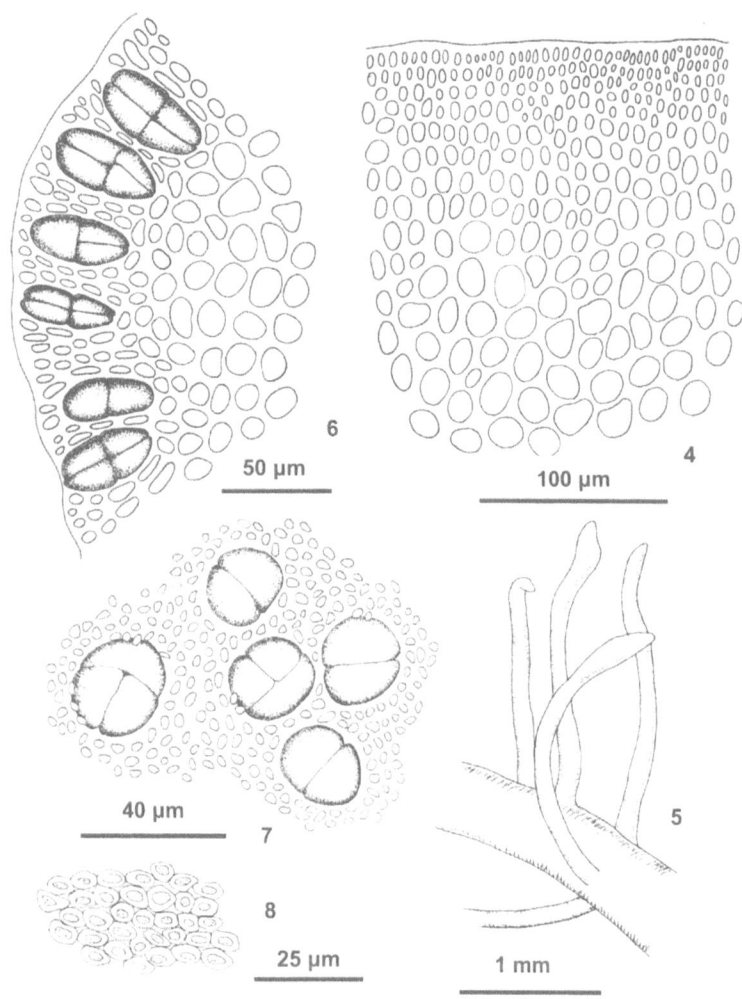

Figures 4–8. Gelidiella acerosa (Forsskål) Feldmann *et* Hamel. **4.** Transection of part of frond (**AST** 76-1364); **5.** Tetrasporangial branchlet (**AST** 76-1364); **6.** Transection of tetrasporangia (**AST** 76-1364); **7.** Surface view of tetrasporangia (**AST** 54-4545); **8.** Surface view of part of cortex cells of frond (**AST** 54-4545).

Basionym: *Gelidium caerulescens* Kützing, Tab. Phyc. 18(1): 19, Pl. 56c, d, 1868.

Synonyms: *Pterocladia tropica* Dawson, Pac. Natur. I: 40, Figures 21 A–D, Figure 22 B, 1959.

Gelidium irregulare Loomis, Allan Hancock Found. Occas. Pap. 24: 6, Pl. 9, Figure 1, Pl. 10, Figures 1–2, Pl. 11, 2–3, 1960.

Pterocladia rigida Loomis, Allan Hancock Found. Occas. Pap. 24: 8, Pl. 12, Figs 1–4, 1960.

Pterocladiella caloglossoides sensu Xia *et* Wang *non P. caloglossoides* (Howe) Santelices. Tax. Econ. Seaweeds 7: 81–86, Figures 1–8, 1999.

Thallus erect, 2–3.6 cm high, flattened axes rising from subcylindrical or compressed creeping axes, attached to the substratum by peg-like holdfasts. Erect axes subcylindrical below, flattened above, 813–868 μm (up to 1 mm) broad, 76–79 μm thick; branching extremely irregular, from simple to alternate, pinnate or secund slightly constricted at base, with acute tips. Blackish, subcartilaginous, adhering slightly to paper on drying.

Transverse section showing numerous rhizoidal filaments aggregated within medullary tissue, medullary cells irregularly ovate or oblong, 13–17 μm × 7–12 μm, cortical cells 1–2 layers, 5–8 μm × 3.3–5 μm.

Tetrasporangia disposed in sori at apex of branches, with sterile margins; in surface view, tetrasporangia irregularly arranged, roundish, 26–33 μm in diam., in transverse section, ovate or obovate, 26–50 μm × 20–26 μm, cruciate. Cystocarps unilateral,

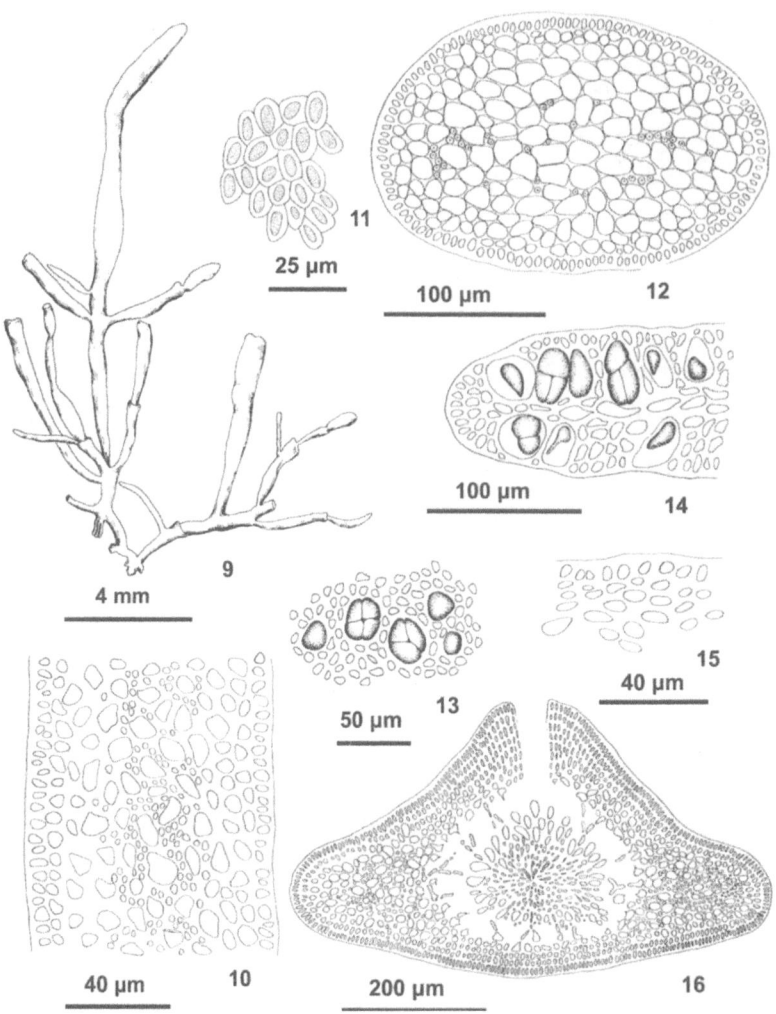

Figures 9–16. *Pterocladiella caerulescens* (Kützing) Santelices. **9.** Habit sketch of frond (**AST** 60-4586); **10.** Transection of erect branch (**AST** 60-4586); **11.** Surface view of part of cortex cells of frond (**AST** 93-0674); **12.** Transection of basal creeping axes (**AST** 93-0674); **13.** Surface view of tetrasporangia (**AST** 93-0674); **14.** Transection of tetrasporangia (**AST** 93-0674); **15.** Longitudinal section of pericarp (**AST** 93-0674); **16.** Longitudinal section of a cystocarp (**AST** 60-4586).

elongate swellings at the apex of branches, with a single ostiole on one surface of the frond and with sterile margins around the cystocarps. In longitudinal section of cystocarp, gonimoblast is usually attached on one side to the cystocarp floor, and produces chains of carposporangia on the remaining three sides. Spermatangia unknown.

Habitat: Growing in intertidal rock pools. Guangdong and Hainan Provinces.

Distribution: Hawaiian Islands, Guam, Vietnam, New Caledonia (Type locality).

Remarks: Santelices (1976) demonstrated extensive morphological variation in Hawaiian populations of *Pterocladiella caerulescens*. This extensive mor-

phological variation was also observed in Chinese materials.

Pterocladiella capillacea **(Gmelin) Santelices** *et* **Hommersand** (Figs 17–27)

Santelices & Hommersand, Phycologia 36: 118, 1997.

Basionym: *Fucus capillaceus* Gmelin, Historia fucorum: 146, Pl. 15, Figure 1, 1768.

Synonyms: *Pterocladia tenuis* Okamura, Jour. Imp. Fish. Inst. (Tokyo) 29(2): 62, Pl. 29, Pl. 30, Figure 3, Pl. 33, Figures 1–3, 1934; Tseng et al. Econ. Seaweeds China: 122, Pl. III: 56, Figures

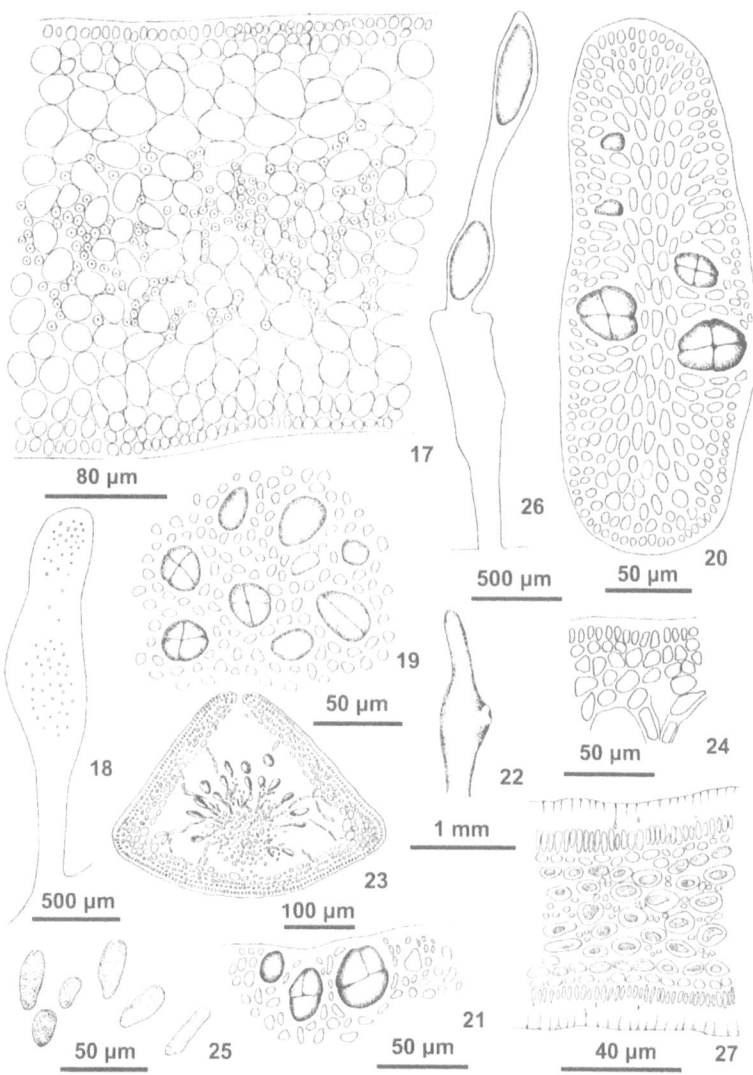

Figures 17–27. Pterocladiella capillacea (Gmelin) Santelices *et* Hommersand. **17.** Transection of part of frond 17 cm from apex (**AST** 64-99); **18.** Tetrasporangial branchlet (**AST** 57-531a); **19.** Surface view of tetrasporangia (**AST** 57-799); **20.** Transection of tetrasporangial branchlet (**AST** 57-799); **21.** Transection of tetrasporangia (**AST** 57-945); **22.** Cystocarpic branchlet (follows Tseng et al., 1962, Figure 28:4); **23.** Longitudinal section of a cystocarp (follows Tseng et al., 1962, Figure28:3); **24.** Longitudinal section of pericarp (**AST** 56-851); **25.** Carposporangia (**AST** 56-851); **26.** Spermatangial branchlet (**AST** 56-595); **27.** Transection of a spermatangial branchlet (**AST** 56-595).

27: 2, 28: 3–4, 1962; Xia, Xia & Zhang, In Tseng, C. K. (ed.), Comm. Seaweeds. China: 68, Pl. 37, Figure 3, 1983.

Thalli purplish red, compressed, 5–15 cm high, consisting of one to several distichously, pinnately branched percurrent axes rising from a holdfast of entangled stolons. The erect axes are slightly flattened below, strongly flattened above, with spatuliform or attenuated apices, up to 0.5 mm in diam. at their bases and up to 1.8 mm broad in the flat parts. Branching pinnately 2–3 times with pyramidal outline, with

opposite or alternate pinnules, 1–2 mm broad, abruptly constricted at the base and with obtuse apices. Cartilaginous, adhering slightly to paper on drying.

In transverse section, rhizoidal filaments scattered only in the middle portion of the central tissue, medullary cells irregularly oblong, 23–30 μm × 10–23 μm; cortical cells 1–2 layers, round or square, 3.3–5 μm.

Tetrasporangial sori in ultimate ramuli, scattered among the cortical layers of frond, circular or ovoid in surface view, 26–30 μm × 23–30 μm, ovoid or

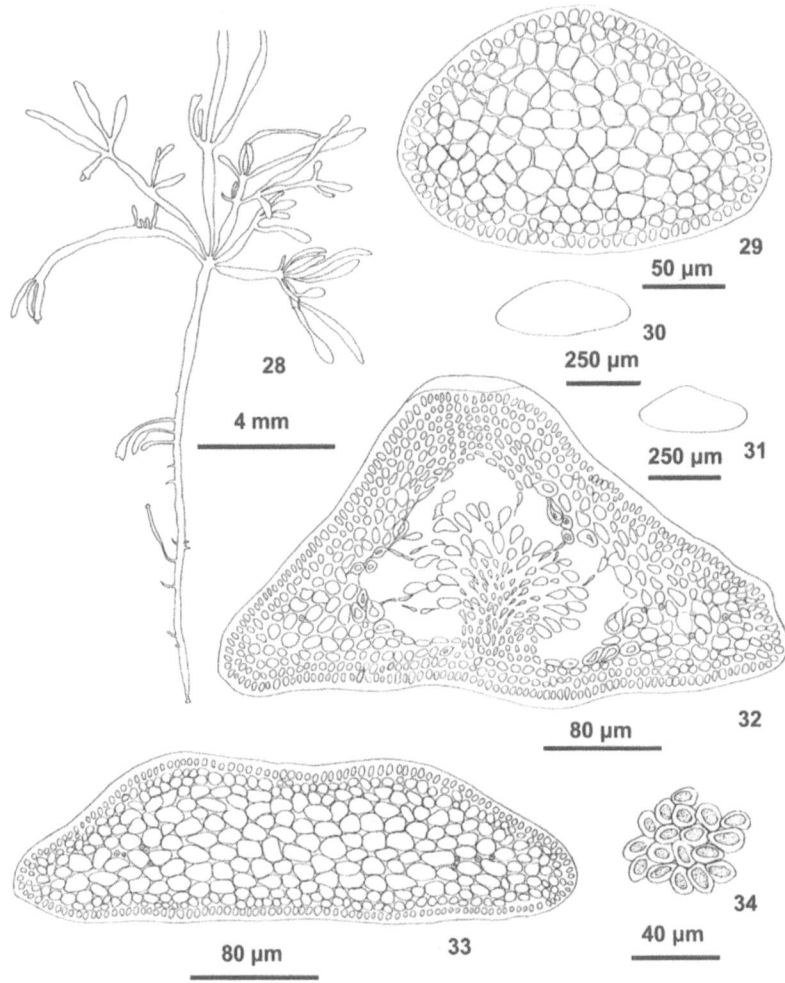

Figures 28–34. Pterocladiella yinggehaiensis Xia *et* Tseng sp. nov. (**AST** 93-0768). **28.** Habit sketch of frond; **29.** Transection of basal creeping axes; **30.** Transection of upper part of erect branch; **31.** Transection of lower part of erect branch; **32.** Longitudinal section of a cystocarp; **33.** Transection of middle axes; **34.** Surface view of part of cortex cells of frond.

circular in transverse section, 33–40 μm × 23–33 μm, cruciately divided, surrounded by modified cortical cells. Cystocarps formed on the median axis of ramuli, swollen on one side, roundish, 249–300 μm × 300–332 μm, slightly rostrate, nonconstricted at base; in longitudinal section, gonimoblast consisting of very small cells, carposporangia oblong or ovoid, 26–40 μm × 13–20 μm; pericarp 33–53 μm thick, consisting of 4–5 cells layer. Spermatangial sori on slightly compressed ramuli, 498-697 μm long and 149 μm in diameter, spermatangia cut off directly from outer cortical cells.

Habitat: Growing on intertidal to subtidal rocks.

Distribution: Common on entire Chinese coast, Mediterranean (Type locality).

Pterocladiella yinggehaiensis **Xia** *et* **Tseng sp. nov.** (Figs 28–34)

Thallus parvus, 1–2 cm altus, dense caespitosus: rami erecti orientes ex axibus 132–145 μm diametro, compressis, repentibus, affixis ad substratum per haptera parva et obtusa. Rami erecti compressi, 316–389 μm lati, irregulariter 3–5(-9)-plo palmati in partibus superis. Fila rhizoidea rara intra medullam. Cystocarpia prope extrema ramorum liberorum.

Holotype: AST 93-0768, Cystocarpic, growing on lower intertidal rocks. Collected by Xia Bangmei, Kuang Mei and Wang Yongqiang at Yinggehai, Hainan Island, Hainan Province, China, September, 21, 1993.

Thallus small, 1–2 cm high, growing as a dense tuft, erect branches rising from compressed creeping axes, attached to the substratum by peg-like holdfasts. Erect axes compressed, 316–389 μm broad, 100–166 μm thick; more densely branched on the middle to upper parts (or rarely on branchlets) for one half their lengths, with branches 3–5 (-9) times palmate, apex obtuse and often broken, slightly constricted at base; rarely proliferous from surface; purple red, cartilaginous.

Transverse section creeping stem 132–145 μm broad, 165–224 μm thick, consisting of a medulla of irregularly angular parenchymatous cells, 11–20 μm × 8–13 μm, one layer of smaller cortical cells, 7–9 μm × 4–7 μm; transverse section erect axes consist of a medulla of irregularly angular parenchymatous cells, 10–17 μm × 7–13 μm, with one layer of smaller cortical cells, 7–10 μm × 4–6 μm, rare rhizoidal filaments found within two edges of compressed medullary tissue.

Cystocarps unilateral, prominently protruding, triangular, 290 μm high, 422 μm broad, rostrate, unconstricted at base; gonimoblast consisting of few small cells, chains of carposporangia radiating from a core of gonimoblast filaments surrounding the central axis, carposporangia obovate, 17–26 μm × 8–10 μm, pericarp 53–59 μm thick, consisting of 5-8 layers of cells, some inner cells elongated, extending to the placenta. Tetrasporangia and spermatangia unknown.

Remarks: Pterocladiella yinggehaiensis is characterized by its small thalli that are more densely branched on the middle to upper parts for one half their lengths. These characteristics are so distinctive that they can be used to separate this species readily from all other species now reported in this genus.

Acknowledgements

This project was supported by a grant for systematic and evolutionary biology, Chinese Academy of Sciences. This is contribution no.3746 from the Institute of Oceanology, Chinese Academy of Sciences. We thank Prof. Karla McDermid of the University of Hawaii for help with the Latin description.

References

Børgesen, F., 1938. Contribution to a South Indian marine algal flora-III. J. Ind. Bot. Soc. 17: 205–242.

Feldmann, J. & G. Hamel, 1934. Observations sur quelques Gelidiacées. Rev. gén. Bot. 62: 422–431.

Kraft, G. T. & I. A. Abbott, 1998. *Gelidiella womersleyana* (Gelidiales, Rhodophyta), a diminutive new species from the Hawaiian Islands. Bot. mar. 41: 51–61, Figures 1–17.

Santelices, B.,1976. Taxonomic and nomenclatural notes on some Gelidiales (Rhodophyta). Phycologia 15: 165–173, Figures 1–33.

Santelices, B., 1998. Taxonomic review of the species of *Pterocladia* (Gelidiales, Rhodophyta). J. appl. Phycol. 10: 237–252, Figures 1–4.

Santelices, B. & M. Hommersand, 1997. *Pterocladiella*, a new genus in the Gelidiaceae (Gelidiales, Rhodophyta). Phycologia 36: 114–119.

Tseng, C. K., C. J. Chang, C. F. Chang, E. Z. Xia, B. M. Xia, M. L. Dong & Z. D. Yang, 1962. Economic Seaweeds of China. Science Press, Beijing: 122, Pl. III: 56, Figures 28: 3–4 [in Chinese].

Xia, B. M. & Y. Q. Wang, 1999. Taxonomic studies on *Pterocladiella* (Gelidiaceae, Gelidiales, Rhodophyta) from China. In Abbott, I. A. (ed.), Taxonomy of Economic Seaweeds: with Reference to Some Pacific Species. California Sea Grant College, Univ. Calif. La Jolla, CA: 81–86, Figures 1–13.

Hydrobiologia **512:** 209–214, 2004.
P.O. Ang, Jr. (ed.), Asian Pacific Phycology in the 21st Century: Prospects and Challenges.
© 2004 *Kluwer Academic Publishers.*

Some observations on harmful algal bloom (HAB) events along the coast of Guangdong, southern China in 1998

Yuzao Qi[1,*], Jufang Chen[1], Zhaohui Wang[1], Ning Xu[1], Yan Wang[1], Pingping Shen[1],
Songhui Lu[1] & I. J. Hodgkiss[2]
[1]*Institute of Hydrobiology, Jinan University, Guangzhou, 510632, China*
[2]*Department of Ecology & Biodiversity, The University of Hong Kong, Pokfulam Road, Hong Kong SAR, China*
Author for correspondence: E-mail: tqyz@jnu.edu.cn

Key words: Guangdong, harmful algal bloom, Hong Kong, southern China

Abstract

The year 1998 was an unusual year for Guangdong Province and Hong Kong, both in southern China, as the frequency and intensity of harmful algal blooms (HAB) were much higher than usual. This paper describes the causative organisms found associated with these blooms and speculates on the possible causes of these blooms, including the effects of increased temperature, reduced salinity, eutrophication and meteorological and oceanographic events on the initiation and spread of these blooms.

Introduction

The occurrence of red tides in the South China Sea (SCS) has been monitored over the last decade (Qi et al., 1992, 1993; Hodgkiss & Yim, 1995; Liang, 1996; Hodgkiss & Ho, 1997; Hodgkiss & Lu, 2004). In about 50% of these cases, these red tide events were due to the dinoflagellate *Noctiluca scintillans*. However, other causative organisms like *Skeletonema costatum* (Grev.) Cleve, *Pseudo-nitzschia pungens* (Grunow *ex* Cleve) Hasle, *Chaetoceros* spp. and *Rhizosolenia longisima* were also often found to occur in the bloom. In 1998, massive harmful algal blooms (HAB) occurred along the coast of Guangdong Province and Hong Kong in southern China. These HABs differed from those occurred in the previous years in that the causative species were diverse (Table 1). They were toxic or potentially toxic (Lu & Hodgkiss, 1999, 2004; Yang et al., 2000a,b) and inflicted significant negative impacts on the society. They devastated aquaculture and destroyed natural marine ecosystems, with a tremendous economic loss estimated at over 0.3 billion Chinese yuan (about 35 million US$).

This paper gives an overview of HABs in the coastal waters of Guangdong and Hong Kong in 1998, and explores the possible initiating factors to explain why so many HAB events occurred in this area at this time.

Materials and methods

Water sampling

The sea areas covered in this investigation included mainly the coastal waters of Guangdong and Hong Kong (Fig. 1). For every area that experienced algal bloom, samples were taken at every 3- or 6-day interval using a 20-μm phytoplankton net and a 5-l water sampler. Phytoplankton were enumerated in the laboratory using a Sedgwick–Rafter counting cell.

Water quality analysis

Water quality analyses, including chlorophyll *a*, dissolved inorganic nitrogen (DIN), dissolved inorganic phosphorus (DIP), and dissolved oxygen (DO) concentrations and water temperature (*T*) were conducted

Table 1. HABs on the coasts of Guangdong and Hong Kong in 1998

Date	Causative species	Area	Effect
February	*Phaeocystis globosa* Scherffel	Raoping, Guangdong	Fish kill
March 19	*Gyrodinium* and *Gymnodinium* spp.	Hong Kong	Fish kill
March 24	*Gymnodinium mikimotoi* Miyake et Kominami ex Oda	Dapeng Bay, Guangdong	Fish kill
April	*Gymnodinium mikimotoi*	Huidong, Guangdong	Fish kill
April	*Gymnodinium* sp.	Yangjiang, Guangdong	Fish kill
April 9–17	*Gymnodinium mikimotoi*	Zhuhai, Guangdong	Fish kill
April 22–May 2	*Gyrodinium instriatum* Freudenthal et Lee	Shenzhen Bay, Guangdong	None
April 23	*Gyrodinium aureolum* Hurlbert	Shenzhen Bay, Guangdong	Fish kill
May 2–5	*Gymnodinium mikimotoi*	Daya Bay, Guangdong	Fish kill
Sept. 5–9	*Scrippsiella trochoidea* (Stein) Loeblich III	Daya Bay, Guangdong	None
Sept. 29–Oct. 3	*Ceratium furca* Ehrenberg (Claparède et Lachmann)	Dapeng Bay, Guangdong	Beach closed
Nov. 7	*Noctiluca scintillans* (Macartney) Ehrenberg	Shanwei, Guangdong	None
Nov. 8–15	*Mesodinium rubrum* Leegaard	Daya Bay	None

following the standard protocol prescribed in 'Regulation of Ocean Investigation' by Chinese State Oceanic Administration (SOA), 1989.

Morphological observation of causative organisms

Morphological observation of the causative organisms of HAB was carried out using an Olympus BH-2 light microscope. For more detailed observation, scanning electron microscopic studies were conducted. DNA sequences of doubtful species were made to help in the species identification.

Results and discussion

Major causative species of HABs in 1998

Red tide causative species in 1998 were very different from those in the previous outbreaks. Some were new records, some were rare, some were probably new species (Yang et al., 2000a,b), but most were dinoflagellates.

Phaeocystis globosa *Scherffel (Fig. 2)*
In September 1997, a Haptophycean species, *Phaeocystis globosa*, bloomed in Quanzhou Bay, Fujian province (part of East China) and then spread down south to the Guangdong coast. The bloom covered an area over 3000 km^2 in size and lasted for about 6 months. It resulted in massive fish mortality and devastated the caged-fish aquaculture industry. The main species of fish killed were: *Scombermorus guttatus* (Bloch et Schneider), *Pampus chinensis* (Euphrasen), *Lutjianus chrysotaenia* (Bleeker) and *Pagrosomus major* (Temminck et Schlegel) [*Pagrus major*]. The estimated economic loss was more than 60 million Chinese yuan (about 7.5 million US$).

Phaeocystis colonies have long been noticed by local fishermen and the term 'red bubble' was used to describe these colonies. In Raoping county water area, the colonies began to appear in mid-October and the colony numbers reached a maximum on 20 November 1998. The average level of chlorophyll *a* near cage-fishing area was 10.2 μg l^{-1}. A large amount of foam accumulated on beaches, causing a great nuisance smell. Brownish colonies in the sea were big enough to be seen easily (the biggest one was over 30 mm in diameter). Both morphological forms – free-living motile cell and colony – have been identified in the natural samples. The motile cells possess two flagella and a haptonema, which is difficult to be observed under the light-microscope. Their sizes are about 3–5 μm

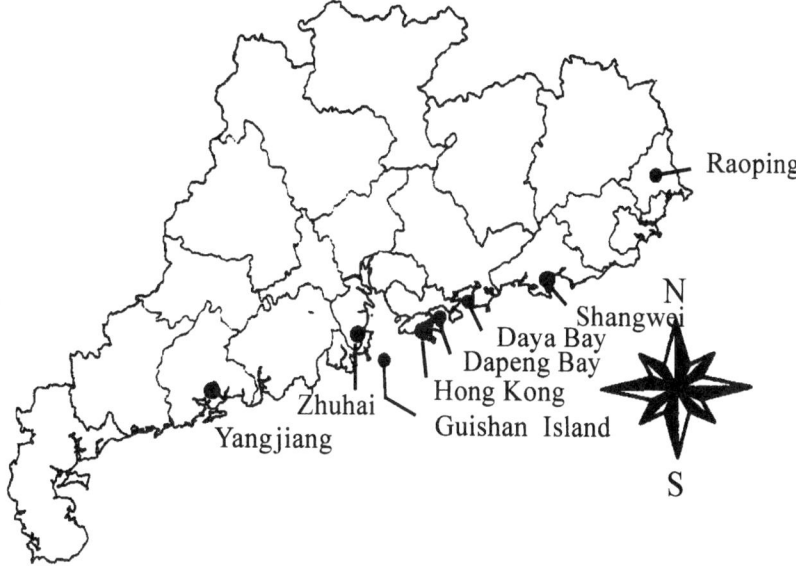

Figure 1. Map of Guangdong Province, China showing sites where HABs occurred in 1998.

(Fig. 2). Each cell contains about two to three parietal yellow-green chloroplasts. Colonies are shown to vary widely in size ranging from 20 μm to over 30 mm. These are composed of thousands of cells embedded in a mucilaginous matrix. Individual cells are distributed within the gel matrix of the homogeneous spheres.

Identification of this causative species was difficult. It closely resembled *Phaeocystis pouchetii* Lagerhein in colonial form, but was not lobed when it became ripe. The size was also bigger than that originally diagnosed for *P. pouchetii* (2 mm in diameter) (Sournia, 1988). DNA sequences confirmed that this species was *P. globosa*. The DNA sequencing analysis gave an interesting result and revealed that the species in China is somewhat different from the European and American strains (Medlin & Lange, 1994), showing that this strain is endemic.

Gymnodinium mikimotoi *Miyake et Kominami ex Oda (Fig. 2)*

The blooms caused by *Gymnodinium mikimotoi* in March to April stretched over almost all coastal areas across the Pearl River estuary from Huidong, east of Guangdong, to Yangjiang, west of Guangdong (Fig. 1). The highest density of *G. mikimotoi* was up to 2.08×10^6 cells l^{-1} in Guishan, Zhuhai. The bloom inflicted a loss of over 0.1 billion Chinese Yuan in aquaculture. A large number of fish, such as: *Seriola* sp., *Pagrosomus major*, *Epinephelus epistictus* (Temminck et Schlegel) were killed.

The causative organism of the red tide which occurred at the same time in Hong Kong was identified to be *Gyrodinium aureolum* (Yang & Hodgkiss, 1999). It is believed by some workers (e.g., Hallegraeff et al., 1995) that *Gymnodinium mikimotoi* and *Gyrodinium aureolum* are the same species. Whether or not these two organisms are the same remains to be verified in more details. Furthermore, Yang & Hodgkiss (1999) reported that another species occurred in Hong Kong in March and April is a new species of *Karenia*. They later named it *Karenia digitata* (Yang et al., 2000b). Also, the *K. digitata* bloom devastated caged fish aquaculture, resulted in over HK$ 0.3 billion (about 38 million US$) economic loses. These authors also found another new species, *Karenia longicanalis*, in mid-May, 1998 in Victoria Harbor, Hong Kong (Yang et al., 2000a). Further studies are needed to verify the identification of species from Hong Kong and the whole of Guangdong coast.

Gyrodinium instriatum *Freudenthal et Lee (Fig. 2)*

This species was not recorded previously in China, but has been noted as a red tide species in Japan (Fukuyo et al., 1990). A bloom caused by this species occurred in Shenzhen Bay from 23 April to 3 May 1998, covering an area of 200 km^2. High cell concentrations between 10^5–10^6 cells l^{-1} were observed at the caged fish area and the outer estuary (Fig. 3). The highest cell density was 3.8×10^6 cells l^{-1}. Although the nutrient contents were relatively high in the inner estuary of the Gulf of Shenzhen River, phytoplankton biomass was

212

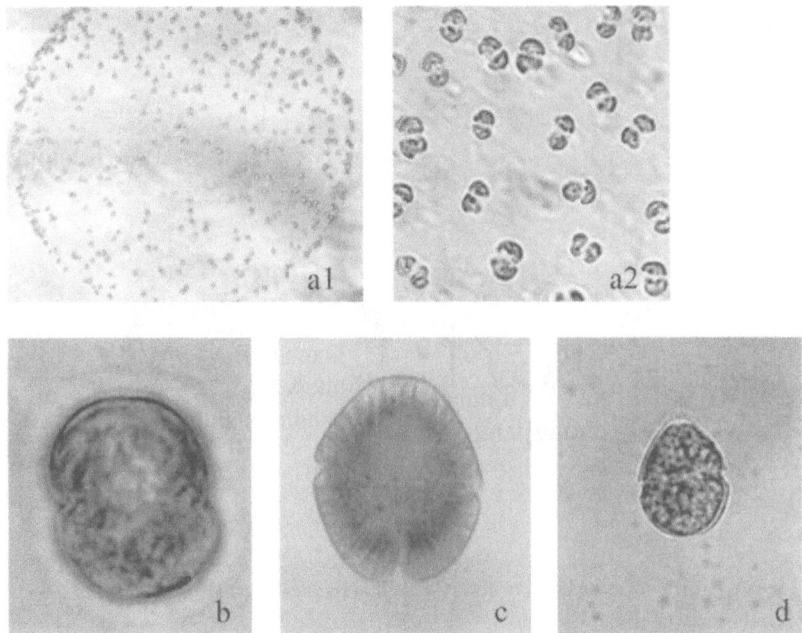

Figure 2. Representative causative bloom organisms. (a) *Phaeocystis globosa*; (a1) colony; (a2) pair cells in a colony; (b) *Gymnodinium mikimotoi*; (c) *Gyrodinium instriatum*; (d) *Scrippsiella trochoidea*.

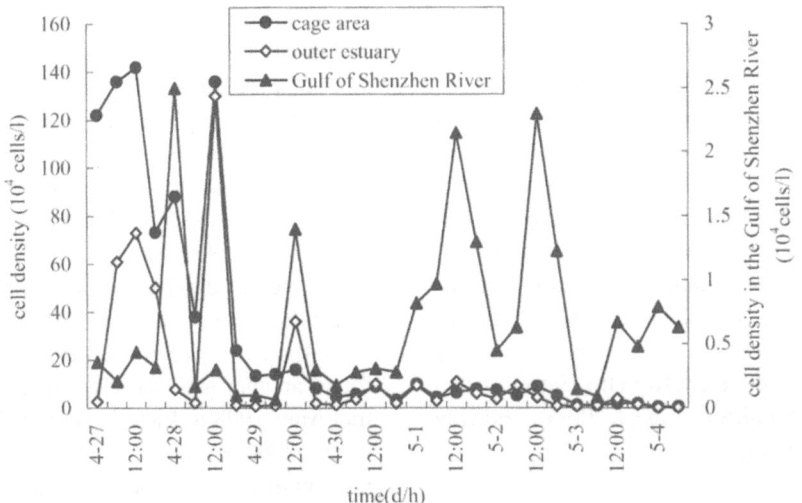

Figure 3. The quantitative variation of cell densities of *Gyrodinium instriatum* in Shenzhen Bay from 27 April to 4 May 1998 during the red tide bloom.

normally low due to heavy suspension of sediments. So far, there is no report showing this to be a toxic species and it did not cause any fish mortality this time.

Scrippsiella trochoidea *(Stein) Loeblich III (Fig. 2)*

This is a very common non-toxic species which often forms blooms in the coast of Guangdong. From 5 to 8 September 1998, a bloom of this species occurred in Daya Bay. The cell density was up to 6.3×10^5 cells l^{-1} (Wang et al., 1998).

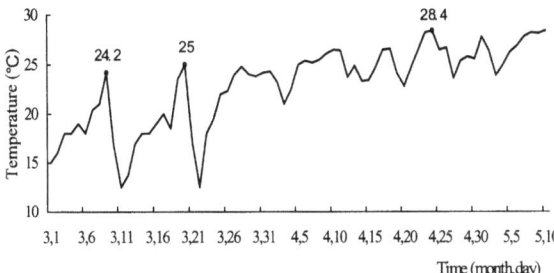

Figure 4. Daily changes of air temperature in Shenzhen City from March to May, 1998.

Suspected causes initiating algal blooms in 1998

Many factors have been speculated to be responsible for initiating algal blooms. Among these are: (1) climate change and temperature, (2) meteorological and oceanographic features and (3) anthropogenic influences in the form of excess nutrient loading.

Climate change and temperature

The year 1998 was an El Niño year, and an unusual climate pattern was observed. For example, there was a high of up to 32 °C air temperature in Guangdong in March, but a low of 18 °C in May. Temperature is thought to be one of the most important factors initiating blooms (Liang & Qian, 1991; Qian et al., 1991). Algal bloom that occurred in March and April 1998 apparently coincided with a rise in temperature (Fig. 4). When the temperature increased from 21 to 25 °C, a *G. aureolum* bloom was initiated in Hong Kong waters on 9 March and then on 21 March. Another temperature variation apparently induced a *G. mikimotoi* bloom in Nan-au, Dapeng Bay, east of Hong Kong. Interestingly, on 24 April, when the temperature rose from 22 to 28 °C, a *Gyr. instriatum* bloom was observed in Shenzhen Bay. The northeast monsoon changes to southwest monsoon from March to May. During this period, the weather gradually becomes warmer. This period is also the beginning of the rainy season, with the wind velocity generally becoming weakened and the air pressure becoming lower. It is during this period that algal bloom could easily occur (Liang & Qian, 1991).

Precipitation also played an important role in the algal bloom in 1998. There were few rainfalls from March to early April in Guangdong in 1998. Rains became heavier and more frequent after mid-April. Heavy rain reduced the salinity of seawater. There was a heavy rainfall (28.3 mm) on 12 April, 1998 in Guishan Island area before the *G. mikimotoi* bloom.

Marked decrease in the salinity of the surface layer by heavy rain was reported to be the important factor causing huge bloom of *G. nagasakiense*, a relative species of *G. mikimotoi*, in Japan (Yamaguchi, 1994). Similarly, *G. mikimotoi* bloom often occurred after heavy rain in Korea (Cho, 1981). In Shenzhen Bay, a heavy downpour from 26 to 27 April, 1998 recorded 126 mm of precipitation. This rain greatly reduced the salinity of the water around Shenzhen Bay to brackish level, and together with high nutrient levels and high temperature, induced the outbreak of *Gyr. instriatum* bloom.

Meteorological and oceanographic features

Effects of meteorological and oceanographic features on the occurrence of HABs are relatively poorly studied in China. This could be another key factor inducing HAB events.

Bloom distribution often depends on wind direction or flow of ocean currents. There are arguments as to where the HABs in 1998 originated, either in mainland or in Hong Kong waters. However, it is agreed that bloom movement should have been influenced by wind direction and current movement in either case, and that monsoon wind is an important element in determining the bloom direction in coastal waters of Guangdong in the spring. The spring monsoon blows in a southeast direction and this direction changes to northwest in winter. So, the HABs that occurred in the spring of 1998 probably started in Hong Kong waters, then moved to Dapeng Bay of mainland waters. Algal bloom expansion also followed the same phenomenon in Guishan, Zhuhai. *G. mikimotoi* first appeared on 8 April around Hong Kong waters and then spread to Guishan Island on 10 April. It was carried by current movement, generated from the southeast wind.

Anthropogenic impacts and excessive nutrient loading

Many studies have shown that water pollution is a key factor inducing HABs (Lancelot et al., 1987; Anderson, 1997). Wastes and raw sewage discharging into the sea without treatment often cause nutrient overloading which then easily induces bloom outbreaks. For instance, there were 2.8 billion t of sewage discharged into the estuary of the Pearl River in 1997, among which just 10% were given primary treatment. This causes excess input of nitrate and/or phosphate (eutrophication) and often results in HABs.

Another source of excessive nutrients is mariculture. Mariculture has increased dramatically along the coasts of China. There are now over 110 000 fish-cages in Guangdong province alone. Intensive aquaculture causes self-pollution as a result of excess feeding, fish feces and aging-water. A good example to elucidate the nutrient situation in caged fish areas is Ya-qian Bay, a small bay in Daya Bay (Fig. 1). The size of the bay is 23 500 m^2. Fish production was 132 t in 1997 and the total feed used amounted to 1056 t. There was a discharge of 48.9–131.8 kg of nitrogen into the sea for every ton of fish cultured. Within a year, 6.5–17.4 t of nitrogen in total would have been discharged into the sea. This figure does not include nitrogen from fish feces. This example clearly indicates that intensive aquaculture causes self-pollution. Long term aquaculture in the same area, over-crowded cage arrangement and intensive fish culture result in eutrophication of the marine aquaculture area, and thus offers a suitable environment for algae to grow and form blooms.

When a *G. mikimotoi* bloom occurred in April in Guishan, Zhuhai water (at the mouth of the Pearl River), the total inorganic nitrogen in the water was 211 μg l^{-1}, but the total inorganic phosphorus level was 7 μg l^{-1}. Meanwhile, when a *Gyr. instriatum* bloom happened in Shenzhen Bay at the end of April, the nitrogen level was up to 977 μg l^{-1}. Moreover, at the mouth of the Shenzhen River, the ammonia nitrogen level was 4500 μg l^{-1}. There are many examples indicating that nutrients are a key factor in initiating algal blooms (Liang, 1996).

HAB formation is a complicated phenomenon. We need to put more effort into exploring its mechanisms. There is a need to synthesize approaches from biology, ecology, oceanography and meteorology.

Acknowledgements

This research project was supported by a Chinese NSF Grant, No. 39790110; by Department of Education of Guangdong Province; by Red tide Key Project, Department of Science and Technology, Guangdong Province, China; by Chinese National Basic Research Priorities Programme (973), No. 2001CB409701. Thanks are due to Dr Donald M. Anderson of the Woods Hole Institute of Oceanography, U.S.A., for his generous guidance in this research.

References

Anderson, D. A., 1997. Turning back the harmful red tide. Nature 388: 513–514.

Cho, C. H., 1981. On the *Gymnodinium* red tide in Jinhae Bay. Bull. Korean Fish Soc. 14: 227–232.

Fukuyo, Y., H. Takano, M. Chihara & K. Matswoka, 1990. Red Tide Organisms in Japan. Uchida Rokakuho.

Hallegraeff, G. M., D. A. Anderson, A. D. Cembella & H. O. Enevoldsen (eds), 1995. Manual on Harmful Marine Microalgae. UNESCO, IOC: 33.

Hodgkiss, I. J. & K. C. Ho, 1997. Are changes in N:P ratios in coastal waters the key to increased red tide blooms? Hydrobiologia 352: 141–147.

Hodgkiss, I. J. & S.-H. Lu, 2004. The effects of nutrients and their ratios on phytoplankton abundance in Junk Bay, Hong Kong. Hydrobiologia 512/Dev. Hydrobiol. 173: 213–227.

Hodgkiss, I. J. & W. W. S. Yim, 1995. A case study of Tolo Harbour, Hong Kong. In McComb, A. J. (ed.), Eutrophic Shallow Estuaries and Lagoons. CRC Press, Baton Rouge, USA: 41–57.

Lancelot, C., G. Billen, A. Sournia, T. Weisse, F. Colum, M. Veldhuis, A. Davis & P. Wassman, 1987. *Phaeocystis* blooms and nutrient enrichment in the continental zone of the North Sea. Ambio 16: 38–46.

Liang, S. (ed.), 1996. Studies on the Environments of Dapeng Bay and Red Tides. Ocean Press, Beijing, China [In Chinese].

Liang, S. & H. L. Qian, 1991. The exploration of relationship between the occurrence frequency of red tide and the period of the monsoon change in Northern South China Sea. Res. Dev. S. China Sea 3: 1–5 [in Chinese].

Lu, S.-H. & I. J. Hodgkiss, 1999. An unusual year for the occurrence of harmful algae. Harmful Algae News 18: 1–3.

Lu, S.-H. & I. J. Hodgkiss, 2004. Harmful algal bloom causative collected from Hong Kong waters. Hydrobiology 511/Dev. Hydrobiol. 173: 229–236.

Medlin, L. K. & M. Lange, 1994. Genetic differentiation among three colony-forming species of *Phaeocystis*: Further evidence for the phylogeny of the Prymnesiophyta. Phycologia 33: 199–207.

Qi, Y. Z., Y. Hong, H. L. Qian, & S. H. Lu, 1992. Problems caused by harmful algal blooms in China. UNESCO, IOC Workshop Report No. 94: 22–24.

Qi, Y. Z., Z. P. Zhang, Y. Hong, S. H. Lu, C. J. Zhua & Y. Q. Li, 1993. Occurrence of red tides on the coasts of China. In Smayda T. J. & Y. Shimizu (eds), Toxic Phytoplankton Blooms in the Sea. World Publishing Corporation, Beijing, China: 43–46.

Qian, H. L., 1991. Study on the red tide along northern coast of South China Sea. J. Jinan University 12: 108–111 [In Chinese].

Qian, H. L., S. Liang & Y. Z. Qi, 1991. Discussion on the relationship between red tides and El Niño phenomenon. J. Jinan University. 12: 112–116 [In Chinese].

Sounia, A.,1988. *Phaeocystis* (Prymnesiophyceae): How many species? Nova Hedwigia 47: 211.

Wang, Z. H., S. H. Lu, J. F. Chen, N. Xu, & Y. Z. Qi, 1998. Taxonomic studies on red tide causative algae on the Guangdong coast, South China Sea. J. Wuhan Bot. Res. 4: 310–314 [In Chinese].

Yamaguchi, M., 1994. Physiological ecology of the red tide flagellate *Gymnodinium nagasakiense* (Dinophyceae) mechanism of the red tide occurrence and its prediction. Bull. Nansci Nat. Fish. Res. Inst. Japan 27: 251–253.

Yang, Z. B. & I. J. Hodgkiss, 1999. Massive fish killing by *Gyrodinium* sp. Harmful Algae News 18: 4–5.

Yang, Z. B., I. J. Hodgkiss & G. Hansen, 2000a. *Karenia longicanalis* sp. nov. (Dinophyceae), a new bloom-forming species isolated from Hong Kong, May 1998. Bot. mar. 44: 67–74.

Yang, Z. B., H. Takayama, K. Matsuoka & I. J. Hodgkiss, 2000b. *Karenia digitata* sp. nov. (Gymnodiniales, Dinophyceae), a new harmful algal bloom species from coastal waters of west Japan and Hong Kong. Phycologia 39: 463–470.

Hydrobiologia **512**: 215–229, 2004.
P.O. Ang, Jr. (ed.), Asian Pacific Phycology in the 21ˢᵗ Century: Prospects and Challenges.
© 2004 *Kluwer Academic Publishers.*

The effects of nutrients and their ratios on phytoplankton abundance in Junk Bay, Hong Kong

I.J. Hodgkiss[1,*] & Songhui Lu[1,2]
[1]*Department of Ecology & Biodiversity, The University of Hong Kong, Pokfulam Road, Hong Kong SAR, China*
[2]*Institute of Hydrobiology, Jinan University, Guangzhou 510632, China*
Author for correspondence; E-mail: hodgkiss@hkucc.hku.hk

Key words: eutrophication, nutrients, N:P ratio, phytoplankton, harmful algal bloom, Junk Bay, Hong Kong

Abstract

Eutrophication has been considered to be undoubtedly one of the key factors stimulating phytoplankton growth, since it involves the enrichment of a water mass with both inorganic and organic nutrients supporting plant growth. Nutrient enrichment as a result of anthropogenic activity occurs in estuaries and coastal waters as well as in lakes and freshwater impounds, and blooms of phytoplankton are one of the effects of such an accelerated process of nutrient enrichment. This paper presents the results of a two-year survey of the nutrients and phytoplankton at 3 stations in Junk Bay, Hong Kong, carried out from 1997 to 1998. The relationships between nitrogen, phosphorus, and their ratio, with phytoplankton abundance have been studied. The results show that the highest nitrogen concentration was in Station 2 which is close to a sewage input, whereas the highest phosphorus concentration was in Station 1 which is close to a landfill area. The mean N:P ratios at the three stations were between 8 and 14. The diatoms were the dominant group during most of the year but it seems that diatoms were more sensitive than dinoflagellates and other algal groups to the increase in nutrients.

Introduction

After almost a century of study, the term 'eutrophication' can be defined in terms of both systems naturally enriched by nutrients, e.g. upwelling in the sea, or so-called natural eutrophication, and environments polluted by anthropogenic activities such as sewage discharge, or so-called cultural eutrophication. Most recent studies have focused on cultural eutrophication. Coastal marine waters, particularly embayments and estuaries, are typically more fertile than the open ocean. The urbanization of coastal marine and estuarine areas results in a dramatic increase in population, so that eutrophication by anthropogenic activities such as urban waste and sewage discharge, increasing use of agricultural fertilizers, freshwater runoff, riverine nutrient inputs, coastline construction, tourism, mariculture, etc., are now the major causes in the world of environmental pollution. Many estuaries and coastal waters are eutrophic because of the large amounts of inorganic nutrients (primarily nitro-

gen and phosphorus) and organic matter they receive from anthropogenic activities. It is a widespread phenomenon in coastal areas and estuaries all around the world. Well-known examples from elsewhere in the world are Chesapeake Bay (Fisher, et al., 1992), the Baltic Sea (Larsson et al., 1985), Narragansett Bay (Nixon, 1997), the Black Sea (Mee, 1992), the Dutch Wadden Sea (De Jonge & Raaphorst, 1995), Mediterranean Sea (Stirn, 1988), the North Sea (Paetsch & Radach, 1997), Skagerrak and Kattegat (Rosenberg et al., 1996), and the Seto Inland Sea (Yamanaka, 1983).

Phytoplankton primary production in both fresh water and marine ecosystems is basically controlled by three factors: nutrient availability, light availability, and the response of the algae to nutrients and light (Kelly & Naguib, 1984). The direct effect of nutrient enrichment is phytoplankton 'population explosions' (or algal blooms). The decomposition of these blooms depletes dissolved oxygen in the water column, resulting in invertebrate, fish and even marine mammal kills (Hallegraeff, 1993). Some algae can produce toxins which can be hazardous to humans through the food web (Yasumoto, 1990; Baden et al., 1993), and the resulting harmful algal blooms (HABs) in coastal waters and embayments have increased in the last few decades (Anderson, 1989; Hallegraeff, 1993). Eutrophication of coastal waters, which has been reported all around the world (Smayda, 1990), is one of the major reasons, at least in the case of some specific blooms, such as *Phaeocystis* (Riegman, et al. 1992), *Chattonella* (Amano, et al. 1998), *Aureococcus anophagefferens* Hargraves *et* Seiburth (brown tide) (Keller & Rice, 1989), *Heterosigma akashiwo* Hada (Honjo, 1993), and *Chrysochromulina polylepis* Manton *et* Parke (Maestrini & Granéli, 1991).

In Hong Kong, Tolo Harbour is a well-known example of the effects of cultural eutrophication. Urbanization of the area since the mid-1970s resulted in mass loadings of livestock wastes and domestic sewage. In the period 1976–1985, nutrient loading increased more than two-fold, and the frequency of red tide occurrences markedly increased during this period (Lam & Ho, 1989).

Junk Bay is one of those enclosed small bays in Hong Kong facing pollution problems. Land development, including house construction and reclamation, is under way in Tseung Kwan O, Tseng Lan Shue, Ma Yau Tong, Sheung Lau Wan, Chik Sha, Tiu Keng Leng and Hang Hau. A new town is being constructed to house more and more residents in the near future. In order to provide land for urbanization, major coastal land reclamation schemes were undertaken in Tiu Keng Leng, which is just beside the western side of the inner bay. To the eastern side of the inner bay, there is an effluent discharge tunnel which receives sewage and runoff. In 1989, the first toxic algal bloom in Hong Kong, caused by *Alexandrium catenella* (Whedon *et* Kofoid) E. Balech, was recorded in Junk Bay. Other harmful algal species were also found in the bay (Lu & Hodgkiss, 1999). Because of the increasing water pollution and nutrient loading, Junk Bay was included in one of the 10 gazetted water control zones set up by the Environmental Protection Department of the Hong Kong Government (EPDHK, 1990).

In addition to sewage effluents, landfill leachates are another source of nutrient enrichment of seawater. A case study conducted in Hong Kong (Chu et al., 1994) revealed that the principal pollutant in the leachate was ammoniacal nitrogen. The ammoniacal nitrogen concentration in leachate samples from Junk Bay reached 594–1610 mg l^{-1} (mean 1040 mg l^{-1}). The concentration of nitrate/nitrite nitrogen and phosphorus were relatively low in the leachates, but still contributed to the enrichment of the seawater.

The objectives of this study were to understand the present status of water enrichment by anthropogenic activities such as land reclamation and effluent discharge, to understand phytoplankton population dynamics and their relationship to nutrient levels and their ratios, and finally to try to find a linkage between algal blooms and nutrient enrichment.

Description of sites studied

Junk Bay is located to the south of the Kowloon Peninsula, facing Hong Kong Island and the South China Sea (Fig. 1). Three representative stations were chosen for the study: S1 (station 1) located in the west side of the inner bay and surrounded by reclamation and a new housing construction area; S2 (station 2) located in the east side of the inner bay and close to an effluent discharge tunnel; and S3 (station 3) is a station located in the outer bay. The depths of the three stations were 6, 8, and 15 m, respectively.

Materials and methods

Sampling

Sampling was undertaken every 10 days (three times a month) from January to December 1997. Water samples were taken from surface (0.5 m below surface) and bottom (0.5 m above bottom) levels from 3 stations. Water samples were collected using a 3-l Van Dorn-type water sampler. To minimize changes in nutrient concentration, all samples were stored in a cooler with enough dry ice immediately after sample collecting, and the nutrients were measured within 3 h of sampling.

Phytoplankton

Water and net phytoplankton samples were collected for quantitative and qualitative analysis respectively. Water samples were collected at each station using a water sampler and 2 l of water were then concentrated to about 50 ml by filtering through a 10 μm mesh size sieve. Net trawls were made using a 10 μm mesh size phytoplankton net at each station. Samples were preserved immediately after sampling using acidic

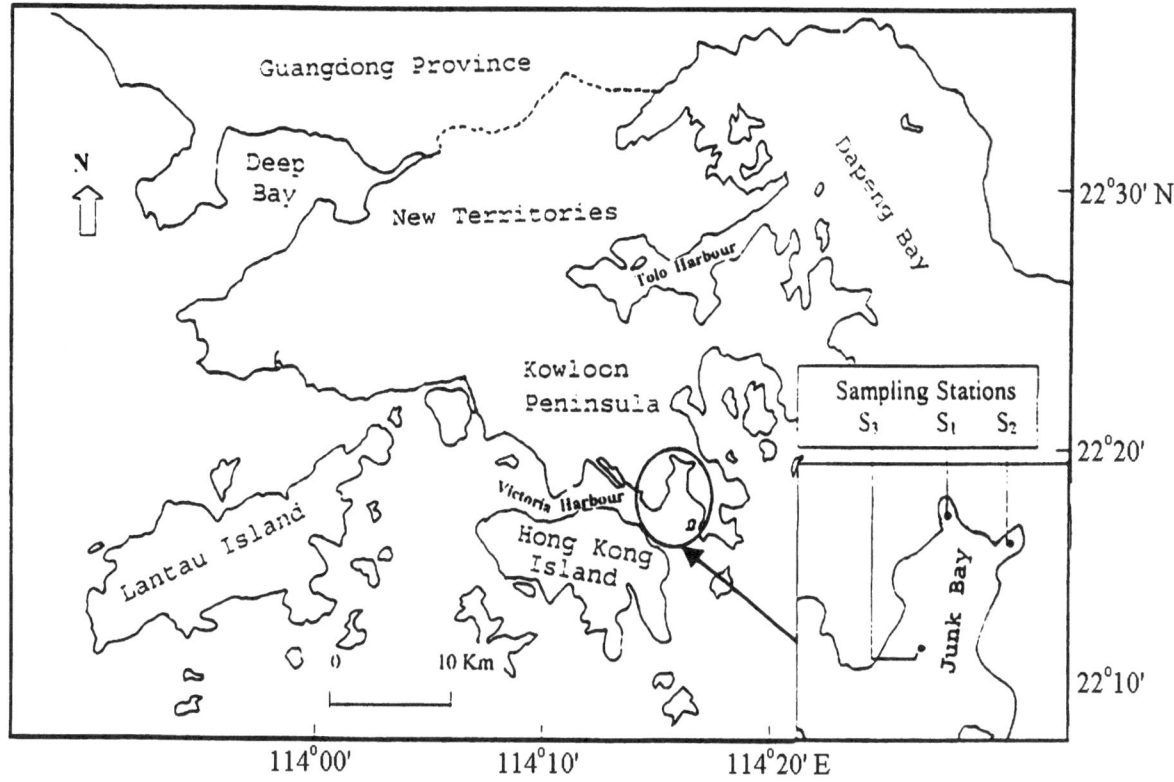

Figure 1. Map showing position of Junk Bay and the sampling stations.

Lugol's solution (Sournia, 1978). Some live phytoplankton samples were also kept for observation of fragile species. Sample processing, storage and concentration followed Sournia (1978). Water samples for quantitative analysis were finally concentrated to 15 ml by sedimentation. A Sedgwick–Rafter counting cell was used for cell counting and a 1 ml concentrated phytoplankton sample was pippetted into the counting chamber and then allowed to stand for 10 min, to permit settling of the phytoplankton before counting with an Olympus inverted microscope (Model IX50-S8F2). The total number of cells of individual species collected at each depth of each station was quantified and the phytoplankton abundance was expressed as number of cells per l sample.

Nutrient analysis

The nutrients in this study were mainly dissolved inorganic nitrogen (DIN), including ammonium-nitrogen (NH_4-N), nitrate-nitrogen (NO_3-N) and nitrite-nitrogen (NO_2-N), and dissolved inorganic phosphorus (DIP), including phosphate-phosphorus (PO_4-P).

The method for analysis of nutrients used in this study was based on the transformation, through a chemical reaction of the substance to be analyzed, to another compound which can be measured colorimetrically within the wavelength range of the visible spectrum. The steps followed Parsons et al. (1984) and Grasshoff (1976). The spectrophotometer employed in this study was a Philips PYE Unicam (Model PU8600). As a general rule, all samples were analyzed as soon as possible after collection and especially when the concentration was expected to be low.

Other Oceanographic parameters such as water depth, water temperature, dissolved oxygen, and salinity were measured *in situ*. The readings of water temperature and dissolved oxygen were taken directly from a YSI Dissolved Oxygen Meter (Model 59) and salinity (in parts per thousand) was measured using an ATAGO Hand Refractometer.

218

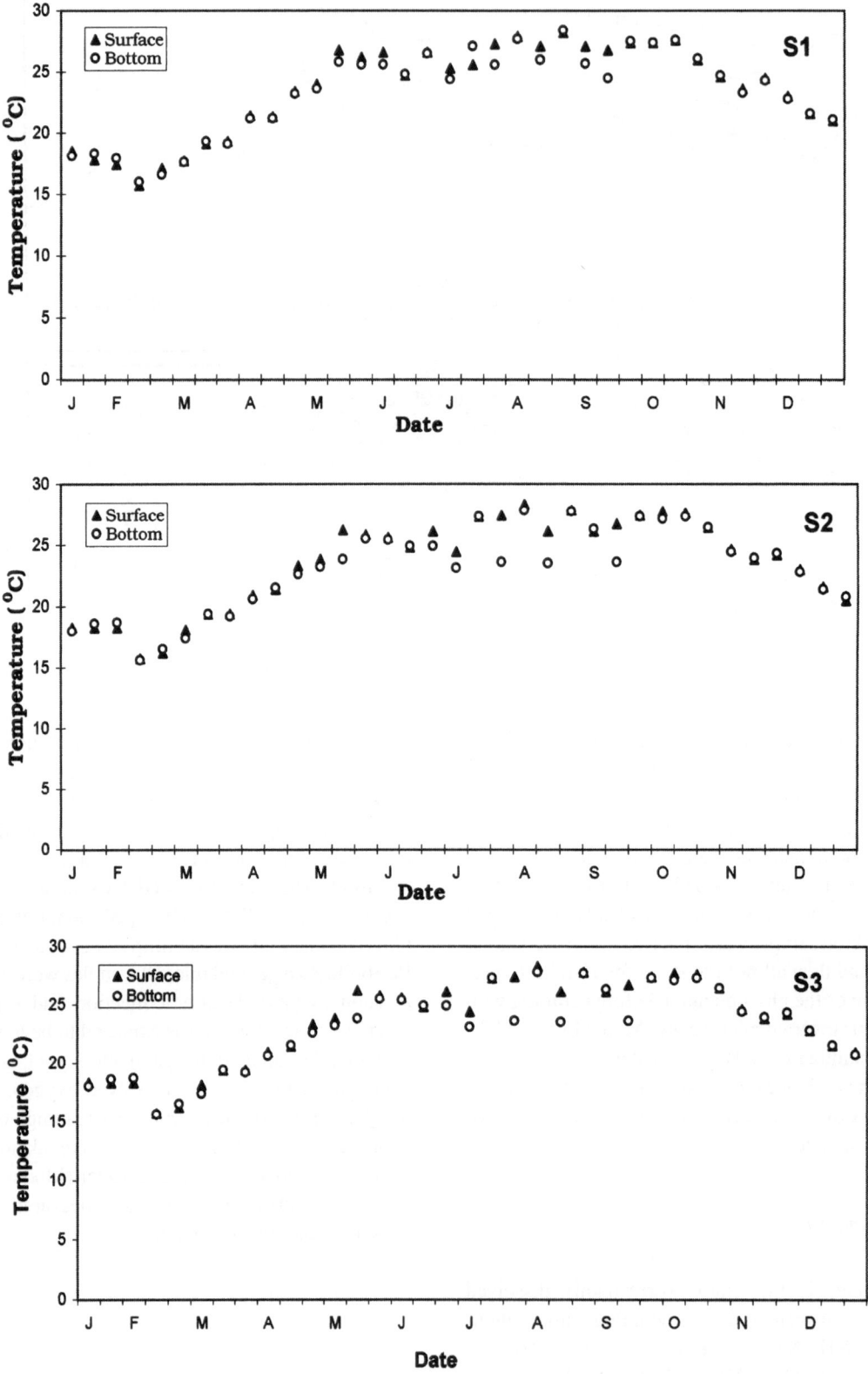

Figure 2. Seasonal variation of water temperature at the 3 sampling stations in Junk Bay in 1997.

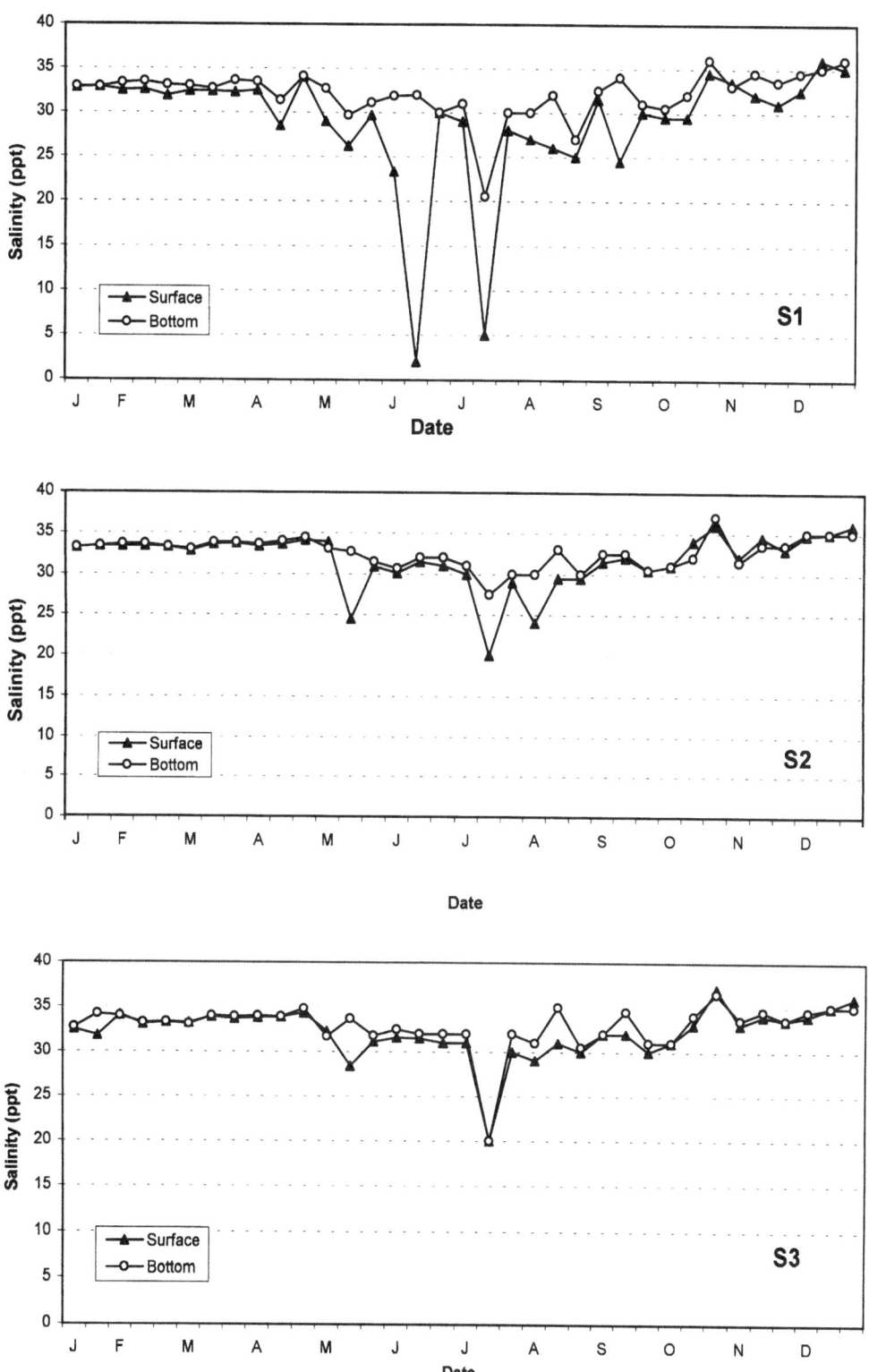

Figure 3. Seasonal variation of salinity at the 3 sampling stations in Junk Bay in 1997.

Results

Physical characters

The annual average water temperature at the 3 sampling stations ranged between 23.11 °C and 24.28 °C. The lowest and highest water temperatures during the study period were 15.6 °C (in March) and 28.9 °C (in August), respectively (Fig. 2). The annual average salinity at the 3 sampling stations ranged from 28.5‰ to 32.7‰. The lowest and highest salinity recorded during the study period was 5.0‰ in the inner bay (station 1) and 37.0‰ in the outer bay (station 3), respectively. The salinity was almost stable, except from June to August during the rainy season (Fig. 3). The annual average dissolved oxygen at the 3 stations ranged between 6.88 mg l^{-1} and 7.69 mg l^{-1}. The lowest and highest values during the study period were 3.02 and 11.94 mg l^{-1}, respectively. The annual average Secchi disc depth at the 3 stations ranged between 1.25 m and 2.96 m. The lowest and highest values during the study period were 0.3 m and 4.8 m, respectively. Because of the relatively shallow water depth (6–15 meters), as well as flushing by tides and currents, no thermal stratification occurred in Junk Bay.

Nutrients and their ratios

The summary of the annual average and the range of concentrations for all nutrients are listed in Table 1. DIN (DIN=NH_4+NO_2+NO_3) in the water column of the three stations ranged from a low of 0.007 (surface-station 3) to a high of 8.579 mg l^{-1} (bottom-station 2). The highest annual average concentration (0.97 mg l^{-1}) was in the surface water of station 2, which was close to an effluent discharge tunnel, and the lowest value was 0.2 mg l^{-1} in the surface water of station 3, which is the offshore station. Figure 4 shows the DIN in the surface and bottom waters of the three stations. It indicates that the DIN was highest from July to September, especially in stations 1 and 2, which are nearshore stations, station 1 receiving land runoff from a landfill construction site and station 2 receiving effluents from a discharge tunnel. This might have resulted from heavy summer rainfall (Fig. 8), which would bring more land runoff (and thus wastes) to these coastal waters. The differences of DIN between surface and bottom water at station 1 and station 2 were significant ($p<0.05$), but were not significant at station 3. Among the three forms of nitrogen (NH_4, NO_2 and NO_3), NH_4-N was the major element. It comprised 65% (station 1) to 84% (station 2) of the total DIN.

Phosphate (PO_4-P) concentrations in Junk Bay ranged from almost non-detectable (station 3) to 0.151 mg l^{-1} (surface-station 2). Annual variation of phosphate at the three sampling stations is shown in Figure 5. The highest annual mean phosphate concentration was in the surface water of station 1 (0.036 mg l^{-1}), and the lowest was in the bottom water at station 3 (0.023 mg l^{-1}) (Table 1). The phosphate concentrations in bottom waters were relatively stable, with average values ranging from 0.020 to 0.027 mg l^{-1}. Statistical analysis showed that the phosphate concentrations in the surface water were significantly higher than at the bottom in station 1 ($P<0.05$), but there were no significant differences at stations 2 and 3.

Overall, comparing the offshore station (station 3) with the nearshore stations (station 1 and station 2), the concentrations of DIN and phosphate in station 3 were significantly lower than at stations 1 and 2 ($P<0.05$), with the highest annual mean DIN concentration in station 2 and highest annual mean phosphate concentration in station 1 (Table 1). DIN concentrations in Junk Bay fluctuated widely (from 0.007 to 8.579 mg l^{-1}), with higher concentration in summer from June to September (Fig. 4). Whereas phosphate concentrations did not vary significantly (except some of those at station 1), and there was no clear seasonal trend (Fig. 5).

With the exception of the surface water at station 2, where the annual average N:P (DIN to PO_4-P) ratio (atomic) was 31.9, the ratios at each depth of each of the other stations were all less than the Redfield Ratio of 16 (Table 1). The annual N:P ratio data (Fig. 6) indicated that the ratios were usually higher than the Redfield ratio in the Summer months of May to September, and less than the Redfield ratio for most time of the other months. The N:P ratio (atomic) shows that Junk Bay, like most other marine waters, is generally nitrogen limited. The only exception was in the surface water at station 2, where 60% of the samples were higher than the Redfield ratio, and were thus phosphorus limited. This coincided with the high inputs of nitrogen at this station.

Phytoplankton abundance

The annual variation in total phytoplankton abundance is given in Figure 7. The highest annual average phytoplankton abundance (1.07×10^7 cells l^{-1}) was in the surface water of station 3, whereas station 2 had the

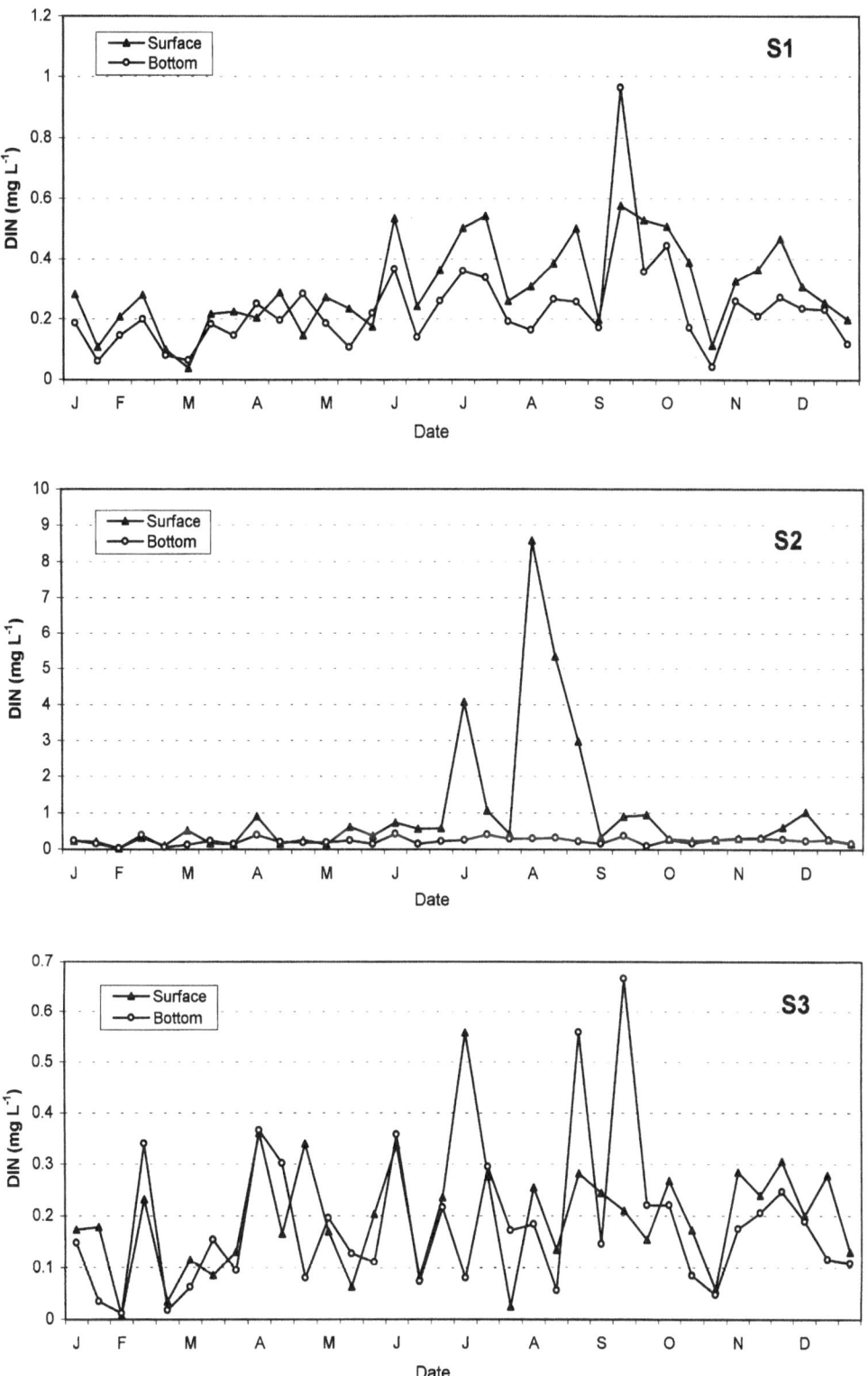

Figure 4. Seasonal variation of dissolved inorganic nitrogen (DIN) at the 3 sampling stations in Junk Bay in 1997.

222

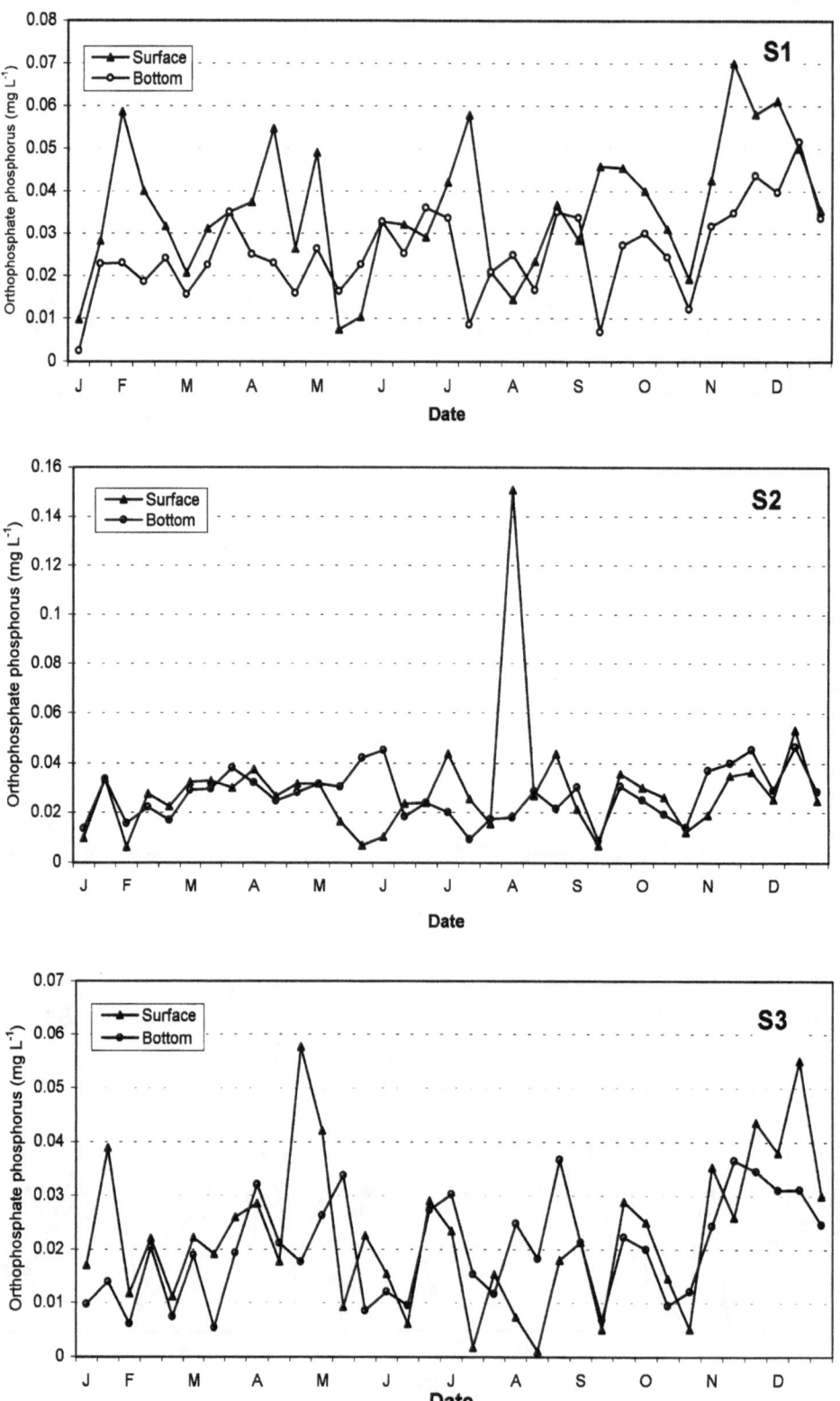

Figure 5. Seasonal variation of phosphate-phosphorus at the 3 sampling stations in Junk Bay in 1997.

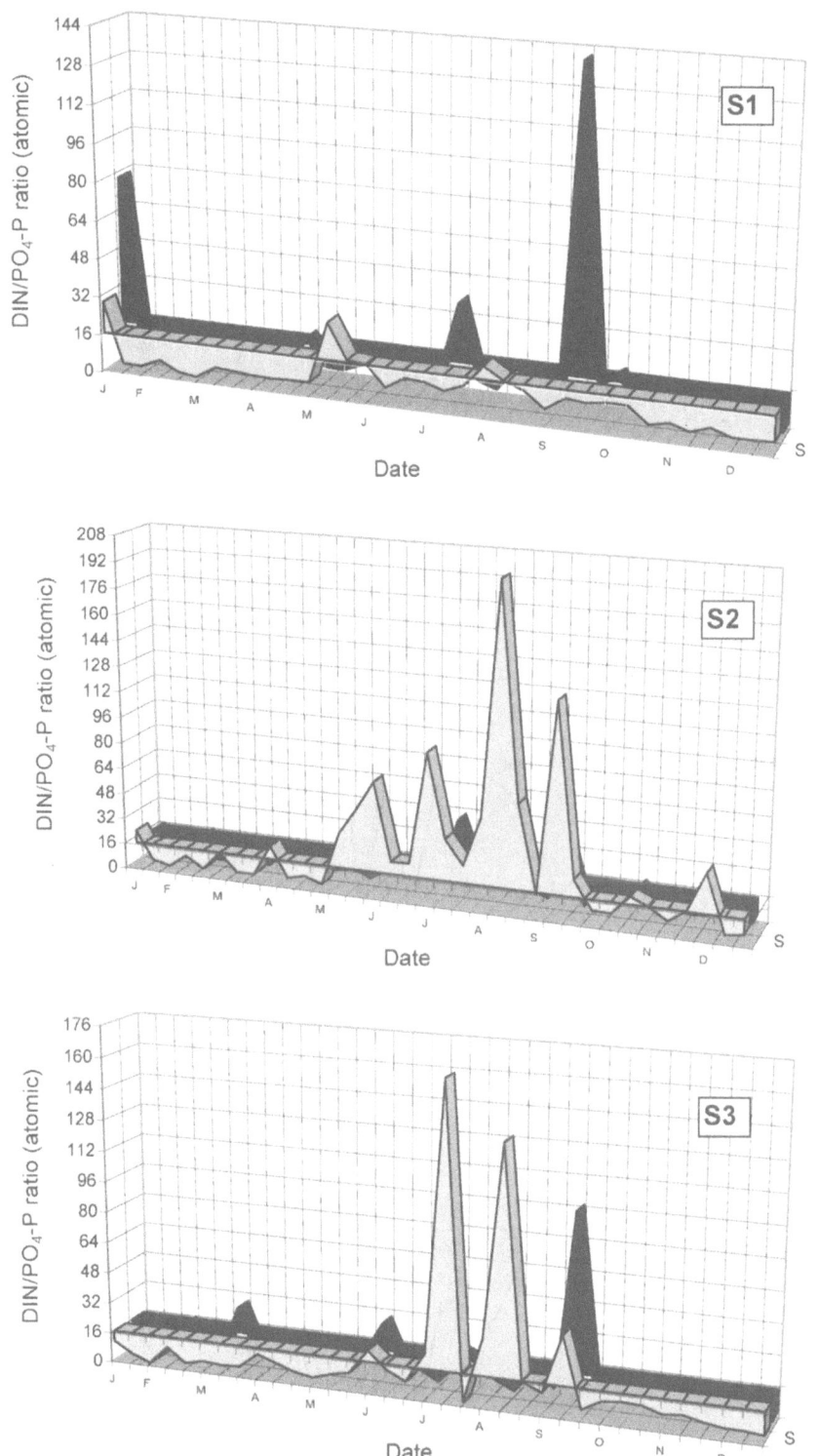

Figure 6. Seasonal variation of N:P ratio (atomic) at the 3 sampling stations in Junk Bay in 1997 (Solid area = bottom water; stippled area = surface water).

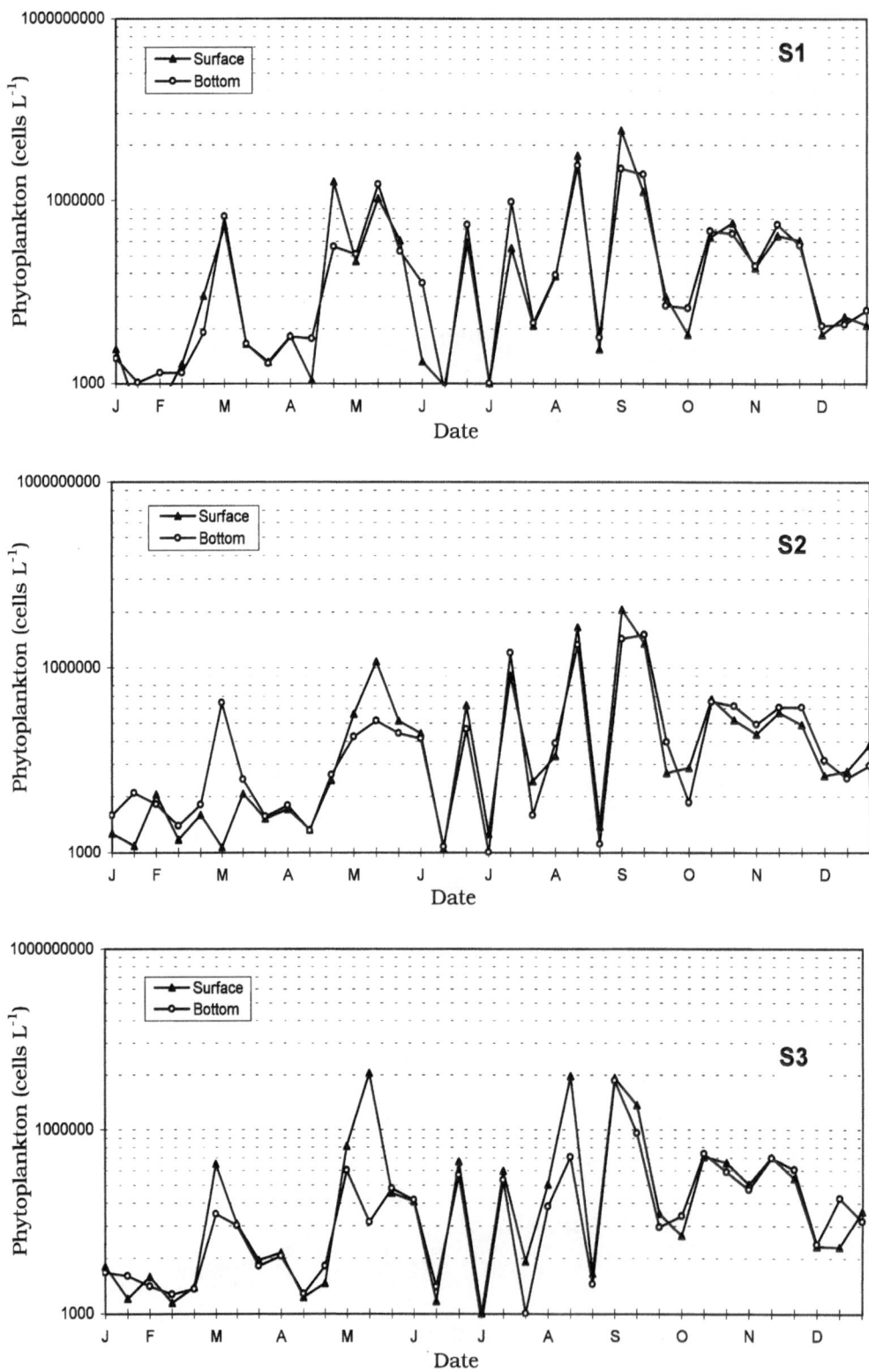

Figure 7. Seasonal variation of phytoplankton numbers at the 3 sampling stations in Junk Bay in 1997.

Table 1. Mean and range (in brackets) of nutrient concentrations and phytoplankton abundance in Junk Bay, Hong Kong

Measurements	Station 1		Station 2		Station 3	
	Surface	Bottom	Surface	Bottom	Surface	Bottom
NH_4-N (mg l^{-1})	0.197	0.156	0.811	0.155	0.151	0.135
	(0.026~0.431)	(0.007~0.60)	(0.014~8.364)	(0.001~0.28)	(0.001~0.485)	(0.001~0.662)
NO_2-N (mg l^{-1})	0.0133	0.0131	0.0186	0.0093	0.0075	0.0067
	(0~0.0374)	(0~0.10)	(0~0.114)	(0~0.044)	(0~0.054)	(0~0.054)
NO_3-N (mg l^{-1})	0.095	0.0616	0.1145	0.0555	0.0392	0.0413
	(0.0003~0.47)	(0.0001~0.35)	(0.0002~0.93)	(0.0001~0.31)	(0.0001~0.213)	(0.0001~0.32)
DIN (mg l^{-1})	0.304	0.232	0.97	0.226	0.20	0.184
	(0.038~0.576)	(0.04~0.963)	(0.134~8.579)	(0.023~0.423)	(0.007~0.558)	(0.011~0.666)
PO_4-P (mg l^{-1})	0.036	0.026	0.03	0.027	0.023	0.020
	(0.007~0.07)	(0.0024~0.052)	(0.006~0.151)	(0.008~0.046)	(0.001~0.058)	(0.005~0.037)
N/P Ratio	10.1	14.6	31.9	10.6	15.14	11.84
	(1.8~31.6)	(2.6~77.6)	(3.4~199.0)	(1.5~46.1)	(0.6~162.5)	(2.0~98)
Phytoplankton abundance (cells l^{-1})	9.64×10^5 (333~ 1.4×10^7)	5.38×10^5 (695~ 3.69×10^6)	7.12×10^5 (1135~ 8.75×10^6)	4.36×10^5 (1000~ 3.39×10^6)	1.07×10^6 (1490~ 8.58×10^6)	3.32×10^5 (1000~ 6.46×10^6)

lowest value (7.12×10^5 cells l^{-1}) among the surface waters of the three stations (compare this with the fact that the highest nitrogen concentration was at this station). Phytoplankton abundance was higher in the summer than in other seasons (Fig. 7). The dominant phytoplankton groups in Junk Bay were diatoms, and then dinoflagellates. The chlorophyceae, cryptophyceae, cynobacteria, dictyophyceae, and raphidophyceae were minor groups. The diatom species were dominant almost year round, usually making up more than 90% of the total abundance. The abundance of dinoflagellate was relatively higher in the Spring than in other seasons.

The dominant taxa during the study period were: *Skeletonema costatum*, (Grev.) Cleve *Asterionella japonica* Cleve, *Chaetoceros* spp. *Cylindrotheca closterium* (Ehrenberg) Lewin *et* Reimann, *Bacillaria paxillifera* (Müller) Hendey, *Pseudonitzschia* spp. *Thalassiosira mala* Takano, *T. rotula*, Meunier *Thalassionema nitzschioides* (Grunow) Grunow *ex* Hustedt, *Prorocentrum micans* Ehrenberg, *P. minimum* (Pavillard) Schiller, *P. triestinum,*Schiller and*Ceratium furca* Ehrenberg (Claparede *et* Lachmann). Toxic species were also recorded (Lu & Hodgkiss, 2004).

Discussion

Eutrophication

The term 'eutrophication' is complicated, and so far there have been about a dozen different definitions. However, it is clear that the most common single factor causing eutrophication in coastal marine ecosystems is an increase in the amounts of nitrogen and phosphorus they receive, and one of its effects may increase the productivity of the ecosystem. Nixon (1995) proposed a definition and trophic classification standard for eutrophication based on organic matter supply. He emphasized that "eutrophication is a process, not a trophic state". Based on this, as well as the results of the present study, we can say that Junk Bay is now experiencing eutrophication. A long term environment monitoring program carried out by the Environmental Protection Department (EPD) of the Hong Kong S.A.R. Government revealed that there was a significantly increasing trend for PO_4-P and NH_4-N over the 10 years from 1988 to 1997 (EPDHK, 1998). Junk Bay is a nitrogen limited water, and thus, the enrichment of nitrogen is the most important contribution to this water. Normally, nitrogen is supplied to coastal waters by riverine and land runoff, precipitation, atmospheric resources (nitrogen fixation), offshore waters (upwelling), and waste effluents. The most important nitrogen inputs to Junk Bay are land runoff and waste effluents. Because no river enters the bay, and the cynobacterial population in the bay is a minor group, riverine input and nitrogen fixation in the bay are not important. Nutrient enrichment of the inner bay was higher than that of the outer bay. As well as groundwater, landfill leachate is an important source of nutrients. One of the surveys conducted in Junk Bay (Cheung et al., 1997) showed the nutrient enrichment contributed by landfill leachate. The present study also revealed that phosphorus was highest at station 1, and this is probably from landfill leachate and runoff from the naked land flushing.

Even though there are no detailed data concerning either inputs of leachate and runoff from landfill construction to station 1, or the effluents to station 2, the two nearshore stations were heavily affected by land runoff and effluents, especially the surface waters and the average N and P values in the surface waters of these two stations were higher than at station 3, the outer bay station. The highest values were all found in summer, when there were heavy rainstorms in this subtropical area (Fig. 8). The heavy rain flushed the naked land of the landfill area in station 1, and a large amount of land runoff, as well as effluents, were discharged from the tunnel in station 2. This resulted in the concentration of N and P in these two stations being even higher. The data also show that the deeper waters were less affected by runoff. The differences between N and P at the three stations were not, however, statistically significant.

Nutrients and phytoplankton

Phytoplankton growth depends largely on the availability of inorganic nutrients. Thus, nutrient enrichment results in an increase of phytoplankton productivity and population selection. A good example is in Tolo Harbour, which is the most heavily polluted embayment in Hong Kong, and where the relationship between eutrophication and phytoplankton has been well studied (Hodgkiss & Chan, 1983, 1987; Chan & Hodgkiss, 1987). The same result was also shown in Tai Tam Bay, Hong Kong (Chan et al., 1991; Chiu et al., 1994).

Junk Bay is another example where high nutrient levels have resulted in high phytoplankton abundance. From the point of view of seasonal dynamics, the highest phytoplankton numbers coincided with the highest nutrient enrichment in summer. One explanation for higher diatom abundance (over 90% all over the year) and lower abundance of other algal groups is that diatoms are more sensitive to nutrient enrich-

ment than the others. Studies have shown that diatoms have high growth rates under nutrient-rich conditions (Eppley, 1977), compared to the generally lower growth rates of dinoflagellates. Experimental study from Sweden has also indicated that river water draining from forest areas (containing rich humic and fulvic acids) can stimulate dinoflagellate bloom such as *Prorocentrum minimum*, whereas the river water draining from agricultural soils (containing rich inorganic N and P) stimulates diatom blooms (Granéli & Moreira, 1990). The higher concentrations of inorganic N and P in Junk Bay may result in the higher abundance of diatoms. Unfortunately, there is a lack of long term phytoplankton data from this bay and so algal population succession and species selection under different nutrient status conditions remains unclear.

Nutrients affect not only phytoplankton numbers, but also toxin production by some algal species. In the case of *Alexandrium tamarense* (Lebour) Balech, *Pseudo-nitzschia pungens f. multiseries* (Hasle) and *Chrysochromulina polylepis* it has been proven that the cells increase their toxin content considerably when grown in a P-deficient (excess nitrogen) medium and in the stationary phase (Anderson et al., 1990; Carlsson et al., 1990; Bates et al., 1991). Laboratory experiments also showed that *Prorocentrum lima* increased its toxin production when grown under nitrogen limited conditions (McLachlan et al., 1994). There were many toxic species recorded in Junk Bay (Lu & Hodgkiss, 2003), although there were no toxic blooms recorded during the study. The production of toxins and their relationship with nutrients needs to be further studied.

N/P ratio, phytoplankton biomass and population composition

Response of the phytoplankton to nutrients is considerably influenced by its physiological state when the nutrients are in short supply. It has been demonstrated that algae increase their uptake rates for NH_4^+ when N starved and for PO_4^{3-} when P starved. Redfield (1958) reported atomic ratios of available nitrogen to phosphorus of 15:1 in seawater, depletion of nitrogen and phosphorus in the ratio of 15:1 during phytoplankton growth, and ratios of 16:1 for laboratory analyses of phytoplankton. This ratio was subsequently called the Redfield ratio. It is now agreed that freshwater (especially lake) phytoplankton growth is limited primarily by phosphorus, and marine phytoplankton growth by nitrogen (Dugdale, 1967; Hecky & Kil-

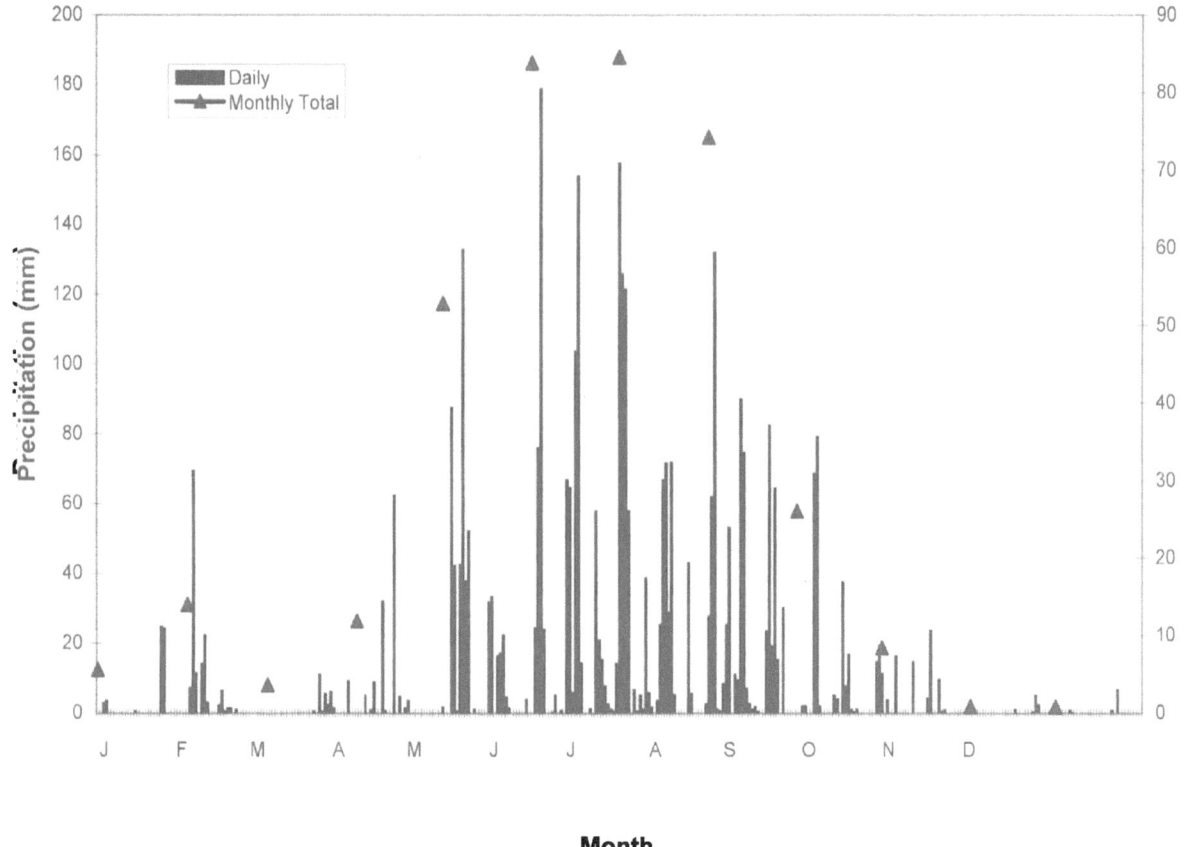

Figure 8. Annual precipitation in the Junk Bay area (from the Hong Kong Observatory).

ham, 1988). This means that N:P values are normally more than the Redfield ratio in freshwater, and less in seawater. Ryther & Dunstan (1971) indicated that 'although there is no indication of any normal or optimal nitrogen to phosphorus ratio in algae, values between 5:1 and 15:1 are most commonly encountered and an average ratio of 10:1 is therefore a reasonable working value'. Hodgkiss & Ho (1997) experimentally determined the optimal N:P ratio for various algal species. Most of the values on their list are within the range 5:1 to 15:1. In Junk Bay, 71%, 26%, and 60% of the samples had N:P ratios between 5:1 and 15:1 at stations 1, 2 and 3, respectively. Conversely 60 percent of the samples at station 2 had an N:P ratio larger than 15:1 and this is probably the reason why the higher nitrogen concentration at station 2 resulted in the lower phytoplankton numbers (phosphorus limited). Comparing with stations 1 and 2, the station 3 (surface water) has a favourable annual mean N:P ratio (15.14). It might be one of the reasons that station 3 has a lower nutrient level but higher phytoplankton abundance.

The nutrient supply and its ratios have a decisive effect on the species composition of the phytoplankton since different algal species have different nutrient requirements. Although the optimal nutrient requirement for each algal species is not well-known, diatoms as a group have an obligate requirement for silica, and many blue-green bacteria can fix molecular nitrogen, making them potentially superior competitors under low nitrogen conditions (Granéli et al., 1990). A good example are the nitrogen fixing cynobacteria, *Nodularia spumigena* Mertens, *Aphanizomenon flos-aquae* (L.) Ralfs *ex* Bornet *et* Flahault *Anabaena lemmermanni* (Skuja) Cronberg *et* Komárek in the Baltic, which usually develop blooms in late summer when DIN is low in surface water (Granéli & Granéli, 1982). On the other hand, dinoflagellates may show stronger competition under nutrient deficient conditions, because of their diurnal migrations to deeper nutrient-rich layers (Granéli & Moreira, 1990). In Junk Bay, the phosphorus level is relatively stable year round, whereas the nitrogen level is highest in the Summer and lowest in the Spring. Diatom numbers were highest in the summer, whereas the population diversity and numbers of dinoflagellates were higher in the Spring, showing, therefore, that their relative abundance is closely related to the nutrient ratio in Junk Bay.

Acknowledgements

We are grateful to Miss Lo Gar Yee for kindly assisting in some sampling and nutrient analysis.

References

Amano, K., M. Watanabe, K. Kohata & S. Harada, 1998. Conditions necessary for *Chattonella antiqua* red tide outbreaks. Limnol. Oceanogr. 43: 117–128.

Anderson, D. M., 1989. Toxic algal blooms and red tides: a global perspective. In Okaichi, T., D. M. Anderson & T. Nemoto (eds), Red Tides: Biology, Environmental Science, and Toxicology. Proceedings of 1st Int. Symp. on Red Tides, Takamatsu, Japan, 10–14 Nov. 1987. Elsevier, New York, Amsterdam, London: 11–16.

Anderson, D. M., D. M. Kulis, J. J. Sullivan, S. Hall & C. Lee, 1990. Dynamics and physiology of saxitoxin production by the dinoflagellate *Alexandrium* spp. Mar. Biol. 104: 511–524.

Baden, D. G. & V. L. Trainer, 1993. Mode of action of toxins of seafood poisoning. In Falconer, I. A. (ed.), Algal Toxins in Seafood and Drinking Water. Academic Press, London, San Diego, New York, Boston, Sydney, Tokyo, Toronto: 49–74.

Bates, S. S., A. S. W. de Freitas, J. E. Milley, R. Pocklington, M. A. Quilliam, J. C. Smith & J. Worms, 1991. Controls on domoic acid production by the diatom *Nitzschia pungens f. multiseries* in culture: nutrients and irradiance. Can. J. Fish. aquat. Sci. 48: 1136–1144.

Carlsson, P., E. Granéli & P. Olsson, 1990. Grazer elimination through poisoning: one of the mechanisms behind *Chrysochromulina polylepis* bloom? In Granéli E., L. Edler & D. M. Anderson (eds), Toxic Marine Phytoplankton. Elsevier, New York, Amsterdam, London: 116–122.

Chan, B. S. S. & I. J. Hodgkiss, 1987. Phytoplankton productivity in Tolo Harbour. Asian mar. Biol. 4: 79–90.

Chan, B. S. S., M. C. Chiu & I. J. Hodgkiss, 1991. Plankton dynamics and primary productivity of Tai Tam Bay, Hong Kong. Asian mar. Biol. 8: 169–192.

Cheung, K. C., L. M. Chu & M. H. Wong, 1997. Ammonia stripping as a pretreatment for landfill leachate. Wat. Air Soil Pollut. 94: 209–221.

Chiu, M. C., I. J. Hodgkiss & B. S. S. Chan, 1994. Ecological studies of phytoplankton in Tai Tam Bay, Hong Kong. Hydrobiologia 273: 81–94.

Chu, L. M., K. C. Cheung & M. H. Wong, 1994. Variations in the chemical properties of landfill leachate. Envir. Manage. 18: 105–117.

De Jonge, V. N. & W. Van Raaphorst, 1995. Eutrophication of the Dutch Wadden Sea (western Europe), an estuarine area controlled by the River Rhine. In McComb, A. J. (ed.), Eutrophic Shallow Estuaries and Lagoons. CRC Press. 129–149.

Dugdale, R. C., 1967. Nutrient limitation in the sea: dynamics, identification, and significance. Limnol. Oceanogr. 12: 685–695.

EPDHK (Environmental Protection Department, Hong Kong), 1990. Marine Water Quality in Hong Kong-1989. Water Policy Group, Environmental Protection Department of Hong Kong Government, Hong Kong.

EPDHK (Environmental Protection Department, Hong Kong), 1998. Marine Water Quality in Hong Kong in 1997. Hong Kong S.A.R. Government, Hong Kong.

Eppley, R. W., 1977. The growth and culture of diatoms. In Werner, D. (ed.), The Biology of Diatoms. Bot. Monogr. 13: 24–64.

Fisher, T. R., E. R. Peele, J. W. Ammerman & L. W. Harding Jr., 1992. Nutrient limitation of phytoplankton in Chesapeake Bay. Mar. Ecol. Prog. Ser. 82: 51–63.

Granéli, E. & W. Granéli, 1982. Eutrophication and dinoflagellate blooms in Swedish coastal waters – possible causes and countermeasures. In Figueiredo, M. R. C., N. L. Chao & W. Kirby-Smith (eds), Proceedings of the International Symposium on Utilization of Coastal Ecosystems: Planning, Pollution and Productivity, 21–27 Nov. 1982, Rio Grande, Brazil. Editora Da Furg.: 261–282.

Granéli, E. & M. O. Moreira, 1990. Effects of river water of different origin on the growth of marine dinoflagellates and diatoms in laboratory cultures. J. exp. mar. Biol. Ecol.136: 89–106.

Granéli, E., K. Wallström, U. Larsson, W. Granéli & R. Elmgren, 1990. Nutrient limitation of primary production in the Baltic Sea area. Ambio 19: 142–151.

Grasshoff, K. (ed.), 1976. Methods of Seawater Analysis. Verlag Chemie, Weinheim, New York.

Hallegraeff, G. M., 1993. A review of harmful algal blooms and their apparent global increase. Phycologia 32: 79–99.

Hecky, R. & P. Kilham, 1988. Nutrient limitation of phytoplankton in freshwater and marine environments: a review of recent evidence on the effects of enrichment. Limnol. Oceanogr. 33: 796–822.

Hodgkiss, I. J. & B. S. S. Chan, 1983. Pollution studies on Tolo Harbour, Hong Kong. Mar. Envir. Res. 10: 1–44.

Hodgkiss, I. J. & B. S. S. Chan, 1987. Phytoplankton dynamics in Tolo Harbour. Asian mar. Biol. 4: 103–112.

Hodgkiss, I. J. & K. C. Ho, 1997. Are changes in N:P ratios in coastal water the key to increased red tide blooms? Hydrobiologia 352: 141–147.

Honjo, T., 1993. Overview on bloom dynamics and physiological ecology of *Heterosigma akashiwo*. In Smayda, T. J. & Y. Shimizu (eds), Toxic Phytoplankton Blooms in the Sea. Elsevier, Amsterdam, London, New York, Tokyo: 33–41.

Keller, A. A. & R. L. Rice, 1989. Effects of nutrient enrichment on natural populations of the brown tide phytoplankton *Aureococcus anophagefferens* (Chrysophyceae). J. Phycol. 25: 636–646

Kelly, M. & M. Naguib, 1984. Eutrophication in coastal marine areas and lagoons: a case study of 'Lac de Tunis'. Unesco Rep. mar. Sci. 29: 1–41.

Lam, C. W. Y. & K. C. Ho, 1989. Red tides in Tolo Harbour, Hong Kong. In Okaichi, T. D. M. Anderson & T. Nemoto (eds), Red Tides: Biology, Environmental Science and Toxicity, Proc. 1st Int. Symp. On Red Tides, Japan, 1987. Elsevier, New York, Amsterdam, London: 49–52.

Larsson, U., R. Elmgren & F. Wulff, 1985. Eutrophication and the Baltic Sea: Causes and consequences. Ambio 14: 9–14.

Lu, S. H. & I. J. Hodgkiss, 1999. An unusual year for the occurrence of harmful algae. Harmful Algal News 18: 1 & 3.

Lu, S. H. & I. J. Hodgkiss, 2004. Harmful algal bloom causative collected from Hong Kong waters. Hydrobiologia 511/Dev. Hydrobiol. 512: 229–236.

McLachlan, J. L., J. C. Marr, A. Conlon-Kelly & A. Adamson, 1994. Effects of nitrogen concentration and cold temperature on DSP-toxin concentrations in the dinoflagellate *Prorocentrum lima* (Prorocentrales, Dinophyceae). Natural Toxins 2: 263–270.

Maestrini, S. Y. & E. Granéli, 1991. Environmental conditions and ecophysiological mechanisms which led to the 1988 *Chrysochromulina polylepis* bloom: an hypothesis. Oceanol. Acta. 14: 397–413.

Mee, L. D., 1992. The Black Sea in crisis: a need for concerted international action. Ambio 21: 278–286.

Nixon, S. W., 1995. Coastal marine eutrophication: a definition, social causes, and future concerns. Ophelia 41: 199–219.

Nixon, S. W., 1997. Prehistoric nutrient inputs and productivity in Narragansett Bay. Estuaries 20: 253–261.

Paetsch, J. & G. Radach, 1997. Long-term simulation of the eutrophication of the North Sea: temporal development of nutrients, chlorophyll and primary production in comparison to observations. J. Sea Res. 38: 275–310.

Parsons, T. R., Y. Marita & C. M. Lalli (eds), 1984. A Manual of Chemical and Biological Methods for Seawater Analysis. Pergamon Press, Oxford.

Redfield, B. C., 1958. The biology control of chemical factors in the environment. Am. Sci. 46: 205–221.

Riegman, R., A. A. M. Noordeloos & G. C. Cadée, 1992. *Phaeocystis* blooms and eutrophication of the continental coastal zones of the North Sea. Mar. Biol. 112: 479–484.

Rosenberg, R., I. Cato, L. Foerlin, K. Grip & J. Rodhe, 1996. Marine environment quality assessment of the Skagerrak-Kattegat. J. Sea Res. 35: 1–8.

Ryther, J. H. & W. M. Dunstan, 1971. Nitrogen, phosphorus, and eutrophication in the coastal marine environment. Science 171: 1008–1013.

Smayda, T. J., 1990. Novel and nuisance phytoplankton blooms in the sea: evidence for a global epidemic. In Granéli, E., L. Edler & D. M. Anderson (eds), Toxic Marine Phytoplankton. Elsevier, New York, Amsterdam, London: 29–40.

Sournia, A., (ed.), 1978. Phytoplanton Manual. UNESCO: Monographs on oceanographic methodology 6. 1–337.

Stirn, J., 1988. Eutrophication in the Mediterranean. In UNESCO (ed.), Eutrophication in the Mediterranean Sea: Receiving Capacity and Monitoring of Long-term Effects. Report and Proceedings of a Scientific Workshop, Bologna, Italy, 2–6 March 1987. UNESCO Reports in Marine Science 49: 161–188.

Yamanaka, Y., 1983. Present state and problems of water pollution in Japan. Bull Jap. Soc. Fish. Oceanogr. 43: 77–80.

Yasumoto, T., 1990. Marine microorganisms toxins-an overview. In Granéli, E., L. Edler & D. M. Anderson (eds), Toxic Marine phytoplankton, Proc. 4th Int. Conf. on Toxic Marine Phytoplankton, Sweden, 1989. Elsevier, New York, Amsterdam, London: 3–10.

Hydrobiologia **512**: 231–238, 2004.
P.O. Ang, Jr. (ed.), Asian Pacific Phycology in the 21ˢᵗ Century: Prospects and Challenges.
© 2004 *Kluwer Academic Publishers.*

Harmful algal bloom causative collected from Hong Kong waters

Songhui Lu[1,2] & I. J. Hodgkiss[1,*]
[1]*Department of Ecology & Biodiversity, The University of Hong Kong, Pokfulam Road, Hong Kong SAR, China*
[2]*Institute of Hydrobiology, Jinan University, Guangzhou 510632, China*
Author for correspondence; E-mail: hodgkiss@hkucc.hku.hk

Key words: dinoflagellates, harmful algal blooms, red tides, harmful species, Junk Bay, Hong Kong

Abstract

Harmful algal blooms (HABs) have increased globally in recent years. In Hong Kong, a record algal bloom, caused by *Gymnodinium mikimotoi* and *Gyrodinium* sp. HK'98 (subsequently described as *Karenia digitata*) occurred in March and April 1998. Almost all fishes died in the affected cages, and the estimated economic loss caused by the HAB was HK$315 000 000 (equivalent to US $40 000 000). Most of the known toxic or harmful algal species are dinoflagellates. Some common dinoflagellate species such as *Ceratium furca, Gonyaulax polygramma, Noctiluca scintillans, Heterocapsa triquetra, Prorocentrum minimum, Prorocentrum sigmoides,* and *Prorocentrum triestinum* frequently bloom in Hong Kong waters. Others, such as *Alexandrium catenella, Alexandrium tamarense, Gymnodinium mikimotoi, Gymnodinium* cf. *breve, Gymnodinium catenatum, Dinophysis caudata, Dinophysis acuminata,* and *Gambierdiscus toxicus* bloom only occasionally, but their toxic effects or potentially toxic and harmful effects are very significant. Some important toxic and harmful, or potentially toxic and harmful dinoflagellate species are described. Among them, *Gambierdiscus toxicus*, a potential ciguatera fish poison producing species, and *Gymnodinium* cf. *breve*, a neurological shellfish poison producing species were new records from Hong Kong waters.

Introduction

Sournia (1995) summarized that out of a total number of between 3365 and 4024 marine phytoplankton species, about 184-267 species (~6%) are responsible for algal blooms (or red tides), including diatoms, dinoflagellates, raphidophytes, prymnesiophytes, and silicoflagellates, and about 60–78 species (~2%) are toxic. Among these 60–78 toxic species, 45–57 are dinoflagellates. It is thus clear that most of the toxic species (73–75%) are dinoflagellates.

Dinoflagellates are the most important group of marine phytoplankton producing both toxic and harmful algal blooms (Steidinger, 1983, 1993; Anderson, 1989; Hallegraeff, 1993). The negative effects of dinoflagellates include the biotoxins, physical damage, and the anoxia or hypoxia they produce. So far, five major groups of toxins produced by marine phytoplankton and their effects have been identified. They are amnesic shellfish poisoning (ASP) (Bates et al., 1989);

diarrhetic shellfish poisoning (DSP) (Yasumoto et al., 1980); neurological shellfish poisoning (NSP) (Baden & Mende, 1982); paralytic shellfish poisoning (PSP) (Shimizu, 1979); and ciguatera fish poisoning (CFP) (Yasumoto et al., 1977). Of these five groups of toxins, four (DSP, NSP, PSP, CFP) are mainly produced by dinoflagellates (Shumway, 1990; Yasumoto, 1990). Beside toxins, mass accumulation of dinoflagellate cells can also cause physical damage or clog fish and invertebrate gills, for example *Noctiluca scintillans, Gymnodinium mikimotoi, Gymnodinium sanguineum,* and *Cochlodinium polykrikoides*; or cause depletion of dissolved oxygen, such as *Gonyaulax polygramma,* and *Scrippsiella trochoidea* (Hallegraeff, 1993).

In Hong Kong, as elsewhere, dinoflagellate blooms are the most harmful. The first harmful algal bloom (HAB) caused by a dinoflagellate (*Noctiluca scintillans*) was recorded in the south bays of Hong Kong (Morton & Twentyman, 1971). In 1988, a continuous bloom of *Gonyaulax polygramma* lasted for three and

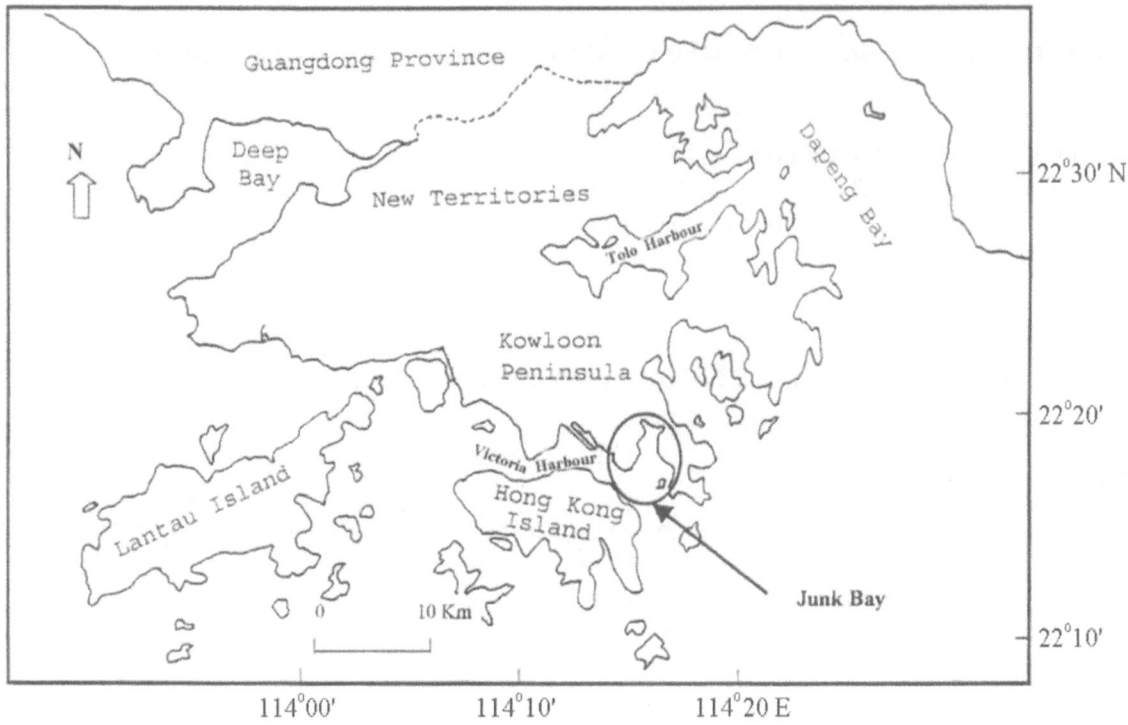

Figure 1. Map of Hong Kong and location of Junk Bay.

a half months from early February to May in Tolo Harbour. The collapse of the bloom in May caused the whole of the Tolo Harbour waterbody to become anoxic, resulting in a massive fish kill in the fish culture zone (Lam & Yip, 1990). The first toxin-related HAB caused by *Alexandrium catenella* was recorded in inner Junk Bay in 1989 (Ho & Hodgkiss, 1993). The toxicity of PSP in green-lipped mussel increased from 2280 MU kg^{-1} meat when the bloom was first detected to 13500 MU kg^{-1} meat within a week (EPDHK, 1990).

A review of HAB occurrences in Hong Kong from 1975 to 1986 (Wong, 1989) showed that of a total of 26 identified HAB causative species, 17 were dinoflagellates (65%). In a total of 133 HAB incidents, 102 were caused by dinoflagellates (77%), and almost all of the fish kills were caused by dinoflagellates.

In Spring 1998, a record bloom of *Gymnodinium/Karenia* hit Hong Kong waters. About two thirds of the mariculture farms (estimates are 1000 out of 1500) were affected. Almost all fish died in the affected cages, and the estimated economic loss was HK $315 000 000 (equivalent to US $40 000 000). The causative species were *Gymnodinium mikimotoi* and a new dinoflagellate species, *Karenia digitata* (Yang

et al., 2000) and, since then, more and more harmful algal species have been found (Lu & Hodgkiss, 1999b).

Hong Kong is one of the worst HAB affected areas in the world, with a high diversity of harmful and toxic algal species, especially dinoflagellates. The objective of this study was to investigate the morphology and diversity of toxic and harmful, or potentially toxic and harmful species in Hong Kong waters, with attention focused on Junk Bay, where the first PSP event happened.

Materials and methods

Hong Kong is located in the north-eastern part of the subtropical South China Sea. It is surrounded by sea to the south, east, and west. Junk Bay is located on the south of the Kowloon Peninsula, facing Hong Kong Island (Fig. 1). Phytoplankton samples were taken regularly (three times a month from 1997 to 1998) from Junk Bay, and occasionally from other sites in Hong Kong waters.

Net and water samples were taken at each sampling. Net samples were collected using a 10 μm

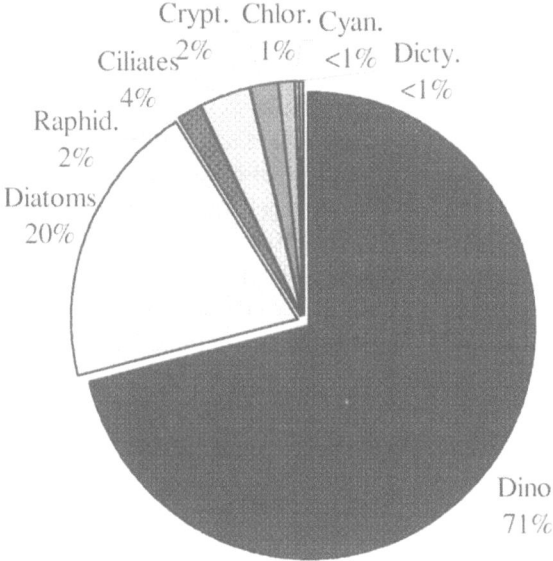

Figure 2. Percentage contributions of different taxonomic groups to algal blooms recorded in Hong Kong waters from 1975 to 1998 (from the red tide database, Agriculture, Fisheries and Conservation Department, Government of the Hong Kong SAR).

Table 1. Dinoflagellate species recorded in Hong Kong which have had the harmful impacts indicated elsewhere

Species	Impacts
Alexandrium catenella (Whedon et Kofoid) Balech *	Toxic
Alexandrium tamarense (Labour) Balech*	Toxic
Amphidinium carterae Hulburt	Toxic
Ceratium furca Ehrenberg (Claparède *et* Lachmann)*	Fish kills
Ceratium fusus (Ehrenberg) Dujardin	Fish kills
Cochlodinium polykrikoides Margelef*	Fish kills
Dinophysis acuminata Claparède *et* Lachmann	Toxic
Dinophysis acuta Ehrenberg	Toxic
Dinophysis caudata Saviller-Kent	Toxic
Dinophysis fortii Pavillard	Toxic
Dinophysis tripos Gourret	Toxic
Gambierdiscus toxicus Adachi *et* Fukuyo	Toxic
Gonyaulax polygramma Stein*	Fish kills
Gymnodinium cf. *breve* Davis	Toxic
Gymnodinium catenatum Graham	Toxic
Gymnodinium mikimotoi Miyake et Kominami *ex* Oda*	Fish kills
Gymnodinium sanguineum Hirasaka*	Fish kills
Gyrodinium instriatum Freudenthal *et* Lee*	Fish kills
Heterocapsa triquetra (Ehrenberg) Balech*	Fish kills
Karenia digitata Yang, Takayama, Matsuoka *et* Hodgkiss*	Fish kills
Noctiluca scintillans (Macartney) Ehrenberg*	Fish kills
Prorocentrum dentatum Stein*	Fish kills
Prorocentrum micans Ehrenberg*	Fish kills
Prorocentrum minimum (Pavilard) Schiller*	Fish kills
Prorocentrum sigmoides Böhm*	Fish kills
Prorocentrum triestinum Schiller*	Fish kills
Scrippsiella trochoidea (Stein) Loeblich*	Fish kills

* Those species which have caused blooms (with or without negative impacts) in Hong Kong waters.

mesh size phytoplankton net. The samples were preserved using Lugol's solution. Some live samples were collected and taken back to the laboratory immediately without preservation after each sampling, and the live samples were used for isolation and cultivation of some important species. Preserved samples were stored for subsequent quantitative studies.

For most of the armored dinoflagellates, preserved net samples were good for observation under both light and electron microscopes. For unarmored and some fragile armored dinoflagellates, live water samples were observed under the light microscope, and then single cells of each species were picked out and cultivated.

'K' (Keller et al., 1987) and 'f/2' (Guillard, 1975) media were applied for algal culture. The light dark cycle was 12L:12D. Culture temperature was different for different species. *Gambierdiscus toxicus*, a tropical species, was kept at 25 °C. For all other species, the culture temperature was 20 °C.

Preparation of dinoflagellates for scanning electron microscopy (SEM) followed methods recommended by Takayama (1985) and Truby (1997). The combined fixation technique was applied to unarmored dinoflagellates. Fixed dinoflagellate specimens were washed and dehydrated, followed by critical-point-drying and sputter-coating with gold-palladium. They were fi-

nally examined on a Leica S-440 Scanning Electron Microscope.

Results

A total of 27 harmful, or potentially harmful dinoflagellate species were identified during the study (Table 1). Of these, 17 species were harmful and caused toxicity or fish kill events in Hong Kong waters before or during the study period. Ten species, which previously had not caused HABs in Hong Kong, are potentially harmful since they have been proven to be toxic or harmful elsewhere in the world.

Newly recorded harmful dinoflagellate species

Gymnodinium cf. *breve* (Plate 1), a potentially NSP producing species, and *Gambierdiscus toxicus* (Plate 2), a potentially CFP producing species (Lu & Hodgkiss, 1999a) were new records of harmful algal species for Hong Kong waters. Both species were collected from plankton samples in Junk Bay.

Harmful dinoflagellate species

The harmful event caused by *Gymnodinium mikimotoi* and *Karenia digitata* during the study period was the biggest in the historical record of HABs in Hong Kong, both in terms of the area it affected and the economic losses which resulted. The most common harmful dinoflagellate species which were frequently observed in relatively higher density in Hong Kong waters were: *Noctiluca scintillans, Prorocentrum micans, Prorocentrum minimum, Ceratium furca, Gonyaulax polygramma, Scrippsiella trochoidea*, and *Gymnodinium mikimotoi*.

Potentially harmful dinoflagellate species

So far in Hong Kong only PSP toxins have been confirmed in shellfish from local waters. The causative species was believed to be *Alexandrium catenella* (Ho & Hodgkiss, 1993). However, other potentially important PSP producing species like *Alexandrium tamarense* and *Gymnodinium catenatum* were frequently observed during the present study.

In addition, DSP producing species such as *Dinophysis acuminata, D. acuta, D. caudata* and *D. fortii*; the NSP producing species *Gymnodinium* cf. *breve*, and the CFP producing species *Gambierdiscus toxicus* were also observed during this study.

Discussion

Dinoflagellates and HABs in Hong Kong

Hong Kong is believed to be one of the most severely harmful algal bloom affected areas in the world (Smayda, 1990). It is true that there are all kinds of potentially harmful algal species present, which have been linked to DSP, NSP, DSP, CFP, and ASP (produced by diatoms), as well as fish kill species. Amongst the HAB causative species in Hong Kong waters, dinoflagellates play a very important role. A

red tide database from the Agriculture, Fisheries and Conservation Department of the Hong Kong Government showed that a total of 534 identified algal blooms have been recorded from 1975 to July 1998. Of these, 377 (71%) were caused by dinoflagellates (Fig. 2). Twenty six (42%) of the total 62 species responsible for the blooms were dinoflagellates (Fig. 3). Most importantly, out of a total of 50 fish kill incidents during this time, 48 were caused by dinoflagellates, and another two were caused by combinations of diatoms and dinoflagellates. So dinoflagellates caused 98% of the fish kill incidents.

Potential harmful dinoflagellate species and harmful algal blooms

In Hong Kong, only the negative effects of PSP and CFP, and fish kills, have been reported. It has been suggested that the CFP was from imported reef fishes. Dinoflagellates containing the toxins responsible for NSP and DSP have not been reported previously.

Ciguatera toxins originate in benthic, epiphytic dinoflagellates, mainly *Gambierdiscus toxicus* (Yasumoto et al., 1977), which are grazed by herbivorous reef fish, the toxins then being transferred to carnivorous reef fish. Humans become intoxicated by eating these toxic reef fish. Ciguatera poisoning occurs throughout the Caribbean and tropical Pacific regions. There are about 50 000 people poisoned each year throughout the world (Steidinger, 1993). *Gambierdiscus toxicus* lives epiphytically on red, brown and green seaweeds, and also lives free in sediments and coral rubble, occasionally appearing in plankton suspended by current.

Gambierdiscus toxicus was first observed in planktonic samples from Junk Bay in 1997 (Lu & Hodgkiss, 1999b). Fortunately the cell number was quiet low. The toxicity of the species has not been tested. In Hong Kong, the occurrence of fish contaminated with CFP is common, and many people have been affected (Lu & Hodgkiss, 1999a). Though ciguatera is not linked in any way to local harmful algal blooms, and it has been proven that it was a result of the consumption of imported tropical coral reef fish, the occurrence of *Gambierdiscus toxicus* is still a potential source for ciguatera. Therefore, more attention should be paid to this species, and more studies should be carried out.

Gymnodinium breve originated NSP incidents have only been reported in the Gulf of Mexico (Steidinger, 1993), and more recently in New Zealand (Chang et al., 1995). The impacts of NSP are extremely severe,

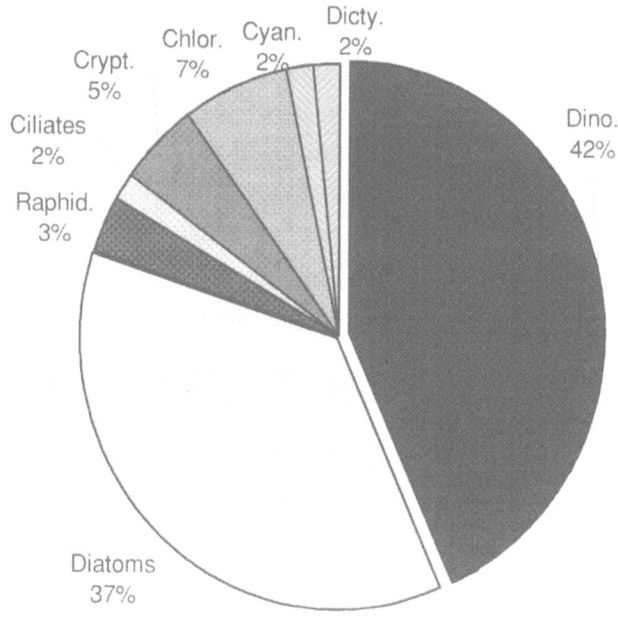

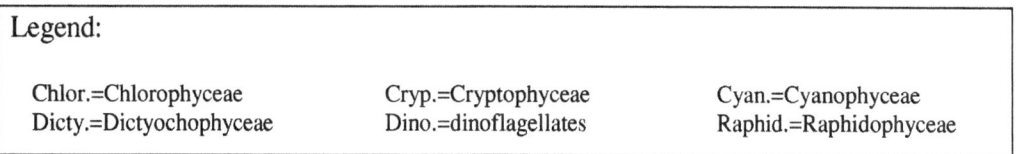

Legend:

Chlor.=Chlorophyceae   Cryp.=Cryptophyceae   Cyan.=Cyanophyceae
Dicty.=Dictyochophyceae   Dino.=dinoflagellates   Raphid.=Raphidophyceae

Figure 3. Percentage contributions of different causative species to the algal blooms recorded in Hong Kong waters from 1975–1998 (from the red tide database, Agriculture, Fisheries and Conservation Department, Government of the Hong Kong SAR).

causing massive fish kills as well as mortalities of marine animals such as whales, porpoises, manatees, dolphins, sea turtles, and sea birds through food chains (Anderson & White, 1989; Landsberg & Steidinger, 1998).

The distribution of *Gymnodinium breve* is very restricted. It has only been confirmed in the Gulf of Mexico, the southeast coast of the United States and the West Indies. However, *G. breve*-like cells have been recorded from Japanese, European, Australian, and New Zealand waters (Steidinger & Tangen, 1997). The *Gymnodinium breve*-like species found in Hong Kong waters will undoubtedly spread and so, the toxicity and population dynamics of the species should be studied.

In addition to *Alexandrium catenella*, several other *Alexandrium* species as well as *Gymnodinium catenatum,* which are potential PSP producing species,

have been observed in the present study. The chain-forming species of *Gymnodinium catenatum* associated with PSP was first found in 1976 in northwest Spain (Anderson et al., 1989), and the first death (of three people from eating toxic oysters and coquina contaminated by *Gymnodinium catenatum*) was reported in 1979 in Mexico (Cortes-Altamirano, 1987). This species is the most important source of PSP on the Iberian coast, causing severe problems to shellfish fisheries (Anderson, 1989). The species and incidents have also been reported in the Gulf of Mexico, Gulf of California, South America, Italy, Japan, the Philippines and Tasmania. Though this species is present in Hong Kong waters, it has never bloomed, and there is no evidence that it has caused problems in local waters.

236

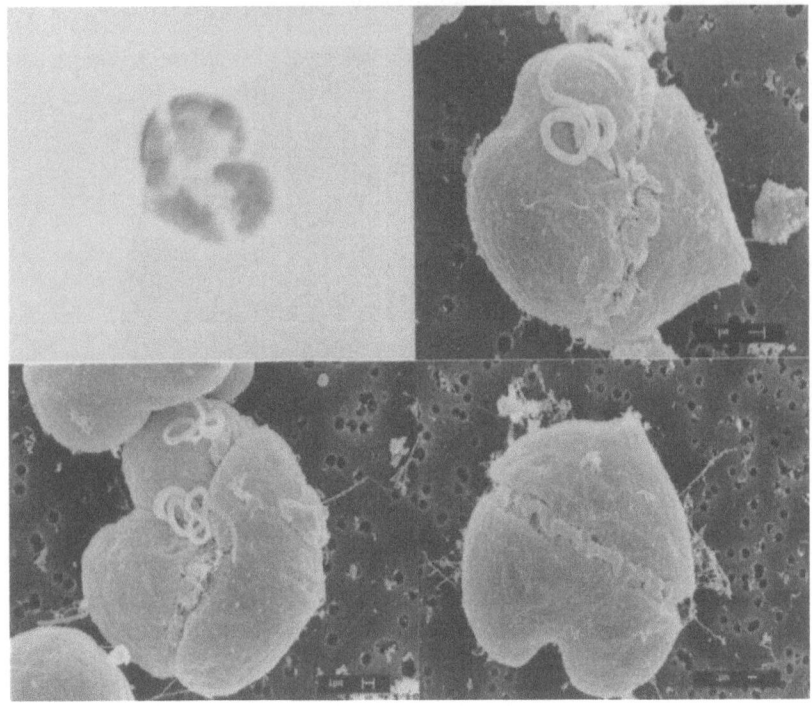

Plate 1. Gymnodinium cf. *breve.*

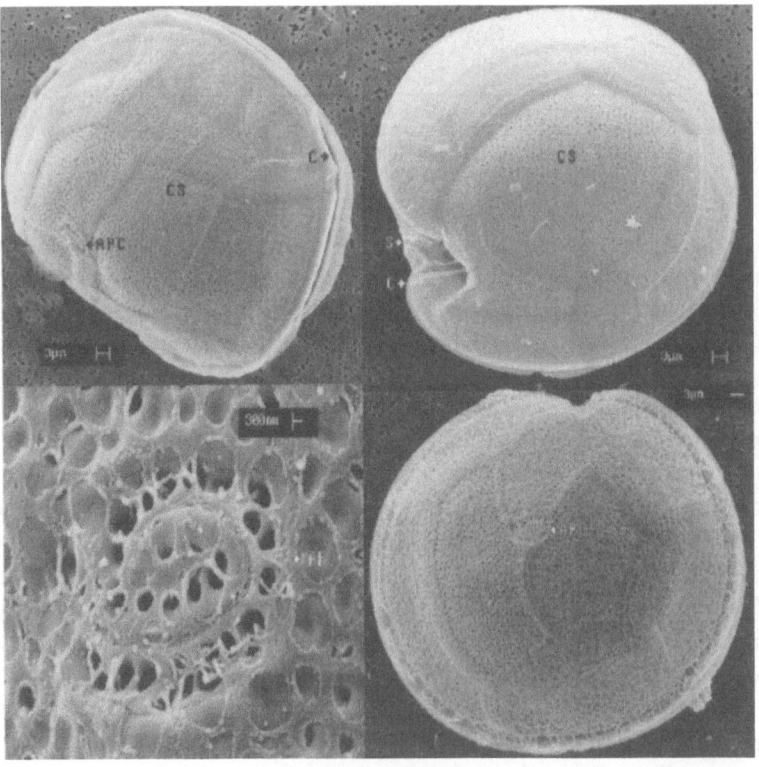

Plate 2. Gambierdiscus toxicus.

Fish kill dinoflagellates

According to historical HAB records of fish kill events, three groups of dinoflagellates have been identified as responsible for the incidents: the *Gymnodinium/Karenia* group, the *Prorocentrum* group and the *Noctiluca/Gonyaulax* group. These group were separated on the basis of the fish kill (or potential fish kill) capabilities of the species, and the frequency with which they bloomed in Hong Kong waters.

Gymnodinium/Karenia *group*

This group of species is the most harmful in Hong Kong waters and it has caused the majority of fish kill incidents. The species involved are *Gymnodinium mikimotoi, Karenia digitata,* and some other small unidentified *Gymnodinium* species. *Gymnodinium sanguineum,* and *Gyrodinium instriatum* might also be included in this group because of their high density during the present study, though they have never caused fish kills in Hong Kong waters. This group of species was abundant in spring, and most of the fish kill blooms happened in spring.

Prorocentrum *group*

This group of species includes *Prorocentrum dentatum, P. micans, P. minimun, P. sigmoides,* and *P. triestinum.* It is the second largest group responsible for fish kills in Hong Kong. The fish kills caused by this group of species were due to hypoxia or anoxia. The *Prorocentrum* species can be observed year round with highest cell density in spring.

Noctiluca/Gonyaulax *group*

The species in this group are mainly *Noctiluca scintillans* and *Gonyaulax polygramma. Noctiluca scintillans* is the most common bloom causative species in Hong Kong waters (Wong, 1989). Although some toxic substance has been reported to be produced by *Noctiluca scintillans* (Okaichi & Nishio, 1976), so far only one fish kill incident has been reported. *Gonyaulax polygramma* is another major causative species in Hong Kong. In 1988, a HAB caused by this species lasted as long as three months (Lam & Yip, 1990), which is the longest lasting bloom in Hong Kong. So far two fish kill incidents were caused by this species. The characteristics of this group of species are high frequency of blooms and low ability to kill fishes. *Ceratium furca* and *Scrippsiella trochoidea* should also be included in this group.

Acknowledgements

We are grateful to the Agriculture, Fisheries and Conservation Department, Government of the Hong Kong SAR, for providing access to their database on Hong Kong red tides. We would like to thank Mr Lee Wing Sang for his kind help in using the Scanning Electron Microscope, and Ms Maria G.Y. Lo for her assistance in collecting some samples.

References

Anderson, D. M., 1989. Toxic algal blooms and red tides: a global perspective. In Ochaichi, R., D. M. Anderson & T. Nemoto (eds), Red Tides: Biology, Environmental Science and Toxicology. Proc. First Int. Symp. Red Tides, Japan, 1987. Elsevier, New York: 11–16.

Anderson, D. M. & A. W. White (eds), 1989. Toxic Dinoflagellates and Marine Mammal Mortalities. Woods Hole Oceanographic Institution Tech. Rept. WHOI-89-36 (CRC-89-6).

Anderson, D. M., J. J. Sullivan & B. Reguera, 1989. Paralytic shellfish poisoning in northwest Spain: the toxicity of the dinoflagellate *Gymnodinium catenatum.* Toxicon 27: 665–674.

Baden, D. G. & T. J. Mende, 1982. Toxicity of two toxins from the Florida red tide dinoflagellate, *Ptychodiscus brevis.* Toxicon 20: 457–461.

Bates, S. S., C. J. Bird & A. S. W. De Freitas, 1989. Pennate diatom *Nitzschia pungens* as the primary source of domoic acid, a toxin in shellfish from eastern Prince Edward Island, Canada. Can. J. Fish. aquat. Sci. 46: 1203–1215.

Chang, F. H., L. Mackenzie, D. Till, D. Hannah & L. Rhodes, 1995. The first toxic shellfish outbreaks and the associated phytoplankton blooms in early 1993 in New Zealand. In Lassus, P., G. Arzul, E. Erard, P. Gentien & C. Marcaillou (eds), Harmful Marine Algal Blooms, Paris, Lavoisier: 145–150.

Cortes-Altamirano, R., 1987. Observaciones de mareas rojas en la Bahia de Mazatlan, Sinaloa, Mexico. Cienc. Mar. 13: 1–19.

EPDHK (Environmental Protection Department, Hong Kong), 1990. Marine Water Quality in Hong Kong-1989. Water Policy Group, Environmental Protection Department of Hong Kong Government, Hong Kong.

Guillard, R. R. L., 1975. Culture of phytoplankton for feeding marine invertebrate animals. In Smith, W. L. & M. H. Chanley (eds), Culture of Marine Invertebrate Animals. Plenum Press, New York: 29–60.

Hallegraeff, G. M., 1993. A review of harmful algal blooms and their apparent global increase. Phycologia 32: 79–99.

Ho, K. C. & I. J. Hodgkiss, 1993. Characteristics of red tides caused by *Alexandrium catenella* (Whedon and Kofoid) Balech in Hong Kong. In Smayda, T. J. & Y. Shimizu (eds), Toxic Phytoplankton Blooms in the Sea. Elsevier, Amsterdam, London: 263–268.

Keller, M. D., R. C. Selvin, W. Claus & R. R. L. Guillard, 1987. Media for the culture of oceanic ultraphytoplankton. J. Phycol. 23: 633–638.

Lam, C. W. Y. & S. S. Y. Yip, 1990. A three-month red tide event in Hong Kong. In Granéli, E., B. Sundström, L. Edler & D.M. Anderson (eds), Toxic Marine Phytoplankton. Elsevier, New York, Amsterdam, London: 481–486.

Landsberg, J. H. & K. A. Steidinger, 1998. A historical review of *Gymnodinium breve* red tides implicated in mass mortalities of the manatee (*Chichechus manatus latirostris*) in Florida, U.S.A. In Reguera, B., J. Blanco, M. Fernandez & T. Wyatt (eds), Harmful Algae, Xunta de Galicia and Intergovernmental Oceanographic Commission of UNESCO: 97–100.

Lu, S. H. & I. J. Hodgkiss, 1999a. *Gambierdiscus toxicus*, a ciguatera fish poisoning producing species found in Hong Kong waters. In 1st Conference om Harmful Algae – Management and Mitigation, 10–14 May, 1999, Philippines (Abstract).

Lu, S. H. & I. J. Hodgkiss, 1999b. An unusual year for the occurrence of harmful algae. Harmful Algal News 18: 1 & 3.

Morton, B. & P. R. Twentyman, 1971. The occurrence and toxicity of a red tide caused by *Noctiluca scintillans* (Macartney) Ehrenb. in the coastal water of Hong Kong. Environ. Res. 4: 544–557.

Okaichi, T. & S. Nishio, 1976. Identification of ammonia as the toxic principle of red tide *Noctiluca miliaris*. Bull. Plankton Soc. Jap. 23: 75–80.

Shimizu, Y., 1979. Developments in the study of paralytic shellfish toxins. In Taylor, D. L. & H. H. Seliger (eds), Toxic Dinoflagellate Blooms. Elsevier, New York: 453–455.

Shumway, S. E., 1990. A review of the effects of algal blooms on shellfish and aquaculture. J. World Aquacult. Soc. 21: 65–104.

Smayda, T. J., 1990. Novel and nuisance phytoplankton blooms in the sea: evidence for a global epidemic. In Granéli, E., B. Sundström, L. Edler & D. M. Anderson (eds), Toxic Marine Phytoplankton. Elsevier, New York, Amsterdam, London: 29–40.

Sournia, A., 1995. Red tide and toxic marine phytoplankton of the world ocean: An inquiry into biodiversity. In Lassus, P., G. Arzul, E. Erard, P. Gentien & C. Marcaillou (eds), Harmful Marine Algal Blooms. Proc. 6th Int. Conf. on Toxic Marine Phytoplankton, France, 1993. Lavoisier: 103–112.

Steidinger, K. A., 1983. A re-evaluation of toxic dinoflagellate biology and ecology. Prog. in Phycol. Res. 2: 147–188.

Steidinger, K. A., 1993. Some taxonomic and biologic aspects of toxic dinoflagellates. In Falconer, I. A. (ed.), Algal Toxins in Sea-food and Drinking Water. Academic Press, London, San Diego: 1–28.

Steidinger, K. A. & K. Tangen, 1997. Dinoflagellates. In Tomas, C. R. (ed.), Identifying Marine Phytoplankton. Academic Press, San Diego, New York: 387–584.

Takayama, H., 1985. Apical grooves of unarmored dinoflagellates. Bull. Plankton Soc. Japan. 32: 129–140.

Truby, E. W., 1997. Preparation of single-celled marine dinoflagellates for electron microscopy. Microscopy Res. Tech. 36: 337–340.

Wong, P. S., 1989. The occurrence and distribution of red tides in Hong Kong – applications in red tide management. In Okaichi, T., D. M. Anderson & T. Nemoto (eds), Red Tides: Biology, Environmental Science and Toxicity, Proc. 1st Int. Symp. On Red Tides, Japan, 1987. Elsevier, New York, Amsterdam, London: 125–128.

Yang, Z. B., H. Takayama, K. Matsuoka & I. J. Hodgkiss, 2000. *Karenia digitata* sp. nov. (Gymnodiniales, Dinophyceae), a new harmful bloom species from the coastal waters of west Japan and Hong Kong. Phycologia 39: 463–470.

Yasumoto, T., 1990. Marine microorganisms toxins – an overview. In Granéli, E., L. Edler & D. M. Anderson (eds), Toxic Marine Phytoplankton, Proc. 4th Int. Conf. on Toxic Marine Phytoplankton, Sweden, 1989. Elsevier, New York, Amsterdam, London: 3–10.

Yasumoto, T., I. Nakajima, R. Bagnis & R. Adachi, 1977. Finding of a dinoflagellate as a likely culprit of ciguatera. Bull. Jap. Soc. Sci. Fish. 43: 1021–1026.

Yasumoto, T., Y. Oshima, W. Sugawara, Y. Fukuyo, H. Oguri, T. Igarashi & N. Fujita, 1980. Identification of *Dinophysis fortii* as the causative organism of diarrhetic shellfish poisoning. Bull. Jap. Soc. Sci. Fish. 46: 1405–1411.

Hydrobiologia **512**: 239–245, 2004.
P.O. Ang, Jr. (ed.), Asian Pacific Phycology in the 21ˢᵗ Century: Prospects and Challenges.
© 2004 *Kluwer Academic Publishers.*

Crossing test among floating *Ulva* thalli forming 'green tide' in Japan

Masanori Hiraoka[1,5], Masao Ohno[2,*], Shigeo Kawaguchi[3] & Goro Yoshida[4]
[1]*Marine Algae Research Co. Ltd., 3-9-4 Minatozaka, Shingu, Kasuya, Fukuoka 811-0114, Japan*
[2]*Usa Marine Biological Institute, Kochi University, Usa, Tosa City, Kochi 781-1164, Japan*
[3]*Department of Fisheries, Faculty of Agriculture, Kyushu University, Fukuoka 812-0053, Japan*
[4]*National Research Institute of Fisheries and Environment of Inland Sea, 2-17-5 Maruishi, Ohno, Saeki, Hiroshima 739-0452, Japan*
[5]*Present address: Kochi Prefectural Deep Seawater Laboratory (NEDO fellow), Maruyama, Murotomisaki, Muroto, Kochi 781-7101, Japan*
*Author for correspondence; E-mail: mohno@cc.kochi-u.ac.jp

Key words: crossing test, 'green tide', induced maturation, punching method, *Ulva*

Abstract

Crossing tests were made to determine the relationship between the identified *Ulva pertusa*, which commonly grows in Japan as an attached form on exposed rocks, and the floating *Ulva* forming "green tide" inside calm bays. The floating *Ulva* thalli were collected from five major green tide sites in Japan (Yokohama, Mikawa, Miyajima, Kochi and Hakata). Reproductive maturation was induced in *U. pertusa* and the floating thalli from each site. Mating between induced gametes was observed. It is therefore believed that the floating thalli from Yokohama, Mikawa and Miyajima were mainly *U. pertusa*, while those from Kochi and Hakata were of a different species (*Ulva* sp.1). Furthermore, the *Ulva* species found in Mikawa is also a species (*Ulva* sp.2) different from both *U. pertusa* and *Ulva* sp.1.

Introduction

Recently, there has been an increasing number of worldwide reports on the formation of huge mass of benthic algae due to eutrophication. These excessive growths largely consisted of various species of green benthic algae such as *Enteromorpha, Ulva, Chaetomorpha* and *Cladophora*, and are termed 'green tides' (Fletcher, 1996). In Japan, green tides caused by *Ulva* spp. have been reported since 1970s in many coastal areas where nutrient levels were higher than normal. The accumulation of *Ulva* spp. has caused serious economic and ecological problems (Fig. 1). Green tide phenomenon occurs in the southern and western Pacific coasts around Tokyo, Inland Sea, and the islands of Shikoku and Kyushu. All these areas are affected by the warm Kuroshio Current. There are no reports of green tides in the coasts facing Japan Sea and in the northern Pacific coasts, which are affected by the cold currents.

The conditions which led to green tides in Japan include high nutrient fluxes and the lack of strong currents or wave actions in relatively shallow water (Ohno, 1999). These conditions are similar to the ones reported from other countries (Fletcher, 1996). The development of green tides is most significant at sites with muddy, silty or sandy bottoms. It is a unique characteristic of *Ulva* spp. that they can grow without the need of any attachment to the substratum.

Arasaki (1984) surveyed several green tide phenomena in Japan and found that the *Ulva* species forming green tide were not made up of *Ulva pertusa* Kjellman, a very common species that grows on rocks in lower intertidal zone along almost the entire coast of Japan, but were made up of other species. He thought that these foreign *Ulva* spp. were introduced to Japan via large ships. On the other hand, Migita (1985) described that the green tide at Ohmura bay in western Japan was formed by a sterile mutant of *U. pertusa*. This *Ulva* strain was subsequently recognized as a sterile mutant of *U. pertusa*, which was

240

Figure 1. Gathering of *Ulva* biomass in Yokohama. The government of this city spends more than 40 million yen every year to eliminate accumulation of this alga on the beach.

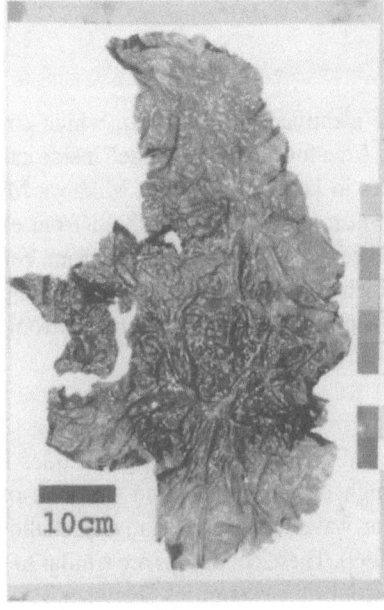

Figure 2. Ulva from Hakata collected from the floating population.

Figure 3. Collection sites of *Ulva* in Japan where green tides are observed.

Table 1. Date and number of collected samples

Collection site	Collection date	Sample number
Hakata	August 28, 1997	5
Kochi	September 19, 1997	25
Mikawa	June 9, 1998	8
Yokohama	June 10, 1998	11
Miyajima	April 19, 1999	26

utilized for some experiments in nearby fishery institutes. There are thus two opinions concerning the identity of *Ulva* species forming green tides in Japan. The taxonomic status of these species remains uncertain. In the present study, to determine the relationship between the identified *Ulva pertusa* and the floating *Ulva* forming 'green tide' and to evaluate the validity of the two opinions, we made crossing tests between the attached *U. pertusa* and the floating *Ulva* spp. from five sites in Japan where green tides have largely occurred using an artificial technique to induce reproductive maturation.

Materials and methods

Well-developed thalli from floating populations of *Ulva* spp. (Fig. 2) were collected from Yokohama, Mikawa, Miyajima, Kochi and Hakata, Japan, where significant green tides occurred (Fig. 3). The collection date and sample number in each site are shown in Table 1. After collections of these samples, we found out that there were two morphotypes in color and texture. One morphotype was dark green color with comparatively thick thallus (40–100 μm). Other type was light green color and thin thallus (30–50 μm). All samples from Yokohama and Miyajima and five samples from Mikawa belonged to the former. The samples from Kochi and Hakata and three others from Mikawa belonged to the latter. Typical morphotypes in color and texture were selected for crossing tests. The attached *Ulva pertusa* with rhizoid was collected

Table 2. Percentage frequency of zooid formation (reproductive maturation rate) over 10 days after excision in 100 disks (1.2 mm diam.) obtained from floating *Ulva* and attached *Ulva pertusa* (male and female gametophytes)

Sample no.	Zooid	Reproductive maturation rate (%)										
		Days after excision of thallus disks										
		1	2	3	4	5	6	7	8	9	10	Total
Yokohama 1	Gamete	0	0	1	0	4	0	0	0	0	0	5
Yokohama 2	gamete	0	0	3	0	48	30	13	6	–	–	100
Yokohama 4	Zoospore	0	0	45	10	22	1	11	11	–	–	100
Mikawa 1	Gamete	0	0	26	0	0	16	0	18	9	3	72
Mikawa 2	Zoospore	0	0	33	0	0	17	0	1	13	2	66
Mikawa 6	Zoospore	0	0	41	0	0	17	0	11	3	0	72
Mikawa 7	Zoospore	0	0	19	0	0	0	0	0	0	0	19
Miyajima 1	Zoospore	0	0	45	4	0	0	0	0	0	0	49
Miyajima 3	Zoospore	0	0	49	6	0	0	0	0	0	0	55
Miyajima 4	Zoospore	0	0	2	0	0	0	0	0	0	0	2
Kochi 2	Zoospore	0	0	1	15	0	0	7	0	0	0	23
Kochi 6	Zoospore	0	0	0	4	0	0	30	0	0	0	34
Kochi 11	Zoospore	0	0	7	1	0	0	0	0	0	0	8
Kochi 22	Zoospore	0	0	68	10	0	0	21	0	0	0	99
Hakata 6	Zoospore	0	0	0	11	4	0	0	0	0	0	15
Hakata 9	Zoospore	0	0	0	15	0	0	0	0	0	0	15
U. pertusa	Gamete($\male$)	0	100	–	–	–	–	–	–	–	–	100
U. pertusa	Gamete($\female$)	0	100	–	–	–	–	–	–	–	–	100

from rocky shores of Kochi. Samples agreeing well with the description of Kjellman (1897) were selected (Fig. 4a–d). Male and female gametes released from the attached *U. pertusa* formed zygotes, which were isolated and cultured up to sporophytes. Male and female gametophytes were established from zoospores released from the sporophytes for crossing test.

In order to obtain zooids (viz. male gametes, female gametes or zoospores), samples of *Ulva* spp. were induced using the 'punching method' described in Hiraoka & Enomoto (1998) and Hiraoka et al. (1998). This method involved the excision of small thallus disks of 1–2 mm diameter and the incubation of these disks in PES medium at 20 °C, 12:12 h L:D cycle and fluorescent light at 100 μmol photons m^{-2} s^{-1} (Hiraoka & Enomoto, 1998). Under such conditions, zooids can normally be formed within several days. Quadriflagellate zoospores were microscopically distinguished from biflagellate gametes. When the floating *Ulva* samples did not release gametes but zoospores, these could not be used directly for crossing tests. Therefore, the zoospores were cultured up to their gametophyte stage, of which gamete formation was induced for crossing test. Successful mating

of gametes released from attached *U. pertusa* and the floating *Ulva* was also determined microscopically.

Results and discussion

Ulva spp. in this study produced either quadriflagellate zoospores or biflagellate gametes. The percentage frequency of daily zooid formation is shown in Table 2. Zooid formation in cultured *Ulva pertusa* occurred on the second day after excision. However, it took 3 or more days for zooid formation to occur in the floating thalli. Depending on individuals, a large variation in the degree of zooid formation in the floating thalli was observed. Using the punching method, the zooid formation in the floating thallus occurred in less than 50% of the disks within 3 days (except for Kochi 22), while zooid formation in established male and female gametophyte of *U. pertusa* occurred in 100% of disks 2 days after excision. In some of the floating thalli zooids could not be induced at all by this method. Zooid formation in *Ulva* spp. would normally occur when the thallus is fragmented and the inhibitors of zooid formation leaked out from the fragmented tissue (Nordby, 1977; Stratmann et al., 1996). The difference in maturation time and rate between attached and

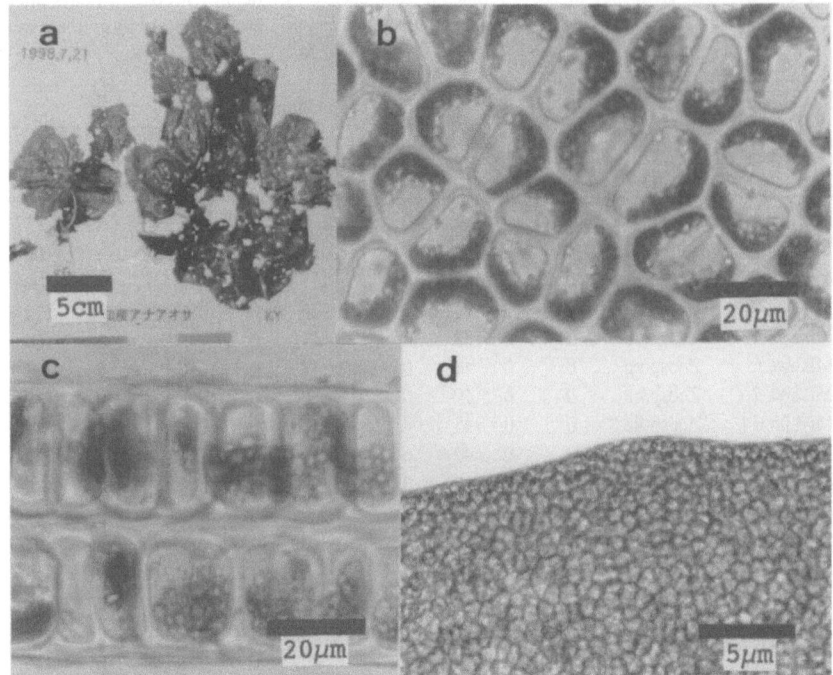

Figure 4. Morphological characteristic of attached *Ulva pertusa*. (a) Two gametophytes of different sizes with rhizoidal holdfast. (b) Surface view of cells from apical region. (c) Cross section from apical region. (d) Marginal part of apical region showing an absence of tooth-like protuberances.

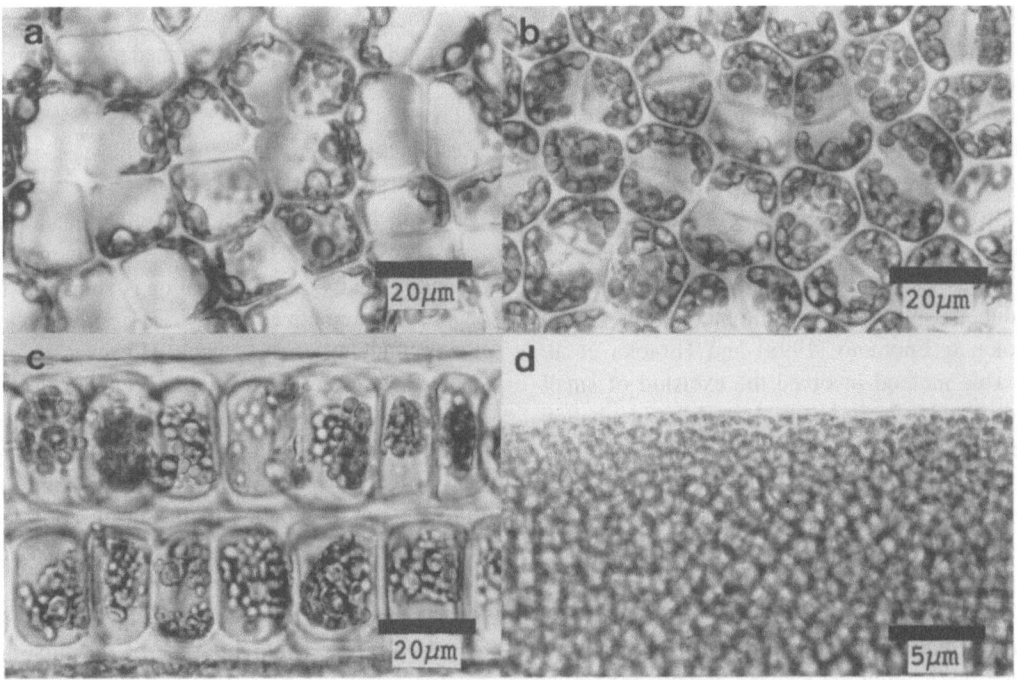

Figure 5. Morphological characteristic of floating *Ulva* crossed with *Ulva pertusa*. (a) Surface view showing cells with 1–3 pyrenoids. (b) Surface view showing cells with many small granules. (c) Cross section. (d) Surface view of marginal part.

Table 3. Crossing tests between floating *Ulva* and attached *Ulva pertusa* thalli. All samples from Yokohama and Miyajima and Mikawa 1, 2 and 6 were dark green color with comparatively thick thallus. Samples from Kochi and Hakata and Mikawa 7 were light green color with thin thallus. Male and female gametophytes were established from the zoospores released from the floating *Ulva* during *in vitro* culture. Success and failure of copulation between male and female gametes are indicated by + and − respectively

Sample no.	Zooid	Sex of gamete from cultured Gametophyte	*Ulva pertusa* attached to rock		Floating *Ulva* from Kochi	
			Male	Female	Male	Female
Yokohama 1	Gamete		+	−	−	−
Yokohama 2	Gamete		−	+	−	−
Yokohama 4	Zoospore	Male	−	+	−	−
		Female	+	−	−	−
Mikawa 1	Gamete		+	−	−	−
Mikawa 2	Zoospore	Male	−	+	−	−
		Female	+	−	−	−
Mikawa 6	Zoospore	Male	−	+	−	−
		Female	+	−	−	−
Mikawa 7	Zoospore	Male	−	−	−	−
		Female	−	−	−	−
Miyajima 1	Zoospore	Male	−	+	−	−
		Female	+	−	−	−
Miyajima 3	Zoospore	Male	−	+	−	−
		Female	+	−	−	−
Kochi 22	Zoospore	Male	−	−	−	+
		Female	−	−	+	−
Hakata 9	Zoospore	Male	−	−	−	+
		Female	−	−	+	−

Table 4. *Ulva* species occurring in each sites determined by crossing test

Collection site	*Ulva* species
Yokohama	*Ulva pertusa*
Mikawa	*Ulva pertusa*, *Ulva* sp. 2
Miyajima	*Ulva pertusa*
Kochi	*Ulva* sp.1
Hakata	*Ulva* sp.1

floating *Ulva* spp. may depend on the concentration or distribution of inhibitors in the thalli.

Because zooid formation in floating thalli was normally delayed as mentioned above, male and female gametophytes of *U. pertusa* had to be induced to form gametes everyday in order to ensure the availability of gametes of both *U. pertusa* and the floating *Ulva* for crossing tests. The release of zooids from mature disks occurred in the morning both in *U. pertusa* and the floating *Ulva*. The results of crossing test are shown in Table 3 and summarized in Table 4. Dark green and thick morphotype *Ulva* spp. from Yokohama, Mikawa and Miyajima successfully crossed with *U. pertusa*, but the light green and thin morphotype from Kochi and Hakata did not. The thin morphotypes from Kochi and Hakata crossed with each other, the resulting offspring is designated herein as *Ulva* sp.1. The light green and thin *Ulva* morphotype from Mikawa (Mikawa 7) did not cross successfully with either *U. pertusa* or *Ulva* sp.1 and was considered as a different species, designated herein as *Ulva* sp.2.

The floating *Ulva* spp., which crossed with attached *U. pertusa*, have rounded cells in surface view. In many cases the chloroplasts are pressed against the cell walls (parietal) in surface view and have 1–3 pyrenoids (Fig. 5a). These may be replaced by numerous small granules (Fig. 5b). The marginal part of the thallus is comparatively thick (40–100 μm; Fig. 5c) without tooth-like protuberances in the margin of the thallus (Fig. 5d). These characters are in agreement with those of typical *U. pertusa* with rhizoidal holdfast (Fig. 4b–d; Kamiya et al., 1993). These morphological similarities also support the suggestion that the floating *Ulva* spp. have been derived from *U. pertusa*.

Hiraoka et al. (1998) have earlier suggested that the floating *Ulva* in Hakata is a different taxon from *U. pertusa* based on crossing test. In this report, we suggest that the floating *Ulva* from Kochi, which crossed with the ones from Hakata, belongs to the same unidentified *Ulva* (*Ulva* sp.1). In our field observations, we found some *Ulva* thalli attached to rock which possibly crossed with *Ulva* sp.1. The samples obtained on this species were morphologically similar to *Monostroma* sp. for their thin, easily torn blade and light green color. These characteristics, however, are more in agreement with those of the floating thalli of *Ulva* sp.1. We believe that these attached thalli belong to *Ulva* sp.1, and the floating type of *Ulva* sp.1 could have been derived from the attached type. The cells of *Ulva* sp.1 are polygonal in surface view (Fig. 6a). The thallus is comparatively thin, only 30–50 μm thick (Fig. 6b). In many cases, tooth-like protuberances are found microscopically in the thallus margin (Fig. 6c). These characters separate *Ulva* sp.1 from *U. pertusa*.

Ulva sp.2, which did not cross with *U. pertusa* and *Ulva* sp.1, is similar to *Ulva* sp.1 in having thin thallus, being easily torn, and light green in color and having polygonal cells in its surface view (Fig. 7a). *Ulva* sp.2 is different from *Ulva* sp.1 in having cells at the marginal part of the thallus being oval in shape in transverse sections (Fig. 7b). The microscopic tooth-

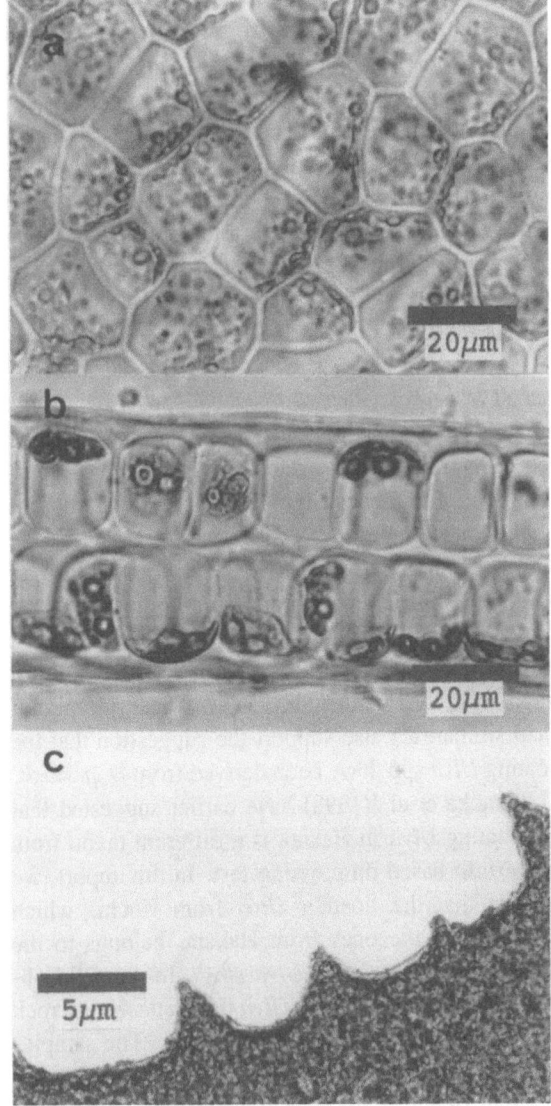

Figure 6. Morphological characteristic of *Ulva* sp.1. (a) Surface view of polygonal cells. (b) Cross section. (c) Marginal part showing tooth-like protuberances of polygonal cells.

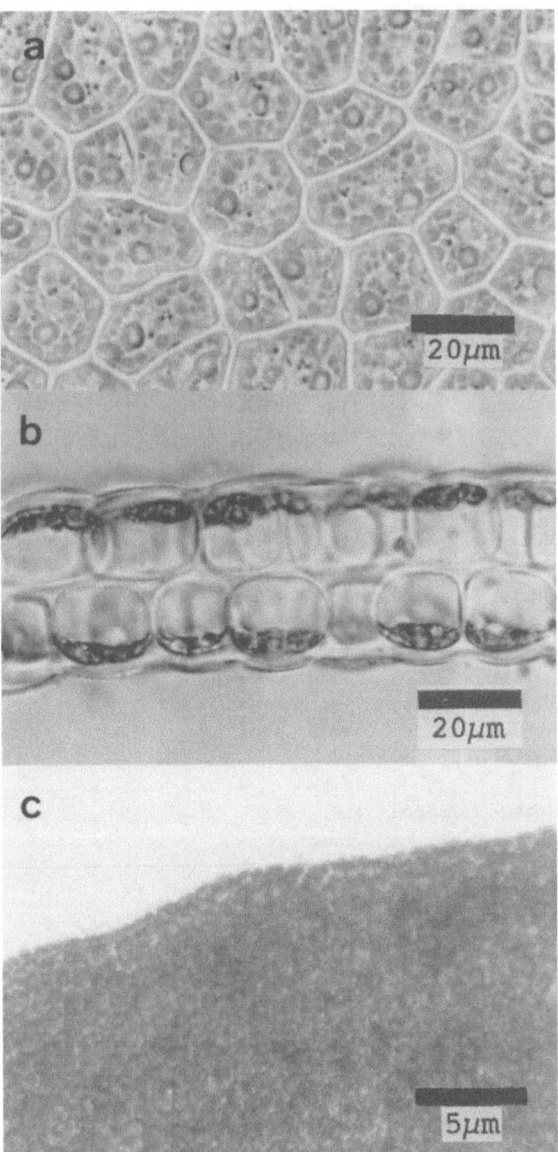

Figure 7. Morphological characteristic of *Ulva* sp.2. a: Surface view of polygonal cells. b: Cross section. c: Marginal part showing an absence of tooth-like protuberances.

like protuberances could not be found in the margin of *Ulva* sp.2 thallus (Fig. 7c). A more detailed description of *Ulva* sp.1 and *Ulva* sp.2 will be presented elsewhere.

Recently, it has been reported that algae forming the green tides of Brittany, France belong to a new species *Ulva armoricana* Dion, de Reviers & Coat (1998). *U. armoricana* is very similar to *Ulva* sp.1 in having a thin, easily torn blade, and polygonal cells in surface view. However, the frequency of tooth-like protuberances at the blade margin of the blade seems

to be higher in *U. armoricana* than in *Ulva* sp.1. At present, crossing tests and morphological comparisons between *Ulva* sp.1 and *U. armoricana* from France are being conducted in our laboratory in order to determine the relationship of French and Japanese green tide *Ulva* species.

Based on the summary of crossing tests shown in Table 4, *Ulva* populations occurring in Yokohama, Mikawa and Miyajima could be those of *U. pertusa*, while those in Kochi and Hakata were of a different species (*Ulva* sp.1). Yokohama, Mikawa and Miyajima are located in the temperate zone of Japan and are influenced only weakly by the Kuroshio Current. On the other hand, Kochi and Hakata are in the warm temperate zone and are affected more strongly by the Kuroshio (Tokuda et al., 1994). When the relationship between the marine flora and the distribution of floating *Ulva* is considered, it would be found that *U. pertusa* occurs only in temperate waters while *Ulva* sp.1 occurs in warm temperate waters affected more strongly by the Kuroshio Current. In Mikawa, *Ulva* sp.2 was collected with *U. pertusa*. This indicates that it is possible for more than two species of *Ulva* to occur in the same water. In this study, only one collection was made in each site. Therefore, there remains a possibility that more than two *Ulva* species may be present in the same site if more collections are to be made at different times in a year.

Acknowledgments

Dr D. B. Largo, Department of Biology, University of San Carlos, Philippines is thanked for critical reading of the manuscript.

References

Arasaki, S., 1984. A new aspect of *Ulva* vegetation along the Japanese coast. Hydrobiologia 116/117: 229–232.

Dion, P., B. de Reviers & G. Coat, 1998. *Ulva armoricana* sp. nov. (Ulvales, Chlorophyta) from the coasts of Brittany (France). I. Morphological identification. Eur. J. Phycol. 33: 73–80.

Fletcher, R. T., 1996. The occurrence of 'green tide'. In Schramm, W. & P. H. Nienhuis (eds), Marine Benthic Vegetation – Recent Changes and the Effects of Eutrophication. Springer Verlag, Berlin: 7–43.

Hiraoka, M. & S. Enomoto, 1998. The induction of reproductive cell formation of *Ulva pertusa* Kjellman (Ulvales, Ulvophyceae). Phycol. Res. 46: 199–203.

Hiraoka, M., M. Ohno & S. Kawaguchi, 1998. A note on the reproductive isolation of two types of *Ulva* (Chlorophyta) growing in Hakata Bay. Jap. J. Phycol. (Sôrui) 46: 161–165.

Kamiya, M., K. Doi, Y. Hara & M. Chihara, 1993. Taxonomic studies on *Ulva pertusa* (Ulvophyceae). I. Morphological study. Jap. J. Phycol. (Sôrui) 41: 191–198.

Kjellman, F. R., 1897. Marina Chlorophyceer från Japan. Bihang Till K. Sevenska Vet.–Akad. Handlingar. 23: III. 1–44, Pl. 3.

Migita, S., 1985. The sterile mutant of *Ulva pertusa* Kjellman from Ohmura Bay. Bull. Fac. Fish. Nagasaki Univ. 57: 33–37.

Nordby, Ø., 1977. Optimal conditions for meiotic spore formation in *Ulva mutabilis* Føyn. Bot. mar. 20: 19–28.

Ohno, M., 1999. *Ulva* and excessive growth. In Notoya, M. (ed.), The Utilization and Environmental Remediation in *Ulva*. Seizandou, Tokyo: 1–11.

Stratmann, J., G. Paputsoglu & W. Oertel, 1996. Differentiation of *Ulva mutabilis* (Chlorophyta) gametangia and gamete release are controlled by extracellular inhibitors. J. Phycol. 32: 1009–1021.

Tokuda, H., S. Kawashima, M. Ohno & H. Ogawa (eds), 1994. A Photographic Guide, Seaweeds of Japan. Midori Shobo Co., Ltd., Japan.

Hydrobiologia **512**: 247–253, 2004.
P.O. Ang, Jr. (ed.), Asian Pacific Phycology in the 21ˢᵗ Century: Prospects and Challenges.
© 2004 *Kluwer Academic Publishers.*

Taxonomic and ecological profile of 'green tide' species of *Ulva* (Ulvales, Chlorophyta) in central Philippines

Danilo B. Largo[1], Jose Sembrano[1], Masanori Hiraoka[2] & Masao Ohno[3]
[1]*Department of Biology, University of San Carlos, Cebu City 6000, Philippines*
[2]*Marine Greens Co., Ltd, Matsuyama, Japan*
[3]*Usa Marine Biological Institute, Kochi University, Usa-cho, Tosa 781-1164, Kochi, Japan*
E-mail: largodb@yahoo.com

Key words: algal biomass, eutrophication, 'green tide', Philippines, *Ulva lactuca, Ulva reticulata*

Abstract

Ulva spp. are common in the intertidal zones of the Philippines, but, at certain times, could over-proliferate producing blooms or 'green tide' in some protected bays. In Mactan Island (Cebu), central Philippines, at least two species constitute the *Ulva* population, either as free-living or attached form. The one referred to in the literature as '*Ulva lactuca*' mainly consists of free-living population while the species referred to as *Ulva reticulata* consists mainly of attached population. Based on morphological and physiological characteristics, '*U. lactuca*' differs much from the descriptions of the species from its type locality in Europe in having a crumpled texture of blade, presence of tooth-like protuberances at the margins, thinner thallus (40–50 µm) and more pyrenoids per cell (two to four). The species referred to as '*U. lactuca*' in the Philippines therefore is a different species. Two morphotypes consisted the '*U. lactuca*' population from Mactan – a thick thallus and a thin thallus type. However, both morphotypes cultured under the same condition in the laboratory could transform into the same thin-thallus type observed in the field. 'Green tide' caused by '*U. lactuca*' occur almost regularly in Station 1 of Mactan Island, reaching an average biomass of up to 2.6 kg wet wt m^{-2} (or 0.5 kg dry wt m^{-2}). *Ulva reticulata*, although was less abundant in the rocky tidal zone at most times, reaching an average biomass of only up to 0.15 kg wet wt m^{-2} (or 0.03 kg dry wt m^{-2}) had caused green tide in Station 2 around February–March. Reproductive structures were not observed in both *Ulva* species during the survey period suggesting that vegetative fragmentation is the main mode of propagation. Vegetative tissues excised from the thallus can be induced to release biflagellated large and small zooids.

Introduction

Species of the green algal genus *Ulva* (Ulvaceae, Chlorophyta) have been used since the 1970s in some countries as biofilters to remove nitrogen and phosphorus pollutants–chemicals which are often found in domestic wastewater and wastes from aquaculture facilities (Cohen & Neori, 1991; Neori et al., 1991; Jimenez del Rio et al., 1996). Thus, *Ulva* has been tagged as 'pollution indicator' due to its biomass accumulation in highly polluted waters (Morand et al., 1991). In the island of Cebu, in central Philippines, 'green tide' or algal proliferation caused by

this green algal genus, is becoming a common phenomenon especially in areas associated with dense human population.

Based on the taxonomic guides of Trono & Ganzon-Fortes (1988) and Calumpong & Meñez (1997), *Ulva* in the Philippines consists of two species: '*U. lactuca*' and *U. reticulata* (Forsskål. The former is described as having 'thin, glossy broad sheets with lobed undulating margin' and the latter as having 'highly perforated thallus'. Free-living thalli of species identified as '*U. lactuca*' in the Philippines has, at least, two different morphotypes observed in Mactan Island. These morphotypes could be ecolo-

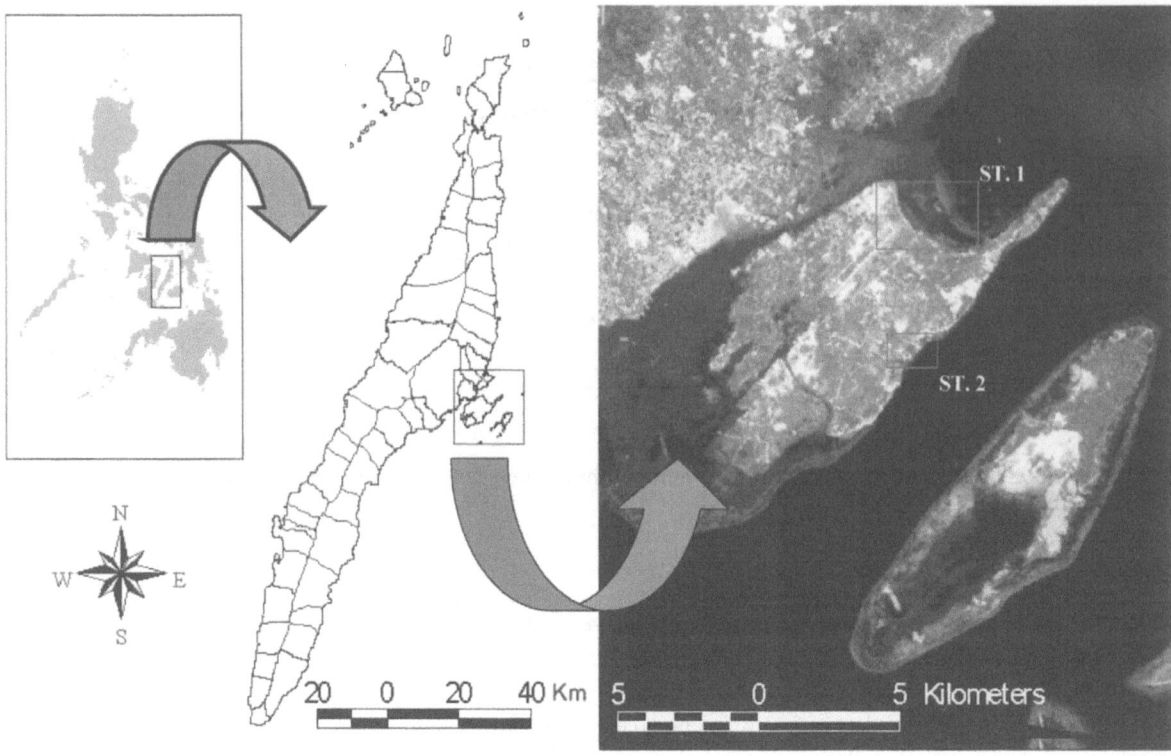

Figure 1. Map of the study area where the two monitoring stations of *Ulva* were established.

gical variants of the same species or could be entirely different species. The purpose of this study, therefore, is to provide baseline taxonomic and ecological data of *Ulva* in the Philippines and to document the occurrence of 'green tide' particularly in the increasingly industrialized island of Mactan (Cebu), in central Philippines.

Materials and methods

Description of the monitoring stations

The monitoring of *Ulva* was conducted from December 1998 to January–July 1999 mainly in the intertidal areas of Mactan Island, Cebu. Monthly biomass of '*U. lactuca*' and *U. reticulata* was obtained at two stations within Mactan Island. Station 1 was established within Magellan Bay, in the northeastern side of the island (across the foreshore light guides of the Cebu International Airport runway) while Station 2 was located in the mid-eastern portion of Mactan Island (Fig. 1).

Morphological and reproductive examination

Fresh materials of both species from the field were obtained monthly for morphological and reproductive examinations. However, this study gave more attention to '*U. lactuca*' due to its taxonomic uncertainty. Cell size and shape in both surface and transverse views were determined; the number of pyrenoids and chloroplast concentration of the two morphotypes were noted as well. To induce zooid (spore/gamete) release, thalli pieces, approximately 3–4 mm diam., were excised by punching on vegetative thalli using the method described in Hiraoka & Enomoto (1998).

Measurements of biomass and conditions of surrounding waters

Biomass estimates were obtained monthly from 10 randomly placed 0.25-m^2 quadrats. Samples were oven-dried at 60 °C for 48 h. For convenience, *Ulva* was segregated into two morphologically distinct species as '*U. lactuca*' and *U. reticulata*, based on their general description by Trono & Ganzon-Fortes (1988) and by Calumpong & Meñez (1997). As part of a general description of the study area, water temper-

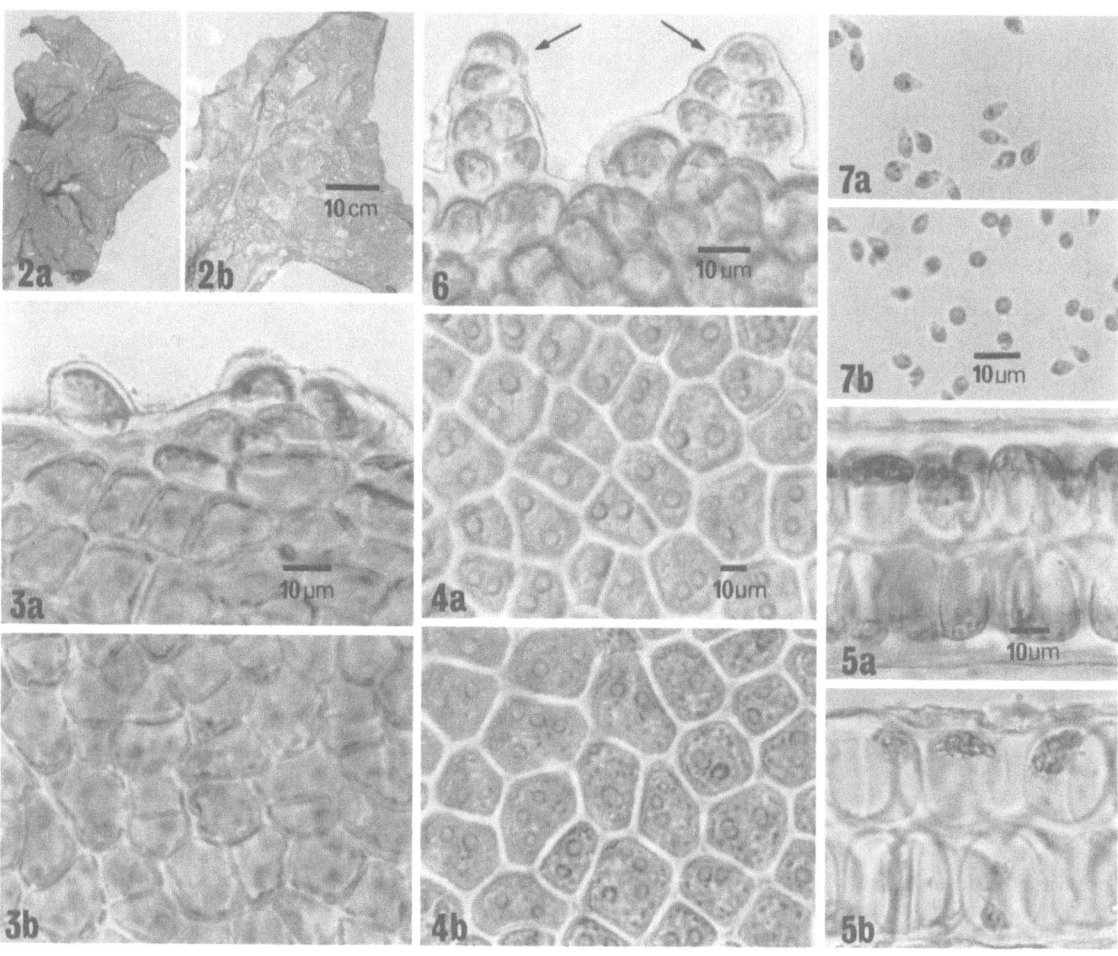

Figures 2–7. Morphological characteristics of '*Ulva lactuca*'. (2) Two morphotypes observed in the study area consisted of thick (2a) and thin (2b) thalli. (3) Surface view of thallus showing quadrangular marginal cells (3a) and polygonal middle cells (3b). (4) Surface view of thallus showing distinct pyrenoids ranging in number from mostly one to three per cell (4a) to mostly two to four per cell (4b). (5) Transverse section showing dense (5a) and sparse (5b) chloroplast concentrations. (6) Marginal tooth-like protuberances consisting of two-cell rows and an apical cell. (7) Release of large (7a) and small (7b) biflagellated zooids induced from excised tissues (by punching method).

ature and salinity were determined using fluid-filled thermometer and refractometer, respectively.

Growth rate measurements of 'U. lactuca'

The growth rate of '*U. lactuca*' was determined at different seawater temperatures – 18, 20, 22, 25, 28, and 30 °C, using a re-circulating culture system (Aquatron by Ohno, 1977). The increase in size of '*U. lactuca*' thalli was determined (in relative units) on a weekly basis using a computer image scan of 12 pressed samples obtained from the incubated materials. The growth rate based on 12 samples was computed as percentage increase per day using the formula described in Ohno et al. (1994). Except for temperature and weekly changes of seawater, condi-

tion for incubation was made constant at $100 \ \mu E \ m^{-2} \ s^{-1}$ (natural light+fluorescence light) at 12:12 h L/D photoperiod. Incubation period consisted of 7 days in each temperature regime.

Results and discussion

Ulva *morphotypes*

Ulva reticulata has highly perforated thallus in contrast to '*U. lactuca*' which has foliose thallus. However, the latter may also become perforated due to age. *Ulva* populations observed in Mactan Island were either free-living or attached, usually on hard substrate. *Ulva reticulata* mainly comprised the attached

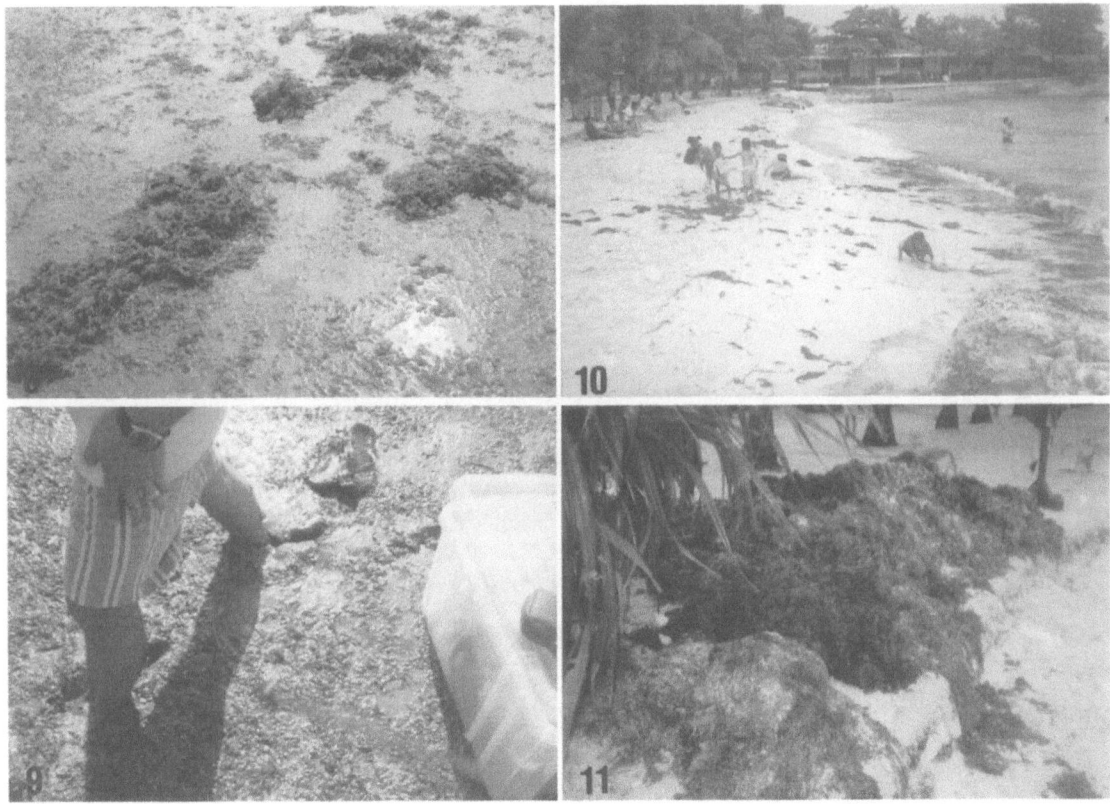

Figures 8–11. (8) Massive accumulation of free-living '*Ulva lactuca*' during a 'green tide' event in the intertidal zone of Station 1 (northeastern side of Mactan Island). (9) *U. lactuca* green tide over 30–50 cm thick and dark mud layer characteristic of Station 1. (10) Green tide caused by *Ulva reticulata* occurring in a beach resort in Station 2 (mid-eastern side of Mactan Island) on February–March 1999. (11) Harvesting of *U. reticulata* biomass mixed with some other algal species heaped up on a beach resort during a green tide event in Station 2.

population of *Ulva* in the sandy-rocky substratum of Station 2 in Mactan Island. However, non-reticulate thalli of *Ulva* were also, but not often, observed in this station. '*U. lactuca*', on the other hand, almost exclusively dominated the free-living population found in the eutrophied waters of Station 1 and was observed to be of two morphotypes: thick (dark green; Fig. 2a) and thin (light green color; Fig. 2b). Transverse sections made from these thalli show variation in cell shapes and concentrations of chloroplast. Vegetative thalli of '*U. lactuca*' consisted of quadrangular, near the margin (Fig. 3a), to polygonal cells in midthallus (Fig. 3b). Pyrenoids were distinct in surface view, varying in number from 1 to 4 (often more in thick type; Fig. 4a,b). In transverse view, cells of thick-thallus type were generally elliptical while those of thin-thallus type were shorter and round. Chloroplasts also varied in concentration from dense (thick type; Fig. 5a) to sparse (thin type; Fig. 5b). When these two morphotypes were grown in similar culture conditions, their cell size and color became similar in character, reveal-

ing their morphological plasticity. Protuberances, consisting of one to two cells, marked the thallus margin (Fig. 6). The appearance of these marginal protrusions is strikingly similar to those in *Ulva armoricana* Dion, de Reviers *et* Coat described in Dion et al. (1998) paper. *Ulva lactuca* is cited by these authors (Dion et al., 1998) as having a north European distribution. Based on this account, and the difference in thallus characters of the Philippine morphotype with that of *U. lactuca* described from Europe (e.g., no tooth-like protuberances), it is suggested that species referred to as '*U. lactuca*' in the Philippines is a different species. While it is closely similar to *U. armoricana* in most features especially in having, at least, a crumpled texture of blade and presence of tooth-like protuberances (Table 1), a detailed eco-physiological and molecular-based taxonomic examination is necessary to unequivocally identify '*U. lactuca*' as entirely different species.

Table 1. Morphological characteristics of Philippine morphotype of '*Ulva lactuca*' with respect to European morphotypes of *U. lactuca* and *U. armoricana*, a closely similar species

Morphotypes	Author	Crumpled texture of blade	Tooth-like protuberances	Cell shape and size (μm)		Thickness in mid- and apical region	Pyrenoid number per cell
				In surface view (mid-section)	Transverse section (μm) (rhizoidal region)		
U. lactuca							
European type	Dion et al. (1998)	No	No	Mostly polygonal	Cylindrical	50–90	Mostly 1
Philippine type (Mactan, Cebu)	This study	Yes	Yes	Quadrangular to polygonal	Oval	40–50	1–4 (2–3 most)
U. armoricana	Dion et al. (1998)	Yes	Yes	Polygonal to quadrangular	Oval or spindle-shaped, with tapered ends	30–55 (80 in winter)	Mostly 1–2

Spore/gamete release induction

The materials of both *Ulva* species, as examined microscopically, were all vegetative during the study period, suggesting fragmentation to be the main mode of propagation in these species. Release of zooids in excised pieces was successfully induced in '*U. lactuca*' using the punching method described by Hiraoka & Enomoto (1998). Large (+) and small (−) biflagellate spore/gametes were released within 3 days from the excised tissues of '*U. lactuca*' by this method (Fig. 7).

Observations on Ulva *habitat*

The two monitoring sites were selected based on the contrast of their substrate and water conditions. The sampling stations were up to 1–1.5 m deep at high tide but bare, except in tide pools, during low tide. Water temperature in the intertidal zone ranged from 26.4 to 31 °C during high tide but may increase up to 38 °C in the shallow tide pools. In Station 1 (within Magellan Bay), 'green tide' caused by massive accumulation of '*U. lactuca*' in the intertidal zone was observed (Fig. 8). The waters in this area are highly eutrophied by sources from Cebu harbor and from a river fronting Magellan Bay (see map in Fig. 1). The layer of muddy sediment in this area could be as deep as 30–50 cm; Fig. 9). On more consolidated substrate, the upper mud layer is less than 30 cm which results to the poor algal diversity in this area. Confinement of the *U. 'lactuca'* population within the shallow zone could be a result of limited water circulation within the bay. Effluents from households and industrial firms discharged directly into rivers and streams could have exacerbated this bloom. Although no data is available regarding the extent of eutrophication of waters around Metro Cebu, based on analysis of water samples from 5 major rivers which practically empty into the surrounding waters, dissolved oxygen, BOD, coliform bacteria and heavy metal pollutants all indicated highly polluted waters (DENR 7, 1993). Owing to the potential of *Ulva* as a biofilter of nutrients, and possibly toxicants, such as heavy metals and other xenobiotic pollutants, the removal of *Ulva* biomass could be a possible remedy to reduce input of nutrients in this place.

In Station 2, located in the mid-eastern side of Mactan Island (fronting a beach resort) contrasted that of Station 1 in the northern side, in having still relatively pristine waters, this being far from the influence of the Cebu harbor and water run-off from mainland Cebu. This area is well-flushed by tidal current from a relatively open sea. *Ulva reticulata* predominated the month of January (during a northeast monsoon season) causing 'green tide' in the shallow waters of Station 2. The wave-cast materials heaped up on the beach in large amount causing nuisance to swimmers (Figs 10 and 11). The plants were mainly in their young vegetative stage.

Ulva *biomass*

The average biomass of *Ulva* was highest in March 1999. Biomass of '*U. lactuca*' in Station 1 ranged from 0.15 to 2.6 kg wet wt m^{-2} (0.02–0.45 kg dry wt m^{-2}; Fig. 12). As its growth declined to a low standing crop towards July, its blades became increasingly perforated and fragile. This was accompanied by the appearance of another green alga, *Enteromorpha intestinalis* (L.) Nees and occasional appearance of young *U. reticulata*.

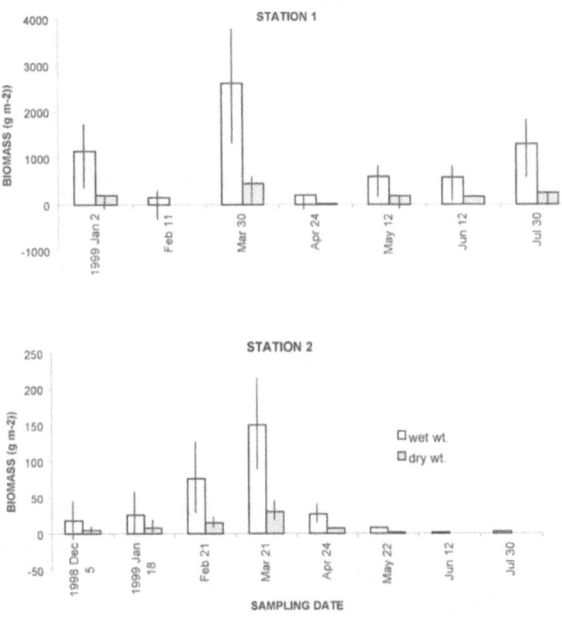

Figure 12. Monthly biomass of *Ulva* in green tide areas of Station 1 (above panel) and Station 2 (bottom panel) from December/January 1998/1999 to July 1999 as determined by quadrat sampling. *Ulva* population in Station 1 consisted almost exclusively of '*U. lactuca*' while Station 2 was predominated by '*U. reticulata*' (with few attached *Ulva lactuca*) all throughout the study period. Vertical lines represent standard deviation (*n*=10).

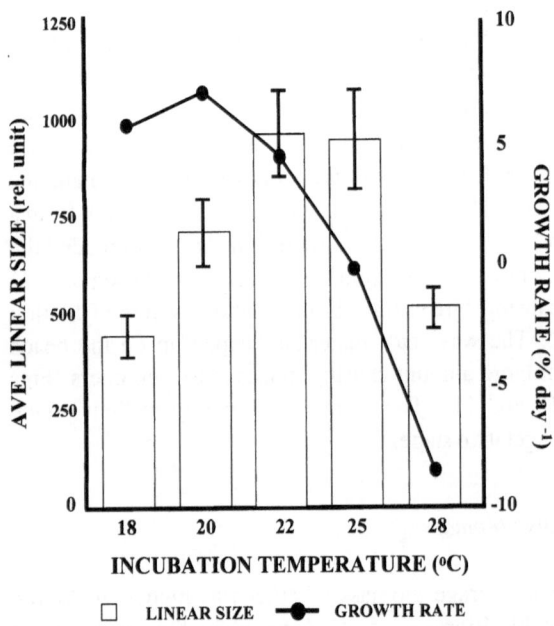

Figure 13. Relative size and growth rates of '*Ulva lactuca*' under different incubation temperatures in an aquatron culture system maintained at a light intensity of 100 μE m^{-2} s^{-1} under 12:12 h L/D photoperiod. Note maximum relative size and growth rate of this species at 20 and 22 °C, respectively. Incubation period was 7 days in each temperature regime. Vertical lines represent standard deviations (*n*=12).

In Station 2, the biomass of *U. reticulata*, which is a predominant species, increased dramatically from December to March where it attained its highest of 0.15 kg wet wt m^{-2} (0.03 kg dry wt m^{-2}; Fig. 12) during the study period. Highest values were recorded between February and March and were responsible for the unialgal bloom in this area during February–March (see Figs 10 and 11).

Growth rate of 'U. lactuca' *(no data for* U. reticulata*)*

'*U. lactuca*' grown in aquatron culture system at different temperature regimes showed highest growth rate of approximately 7% day^{-1} at 20 °C. However, highest increase in size was only observed at 22 °C (Fig. 13). Temperatures over 22 °C resulted in the decline of both growth rate and biomass under the aquatron condition. Highest biomass of '*U. lactuca*' and *U. reticulata* in the two stations was observed in February–March when *in situ* temperatures were generally low.

'*U. lactuca*' and *U. reticulata* showed distinct population dominance in two stations in Mactan (Cebu), Philippines suggesting both species to vary in their nutrient requirements, the former being more adopted to more eutrophied waters while the latter in low nutrient concentration. Future studies on *Ulva* in this area, therefore, should be focused on its nutrient uptake variability.

Acknowledgements

The first author is grateful for the assistance rendered by his Advance Phycology students in the University of San Carlos during some field samplings.

References

Calumpong, H. & E. G. Meñez, 1997. Field Guide to the Common Mangroves, Seagrasses and Algae of the Philippines. Bookmark, Inc., Makati City, Philippines.

Cohen, T. & A. Neori, 1991. *Ulva lactuca* biofilters for marine fishpond effluents. Bot. mar. 34: 475–482.

DENR (Department of Environment and Natural Resources), 1993. Environmental Profile of Metro Cebu. A Technical Report of the DENR 7, Cebu City.

Dion, P., B. de Reviers & G. Coat, 1998. *Ulva armoricana* sp. nov. (Ulvales, Chlorophyta) from the coasts of Brittany (France). I. Morphological identification. Eur. J. Phycol. 33: 73–80.

Hiraoka, M. & S. Enomoto, 1998. The induction of reproductive cell formation of *Ulva pertusa* Kjellman (Ulvales, Ulvophyceae). Phycol. Res. 46: 199–203.

Jimenez del Rio, M., Z. Ramazanov & G. Garcia-Reina, 1996. *Ulva rigida* (Ulvales, Chlorophyta) tank culture as biofilters for dissolved inorganic nitrogen from fishpond effluents. Hydrobiologia 326/327: 61–66.

Morand, P., B. Carpenter, R. H. Charlier, J. Maze, M. Orlandini, B. A. Plunkett & J. de Waart, 1991. Bioconversion of seaweeds. In Guiry, M. D. & G. Blunden (eds), Seaweed Resources in Europe: Uses and Potential. John Wiley and Sons Ltd., Chichester: 95–148.

Neori, A., I. Cohen & H. Gordin, 1991. *Ulva lactuca* biofilters for marine fishpond effluents. II. Growth rate, yield and C:N ratio. Bot. mar. 34: 483–489.

Ohno, M., 1977. Effect of temperature on the growth rate of seaweeds in an aquatron culture system. Bull. Jpn. Soc. Phycol. 25 (Suppl): 257–263.

Ohno, M., D. B. Largo & T. Ikumoto, 1994. Growth rate, carrageenan yield and gel properties of cultured kappa-carrageenan producing red alga *Kappaphycus alvarezii* (Doty) Doty in the subtropical waters of Shikoku, Japan. J. appl. Phycol. 6: 1–5.

Trono, G. C. & E. T. Ganzon-Fortes, 1988. Philippine Seaweeds. Technology and Livelihood Resource Center. National Bookstore, Manila, Philippines.

Hydrobiologia **512**: 255–262, 2004.
P.O. Ang, Jr. (ed.), Asian Pacific Phycology in the 21st Century: Prospects and Challenges.
© 2004 *Kluwer Academic Publishers.*

Inorganic ion compositions in brown algae, with special reference to sulfuric acid ion accumulations

Hideaki Sasaki[1], Hironao Kataoka[2], Akio Murakami[3] & Hiroshi Kawai[3,*]
[1] *Graduate School of Science and Technology, Kobe University, Nadaku, Kobe 657-8501, Japan*
[2] *Institute of Genetic Ecology, Tohoku University, Katahira, Sendai 980-8577, Japan*
[3] *Kobe University Research Center for Inland Seas, Rokkodai, Nadaku, Kobe 657-8501 Japan*
Author for correspondence; Fax: 81-78-803-5710. E-mail: kawai@kobe-u.ac.jp

Key words: brown algae, Dictyotales, inorganic ions, ion chromatography, pH, sulfuric acid

Abstract

Cellular pH estimated from cell extract pH and the ion compositions of major inorganic ions (Na^+, NH_4^+, K^+, Mg^{2+}, Ca^{2+}, Cl^-, Br^-, NO_3^-, SO_4^{2-}) were studied by ion chromatography in 61 species of 10 orders (Dictyotales, Desmarestiales, Ectocarpales, Chordariales, Scytosiphonales, Dictyosiphonales, Cutleriales, Sporochnales, Laminariales and Fucales) of Phaeophyceae. Three species in the order Dictyotales, *Dictyopteris* sp., *Spatoglossum solierii* (Chauv.) Kützing and *Zonaria stipitata* Tanaka et K. Nozawa, were newly found to be highly acidic (pH 0.6 and 1.4 within cells), in addition to previously reported dictyotalean species, *Dictyopteris latiuscula* (Okamura) Okamura, *D. prolifera* (Okamura) Okamura, *D. repens* (Okamura) Børgesen and *Spatoglossum crassum* J. Tanaka. They all contained high concentrations of SO_4^{2-} perhaps within the vacuoles. Furthermore, *Delamarea attanuata* (Kjellman) Rosenvinge (Dictyosiphonales) and *Thalassiophyllum clathrus* (Gmel.) P. et R. (Laminariales) were shown to contain relatively high concentrations of SO_4^{2-} balanced by relatively high concentrations of Ca^{2+}.

Introduction

Species of the order Desmarestiales (Phaeophyceae) are well known to accumulate highly acidic substances in vegetative cells of the sporophytes. The pH of the cell extracts have been shown to be as low as 2.0 (Wirth & Rigg, 1937; Eppley & Bovell, 1958; Schiff, 1962; McClintock et al., 1982; Sasaki et al., 1999), and the cellular pH has been estimated to be ca. 0.5–0.9 (McClintock et al., 1982; Sasaki et al., 1999). The acidic substance was thought to be sulfuric acid stored in vacuoles of the sporophytic cells (Eppley & Bovell, 1958; McClintock et al., 1982). Recently, Sasaki et al. (1999) reported a similar phenomenon in some species of Dictyotales; four species of *Dictyopteris* and *Spatoglossum* were found to be highly acidic due to high concentrations of cellular SO_4^{2-}. The authors showed, by ion chromatography of major inorganic ions, that the sum of metal cations was significantly lower than that of anions, and the extremely low pH

indicated that the excess of anions is balanced by protons (H^+). On the other hand, Sasaki et al. (1999) also found that some non-acidic dictyotalean species (*Dictyota dichotoma* (Hudson) Lamouroux, *Padina minor* Yamada, etc.) contained relatively high concentrations of SO_4^{2-}, which was balanced by high concentrations of Mg^{2+} and so did not cause high acidity. The authors therefore suggested that the inorganic ion composition could be rather variable depending on the species.

In this work, we aimed to extend the survey of inorganic ion compositions to cover additional brown algae in other orders, and to explore the diversity of inorganic ion compositions in brown algae.

Materials and methods

Selected members of the following orders were examined for cell extract pH and inorganic ion compositions by ion chromatography: Dictyotales, Desmarestiales, Ectocarpales, Chordariales, Scytosiphonales,

Table 1. Taxonomic list of species examined in the present survey and their origins

Species	Collection sites and dates
Dictyotales	
Dictyopteris divaricata (Okamura) Okamura	Sasaki et al. (1999)
D. latiuscula (Okamura) Okamura	Sasaki et al. (1999), Yamaguchi Pref. (13 June 1997)
D. prolifera (Okamura) Okamura	Sasaki et al. (1999), Awaji Isl., Hyogo Pref. (10 Jan., 7 July, 28 Oct., 12 Dec., 1997, 27 Feb., 28 Apr., 17 July 1998)
D. repens (Okamura) Børgesen	Sasaki et al. (1999)
D. undulata Holmes	Sasaki et al. (1999), Awaji Isl., Hyogo Pref. (17 Dec. 1998); Kinosaki, Hyogo Pref. (11 Oct. 1998)
Dictyopteris sp.	Ishigaki Isl., Okinawa Pref. (21 Apr. 1999)
Dictyota dichotoma (Hudson) Lamouroux	Sasaki et al. (1999), Awaji Isl., Hyogo Pref. (3 July, 12 Dec. 1997, 28 Apr., 17 July, 3 Dec. 1998)
D. divaricata Lamouroux	Sasaki et al. (1999)
D. linearis (C. Agardh) Greville	Sasaki et al. (1999)
Dictyotopsis propagulifera Troll	Sasaki et al. (1999)
Dilophus okamurae Dawson	Sasaki et al. (1999), Awaji Isl., Hyogo Pref. (17 Dec. 1998); Seto, Wakayama Pref. (10 May 1997)
Distromium decumbens (Okamura) Levring	Sasaki et al. (1999), Kasumi, Hyogo Pref. (11 Oct. 1998)
Lobophora variegata (Lamouroux) Womersley	Sasaki et al. (1999)
Pachydictyon coriaceum (Holmes) Okamura	Sasaki et al. (1999)
Padina arborescens Holmes	Sasaki et al. (1999), Awaji Isl., Hyogo Pref. (3 Dec. 1998)
P. crassa Yamada	Sasaki et al. (1999)
P. minor Yamada	Sasaki et al. (1999)
Padina sp.	Sasaki et al. (1999)
Spatoglossum crassum J. Tanaka	Sasaki et al. (1999), Awaji Isl., Hyogo Pref. (10 Jan. 1997)
S. pacificum Yendo	Sasaki et al. (1999), Awaji Isl., Hyogo Pref. (27 Aug. 1997)
S. solierii (Chauv.) Kützing	Tarifa Island, Cadiz, Spain (28 Aug. 1999)
Stypopodium zonale (Lamouroux) Papenfuss	Sasaki et al. (1999)
Taonia leunebackerae Farlow	Sasaki et al. (1999)
Zonaria diesingiana J. Agardh	Sasaki et al. (1999)
Z. stipitata Tanaka et K. Nozawa	Kerama Islands, Okinawa Pref. (8 Mar. 1999)
Desmarestiales	
Desmarestia aculeata (L.) Lamouroux	Sasaki et al. (1999)
D. latifrons Kützing	Kawai culture
D. ligulata (Stackhouse) Lamouroux	Sasaki et al. (1999)
D. tabacoides Okamura	Sasaki et al. (1999), Maiko, Hyogo Pref. (3 Mar. 1999)
D. viridis (Müller) Lamouroux	Sasaki et al. (1999)
Ectocarpales	
Bodanella lauterbornii Zimmermann	D.G. Müller culture
Chordariales	
Papenfussiella kuromo (Yendo) Inagaki	Awaji Isl., Hyogo Pref. (17 Dec. 1998)
Saundersella simplex (Saunders) Kylin	Abacha Bay, Kamchatka, Russia (24 July 1998)
Ishige okamurae Yendo	Awaji Isl., Hyogo Pref. (17 Dec. 1998)
I. sinicola (Setchell et Gardner) Chihara	Awaji Isl., Hyogo Pref. (17 Dec. 1998)
Leathesia difformis (Linnaeus) Areschoug	Awaji Isl., Hyogo Pref. (1 July 1999)
Scytosiphonales	
Colpomenia sinuosa (Mertens *ex* Roth) Derbes et Solier	Awaji Isl., Hyogo Pref. (3 Dec. 1998)
Petalonia binghamiae (J. Agardh) Vinogradova	Awaji Isl., Hyogo Pref. (17 Dec. 1998)
Scytosiphon gracilis Kogame.	Awaji Isl., Hyogo Pref. (17 Dec. 1998)

Continued on p. 257

Table 1. Continued

Species	Collection sites and dates
Dictyosiphonales	
Delamarea attenuata (Kjellman) Rosenvinge	Hokkaido Pref. (June 1997)
Stschapovia flagellaris A. D. Zinova	Kawai culture
Dictyosiphon foeniculaceus (Hudson) Greville	Abacha bay, Kamchatka, Russia (28 July 1998)
Cutleriales	
Cutleria cylindrica Okamura	Mijo, Korea (3 March 1998)
C. multifida (Turner) Greville	Ibota, Yamaguchi Pref. (12 June 1997)
Microzonia velutina (Harv.) J. Ag.	D.G. Müller culture
Sporochnales	
Carpomitra costata (Stackhouse) Batters	Cheju Isl., Korea (10 Feb. 1998)
Sporochnus scoparius Harvey	Ibota, Yamaguchi Pref. (12 June 1997)
Laminariales	
Agarum clathratum Dumortier	Muroran, Hokkaido Pref. (28 Feb. 1999)
Chorda filum (Linnaeus) Stackhouse	Ohzuchi, Miyagi Pref. (11 June 1998)
Ecklonia cava Kjellman	Awaji Isl., Hyogo Pref. (3 Dec., 17 Dec. 1998)
Ecklonia kurome Okamura	Takeno, Hyogo Pref. (2 July 1999)
Kjellmaniella crassifolia Miyabe	Muroran, Hokkaido Pref. (28 Feb. 1999)
Thalassiophyllum clathrus (Gmel.) P. et R.	Abacha Bay, Kamchatka, Russia (24 July 1998)
Fucales	
Hizikia fusiformis (Harvey) Okamura	Awaji Isl., Hyogo Pref. (3 Dec. 1998)
Sargassum confusum C. Agardh	Takeno, Hyogo Pref. (2 July 1999)
S. hemiphyllum (Turner) C. Agardh	Awaji Isl., Hyogo Pref. (3 Dec. 1998); Takeno, Hyogo Pref. (11 Oct. 1997)
S. micracanthum (Kützing) Endlicher	Awaji Isl., Hyogo Pref. (1 July 1999)
S. patens C. Agardh	Takeno, Hyogo Pref. (2 July 1999)
S. ringgoldianum ssp. *coreanum* (J. Agardh) Yoshida	Awaji Isl., Hyogo Pref. (17 Dec. 1998)
S. thunbergii (Mertens *ex* Roth) Kuntze	Awaji Isl., Hyogo Pref. (3 Dec. 1998)
Turbinaria ornata (Turner) J. Agardh	Ishigaki Isl., Okinawa Pref. (22 Jan. 1998)

Dictyosiphonales, Cutleriales, Sporochnales, Laminariales and Fucales. Origins of the materials are listed in Table 1, but only additional materials added after Sasaki et al. (1999) are listed for the following taxa: *Dictyopteris divaricata, D. latiuscula, D. prolifera, D. repens, D. undulata, Dictyota dichotoma, D. divaricata, D. linearis, Dictyotopsis propagulifera, Dilophus okamurae, Distromium decumbens, Lobophora variegata, Pachydictyon coriaceum, Padina arborescens, P. crassa, P. minor, Padina* sp., *Spatoglossum crassum, S. pacificum, S. solierii, Stypopodium zonale, Taonia leunebackerae, Zonaria diesingiana* (Dictyotales), *Desmarestia aculeata, D. latifrons, D. ligulata, D. tabacoides, D. viridis* (Desmarestiales). Field materials collected were transported as quickly as possible to the laboratory in plastic containers filled with seawater kept at 10–15 °C, and promptly used for the analyses. Unialgal culture strains were cultured in PESI medium (Tatewaki, 1966) under a 16:8 h LD cycle illuminated with daylight-type white fluorescent lighting of approximately 50 μmol m^{-2}s^{-1} at 10 °C. Sea water samples for ion-chromatography were collected from several localities at Awaji Isl., Hyogo Pref. (3, 17 July, 16 December 1997) and Ibota, Yamaguchi Pref. (12, 13 June 1997), and their data of ion-chromatography ($N = 7$) were averaged.

The pH measurements and ion chromatography of the cell extracts were done following the protocol in Sasaki et al. (1999) except for *Bodanella lauterbornii, Ishige okamurae* and *I. sinicola.* These species were highly tolerant to freshwater treatment because *B. lauterbornii* is a freshwater species, and *Ishige* spp. are upper subtidal species. Therefore, for the extractions of these species, fresh algal tissue was soaked in 1 ml (when the sample weighed less than 60 mg) or 2 ml (when the sample weighed more than 60 mg) boiling distilled water (ca. 100 °C) in a 1.5 ml (or 2.0 ml) microfuge tube for 60 min using a block incubator. The sample tissue was then removed and the extract was passed through an ultra-filtration filter and used for both analyses.

Intracellular pH and concentrations of inorganic ions were estimated from values of the extracts based on the following calibration formula:

$$\text{Estimated intracellular pH} = -\log [\alpha \times \text{dilution factor}],$$

where α is proton concentration (M) of the extract calculated from the pH measurement data, and the dilution factor = [distilled water (g) + sample fresh weight (g) × water content (ml)] / sample fresh weight (g) × water content (ml).

Results

pH measurements

The pH values measured for the algal tissue extracts of various taxa are shown in Table 2. Among Dictyotales, in addition to *Dictyopteris latiuscula*, *D. prolifera*, *D. repens* and *Spatoglossum crassum*, an unidentified species of *Dictyopteris* collected from Ishigaki Island, *Zonaria stipitata* and *Spatoglossum solierii* were newly found to be highly acidic, with an estimated cellular pH of approximately 0.6–1. Among Desmarestiales, *Desmarestia ligulata*, *D. tabacoides* and *D. viridis* showed extremely low intracellular pH, but no other comparably acidic species were found in the eight other orders examined.

Ion chromatography

The ionic compositions of 60 phaeophycean species from 10 orders are shown in Table 2. Seawater contained approximately 360 mM Na^+ and 380 mM Cl^- as primary ions, and 40 mM Mg^{2+} and 26 mM SO_4^{2-} as secondary major ions (Table 2, Fig. 1a).

Among Dictyotales and Desmarestiales, acidic species (*Dictyopteris latiuscula*, *D. prolifera*, *D. repens*, *Dictyopteris* sp., *Spatoglossum crassum*, *S.*

solierii, *Desmarestia ligulata*, *D. tabacoides* and *D. viridis*) contained significantly high concentrations of SO_4^{2-} (ca. 140–340 mM) (Fig. 1b). *Zonaria stipitata* also contained relatively high concentrations of SO_4^{2-} (65.4 mM).

In contrast, most non-acidic species of Dictyotales and Desmarestiales contained only 2.1–49.1 mM SO_4^{2-}, which was similar to that of seawater (26.3 mM) (Fig. 1c). Exceptionally, *Dictyota dichotoma* (78.7 mM), *D. divaricata* (108.9 mM), *Distromium decumbens* (65.8 mM), *Padina arborescens* (67.7 mM), *P. crassa* (175.4 mM), *P. minor* (193.0 mM) contained relatively high concentrations of SO_4^{2-} (Fig. 1d). These species contained relatively high concentration of Mg^{2+} (60.4–184.1 mM) as counter ion. Outside these orders, *Leathesia difformis* (Chordariales), *Sargassum confusum* and *S. patens* (Fucales) included relatively high concentration of Mg^{2+} (61.0–69.0 mM).

As seen in Table 2, the sum of cations and anions are not perfectly equal even in non-acidic species because not all of the inorganic ions and none of the organic ions were measured by ion chromatography. However, they are more or less equal in non-acidic species, compared with the acidic species listed above. A large part of the difference between total anions and cations in acidic species is considered to be protons, which could not be measured by ion chromatography. This high concentration of protons causes the extremely low pH in the acidic species.

Although most other non-acidic species in the other orders contained relatively low concentrations of SO_4^{2-} (2.1–60.4 mM), *Delamarea attenuata* (Dictyosiphonales, 274.7 mM) and *Thalassiophyllum clathrus* (Laminariales, 284.0 mM) contained considerably higher concentrations of SO_4^{2-} (Fig. 1e, f). As counter ions, *Leathesia difformis* and *Dictyosiphon foeniculaceus* contained relatively, high concentration of Mg^{2+} (respectively 66.5 and 97.3 mM). However, *Delamarea attenuata*, *Agarum clathratum* and *Thalassiophyllum clathrus* contained relatively high concentration of Ca^{2+} (76.5–258.9 mM). The only freshwater alga *Bodanella lauterbornii* also contained relatively high concentration of Ca^{2+} (92.1 mM).

Among other minor ions, *Padina* sp. and *Microzonia velutina* (Dictyotales) contained relatively high concentration of NH_4^+ (42.9–44.2 mM), whereas most other species contained less than 10 mM. *Kjellmania crassifolia* (Laminariales) contained high concentration of NO^{3-} (96.5 mM), whereas most other species contained less than 30 mM.

Table 2. Estimated cellular pH and inorganic ion composition in the species examined. Each datum of the inorganic ion represents average concentration ± S.E. (mM)

Species	Extract pH	pH within cells ± S.D.	No. of samples (n)	Na^+	NH_4^+	K^+	Mg^{2+}	Ca^{2+}	Cl^-	Br^-	NO_3^-	SO_4^{2-}	Σ cation	Σ anion	Σ anion/Σ cation	No. of samples (n)
Dictyotales																
Dictyopteris divaricata	5.3–5.7	4.2±0.27	5	77.3	0.8	164.8	5.8	0.9	184.9	22.4	N.D.	28.9	256.4	265.2	1.03	2
D. latiuscula	1.7–3.4	0.7±0.23	54	195.1±22.8	6.7±3.7	178.5±18.5	36.2±3.9	32.4±4.0	201.6±28.7	1.8±1.7	N.D.	187.0±18.4	518.4±35.7	577.4±51.0	1.11	9
D. prolifera	1.7–4.3	0.8±0.52	160	156.6±10.4	5.7±0.6	181.3±12.2	30.4±1.4	25.9±2.0	160.5±14.6	0.8±0.5	0.2±0.1	214.9±7.1	456.3±19.5	591.3±25.2	1.30	62
D. repens	3.4	0.8	1	161.7	16.6	151.1	54.8	121.7	182.4	2.4	44.1	277.5	682.5	783.8	1.15	1
D. undulata	5.4–7.3	4.2±0.79	43	142.8±19.5	1.6±0.7	243.1±17.2	18.5±2.4	4.1±1.3	351.1±22.2	1.2±0.8	1.0±0.4	17.8±2.0	432.8±28.1	388.8±23.9	0.90	20
Dictyopteris sp.	2.9–3.7	1.1±0.31	10	93.6±8.6	2.4±0.8	331.2±23.6	28.7±3.1	28.4±3.5	317.3±19.2	N.D.	10.4±2.5	138.3±18.6	541.3±40.9	614.7±49.5	1.14	10
Dictyota dichotoma	5.6–7.2	4.5±0.39	123	94.8±6.8	1.1±0.4	204.5±12.6	83.4±10.6	5.7±0.7	284.6±12.3	0.6±0.3	0.6±0.2	78.7±8.7	478.2±19.1	443.3±19.1	0.93	76
D. divaricata	6.2–6.5	4.7±0.09	5	57.4	1.2	82.9	86.1	11.1	95.6	0.8	N.D.	108.9	336.0	314.2	0.94	2
D. linearis	5.9–6.4	4.9±0.17	8	133.1	1.9	374.6	17.8	4.1	428.7	20.5	N.D.	34.3	553.1	517.7	0.94	2
Dictyopsis propagulifera	6.9–7.3	5.1±0.00	4	62.3±2.8	2.4±1.6	64.6±6.3	3.0±1.0	2.1±0.9	40.0±5.8	0.1±0.1	35.7±9.8	1.3±0.8	139.5±11.4	78.4±11.0	0.56	4
Dilophus okamurae	5.9–7.3	5.2±0.67	33	148.5±36.1	0.5±0.3	197.4±16.2	13.0±5.0	4.1±3.0	296.5±61.8	1.5±0.7	N.D.	22.0±3.5	380.7±64.6	342.0±65.8	0.90	9
Distromium decumbens	6.6–8.1	5.1±0.84	10	92.7±11.2	3.5±2.8	104.2±23.0	66.8±18.6	14.2±4.0	132.0±16.8	1.1±0.8	N.D.	65.8±15.0	362.3±26.4	264.6±30.7	0.73	6
Lobophora variegata	6.8–8.9	5.5±0.68	15	134.5±12.6	N.D.	97.6±24.8	11.9±4.0	2.5±0.7	175.6±33.1	0.7±0.7	N.D.	23.4±3.4	261.0±33.9	223.2±39.3	0.86	3
Pachydictyon coriaceum	5.1–6.3	4.4±0.46	19	232.0±87.5	0.3±0.3	381.0±95.8	44.7±14.3	8.5±4.2	618.9±161.1	5.6±3.7	0.5±0.5	33.5±11.2	719.8±188.6	692.0±185.9	0.96	4
Padina arborescens	6.6–7.9	5.7±0.53	31	175.3±30.4	N.D.	206.6±23.8	60.4±16.4	9.3±2.3	376.4±49.2	2.5±1.4	1.9±0.8	67.7±13.4	521.1±70.2	516.2±73.3	0.99	15
P. crassa	7.1–8.2	5.8±0.31	27	66.6±5.8	0.9±0.5	137.3±16.0	184.1±44.8	25.0±4.8	175.2±23.2	0.3±0.2	22.8±10.9	175.4±46.6	623.0±106.0	549.2±96.5	0.88	8
P. minor	7.9–8.4	6.5±0.19	22	75.2±12.0	1.5±1.0	98.9±24.1	159.4±37.0	24.2±2.6	110.7±10.6	0.9±0.9	193.0±24.7	542.7±62.1	503.4±57.1	0.93	6	5
Padina sp.	6.4–6.7	4.5±0.56	5	72.3±9.4	44.2±20.0	166.9±25.6	44.0±9.8	4.4±1.1	235.1±21.4	0.2±0.1	16.3±3.0	44.1±4.7	380.4±38.2	339.9±33.0	0.89	17
Spatoglossum crassum	1.4–3.4	0.5±0.31	27	124.5±11.6	3.0±0.5	204.4±24.2	29.1±2.7	28.8±3.4	148.2±13.5	2.2±2.1	1.9±0.8	256.4±18.0	426.2±25.4	664.9±43.6	1.56	4
S. pacificum	5.0–6.6	4.4±0.03	43	92.5±8.6	0.1±0.1	298.9±30.1	9.8±2.3	324.1±38.1	1.0±0.5	5.1±2.4	23.5±2.7	414.9±37.1	377.2±36.6	0.91	10	6
S. solieri	2.2–2.4	0.6±0.17	4	101.3±11.9	3.9±1.2	171.1±24.6	22.8±1.6	14.9±0.4	187.4±7.7	3.9±2.0	0.5±0.3	253.7±36.9	351.7±18.4	699.2±79.0	1.99	2
Stypopodium zonale	4.4–6.2	3.3±0.62	19	49.9±10.7	1.6±0.7	112.0±24.4	184.6±32.7	12.4±2.5	374.7±44.7	0.4±0.4	4.7±1.4	32.1±6.3	557.6±103.1	444.0±54.9	0.80	10
Taonia lennebackerae	6.1–6.2	4.4±0.21	2	72.5	1.2	116.0	31.8	4.4	68.3	1.0	9.8	49.1	262.1	177.4	0.68	5
Zonaria diesingiana	6.7–8.4	5.3±0.58	23	80.0±4.2	0.6±0.4	138.0±35.4	3.2±0.8	0.9±0.5	135.1±26.8	0.6±0.3	5.0±1.9	15.5±1.9	226.9±34.9	171.8±26.6	0.76	
Z. stipitata	2.9–3.9	1.4±0.28	5	78.8±14.5	5.3±1.3	267.7±80.0	15.3±1.8	12.4±1.9	318.8±86.4	N.D.	N.D.	65.4±3.1	407.2±89.1	449.6±83.2	1.10	
Desmarestiales																
Desmarestia aculeata	6.5–6.8	5.1±0.12	8	73.2±1.0	1.3±1.3	147.0±6.2	0.5±0.5	N.D.	212.2±14.2	0.2±0.2	N.D.	2.1±0.7	222.4±7.6	216.7±12.8	0.97	4
D. latifrons	7.1	4.6	1	143.8	8.0	34.7	36.3	3.0	104.1	N.D.	N.D.	11.1	265.0	126.2	0.48	1
D. ligulata	2.0–2.7	0.8±0.17	6	299.7±84.8	2.6±2.6	240.4±58.6	68.3±18.5	68.3±24.1	313.2±83.5	4.1±3.0	N.D.	349.7±112.6	821.8±222.5	1016.7±292.9	1.24	3
D. tabacoides	1.9–2.6	0.7±0.18	18	118.9±16.9	2.8±1.1	98.3±9.9	23.0±2.9	13.1±2.8	110.1±14.2	N.D.	1.3±0.5	175.4±12.9	292.4±35.5	462.2±35.5	1.58	10
D. viridis	1.3–2.4	0.5±0.12	16	86.2±17.9	1.0±0.8	136.5±18.5	23.6±3.7	30.7±6.0	114.8±25.7	14.0±12.4	15.9±15.5	249.1±30.1	332.3±53.9	642.9±88.0	1.93	4
Ectocarpales																
Bodanella lauterbornii	7.0	4.8	1	146.1	15.6	157.8	23.7	92.1	155.4	16.8	45.3	45.3	551.1	308.0	0.56	1

Continued on p. 258

Table 2 Continued

Species	Extract pH	pH within cells ± S.D.	No. of samples (n)	Na+	NH4+	K+	Mg2+	Ca2+	Cl-	Br-	NO3-	SO4^2-	Σ cation	Σ anion	Σ anion / Σ cation	No. of samples (n)
Chordariales																
Papenfussiella kuromo	6.3–6.4	4.5±0.22	4	127.3±7.5	N.D.	193.1±13.3	10.6±0.4	0.3±0.3	267.4±11.6	N.D.	0.6±0.6	8.5±0.9	342.3±19.3	285.1±11.0	0.83	4
Saundersella simplex	6.3	5.1±0.21	2	199.5	N.D.	66.5	19.7	3.6	276.4	1.5	N.D.	11.6	312.6	301.6	0.96	2
Ishige okamurae	5.4–5.8	4.0±0.18	5	293.5±6.6	3.3±1.1	108.5±4.3	34.2±4.4	16.1±3.2	244.2±15.0	1.1±0.9	0.9±0.9	15.6±1.4	505.9±14.2	277.4±18.8	0.55	5
I. sinicola	5.8–6.3	4.4±0.24	5	223.2±25.7	3.9±2.0	76.5±10.5	30.1±4.4	15.3±2.9	179.9±20.5	0.8±0.6	9.2±1.0	10.1±25.2	394.5±39.2	210.1±23.1	0.53	5
Leathesia difformis	6.5–6.9	5.4±0.25	4	222.4±62.5	N.D.	280.1±64.7	66.5±29.5	1.8±1.0	531.5±59.0	4.3±2.6	4.7±2.9	60.4±1.4	639.1±89.8	661.1±103.5	1.03	4
Scytosiphonales																
Colpomenia sinuosa	5.5–6.2	4.0±0.29	4	128.8±21.1	0.8±0.8	433.0±42.4	10.0±2.2	1.5±0.9	550.5±59.5	0.4±0.4	1.9±1.4	11.0±1.8	585.5±57.6	574.8±61.3	0.98	4
Petalonia binghamiae	6.1–6.4	4.7±0.10	4	60.9±5.3	N.D.	200.8±18.4	5.3±0.7	0.2±0.2	183.5±13.8	N.D.	12.8±1.2	37.0±6.5	272.7±21.4	270.3±25.4	0.99	4
Scytosiphon gracilis	6.2–6.3	4.8±0.08	4	98.9±18.7	N.D.	169.7±14.6	6.2±1.1	0.2±0.2	189.9±17.4	N.D.	10.7±2.1	38.3±2.8	281.3±21.4	277.1±21.5	0.99	4
Dictyosiphonales																
Delamarea attenuata	6.3–6.9	4.8±0.21	6	166.0	3.1	244.7	39.5	125.1	217.7	19.4	5.5	274.7	743.0	792.2	1.07	2
Stschapovia flagellaris	6.6–7.1	4.8±0.99	3	151.3±66.0	N.D.	240.1±58.5	8.9±1.9	2.2±2.2	251.8±39.7	N.D.	N.D.	33.0±6.0	413.6±13.0	318.0±51.2	0.77	3
Dictyosiphon foeniculaceus	6.3–6.4	4.4±0.20	3	150.4±14.0	3.9±0.9	183.9±19.4	97.3±13.4	8.6±3.2	387.7±41.9	4.3±2.2	N.D.	55.1±7.0	550.0±64.8	502.1±56.4	0.91	3
Cutleriales																
Cutleria cylindrica	5.7–5.9	3.8±0.144	126.0	7.2	338.6	2.7	3.3	263.8	N.D.	N.D.	20.1	483.8	304.1	0.63	2	
C. multifida	6.5–7.1	5.4±0.42	3	294.9	N.D.	121.0	25.2	5.2	440.7	21.2	N.D.	7.0	827.0	475.9	0.58	3
Microzonia velutina	6.6	5.0±0.12	3	25.9±8.1	42.9±3.4	63.0±3.7	8.7±0.3	1.3±0.6	47.9±8.9	3.9±0.5	31.3±3.9	1.3±0.8	152.0±11.2	85.7±13.4	0.56	3
Sporochnales																
Carpomitra costata	6.8–7.0	4.7±0.15	6	115.0±24.8	8.7±2.2	592.6±95.6	12.9±8.5	17.7±17.7	516.5±157.1	6.9±6.9	N.D.	11.1±1.9	777.6±172.7	546.0±160.0	0.70	3
Sporochnus scoparius	6.5–6.6	5.3±0.20	3	68.5	N.D.	256.3	6.3	N.D.	319.7	9.2	1.9	4.1	337.0	339.0	1.01	1
Laminariales																
Agarum clathratum	6.1–6.3	4.7±0.14	6	154.1±23.4	5.2±2.0	211.7±17.2	26.2±2.1	76.5±14.9	467.4±37.6	N.D.	18.6±4.5	52.4±13.6	576.2±31.9	590.9±38.0	1.03	4
Chorda filum	4.8–5.6	3.8±0.14	14	258.4±38.7	N.D.	354.4±41.7	21.9±3.4	4.9±1.0	461.1±60.4	0.2±0.2	4.9±2.7	15.6±2.2	666.4±85.1	492.5±61.7	0.74	4
Ecklonia cava	5.8–6.4	4.8±0.25	5	168.6±14.7	2.0±2.0	132.5±16.3	5.0±0.6	1.0±0.4	276.0±36.6	0.3±0.2	4.9±2.7	8.3±0.7	315.3±32.7	297.8±40.6	0.94	5
Ecklonia kurome	6.1–6.3	4.6±0.24	4	157.7±10.2	N.D.	282.3±28.2	10.7±1.9	2.8±1.0	386.7±38.4	1.4±0.5	0.8±0.4	7.0±1.2	466.9±40.3	402.9±40.6	0.86	4
Kjellmaniella crassifolia	6.1–6.3	4.9±0.05	6	206.0±17.9	1.2±1.2	226.2±26.4	7.4±1.9	3.0±0.4	320.1±42.3	N.D.	96.5±29.7	9.2±0.8	454.3±10.7	435.1±21.7	0.96	4
Thalassiophyllum clathrus	6.3	4.9±0.06	3	132.9±38.7	N.D.	158.4±32.5	47.4±19.0	258.9±97.9	242.8±67.3	N.D.	N.D.	284.0±88.6	903.9±323.6	810.0±243.3	0.90	3
Fucales																
Hizikia fusiformis	6.3–6.9	5.1±0.17	4	112.3±5.0	N.D.	388.4±22.1	6.0±1.2	0.9±0.6	476.5±23.4	0.2±0.2	1.2±1.2	4.0±1.5	515.0±27.9	485.8±24.0	0.94	4
Sargassum confusum	5.0–5.6	3.7±0.30	4	99.6±5.4	N.D.	296.1±8.0	69.0±3.2	11.3±1.1	498.5±13.0	1.3±1.3	13.2±2.1	3.1±0.3	556.3±10.4	519.3±11.1	0.93	4
S. hemiphyllum	5.8–6.5	4.4±0.29	10	111.9±17.3	0.7±0.7	227.5±52.4	11.0±2.7	0.3±0.3	238.4±46.9	N.D.	N.D. 7.1±1.4	363.0±44.1	252.6±47.2	0.70	4	4
S. micracanthum	5.7–6.1	4.0±0.17	4	165.0±4.7	2.2±2.2	320.0±35.0	39.8±11.0	2.5±2.5	473.4±57.0	0.3±0.3	20.8±8.2	4.5±1.8	571.9±61.8	503.5±67.2	0.88	4
S. patens	5.5–5.8	3.9±0.18	4	93.1±12.1	N.D.	360.1±39.8	61.0±4.4	2.5±1.0	489.2±48.0	12.5±2.0	34.5±2.3	6.3±1.6	580.4±55.7	548.6±45.2	0.95	4
S. ringgoldianum ssp. coreanum	5.8–6.0 4.2±0.31	4	146.4±2	1.2	N.D.	222.1±29.9	18.8±3.8	2.1±0.8	379.8±55.3	N.D.	N.D.	3.6±1.8	410.0±56.3	387.0±55.5	0.94	4
S. thunbergii	6.2–6.9	5.0±0.34	4	98.5±6.3	N.D.	179.7±10.6	4.7±0.5	0.4±0.2	256.9±16.9	0.1±0.1	1.8±0.9	4.2±1.0	288.0±16.8	267.3±18.1	0.93	4
Turbinaria ornata	6.3–7.0	5.1±0.36	13	85.0	N.D.	108.4	1.7	N.D.	138.9	N.D.	N.D.	0.6	197.0	140.0	0.71	2
Sea Water				360.1±41.3	N.D.	7.5±0.8	38.9±5.1	7.8±1.2	379.4±60.2	0.9±0.2	N.D.	26.3±1.5	460.9±54.7	433.1±61.1	0.97	7
Sea Water Medium				441.5±0.9	N.D.	11.2±0.6	50.4±0.6	5.9±1.8	494.7±11.0	N.D.	N.D.	25.9±1.0	565.3±5.1	547.0±12.9	0.94	4
Fresh water medium				2.40±0.02	N.D.	0.02±0.00	0.04±0.00	0.5±0.00	1.02±0.02	N.D.	1.00±0.00	0.12±0.00	3.5±0.02	2.25±0.02	0.64	4

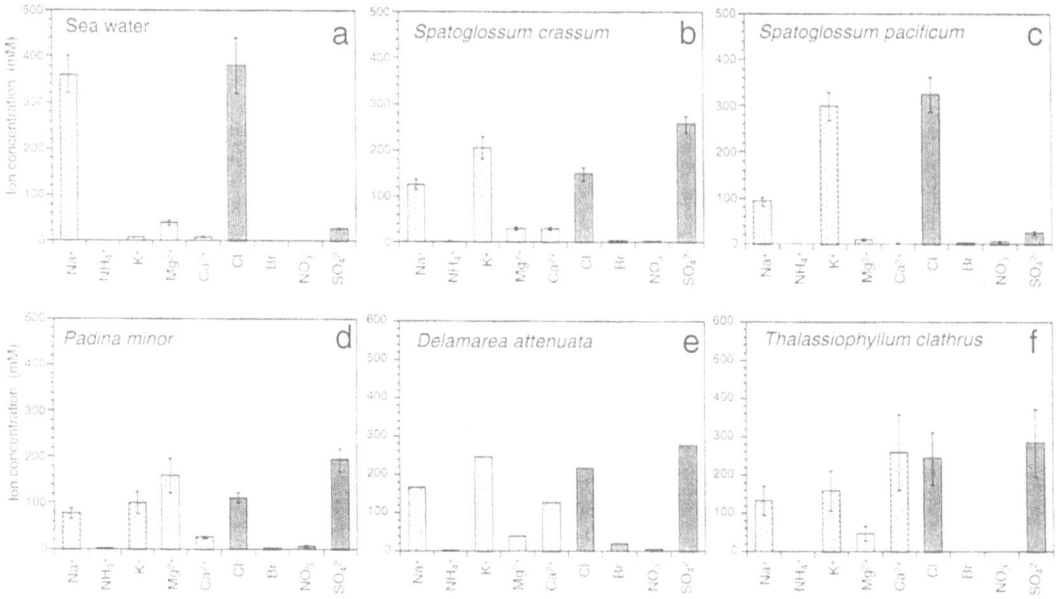

Figure 1. Estimated intracellular ion concentrations (mM) of representative acidic and non-acidic species: *Spatoglossum crassum* (b, acidic with high SO_4^{2-}), *Spatoglossum pacificum* (c, non-acidic), *Padina minor* (d, non-acidic with high SO_4^{2-} and Mg^{2+}), *Delamarea attenuata* (e, non-acidic with high SO_4^{2-} and Ca^{2+}) and *Thalassiophyllum clathrus* (f, non-acidic with high SO_4^{2-} and Ca^{2+}). Ionic compositions of sea water (average of 7 samples) are also shown (a). For number of samples in each species and the data for other species, see Table 2.

Discussion

In the present survey, three dictyotalean species (*Dictyopteris* sp., *Spatoglossum solierii* and *Zonaria stipitata*) were newly found to be highly acidic, accumulating high concentrations of SO_4^{2-} although the concentration was not very high in *Z. stipitata*. Their cellular pH was comparable to previously reported acidic species of Dictyotales and Desmarestiales (McClintock et al., 1982; Sasaki et al., 1999). Therefore, highly acidic species in the order Dictyotales are distributed among three genera (*Dictyopteris*, *Spatoglossum* and *Zonaria*), although not all members of these genera showed high acidity. This situation clearly contrasts with Desmarestiales, in which the monophyly of this character is supported by molecular phylogenetic studies (Peters et al., 1997). The order Dictyotales has been classified into two families based on the number of apical cells (Dictyotaceae with a single apical cell and Zonariaceae with marginal rows of apical cells, J. Agardh 1894), or into two tribes (Dictyoteae Greville and Zonarieae De Toni) in a single family Dictyotaceae (Womersley, 1987). Among these, all genera containing highly acidic species belong to the family/tribe Zonariaceae/Zonarieae. Accumulation of sulfate ion itself is found in much broader systematic groups within Dictyotales (Sasaki et al., 1999) in both of the families/tribes, and other orders (e.g., Dictyosiphonales and Laminariales; present paper). However, high acidity caused by balancing the high concentration of sulfate ion by protons (H^+) might have occurred after the evolutionary divergence of the family/tribe within Dictyotales, as occurred relatively newly in Desmarestiales (Peters et al., 1997). This question would be clarified by molecular phylogenetic analysis of the related genera in Dictyotales that is in progress in our laboratory.

Delamarea attenuata, growing on exposed upper subtidal and lower intertidal rocks in North Pacific and North Atlantic cold waters, has been noted to be relatively delicate and easily damaged when collected and kept in small containers filled with culture medium (H. Kawai, unpublished observations). In such cases, the medium turns yellow from the leakage of some substances from the plants (presumably carotenoids). Such vulnerability of the plants might be explained by the accumulation of a high concentration of sulfate ion. When *Delamarea* plants suffer some environmental changes, sulfate ion might leak to cause further damage to the cytoplasm and the whole cell suffers (sometimes lethal) damage, causing the leakage of yellow substances.

Well-developed vacuoles are one of the most important cellular organelles characterizing plant cells. Their functions have been relatively well studied in higher plant cells, but very little is known for algal cells. Vacuoles of various macro-algal cells have similar structures as those of higher plants and also contribute to production of turgor pressure. However, there could be considerable differences in other functional aspects, because the environmental conditions surrounding the cells are very different. The new finding of accumulation of relatively high concentrations of Ca^{2+} in addition to Mg^{2+}, both functioning as counter ions of SO_4^{2-}, implies that the active accumulation of sulfuric acid itself is not the sole function. It may be a result of complex regulation of various correlated cations and anions in brown algae. As for higher plants, the relationship between SO_4^{2-} and Ca^{2+} has been reported in cultured tobacco cells (Smith, 1978; Jones & Smith, 1981). The transport of SO_4^{2-} was stimulated by Ca^{2+} and increased the ratio of SO_4^{2-} in cell. However, the quantity of intracellular SO_4^{2-} reported was 0.01–10.0 mM, and the quantity of Ca^{2+} used in the experiment was 0.5 mM (Jones & Smith, 1981), much lower than that of accumulated ions in brown algae.

Acknowledgements

We are grateful to Dr Eric Henry for critically reading the manuscript, to Dr D.G. Müller for providing culture materials, and to Dr A. Flores-Moya for his support in collecting *S. solierii*. This study was partly supported by grants-in-aid for Scientific Research (No.10836012 to H. Kawai, No. 05454013 and 10874131 to H. Kataoka) from the Ministry of Education, Science, Sports, and Culture, Japan and the Grant for Collaboration Experiments from the Institute of Genetic Ecology, Tohoku University (1997–1998).

References

Agardh, J. G., 1894. Analecta algogogica. Cont. II. Lunds Univ. Årsskr. 30: 1–98.

Eppley, R. W. & C. R. Bovell, 1958. Sulfuric acid in *Desmarestia*. Biol. Bull. Mar. Biol. Lab. Woods Hole 115: 101–106.

Jones, S. L. & I. K. Smith, 1981. Sulfate transport in cultured tobacco cells. Plant Physiol. 67: 445–448.

McClintock, M., N. Higinbotham, E. G. Uribe & R. E. Cleland, 1982. Active, irreversible accumulation of extreme levels of H_2SO_4 in the brown algae, *Desmarestia*. Plant Physiol. 70: 771–774.

Peters, A. F., M. J. H. van Oppen, C. Wiencke, W. T. Stam & J. L. Olsen, 1997. Phylogeny and historical ecology of the Desmarestiaceae (Phaeophyceae) support a southern hemisphere origin. J. Phycol. 33: 294–309.

Sasaki, H., H. Kataoka, M. Kamiya & H. Kawai, 1999. Accumulation of sulfuric acid in Dictyotales (Phaeophyceae): taxonomic distribution and ion chromatography of cell extracts. J. Phycol. 35: 732–739.

Schiff, J. A., 1962. Sulfur. In Lewin, R. A. (ed.), Physiology and Biochemistry of Algae. Academic Press, New York: 239–246.

Smith, I. K., 1978. Role of calcium in serine transport into tobacco cells. Plant Physiol. 62: 941–948.

Tatewaki, M., 1966. Formation of crustaceous sporophyte with unilocular sporangia in *Scytosiphon lomentaria*. Phycologia 6: 62–66.

Wirth, H. E. & G. B. Rigg, 1937. The acidity of the juice of *Desmarestia*. Am. J. Bot. 24: 68–70.

Womersley, H. B. S., 1987. The marine Benthic Flora of Southern Australia. Part II. South Australian Government Printing Division, Adelaide: 484 pp.

Hydrobiologia **512**: 263–266, 2004.
P.O. Ang, Jr. (ed.), Asian Pacific Phycology in the 21st Century: Prospects and Challenges.
© 2004 *Kluwer Academic Publishers.*

Preliminary studies on the chemical characterization and antihyperlipidemic activity of polysaccharide from the brown alga *Sargassum fusiforme*

Wenjun Mao, Bafang Li, Qianqun Gu, Yuchun Fang & Hongtao Xing
Institute of Marine Drugs and Foods, Ocean University of China, Qingdao, 266003, P.R. China
E-mail: wenjunmqd@hotmail.com

Key words: Sargassum fusiforme, Polysaccharide, IR spectroscopy, alginate, antihyperlipidemic activity, China

Abstract

A polysaccharide (SFP) extracted from the brown alga *Sargassum fusiforme* (Harv.) Setch. was purified by chromatography on DEAE-Sephadex A50 and Sephadex G-100. Studies using paper chromatography (PC), electrophoresis and infrared spectroscopy (IR) indicated that SFP was a kind of alginate with a molecular weight of 16 000 and a molar ratio of mannuronic acid (M) to guluronic acid (G) of 2.75. Pharmacological experiments showed that SFP could markedly decrease the content of total cholesterol (TC), triglyceride (TG) and low density lipoprotein-cholesterol (LDL-C) in the serum of experimental hyperlipidemic rats, and significantly increase the level of high density lipoprotein-cholesterol (HDL-C).

Introduction

China has a rich flora of brown algae. However at present, mainly members of the *Laminariaceae* (kelps) have been commercially exploited. Members of Sargassaceae are of lesser importance. *Sargassum fusiforme* (Harv.) Setch. is distributed along the coasts of Guangdong, Fujian, and Zhejiang provinces in southeastern China. It has been part of the cornucopiae of the traditional Chinese medicine for thousands of years. In the classic Ming Dynasty script, the Pen Tsao Kang Mu, it was recorded as "a salty alga [that] can moisten, let out heat and draw water, therefore it can remove tumor and tuberculosis". Although many investigations have discovered that the bioactivity of seaweed polysaccharide was closely associated with its chemical composition or structure, the relationship between the chemical characterization and the antihyperlipidemic activity of the polysaccharide from *S. fusiforme* has not been reported. The present paper is an attempt to fill this gap.

Materials and methods

Polysaccharide preparation

Individuals of the cultured alga *Sargassum fusiforme* were collected and dried in the sun at the coast of Dongtou county, Zhejiang province, eastern China, in November 1996. The dried seaweed was milled and extracted with water at 80 °C for 10 h. The crude extract was added slowly with 3 volumes of ethanol, and the solution was laid at 4 °C for 24 h. Filtration and centrifugation were conducted to collect precipitate. The precipitate collected was dissolved in distilled water, dialyzed in 30 volumes of water with VISKING dialyze membrane, and freeze-dried successively to get crude polysaccharide. Solution of the crude polysaccharide in water was applied to a column 26 × 100 cm containing DEAE-Sephadex A50 (Pharmacia Co.), and chromatographed using a gradient of 0~2 M NaCl as eluant. The eluates were monitored by UV absorbance at 270 nm. The fractions containing the polysaccharide were purified further by chromatography on Sephadex G-100 (Pharmacia Co.). The major fractions were pooled, concentrated, desalted and lyophilized to obtain polysaccharide (SFP).

Analysis of the polysaccharide

The constituent sugar of SFP was determined by paper chromatography (PC) of the hydrolyzed solution of SFP. Hydrolysis was conducted with 80% sulfuric acid at 20 °C for 18 h. The PC of SFP was developed on XinHua chromatographric paper in the solvent system n-butylalcohol, pyridine, and water (4:1:5). Spots of reducing sugars were visualized by spraying with aniline hydrogen phthalate reagent and subsequently heated at 105 °C for 5 min.

The molar ratio of mannuronic acid to guluronic acid (M/G) was determined by chemical method (Ji et al., 1981). SFP was hydrolyzed with 80% sulfuric acid and neutralized with calcium carbonate. The hydrolyzed solution was concentrated and subjected to Dowex 1 × 8 (200–400 mesh) column chromatography eluted with gradient of 0~2 M acetic acid. The amounts of M and G in the fractions were determined by phenol-sulfuric acid reaction and compared with the curves of standard M (Sigma Co.) and G (Sigma Co.). The cellulose acetate membrane electrophoresis of SFP was performed on DYY-III 8 electrophoresis instrument (China) in 0.1% 1 M pyridine-acetic acid buffer (pH 3.5) at 100 V, 25 °C for 20 min. Chromogenic reaction was carried by spraying toluidine-blue solution on the cellulose acetate membrane at the end of electrophoresis.

The viscosity of SFP was determined by Ubbelohde viscometer with 1% SFP in 0.1 M NaCl at 25 °C (Haug et al., 1962; Moore, 1975), and the molecular weight of SFP was calculated from the intrinsic viscosity [η] by Mark formula (Donnan & Rose, 1950; Ji, 1997).

The Infrared (IR) spectroscopy of SFP was recorded on a Nicolect model 510 FT-IR spectrometer (Mackie, 1971).

Assay of antihyperlipidemic activity

Twenty rats (Wistar strain, weight of body is about 200 ± 20 g) were fed with high fat diet for 30 days to obtain hyperlipidemic rats model. These rats were then divided into two groups randomly (10 rats/group): Hyperlipidemia group and SFP group. Hyperlipidemia group was continuously fed with the high fat diet. SFP group was given SFP by gastrointestinal injection at a daily dose (200 mg kg^{-1}) when raised with the high fat diet. Moreover, a control group (10 rats/group) raised with the basal diet was established. The ingredients of the diets are given in Table 1. The diet and water were supplied *ad libitum* during the experiment.

Table 1. Composition of basal diet and high fat diet for rats used in the experiment

Basal diet ingredients	%	High fat diet ingredients	%
Wheat powder	39.2	Basal diet	82.5
Rice	10.5	Cholesterol	2.0
Bean cake powder	15.1	Cholate	0.5
Corn meal	15.1	Egg yolk powder	5.0
Fish meal	15.1	Lard	10.0
Yeast powder	3.8		
Fish-liver oil	0.7		
Sale	0.5		

The test was conducted for 15 days. Blood samples were taken from the tail region of the rat 12 h after the SFP gastrointestinal injection. The serum obtained by centrifugation was used to determine the level of total cholesterol (TC), triglyceride (TG) and low density lipoprotein-cholesterol (LDL-C) and high density lipoprotein-cholesterol (HDL-C) by enzyme methods. Results were expressed as mean ± SD and differences between groups were compared statistically using Student *t* test.

Results and discussions

Chromatography of polysaccharides from *Sargassum fusiforme* on DEAE-Sephadex A50 (100 × 2.6 cm) is shown in Figure 1. Fraction I was concentrated and further eluted on Sephadex G-100 (100 × 2.6 cm) column with water (Fig. 2). The major fractions were pooled, concentrated *in vacuo* and desalted on a Sephadex G-25 column (100 × 2.6 cm) with water. The eluate was concentrated and freeze-dried to obtain SFP.

The paper chromatogram (PC) of SFP appeared as two brown-yellow spots of which R_f values were 0.24 and 0.28, respectively, corresponding to that of standard guluronic acid and mannuronic acid. The molar ratio of mannuronic acid to guluronic acid (M/G) in SFP resulted from the chemical analysis was 2.75, which was higher than that of the alginate from most of the brown seaweeds, such as, *Laminaria japonica* Aresch. (2.26), *Sargassum pallidum* (Turn.) Ag. (1.26) and *Sargassum hemiphyllum*(Turn.) Ag. (1.06) (Ji & Wang, 1984). The result of electrophoresis showed

Table 2. Mean (± SD) levels of serum total cholesterol (TC), triglyceride (TG), high density lipoprotein-cholesterol (HDL-C) and low density lipoprotein-cholesterol (LDL-C) (10^{-3}M) in control, hyperlipidemia and SFP rats ($n = 10$)

Group	TC	TG	HDL-C	LDL-C
Control	1.51 ± 0.14	0.67 ± 0.13	1.26 ± 0.08	0.96 ± 0.15
Hyperlipidemia	$5.24 \pm 1.20^{\Delta\Delta}$	$1.29 \pm 0.25^{\Delta\Delta}$	$0.63 \pm 0.07^{\Delta\Delta}$	$2.50 \pm 0.23^{\Delta\Delta}$
SFP	$2.12 \pm 0.14^{**}$	$0.68 \pm 0.14^{*}$	$1.09 \pm 0.06^{*}$	$0.80 \pm 0.19^{**}$

Student's *t*-test, $^{*} p < 0.05$, $^{**} p < 0.01$ compared with hyperlipidemia. $^{\Delta\Delta} p < 0.01$ compared with control.

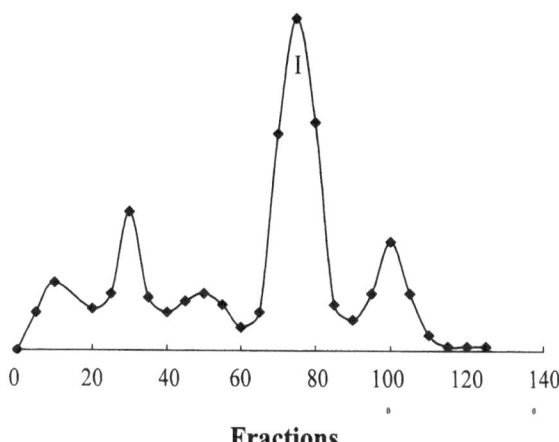

Fractions

Figure 1. Chromatogram of polysaccharides from *Sargassum fusiforme* on DEAE-Sephadex A50 (100 × 2.6 cm).

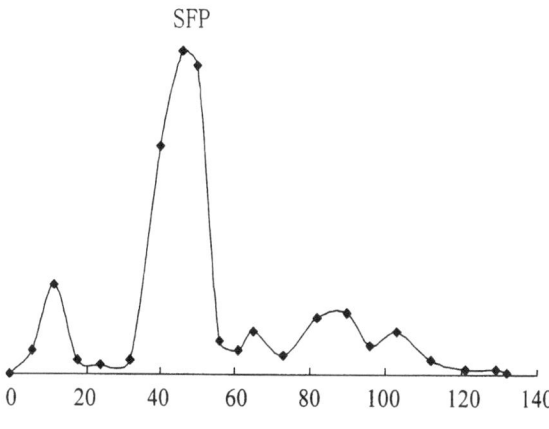

Fractions

Figure 2. Chromatogram of fraction I on Sephadex G-100 (100 × 2.6 cm).

that SFP was an acidic polysaccharide which appeared as a blue spot colored by toluidine-blue on the cellulose acetate membrane. The blue spot migrated from cathode to anode. Using the method of viscometry, the molecular weight of SFP was estimated to be about 16 000.

The absorption peaks of IR for SFP were as follows: 3435 cm^{-1}, the stretch vibration of O-H, existed in the hydrogen bond of the molecules; 2930 cm^{-1}, the stretch vibration of $-$CH; 1614 cm^{-1}, the asymmetric stretch vibration of $-$COO^{-}; 1417 cm^{-1}, the symmetric stretch vibration of $-$COO^{-} and the stretch vibration of C-O within $-$COOH; 1260 cm^{-1}, the stretch vibration of S=O; 1040 cm^{-1}, the stretch vibration of C-O and change angle vibration of O-H; 821 cm^{-1}, the feature absorption of mannuronic acid; 790 cm^{-1}, the feature absorption of guluronic acid. The spectrum of SFP was similar to that of the alginate from *Laminaria japonica* which was used as raw material for alginate production in China. It was found from the IR spectrum of SFP that the intensity of the peaks of mannuronic acid at 821 cm^{-1} and guluronic acid at 790 cm^{-1} was different from that of the alginate from *L. japonica*. The peak of mannuronic acid of SFP was stronger, and the peak of guluronic acid was weaker. It indicated that the M/G ratio in SFP was higher than that of the alginate from *L. japonica*. In addition, there was an asymmetry ring stretch vibration at 940 cm^{-1} for the alginate from *L. japonica,* but this is not present in SFP. It suggested that the properties for substitution groups, hydroxyl and carboxyl groups on the pyran rings of SFP were different from those of the alginate from *L. japonica*.

Table 2 shows that TC, TG, LDL-C levels in serum of the hyperlipidemic rats were significantly higher than those of the control and SFP groups ($p < 0.01$). HDL-C levels in serum of the hyperlipidemic rats was significantly lower than those of the other groups ($p < 0.01$). This indicates that SFP could obviously decrease the content of TC, TG and LDL-C in serum of experimental hyperlipidemic rats, and also significantly increase the level of serum HDL-C (Table 2). The decrease in TC level was 59.54%, TG 47.28%, LDL-C 68%, and the increase in HDL-C level was 42.2%. We deduce from the above findings that SFP has strong antihyperlipidemic activity.

The chemical composition of the polysaccharide from *Sargassum fusiforme* was studied for the first time. We found that it was a kind of acidic polysaccharide-alginate composed of mannuronic acid and guluronic acid with a M/G ratio higher than that of common alginate in China, for example, the alginate from *Laminaria japonica*. Individual seaweed species characteristically differ from each other in their relative proportions (M/G ratio) of these two constituent sugar acids, as well as in their sequencing with the polymer chain structural elements that have a bearing on their physical properties. Also, for a given species, its polysaccharide composition varies considerably depending upon the season, age, and the part of the plant used for extraction. The properties of the substitution groups on pyran rings in the SFP polysaccharide were different from those of common alginate. It could be deduced that the cross-link structures between molecules in SFP were different from those of the latter. These chemical structure characteristics of SFP were similar to those of the polysaccharide from *Sargassum fulvellum* (Turn.) C. Ag. reported by Michio et al. (1984). Though the research showed that the antitumour activities of the polysaccharides from *S. fulvellum* could be associated with its higher M/G ratio, the relation between antihyperlipidemic activity of polysaccharide from *S. fusiforme* and its M/G ratio could not be confirmed by our experiment. We think that the modes of linkages of pyran rings between polysaccharide chains could also be associated with the antihyperlipidemic activities. Further studies on the relationship between chemical structure and antihyperlipidemic activity of the polysaccharide from *S. fusiforme* are currently being undertaken.

References

Donnan F. G. & R. C. Rose, 1950. Osmotic pressure, molecular weight, and viscosity of sodium alginate. Can. J. Res. 28B: 105–113.

Haug A., & O. Smidsrod, 1962. Determination of intrinsic viscosity of alginate. Acta Chem. Scand. 16: 1569–1578.

Ji, M. H., 1997. Chemical composition and structure of algin. In Ji, M. H. (ed.), Seaweed Chemistry. Science Press, Beijing: 231–250.

Ji, M. H. & Y. J. Wang, 1984. Studies on the M:G ratios in alginate. Hydrobiologia 116: 554–556.

Ji, M. H., C. Wenda & H. Lijun, 1981. Determination of uronic acid composition of algin. Oceanol. Limnol. Sin. 12: 533–539.

Mackie, W., 1971. Semi-quantitative estimation of the composition of alginates by infrared spectroscopy. Carbohyd. Res. 20: 413–415.

Michio, F., I. Noriko, Y. Ichiro & N. Terukazu, 1984. Purification and chemical and physical characterization of an antitumour polysaccharide from the brown seaweed *Sargassum fulvellum*. Carbohyd. Res. 125: 97–106.

Moore, W. R., 1975. Viscosity of dilute polymer solutions. Progress in Polymer Science 1: 52–89.

Xu, S. Q., R. L. Bian & X. Chen, 1982. Pharmacology Experiment Methods. Renmin Hygiene Press, Beijing.

Zheng, N. Y., Y. X. Zhang & X. Fan, 1992. Studies on the compositions and sequential structures of uronate residues in alginates from Chinese brown algae *Laminaria* and *Sargassum*. Oceanol. Limnol. Sin. 23: 445–452.

Hydrobiologia **512**: 267–270, 2004.
P.O. Ang, Jr. (ed.), Asian Pacific Phycology in the 21st Century: Prospects and Challenges.
© 2004 *Kluwer Academic Publishers.*

Hepatoprotective effect of seaweeds' methanol extract against carbon tetrachloride-induced poisoning in rats

Chun-Kwan Wong, Vincent E. C. Ooi* & Put O. Ang, Jr.
Department of Biology, The Chinese University of Hong Kong, Shatin, N. T., Hong Kong SAR, China
*Author for correspondence; E-mail: vincent-ooi@cuhk.edu.hk

Key words: Myagropsis, Sargassum, methanol extract of marine macroalgae, hepatoprotective activity, carbon tetrachloride

Abstract

Three species of marine brown macroalgae (seaweeds), *Myagropsis myagroides, Sargassum henslowianum* and *S. siliquastrum* collected from Tung Ping Chau, Hong Kong were studied for their curative effects on hepatotoxicity caused by carbon tetrachloride (CCl_4) in male Sprague-Dawley rats. A single suitable oral dose of 1.25 ml kg^{-1} of 20% CCl_4 was used as a model hepatotoxin to produce significantly elevated levels of serum glutamic pyruvic transaminase (SGPT) and glutamic oxaloacetic transaminase (SGOT). Gavage oral administration of 300 mg kg^{-1} of methanol crude extract from *S. siliquastrum* 6 h post-treatment of CCl_4 significantly reduced the CCl_4-induced acute elevation in the levels of SGPT and SGOT in rats. Similar results, though at a less effective level, were achieved for extracts from *S. henslowianum* and *M. myagroides*. These results indicate that these seaweeds may contain some active principles in their methanol extracts which acted as an antidote against the hepatotoxicity induced by CCl_4. Further investigation is necessary to clarify and characterize the active component(s) in the extracts.

Introduction

Carbon tetrachloride (CCl_4) is one of several chemicals which cause liver injury. It has long been well documented as a hepatotoxin (Recknagel, 1967; Klaassen & Plaa, 1969; Harris et al., 1982). It is introduced into the water mainly as industrial wastes from its primarily use in the manufacture of chlorofluorocarbons (Borzelleca et al., 1990). CCl_4 causes centrilobular necrosis and fatty accumulation in the liver. Early studies have proved that CCl_4-induced hepatotoxicity is catalyzed by cytochrome P-450 in the endoplasmic reticulum of hepatocytes (Recknagel, 1967; Slater, 1984). The present study used CCl_4 as a model hepatotoxin to induce experimental liver injury. Three species of seaweeds, *Myagropsis myagroides* (Mertens *ex* Turner) Fensholt, *Sargassum henslowianum* C. Ag. and *S. siliquastrum* (Turn.) Ag., abundantly found in Tung Ping Chau, an island located in the northeastern part of Hong Kong, were collected and examined for their curative effects against

hepatotoxicity caused by CCl_4 after the gavage oral administration of their methanol extracts in male rats.

Materials and methods

Preparation of methanol extracts of seaweeds

Fresh samples of *Myagropsis myagroides, Sargassum henslowianum* and *S. siliquastrum* were used. For extraction, washed seaweeds were weighed and blended with distilled water. They were kept at 4 °C for 1 d and then filtered through cotton gauze. The filtrates were centrifuged at 23 700 *g* for 20 min. After centrifugation, the water extracted supernatant was stored for other uses and the pellets (residues), which could not dissolve in water, were freeze-dried. The resulting pellets were placed in the Soxhlet apparatus for 6 h by using pure methanol (Ajax Chemicals, Australia). The total amount of methanol extract products was combined and evaporated to dryness under vacuum at 40 °C by using the rotor evaporator to extract water-insoluble extract. The final residue was dissolved in pure methanol and stored in air-tight glass vials for aspiration under nitrogen gas to form dark-green viscous semisolid. The final products were stored in the refrigerator until use.

Experimental protocol

Carbon tetrachloride (CCl_4) induces hepatotoxic effect when taken in suitable dose (1.25 ml kg^{-1}) (Slater, 1966). The CCl_4 (Ajax Chemicals, Australia) was dissolved in corn oil and was introduced into the stomach of the rat by gavage oral administration through an intragastric tube.

Seven to 8 week old male Sprague-Dawley rats, weighing from 150 to 250 g, were provided with tap water and rodent chows *ad libitum*, and housed in a controlled-environment with 12 h day light. These experimental animals were divided into seven groups, each group being consisted of five animals. The first group (the no treatment control group) received no chemical treatment. The second group (the saline vehicle control group) received only the vehicle (6.25 ml kg^{-1}) orally. The vehicle is made of corn oil (Mazola, U.S.A) and normal saline solution (10 ml kg^{-1}). The third group (the CCl_4 toxin treatment group) received the suitable dose of CCl_4 to induce chemical hepatitis followed 6 h later by oral saline administration. The fourth group (the DMSO vehicle control group) was treated similarly to Group 3 except that another vehicle, 25% of dimethyl sulfoxide (DMSO) (Ajax Chemicals, Australia) dissolved in 0.9% (v/v) saline, was administered instead of saline 6 h after exposure to CCl_4 to evaluate its effect on CCl_4-induced hepatotoxicity. DMSO was the vehicle used to carry the seaweed methanol extracts used in the experiments. The last three groups were the treatment groups. They were treated similarly to Group 3 (the CCl_4 toxin treatment group) except that each seaweed extract was individually administered in each group instead of saline to evaluate their curative effects. All animals in each of these treatment groups received one dose of each seaweed extract (300 mg kg^{-1}, dissolved in 10 ml of 25% of DMSO vehicle) respectively.

Biochemical assays

Enzyme activities of serum glutamic pyruvic transaminase (SGPT) and serum glutamic oxaloacetic transaminase (SGOT) in blood serum served as parameters to illustrate the extent of hepatotoxicity in rats.

The animals were anaesthetized with ether 24 h after the last hepatotoxin treatment (CCl_4) and blood (5 ml) was withdrawn from their posterior vena cava with sterile disposable syringes equipped with hypodermic needles. Serum was separated by centrifugation at 1100 g for 15 min. Serum was separated from the cells immediately to prevent interference caused by haemolysis as red blood cells also contain SGOT. The serum was then diluted 10 fold with 0.9% (v/v) saline. The serum enzyme levels of SGPT and SGOT were estimated according to the method of Reitman & Frankel (1957).

Statistical analysis

The Student *t*-test was used to compare the levels of SGPT and SGOT induced by CCl_4 + 25% DMSO (DMSO control group) and the reduced levels due to the effect of individual seaweed extract.

Results

The enzyme assays of the serum transaminases showed that a toxic dose of CCl_4 (1.25 ml kg^{-1}) raised the levels of SGPT in the experimental rats to 1235 $\pm$ 116 (mean $\pm$ S.E.) and that of SGOT to 2119 $\pm$ 89.0 IU l^{-1} (Fig. 1A,B, CCl_4 group). These levels were significantly much higher than those in both the no treatment and saline vehicle control groups with SGPT and SGOT levels being less than 15 and 55 IU l^{-1} respectively (Fig. 1A,B). The DMSO vehicle had no effect on CCl_4-induced hepatotoxicity. The levels of SGPT and SGOT serum enzymes in the DMSO vehicle control group were comparable (Student *t* test, $p > 0.05$) to those of the CCl_4 toxin treatment group (Fig. 1A,B).

In general, methanol extracts of the three species of seaweeds significantly reduced the levels of both the SGPT and SGOT in rats exposed to CCl_4. An exception being that for *M. myagroides*, where its methanol extracts did not reduce the level of SGPT significantly (Fig. 1A). Of the three species, the methanol extracts of *S. siliquastrum* appeared to show the most promising hepatoprotective effect. Experimental animals treated with methanol extracts of this species showed a level of SGPT at 492.4 $\pm$ 42.5 and that of SGOT at 904.8 $\pm$ 101 IU l^{-1}. This is a significant reduction of 60% and 59%, respectively, of their levels in the toxin treatment and DMSO control groups (Fig. 1A,B).

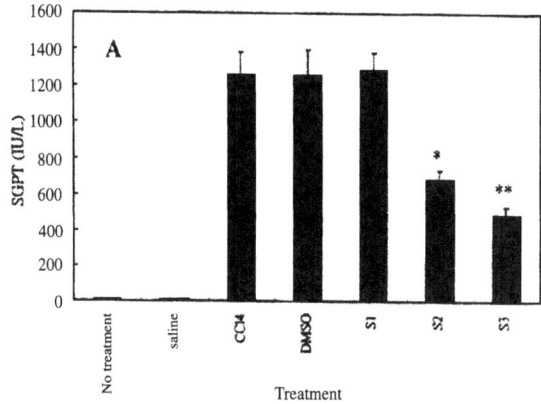

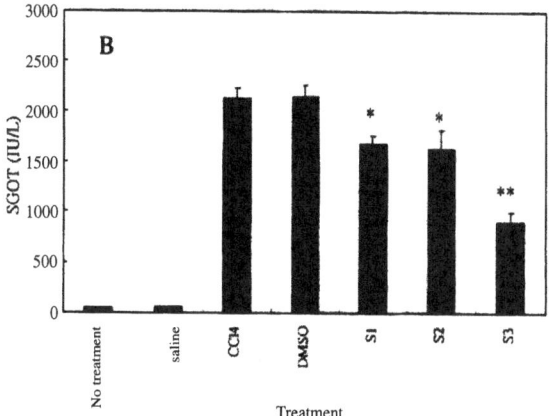

Figure 1. Curative effect of methanol extract (at dosage of 300 mg kg^{-1}) of three species of seaweed on CCl$_4$-induced elevation of (A) SGPT and (B) SGOT activities in rats. Each value represents the mean ± S.E. of 5 treated rats. Values statistically significantly different from those of DMSO control group are indicated by *(Student's *t*-test, $p < 0.05$) and **($p < 0.005$). Treatment groups are as follow: No Treatment control (Group 1); Saline vehicle control (Group 2); CCl$_4$ toxin treatment (Group 3); DMSO vehicle control (Group 4); S1 *Myagropsis myagroides* extract treatment (Group 5); S2 *Sargassum henslowianum* extract treatment (Group 6); and S3 *Sargassum siliquastrum* extract treatment (Group 7).

Discussion

CCl$_4$ is commonly used as a model to study hepatotoxicity (Plaa & Hewitt, 1982; Gilani, & Janbazz, 1995a,b,c). When the liver is injured as a result of the introduction of infectious agents or chemicals, the serum levels of SGPT and SGOT are raised significantly (Recknagel, 1967). The increases in SGPT and SGOT serum levels have been attributed to damage to the structural integrity of the liver (Chenoweth & Hake, 1962). They may be released from the cytoplasm into the blood circulation rapidly after rupture of the plasma membrane and cellular damage (Sallie et al., 1991). At a suitable dose, CCl$_4$ causes exten-

sive necrosis in the liver centrilobular regions around the central veins (Walker et al., 1980). It is generally accepted that CCl$_4$ hepatotoxicity resulted from activation of CCl$_4$ by the respective specific isozyme of the cytochrome P-450 system in the endoplasmic reticulum (ER) of hepatocytes to form the reactive metabolite, trichloromethyl radical (CCl$_3$·), which covalently binds to macromolecules, protein and lipid, and also interact with O$_2$ to yield highly reactive trichloromethylperoxy radical (CCl$_3$O$_2$·). This in turn initiates peroxidative degeneration of membrane lipids of the ER rich in polyunsaturated fatty acids (Slater, 1966; Recknagel, 1967; Klaassen & Plaa, 1969; Packer et al., 1978; Van de Straat et al., 1987). However, the actual mechanisms of how these initial events lead to further degenerative effects, finally bringing about cell necrosis and death, are still obscure (Wang et al., 1996).

Based on the SGPT and SGOT values of the DMSO Control group, the 25% DMSO exhibited no significant influence on CCl$_4$-induced hepatotoxicity. It demonstrated that the dose of DMSO applied could be used as a vehicle for methanol extracts of seaweed in this experiment. Therefore, any resulting effect of seaweed extracts on the serum enzymes was due to the effect of the extracts.

SGPT and SGOT, especially the former, are highly localized in hepatocyte cytosols (Ooi, 1996). The crude methanol extracts of seaweeds probably acted to preserve the structural integrity of the plasma cellular membrane of the hepatocytes to protect it against breakage by the reactive metabolites produced from exposure to CCl$_4$. This prevented further damage to more hepatocytes and hence reduced further leakage of SGPT and SGOT due to cell destruction. This may explain the lower levels of these transaminases observed in rats treated with the seaweed extracts after exposure to the toxin. *S. siliquastrum* extracts appeared to have the best overall curative action, followed by those from *S. henslowianum* and *M. myagroides*. However, the ability of methanol extract of *M. myagroides* to reduce the level of SGOT but not that of SGPT remains difficult to explain.

In contrast to the findings of this present study, the aqueous extracts of *S. henslowianum* showed the most promising hepatoprotective effect against CCl$_4$ exposure than those of the other two seaweeds (Wong et al., 2000). Moreover, those of *S. siliquastrum* showed the least protective effect. It thus appears that both methanol and aqueous extracts of these seaweeds exhibit hepatoprotective effect, but different types of active ingredients are likely involved. Furthermore, it is noted that the methanol extracts of *S. siliquastrum* and *S. henslowianum* at the dose of 300 mg kg^{-1} showed better results than the aqueous extracts of *S. henslowianum* at the same dose. Therefore, it can be assumed that the possible hepatoprotective effect of methanol extracts of these seaweeds is better than that of the aqueous extracts. The same results could also be shown in the histopathological examinations (Wong, 1999). The possible component(s) in methanol extracts may be phenol or polyphenols which are organic in nature (Lee et al., 1996). They are different from those of the aqueous extracts which are likely to be polysaccharides or glycoproteins (Harada et al., 1997). The hepatoprotective activity of the extracts may also be due to their antioxidant properties. In which case, they may act as scavengers of free radicals, such as superoxide and alkoxy radicals, and protect the liver against liver plasma membrane peroxidative degradation or promote cellular mitosis for the repair of damaged liver cells (Ooi, 1996). Some of these processes may involve active binding sites. This was indicated by the saturation of their effects with methanol extracts of seaweed. The active components in the methanol extracts are currently being isolated and further details of the mechanisms involved in their hepatoprotective effects have yet to be elucidated.

Acknowledgements

The authors thank Miss S. N. Lim for technical assistance in preparing the methanol extracts of seaweed. This study was partially supported by an Earmarked Grant from the RGC of Hong Kong.

References

Borzelleca, J. F., T. M. O'Hara, C. Gennings, R. H. Granger, M. A. Sheppard & L. W. Condie. Jr., 1990. Interactions of water contaminants. I. Plasma enzyme activity and response surface methodology following gavage administration of CCl$_4$ and CHCl$_3$ or TCE singly and in combination in the rat. Fundam. Appl. Toxicol. 14: 477–490.

Chenoweth, M. B. & C. L. Hake, 1962. The smaller halogenated aliphatic hydrocarbons. Ann. Rev. Pharmac. 2: 363–398.

Gilani, A. H. & K. H. Janbazz, 1995a. Studies on protective effect of *Cyperus scariosus* extract on acetaminophen and CCl$_4$-induced hepatotoxicity. Phytotherapy Res. 9: 489–494.

Gilani, A. H. & K. H. Janbazz, 1995b. Preventive and curative effects of *Artemisia absinthium* on acetaminophen and CCl$_4$-induced hepatotoxicity. Gen. Pharmac. 26: 309–315.

Gilani, A. H. & K. H. Janbazz, 1995c. Preventive and curative effects of *Berberis aristata* fruit extract on paracetamol and CCl$_4$-induced hepatotoxicity. Gen. Pharmac. 26: 627–631.

Harada, H., T. Noro & Y. Kamei, 1997. Selective antitumor activity *in vitro* from marine algae from Japan coasts. Biol. Pharm. Bull. 20: 541–546.

Harris, R. N., J. H. Ratnayake, V. F. Garry & M. W. Anders, 1982. Interactive hepatotoxicity of CHCl$_3$ and CCl$_4$. Toxicol. Appl. Pharmacol. 63: 281–291.

Klaassen, C. D. & G. L. Plaa, 1969. Comparison of the biochemical alterations elicited in livers from rats treated with CCl$_4$, CHCl$_3$, 1,1,2-trichloroethane and 1,1,1-trichloroethane. Biochem. Pharmacol. 18: 2019–2027.

Lee, J. H., J. C. Park & J. S. Choi, 1996. The antioxidant activity of *Ecklonia stolonifera*. Arch. Pharm. Res. 19: 223–227.

Ooi, V. E. C., 1996. Hepatoprotective effect of some edible mushrooms. Phytotherapy Res. 10: 536–538.

Packer, J. E., T. F. Slater & R. L. Wilson, 1978. Reaction of the CCl$_4$-related peroxy free radical (CCl$_3$OO) with amino acids: pulse radiolysis evidence. Life Sci. 23: 2617–2620.

Plaa, G. L. & W. R. Hewitt, 1982. Quantitative evaluation of indices of hepatotoxicity. In Plaa, G. L. & W. R. Hewitt (eds), Toxicology of the Liver. Raven Press, New York: 103–120.

Recknagel, R. O., 1967. Carbon tetrachloride hepatotoxicity. Pharmacol. Rev. 19: 145–208.

Reitman, S. & S. Frankel, 1957. A colorimetric method for the determination of serum oxaloacetic and glutamic pyruvic transaminases. Am. J. Clin. Pathol. 28: 56–63.

Sallie, R., J. M. Tredgeri & R. William, 1991. Drugs and the liver. Biopharmaceut. Drug Dispos. 12: 251–259.

Slater, T. F., 1966. Necrogenic action of carbon tetrachloride in the rat: a speculative mechanism based on activation. Nature (London) 209: 36–40.

Slater, T. F. 1984. Free-radical mechanisms in tissue injury. Biochem. J. 222: 1–15.

Van de Straat, R., J. Van de Vries, A. J. J. Debets & N. E. Vermueulein, 1987. The mechanism of paracetamol-induced hepatotoxicity by 3,5-dialkyl substitution: the role of glutathion depletion and oxidative stress. Biochem. Pharmac. 36: 2065–2071.

Walker, R. M., W. J. Racz & T. F. Mcelligott, 1980. Acetaminophen-induced hepatotoxicity in mice. Lab. Invest. 42: 181–189.

Wang, D. H., K. Ishii, L. X. Zhen & K. Taketa, 1996. Enhanced liver injury in acatalasemic mice following exposure to carbon tetrachloride. Arch. Toxicol. 70: 189–194.

Wong, C. K., 1999. Protective Effects of Seaweeds Against Liver Injury Caused by Carbon Tetrachloride and Trichloroethylene in Rats. M.Phil. Thesis, The Chinese University of Hong Kong, Hong Kong.

Wong, C. K., V. E. C. Ooi & P. O. Ang. Jr., 2000. Protective effects of seaweeds against liver injury caused by carbon tetrachloride in rats. Chemosphere 41: 173–176.

Hydrobiologia **512**: 271–278, 2004.
P.O. Ang, Jr. (ed.), Asian Pacific Phycology in the 21ˢᵗ Century: Prospects and Challenges.
© 2004 *Kluwer Academic Publishers.*

Nutritional evaluation of protein concentrates isolated from two red seaweeds: *Hypnea charoides* and *Hypnea japonica* in growing rats

K. H. Wong, Peter C. K. Cheung* & Put O. Ang, Jr.
Department of Biology, The Chinese University of Hong Kong, Shatin, New Territories, Hong Kong SAR, China
*Author for correspondence; Tel: (852) 2609-6144. Fax: (852) 2603-5646. E-mail: petercheung@cuhk.edu.hk

Key words: protein concentrates, red seaweeds, *Hypnea charoides*, *Hypnea japonica*, protein quality, growing rats

Abstract

The nutritional values of protein concentrates (PCs) isolated from two subtropical red seaweeds, *Hypnea charoides* Lamouroux and *H. japonica* Tanaka, were evaluated in growing rats. The protein quality of the two seaweed PCs was determined by comparing the net protein ratio (NPR), true protein digestibility (TD), nitrogen balance (NB), biological value (BV), net protein utilization (NPU) and utilizable protein (UP) of the two seaweed PCs diet groups with those of the casein control group. There were no significant differences of NPR and BV in all diet groups. Although the values of TD (ranged from 90.5 to 90.6%), NB (ranged from 108 to 113 mg rat^{-1} day^{-1}), NPU (ranged from 80.1 to 81.3%) and UP (ranged from 80.1 to 81.3%) of these two PCs were significantly lower than those of the casein control, they were comparable to those of other common plant PCs. The growth performance of rats fed the two PCs diets was satisfactory and both PCs had no adverse effect on the weight of their major organs. Together with their good protein quality as mentioned above, the PCs from the two red seaweeds under study could be a potential alternative protein source for human nutrition.

Introduction

The production of plant protein concentrates (PCs) is of growing interest to food industry because of the increasing applications of plant proteins in food especially in developing countries (Akintayo et al., 1998; Sanchez-Vioque et al., 1999). To improve the nutritional quality of the product or for economic reasons, the use of plant PCs in food as functional ingredient is very extensive. For example, whey PCs (Jayaprakasha & Brueckner, 1999) and soybean PCs (Qi et al., 1997) have been widely used as food foaming, emulsifying, water binding and viscosity ingredients. However, these applications in the food trade are almost limited to protein from legumes (Chau et al., 1997; Qi et al., 1997; Sanchez-Vioque et al., 1999) and cereals (Prakash, 1996; Jayaprakasha & Brueckner, 1999), whereas other plant proteins are less used.

People in the Far East and Asia Pacific have a long tradition of consuming seaweeds as part of their diet. In the western countries, the principal uses of seaweeds are as sources of phycocolloids, thickening and gelling agents for various industrial applications including, foods (Darcy-Vrillon, 1993; Mabeau & Fleurence, 1993; Abbott, 1996). Recently in France, seaweeds have been approved for use as vegetable and condiments (Mabeau, 1989). Therefore, seaweeds are becoming a valuable vegetable (fresh or dried) and an important food ingredient in human diet nowadays, even in the western world.

The nutritional potential of seaweeds as food protein sources differs according to species (Fleurence et al., 1999). Seaweeds belonging to the Rhodophyta possess high levels of proteins (10–30% DW) (Darcy-Vrillon, 1993; Mabeau & Fleurence, 1993) comparable to those of edible land vegetables (around 20% DW, Dupin et al., 1992). In some red seaweeds, such as *Palmaria palmata* (L.) Kuntze (dulse) and *Porphyra tenera* Kjellman (nori), the protein contents are 35 and 47% DW, respectively (Morgan et al., 1980; Fujiwara-Arasaki et al., 1984). These levels are even comparable to that of the soybeans (35% DW). However, only a few studies have been undertaken on the quality of seaweed protein (Dam et al., 1986; Ito & Hori, 1989; Amano & Noda, 1990; Fleurence, 1999a) because of the difficulties of extraction and preparation of seaweed PCs. The extraction of seaweed protein by classical procedures is hindered by the presence of large amounts of cell wall polysaccharides, such as the alginates of the brown seaweed or the carrageenans of some red seaweeds. The high content of neutral polysaccharides (e.g. xylans and cellulose) in some red and green seaweeds can also limit the protein accessibility (Fleurence, 1999b). These anionic and neutral polysaccharides are the main impediment during the extraction and purification of seaweed protein (Ochiai et al., 1987; Ito & Hori, 1989; Jordan & Vilter, 1991; Fleurence et al., 1995). Moreover, although *in vitro* seaweed protein degradation by proteolyic enzymes such as pepsin, pancreatin and pronase have been reported previously (Ryu et al., 1982; Fujiwara-Arasaki et al., 1984), the *in vivo* digestibility of seaweed proteins is not well documented (Mabeau & Fleurence, 1993; Fleurence, 1999a).

The extraction procedures for seaweed proteins described in the literature are mainly concerned with the extraction of specific seaweed enzymes such as proteases (Kadokami et al., 1990), peroxides (Sheffield et al., 1993) or carboxylases (Hilditch et al., 1991). In comparison, very little information about the extraction of the total protein fraction from seaweed is available (Fleurence et al., 1995). After comparing with different classical and enzymatic procedures (e.g. using aqueous polymer two-phase system, polysaccharidases, or Tris HCl buffer), Fleurence et al. (1995) reported that the highest yield of seaweed PCs was obtained by the use of NaOH and 2-mercaptoethanol after an initial aqueous extraction.

The inadequacy of bioassay techniques, including protein efficiency ratio (PER), for evaluating protein quality has been recognized (Pellett & Young, 1980; Madi, 1993). Biological indices like net protein ratio (NPR), nitrogen balance (NB), true protein digestibility (TD), biological value (BV), net protein utilization (NPU) as well as utilizable protein (UP), which are widely used in nutritional studies (Kalra & Jood, 1998; Wong & Cheung, 1998), are recommended by the FAO/WHO (1991) for evaluating protein quality. Besides, *in vivo* experiment using the rat balance method (McDonough et al., 1990) is more suitable to predict protein digestibility in humans (FAO/WHO, 1991).

In Hong Kong, the seaweed floras are fairly rich but they are relatively under-utilized (Hodgkiss & Lee, 1983). In general, most Hong Kong seaweeds are mainly used as animal feeds or fertilizers by the coastal villagers (Hodgkiss & Lee, 1983). *Hypnea charoides* Lamouroux and *H. japonica* Tanaka are two subtropical red seaweeds that are very abundant in Hong Kong. The aim of this study was to evaluate the nutritional value of the PCs isolated from these two subtropical red seaweeds in growing rats, as reflected by the computed biological indices mentioned above.

Materials and methods

Sample preparation

Samples of *H. charoides* and *H. japonica* were collected in December of 1997 from A Ma Wan and Lung Lok Shui at Tung Ping Chau, on the northeast of Hong Kong (114° 26′ E; 22° 33′ N). Fresh plants were thoroughly washed with distilled water and their holdfasts and epiphytes were removed. All cleaned seaweeds were then frozen at −70 °C for 24 h and then dried in a freeze-drier (Labconco, MO) for 5 days. All samples were dried to constant weight. The dried samples were pulverized by using a cyclotech mill (Tecator, Höganäs, Sweden) to pass through a screen with an aperture of 0.5 mm. The milled seaweed samples were then stored in air-tight plastic bags in desiccators at room temperature (25 °C) prior to seaweed PCs extraction.

Extraction of seaweed protein concentrates

Seaweed PCs were extracted with the method described by Fleurence et al. (1995) with slight modifications. In brief, seaweed powder was suspended in de-ionzed water (1:20 w/v) to induce cell lysis by osmotic shock in order to facilitate subsequent protein extraction. The suspension was then gently stirred overnight at 35 °C, which is the temperature found to be optimal for seaweed protein solubility (Dua et al., 1993). After incubation, the suspension was centrifuged at $10\,000 \times g$ and 4 °C for 20 min. The supernatant was collected and the pellet was re-suspended in de-ionized water in the presence of 0.5% (v/v) 2-mercaptoethanol (Venkataraman & Shivashankar, 1979). The mixture was then adjusted to pH 12 with 1 M NaOH and gently stirred at room temperature (25 °C) for 2 h before centrifugation at the same conditions mentioned above. The supernatant was collected and combined with the previous supernatant. The combined supernatants were stirred at 0–4 °C and adjusted to pH 7 before precipitation with solid ammonia sulphate. The extraction procedure mentioned above was repeated five times on the residue.

Precipitation of seaweed protein concentrates

Seaweed PCs were precipitated from the supernatant by slowly adding solid ammonia sulphate with stirring until a 85% saturation (60 g/100 ml) was reached (Rosenberg, 1996). The mixture was then allowed to stand for 30 min before centrifugation at the same conditions mentioned before. The pellet (PCs) obtained was dialysed against distilled water until the total dissolved solutes (TDS) (mg 1^{-1}) of the dialysate, measured by its conductivity, was similar to that of the distilled water. The retentates containing seaweed PCs was then freeze-dried, ground to powder and stored in air-tight bags in desiccators before biological evaluation of their protein quality was performed. The percent crude protein of the seaweed PCs was calculated by multiplying percent nitrogen that was determined by a CHNS/O Analyzer (Perkin Elmer 2400, Connecticut, U.S.A.) with a factor of 6.25.

Diet preparation

The control and test diets were prepared according to the AIN-93G purified diet (Reeves et al., 1993) with slight modifications. All diets were formulated to contain identical levels of ingredients. All diets contained 10.0% of protein, as this level of protein is limiting for growth in weanling rat and any effect of seaweed PCs would be more likely to have a significant change on their protein utilization. The isoproteinous level was achieved by taking into account the purity of each protein source. Casein (90.0%, C7078, Sigma Chemical Co., St. Louis, MO) served as the sole protein source for the control group while the sole protein source in the test diets came from their corresponding seaweed PCs (*H. charoides* PCs: 83.1% and *H. japonica* PCs: 85.0%) (Wong & Cheung, 2001). This formulation allowed for constancy of gross energy density of each diet which was calculated by employing the factors: proteins = 4.00 Kcal g^{-1}; carbohydrates = 4.00 Kcal g^{-1}; fats = 9.00 Kcal g^{-1}. The formulations of the semi-purified test diets are shown in Table 1. A total of four diets were investigated: a casein-based control diet, a protein-free diet and two seaweed PCs diets. Each diet was made up in a single batch of about 1 kg and their moisture content was also determined with an infrared moisture analyzer (Mettler LJ 16, Greifensee, Switzerland) at 120 °C.

Rat bioassay

The experiment procedure was similar to the rat balance method described by McDonough et al. (1990). In brief, 16 male, weanling Sprague Dawley rats from the same colony with initial body weight of 50–70 g were obtained from the Laboratory Animal Service Center of The Chinese University of Hong Kong. Animals transported from breeding colony to test laboratory were weighed when received, and fed standardized laboratory rat chow for an acclimation period of 2 days. Four rats were then assigned randomly to each of the four experimental diet groups and were housed in individual metabolic cages kept under the conditions of 18–26 °C, 40–70% relative humidity and with 12 h light /dark cycle. All animals had free access to water and diets.

The rat balance method used in this study consisted of a 5-day preliminary period, during which the rats were allowed to adapt to the diets and experimental conditions, followed by a 5-day balance period. During the balance period, feces, urinary output and spilled food were collected daily and separately for each rat and frozen at −70 °C. At the end of the balance period, total food intake was determined taking into considerations the amount of spilled unconsumed diet. The frozen rat feces and urinary output were freeze-dried, weighed, ground and analyzed for percent nitrogen content by a CHNS/O Analyzer (Perkin Elmer 2400, Connecticut, U.S.A.). The endogenous or metabolic nitrogen loss was determined from the feces of the rats fed the protein-free diet. After completion of the feeding experiment, the rats were deprived of food for 16 h, weighed and then anaesthetized by using diethyl ether. The liver, spleen, kidneys, heart and stomach were rapidly excised and weighed in order to assess the growth performance of the experimental rats. The weight loss of rats fed the protein-free diet was used for computing the NPR.

Biological indices

The data obtained from the animal experiment were used to calculate the parameters such as nitrogen balance (NB, Equation (1)), true protein digestibility (TD, Equation (2)), biological value (BV, Equation (3)), net protein utilization (NPU, Equation (4)), utilizable protein (UP, Equation (5)) and net protein ratio (NPR, Equation (6)) by employing the following formulae recommended by FAO/WHO (1991):

$$NB = I - F - U; \qquad (1)$$

Table 1. Chemical composition of the diets (g/100 g of diet on dried weight basis) used in the experiments

Ingredients	Diet			
	Control	Protein-free	H. charoides	H. japonica
H. charoides PCs	–	–	12.0 (10.0)[1]	–
H. japonica PCs	–	–	–	11.8 (10.0)[1]
Casein[2]	11.1 (10.0)[1]	–	0.00	0.00
Dextrinzed cornstarch[3]	13.2	13.2	13.2	13.2
Corn starch[4]	49.8	59.8	49.8	49.8
Sucrose[5]	10.0	10.0	10.0	10.0
AIN minerals mixed[6]	3.50	3.50	3.50	3.50
Corn oil[7]	7.00	7.00	7.00	7.00
AIN vitamins mixed[8]	1.00	1.00	1.00	1.00
Fiber (cellulose)[9]	5.00	5.00	5.00	5.00
Choline bitartrate[10]	0.20	0.20	0.20	0.20
L-Cystine[11]	0.30	0.30	0.30	0.30
Kcal/100 g of diet	395	395	395	395
Gross energy density (KJ g^{-1})	16.6	16.6	16.6	16.6

[1] Values in parentheses are the 'net' protein added into the diets.
[2] Casein (Product No. C7078, Sigma Chemical Co., St. Louis, MO).
[3] Dextrainzed cornstarch (Catalog No. 160175, Teklad Test Diets, 2826 Latham Dr., Madison, WISC).
[4] Corn starch (Kingsford, CPC/AJI (Hong Kong) Ltd., Hong Hong).
[5] Sucrose (Taikoo Sugar Ltd., Hong Kong).
[6] AIN Mineral Mix (AIN-93G-MX, Nutritional Biochemicals).
[7] Corn oil (CPC International Inc., U.S.A.).
[8] AIN Vitamin Mix (AIN-93G-VX, Nutritional Biochemicals).
[9] Cellulose (Catalog No. 160390, Teklad Test Diets, 2826 Latham Dr., Madison, WISC).
[10] Choline bitartrate (Product No. C1629, Sigma Chemical Co., St. Louis, MO).
[11] L-Cystine (Product No. C8775, Sigma Chemical Co., St. Louis, MO).

$$TD = \frac{I - (F - Fk) \times 100}{I}; \quad (2)$$

$$BV = \frac{I - (F - Fk) - (U - Uk)}{I - (F - Fk)}; \quad (3)$$

$$NPU = \frac{BV \times TD}{100}; \quad (4)$$

$$UP = \frac{NPU \times \text{protein content (g kg}^{-1} \text{ of diet on dry weight basis)}}{100}; \quad (5)$$

$$NPR = \frac{\text{weight gain of test group + weight loss protein-free group}}{\text{weight of test protein consumed}}. \quad (6)$$

The abbreviations used were I (nitrogen intake), F (fecal nitrogen), Fk (metabolic or endogenous fecal nitrogen), U (urinary nitrogen) and Uk (metabolic or endogenous urinary nitrogen). Body weight gain and protein intake were expressed as grams per rat per day while I, F, Fk, U and Uk were expressed as milligrams per rat per day.

Statistical analysis

All data were presented as mean values ± S.D. The results of all mean values were analyzed by one-way ANOVA and Tukey-HSD at $p < 0.05$ (Wilkinson, 1988) to detect significant differences among groups.

Results and discussion

Protein quality of seaweed PCs

Protein quality in food refers to the ability of a protein to support body growth and maintenance (Wardlaw & Insel, 1993). The biological and chemical evaluation of protein quality is a key factor in the search for new protein sources as well as in the development of food proteins (Carias et al., 1998). A major factor determining protein quality is the amino acid profile. Another factor, which is as important as amino acid profile, is protein digestibility. Even with an excellent amino acid profile, a protein would have a low

Table 2. Food intake, protein intake, weight gain, and NPR of rats fed with seaweed PCs diets[1]

Diet	Food intake (g rat^{-1} day^{-1})	Protein intake (g rat^{-1} day^{-1})	Weight gain (g rat^{-1} day^{-1})	NPR[2]
Control	11.6 ± 1.10^a	1.16 ± 0.11^a	5.63 ± 0.63^a	3.81 ± 0.35^a
H. charoides PCs	10.5 ± 0.37^a	1.05 ± 0.04^a	4.63 ± 0.63^{ab}	3.24 ± 0.48^a
H. japonica PCs	10.1 ± 1.02^a	1.02 ± 0.10^a	4.50 ± 0.41^b	3.25 ± 0.10^a

[1] Data are mean values of four determinations $\pm$ SD. Means in columns with different superscripts (a–b) are significantly different from one another ($p < 0.05$, one-way ANOVA, Tukey-HSD).
[2] NPR = Net Protein Ratio.

Table 3. Nitrogen intake, fecal weight, fecal nitrogen and TD of rats fed with seaweed PCs diets[1]

Diet	N intake (mg rat^{-1} day^{-1})	Fecal weight (g rat^{-1} day^{-1})	Fecal N (mg rat^{-1} day^{-1})	TD[2] (%)
Control	185 ± 17.6^a	4.50 ± 0.60^a	14.0 ± 1.94^a	98.2 ± 0.47^a
H. charoides PCs	168 ± 5.75^a	4.27 ± 0.46^a	25.3 ± 1.66^b	90.6 ± 1.16^b
H. japonica PCs	162 ± 16.7^a	4.07 ± 0.32^a	27.8 ± 1.11^b	90.6 ± 4.57^b

[1] Data are mean values of four determinations $\pm$ SD. Means in columns with different superscripts (a–b) are significantly different from one another ($p < 0.05$, one-way ANOVA, Tukey-HSD).
[2] TD = True Protein Digestibility.

nutritional value if its digestibility was low due to its poor bioavailability (Bejosano & Corke, 1998). Our recent work indicated that the amount of essential amino acids of the PCs of *H. charoides* and *H. japonica* accounted for 36.2–38.7% of the total amino acids (Wong & Cheung, 2001). Besides, both *Hypnea* PCs were rich in leucine, valine, threonine, aspartic and glutamic acids but lacked cystine. Except for sulphur-containing amino acids and lysine, the levels of all essential amino acids of these two *Hypnea* PCs were higher than those of FAO/WHO requirement pattern (Wong & Cheung, 2001).

One of the best methods to determine the digestibility of protein is through *in vivo* animal feeding studies (Bejosano & Corke, 1998). A common observation during the evaluation of protein quality *in vivo* is a voluntary reduction in food intake in the animals assigned to the test protein when the quality of this protein is lower than that of control (Carias et al., 1998). In this case, the results may be affected, since the protein consumed is partially utilized as a source of energy and less protein is available to be incorporated in new tissues. As a result, the quality of this protein may be underestimated (Kino & Okumara, 1988; Muramatsu, 1990). Therefore, in order to accurately measure the quality of a protein, the food intake of both the experimental and the control groups should be similar (Carias et al., 1998). In this study, since there were no significant differences of food intake in all different diet groups (Table 2), any effect of seaweed PCs on the biological indices would be reliable and independent of the food intake factor.

As all diets contained the same amount of protein (10%, Table 1), the protein and nitrogen intake of rats fed different diets would be proportional to their corresponding food intake. As a result, no significant differences of protein and N intake were obtained between the seaweed PCs and control diet groups (Tables 2 and 3).

Only the weight gain of *H. japonica* PCs diet group was significantly lower than that of the control group (Table 2). However, when the weight loss of protein-free diet group was considered and the weight gain was expressed as per grams of protein intake (i.e. NPR) (Equation (6)), the NPR of the seaweed PCs diet groups did not significantly differ from that of the control group (Table 2). This implied that the proportion of utilizable seaweed PCs from diet that can turn into new protein tissues was similar to that of the casein control (FAO/WHO, 1991).

According to the National Research Council (U.S.A.), the gross energy density needed for weanling rats to grow is 12.55 KJ g^{-1} of diet (NRCN, 1978). Therefore, the gross energy density of all the diets (16.6 KJ g^{-1} of diet, Table 1) in the present study exceeded the minimal growth value. Besides, it is

Table 4. Urinary nitrogen, NB, BV, NPU and UP in rats fed with seaweed PCs diets[1]

Diet	Urinary N (mg rat^{-1} day^{-1})	NB[2] (mg rat^{-1} day^{-1})	BV[3] (%)	NPU[4] (%)	UP[5] (%)
Control	26.0 ± 1.78^a	145 ± 16.6^a	93.8 ± 1.99^a	92.1 ± 1.88^a	92.1 ± 1.88^a
H. charoides PCs	29.0 ± 5.01^a	113 ± 10.0^b	89.7 ± 3.77^a	81.3 ± 4.29^b	81.3 ± 4.29^b
H. japonica PCs	26.2 ± 2.08^a	108 ± 18.8^b	90.6 ± 3.47^a	80.1 ± 4.79^b	80.1 ± 4.79^b

[1]Data are mean values of four determinations $\pm$ SD. Means in columns with different superscripts (a–b) are significantly different from one another ($p < 0.05$, one-way ANOVA, Tukey-HSD).
[2]NB = Nitrogen Balance.
[3]BV = Biological Value.
[4]NPU = Net Protein Utilization.
[5]UP = Utilizable Protein.

valuable to note that, the energy density of all diets still exceed the minimal growth value of rat, even though the energy supply is only based on carbohydrate and fat (14.9 KJ g^{-1} diet). This indicated that apart from the protein, all experimental diets could provide sufficient adequate energy from other energy sources for weanling rats to grow. Therefore, dietary proteins are unlikely to break down and release energy to meet the body's need. As a result, the evaluation of protein quality on the test protein would be more accurate, reliable and meaningful (Whitney et al., 1991; Wardlaw & Insel, 1993).

Although the fecal weight and urinary N of seaweed PCs diet groups were not significantly different from those of the control group (Tables 3 and 4), the fecal nitrogen loss of the seaweed PCs diet groups was significantly higher (about 2 folds) than that of the control group (Table 3). Because of this significantly higher fecal nitrogen excretion, significantly lower NB and TD values of seaweed PCs diet groups were obtained as compared with those of the control (Equations (1) and (2)) (Tables 3 and 4). All NB was positive (positive N balance) indicating that there is extra nitrogen for the weanling rats to grow (synthesis new protein tissues) although the amount of this extra nitrogen in control group was significantly higher. Besides, N was mainly lost in feces rather than urine. Furthermore, the significantly lower TD values of seaweed PCs implied that their percentage of intake N that can be absorbed were lower than that of the control (FAO/WHO, 1991).

Comparing with the seaweed PC diet groups, the control group not only possessed significantly higher apparent nitrogen retention (i.e. NB) (Table 4), but also exhibited the significantly highest level of absorbed nitrogen (i.e. TD) (Table 3). As a result, no

significant differences between the BV of the seaweed PCs and control diet groups were obtained (Equation (3)) (Table 4). Besides, this suggests that the proportion of absorbed nitrogen that is retained for maintenance or growth in both seaweed PCs and control diet groups were similar (FAO/WHO, 1991).

Table 4 also shows that the NPU and UP of the seaweed PCs diets were significantly lower than those of the control, implying that the proportion of nitrogen intake retained for growth or maintenance as well as the maximum amount of protein that can be utilized in seaweed PCs diet groups were lower than those of the control (FAO/WHO, 1991). Although there were no significant differences in nitrogen intake by all diets groups (Table 3), the result of NPU and UP may probably be due to the significantly higher level of NB in the control group (Equations (4) and (5)) (Table 4).

In conclusion, the overall protein quality of the two seaweed PCs was lower than that of the casein control. This was indicated by their significantly lower levels of true protein digestibility (TD), nitrogen balance (NB), net protein utilization (NPU) and utilizable protein (UP). The intake nitrogen was mainly lost in the feces and not in the urine. This suggested that the lower protein quality of the seaweed PCs would be mainly due to their relatively lower digestibility rather than limited supply of essential amino acids (which resulted in additional N lost in urine) (Whitney et al., 1991; Wardlaw & Insel, 1993). However, the nitrogen retention (for growth and maintenance) of absorbed N from seaweed PCs was similar to that of absorbed casein N, since there were no significant differences in BV of all diets even though the absorbed nitrogen of seaweed PCs was significantly lower than that of the casein. Besides, the protein quality of the two seaweed PCs was similar, as no significant differences

Table 5. Relative weight (g/100 g of body weight) of liver, kidney, spleen, stomach and heart of rats fed with seaweed PCs diets[1]

Diet	Liver	Kidney	Spleen	Stomach	Heart
Control	4.15 ± 0.32^a	0.81 ± 0.02^a	0.22 ± 0.03^{ab}	0.59 ± 0.01^a	0.48 ± 0.03^a
H. charoides PCs	3.68 ± 0.11^a	0.78 ± 0.04^a	0.18 ± 0.01^a	0.59 ± 0.05^a	0.48 ± 0.06^a
H. japonica PCs	3.81 ± 0.39^a	0.78 ± 0.06^a	0.18 ± 0.01^a	0.61 ± 0.02^a	0.51 ± 0.03^a

[1] Data are mean values of four determinations $\pm$ SD. Means in columns with same superscripts do not differ significantly from each other ($p > 0.05$, one-way ANOVA, Tukey-HSD).

were found in all the biological indices between the two seaweed PCs.

The weight of major organs

Variations in the weight and appearance of livers in rats fed with different diets were related to the difference in the amount of cholesterol and oil added into the diets (Beynen et al., 1986; Dadai et al., 1996). The absence of significant differences in the weight of liver of all different diets groups (Table 5) may be explained by the fact that the amount of corn oil added in each test diet was identical (5%, Table 1). Besides, the weights of kidney, spleen, stomach and heart of seaweed PCs diet groups also were not significantly different from those of the control group (Table 5). This suggested that consumption of these two seaweed PCs would not cause any adverse effect on the weight of these major organs.

The protein quality of the seaweed PCs was comparable to that of common plant PCs obtained from legumes and cereals. Rozan et al. (1997) reported that like seaweed PCs diet groups, the N intake of soybean PCs diet group was similar to that of casein control group and the fecal N of the soybean and sweet lupine seed PCs diet groups were significantly lower than that of the casein control group. Besides, the TD of the seaweed PCs diets (over 90%, Table 3) was comparable to that of pea PCs (92.6%), faba bean PCs (93.0%), soybean PCs (81.2%) (Fernández-Quintela et al., 1998) and rice bran PCs (84.1%) (Prakash, 1996). Furthermore, similar to the seaweed PCs diet groups, the NB and NPU of rats fed with pea, faba and soybean PCs were significantly lower than that of the casein control (Fernández-Quintela et al., 1998). On the contrary, however, a reduction in the weight of liver and spleen has been reported in rats fed with pea and soybean PCs (Fernández-Quintela et al., 1998).

Although the protein quality of the two seaweed PCs was lower than that of the casein control based

on the biological indices mentioned, it was comparable to that of other common plant PCs. Besides, both seaweed PCs had no adverse effect on the weight of some major organs and together with their good protein quality; they had the potential to be developed into a new protein source. Further investigation on the functional properties of these two seaweed PCs is currently underway and should provide a better appreciation of the potential of seaweed as a new protein source in animal and human diets.

Acknowledgements

We acknowledge the technical assistance of Mr C. C. Li. This project was funded by the Research Grants Council of the Hong Kong SAR Government.

References

Abbott, I. A., 1996. Ethnobotany of seaweeds: clues to uses of seaweeds. Hydrobiology 326: 15–20.

Akintayo, E. T., K. O. Esuoso & A. A. Oshodi, 1998. Emulsifying properties of some legume proteins. Int. J. Food Sci. Technol. 33: 239–246.

Amano, H. & H. Noda, 1990. Proteins of protoplast from red alga *Porphyra yezoensis*. Bull. Japan. Soc. Sci. Fish. 56: 1859–1864.

Bejosano, F. P. & H. Corke, 1998. Protein quality evaluation of *Amaranthus* wholemeal flours and protein concentrates. J. Sci. Food Agric. 76: 100–106.

Beynen, A. C., A. G. Lemmens, J. J. De Bruije, M. B. Katan & L. F. M. Zutphan, 1986. Interaction of dietary cholesterol with cholate in rats: effect of serum cholesterol, liver cholesterol and liver function. Nutr. Rep. Int. 34: 557–563.

Carias, D., A. M. Cioccia & P. Hevia, 1998. Effect of food intake on protein quality measured in chicks by traditional or biochemical methods. J. Sci. Food Agric. 78: 479–485.

Chau, C. F., P. C. K. Cheung & Y. S. Wong, 1997. Functional properties of protein concentrates from three Chinese indigenous legume seeds. J. agric. Food Chem. 45: 2500–2503.

Dadai, F. D., A. F. Walker, I. E. Sambrook, V. A. Welch & R. W. Owen, 1996. Comparative effects on blood lipids and fecal steroids of five legumes species incorporated into a semi-purified, hypercholesterolemic rats diet. Br. J. Nutr. 75: 557–571.

Dam, R., S. Lee, P. C. Fry & H. Fox, 1986. Utilization of algae as a protein source for humans. J. Nutr. 65: 376–382.

Darcy-Vrillon, B., 1993. Nutritional aspects of the developing use of marine macroalgae for the human food industry. Int. J. Food. Sci. Nutr. 44: 23–35.

Dua, S., M. Kaur & A. S. Ahluwalia, 1993. Functional properties of two pollutant grown green algae. J. Food Sci. Technol. 30: 25–28.

Dupin, H., J. L. Cuq, M. Malewiak, C. Leynaud-Rouaud & A. M. Berthier, 1992. Alimentation et nutrition humanies Paris. Editions Sociales Francaises: 8 pp.

FAO/WHO, 1991. Protein quality evaluation. Report of joint FAO/WHO expert consultation, Besthesda, Md., 4–8 December, 1989. FAO/WHO, Rome, Italy.

Fernández-Quintela, A., A. S. del Barrio, M. T. Macarulla & J. A. Martinez, 1998. Nutritional evaluation and metablic effects in rats of protein isolates obtained from seeds of three legume species. J. Sci. Food Agric. 78: 251–260.

Fleurence, J., 1999a. Seaweed proteins: Biochemical, nutritional aspects and potential uses. Trends Food Sci. Technol. 10: 25–28.

Fleurence, J., 1999b. The enzymatic degradation of algal cell walls: a useful approach for improving protein accessibility? J. app. Phycol. 11: 313–314.

Fleurence, J., E. Chenard & M. Lucon, 1999. Determination of the nutritional value of proteina obtained from Ulva armoricana. J. appl. Phycol. 11: 231–239.

Fleurence, J., C. Le Coeur, S. Mabeau, M. Maurice & A. Landrein, 1995. Comparison of different extractive procedures for proteins from the edible seaweeds Ulva rigida and Ulva rotundata. J. appl. Phycol. 7: 577–582.

Fujiwara-Arasaki, T., N. Mino & M. Kuroda, 1984. The protein value in human nutrition of edible marine algae in Japan. Hydrobiologia 116/117: 513–516.

Hilditch, C. M., P. B. Jones, P. Balding, A. J. Simth & L. J. Rogers, 1991. Ribulose biphosphate carboxylases from marcoalgae: proteolysis during extraction and properties of the enzyme from Porphyra umbilicalis. Phytochemistry 3: 745–750.

Hodgkiss, I. J. & K. Y. Lee, 1983. Hong Kong Seaweeds. The Urban Council of Hong Kong, Hong Kong.

Ito, K. & K. Hori, 1989. Seaweed: Chemical composition and potential uses. Food Rev. Int. 5: 101–144.

Jayaprakasha, H. M. & H. Brueckner, 1999. Whey protein concentrate: a potential functional ingredient for food industry. J. Food Sci. Technol. 36: 189–204.

Jordan, P. & H. Vilter, 1991. Extraction of proteins from material rich in anionic mucilages: Partition and fractionation of vanadate-dependent bromoperoxidases from the brown algae Laminaria digitata and L. saccharina in aqueous polymer two-phase system. Biochem. Biophys. Acta. 1073: 98–106.

Kadokami, K., N. Yoshida, K. Mizusaki, K. Noda & S. Makisumi, 1990. Some properties of trypsin-like proteases extracted from seaweed Codium fragile and their purification. Mar. Biol. 107: 513–517.

Kalra, S. & S. Jood, 1998. Biological evaluation of protein quality of barley. Food Chem. 61: 35–39.

Kino, K. & J. Okumara, 1988. Evaluation of nutritional quality for amino acid mixtures deficient in some essential amino acids in chicks under an equalized feeding condition. Nutr. Rep. Int. 138: 239–247.

Mabeau, S., 1989. La filiere algue francaise en 1988: atouts et point de blocage. Oceanis 15: 673–692.

Mabeau, S. & J. Fleurence, 1993. Seaweed in food products: biochemical and nutritional aspects. Trends Food Sci. Technol. 4: 103–107.

Madi, R.,1993. Evolution of protein quality. Am. Assoc. Cereal Chem. 38: 576–577.

McDonough, F. E., F. H. Steinke, G. Sarwar, B. O. Eggum, R. Bressani, P. J. Huth, W. E. Barbeau, G. V. Mitchell & J. G. Phillips, 1990. In vivo rats assay for true protein digestibility: collaborative study. J. Assoc. Off. Anal. Chem. 73: 801–805.

Morgan, K. C., J. L. C. Wright, & F. J. Simpson, 1980. Review of chemical constituents of the red algae Palmaria palmata. Econ. Bot. 34: 27–50.

Muramatsu, T., 1990. Nutrition and whole body protein turnover in the chickens in relation to mammalian species. Nutr. Rep. Int. 3: 211–228.

NRCN, 1978. National Research Council Nutrient Requirements of Laboratory Rats. In Nutrient Requirements of Laboratory Animals. National Academy of Sciences, Washington, D. C.

Ochiai, Y., T. Katsuragi & K. Hashimoto, 1987. Proteins in three seaweeds,: "Aosa" Ulva lactuca, "Arame" Eisenia bicyclis, and "Makusa" Gelidium amansii. Bull. Jap. Soc. Sci. Fish. 53: 1051–1055.

Pellett, P. L. & V. R. Young, 1980. Nutritional evaluation of protein foods. The United Nation University, Tokyo, Japan 47: 77–78.

Prakash, J., 1996. Rice bran proteins: Properties and food uses. Crit. Rev. Food. Sci. Nutr. 36: 537–552.

Qi, M., N. S. Hettiarachchy & U. Kalapathy. 1997. Solubility and emulsifying properties of soy protein isolates modified by pancreatin. J. Food Sci. 62: 1110–1115.

Reeves, P. G., F. H. Nielsen & G. C. Fahey, 1993. AIN-93 purified diets for laboratory rodents: final report of the American Institute of Nutrition Ad Hoc Writing Committee on the reformulation of the AIN-76A rodent diet. J. Nutr. 123: 1939–1951.

Rosenberg, I. M., 1996. Getting started with protein purification. In Protein Analysis and Purification Benchtop Techniques. Birkhäuser, Boston, U.S.A.: 99–132.

Rozan, P., R. Lamghari, M. Linder, C. Villaume, J. Fanni, M. Parmentier & L. Méjean, 1997. In vivo and in vitro digestibility of soybean, lupine and rapeseed meal proteins after various technological processes. J. Agric. Food Chem. 45: 1762–1769.

Ryu, H. S., L. D. Satterlee & K. H. Lee, 1982. Nitrogen conversion factors and in vitro protein digestibility of some seaweeds. Bull. Korean Fish. Soc. 15: 263–270.

Sanchez-Vioque, R., J. Clemente, J. Vioque, J. Bautista & F. Millan, 1999. Protein isolates from chickpea (Cicer arietinum L.): chemical composition, functional properties and protein characterization. Food Chem. 64: 237–243.

Sheffield, D. J., T. Harry, A. J. Simth & L. J. Rogers, 1993. Purification and characterization of the vanadium bromoperoxidase from the marcoalgae Corallina officinallis. Phytochemistry 32: 21–26.

Venkataraman, L. V. & S. Shivashankar, 1979. Studies on the extractability of proteins from the alga Scenedesmus acutus. Arch. Hydrobiol. 56: 114–126.

Wardlaw, G. M. & P. M. Insel, 1993. Perspective in Nutrition. Mosby-Year Book Inc., St. Louis, Missouri.

Whitney, E. N., C. B. Cataldo & S. R. Rolfes, 1991. Understanding Normal and Clinical Nutrition. West Publishing Company, St. Paul, MN.

Wilkinson, L., 1988. SYSTAT: The System for Statistics. Evanston, IL.

Wong, K. H. & P. C. K. Cheung, 1998. Nutritional assessment of three Chinese indigenous legumes in growing rats. Nutr. Res. 18: 1573–1580.

Wong, K. H. & P. C. K. Cheung, 2001. Nutritional evaluation of some red and green seaweeds part II. In vitro protein digestibility and amino acid profiles of protein concentrates. Food Chem. 72: 11–17.